Painting and Decorating

this Book belongs to: Gerry.

. . . But seeing others work and practising with your hand will make you perceive better than seeing it merely written . . .

From *The Book of the Art of Cennino Cennini* — 15th century

Also by J. H. GOODIER:

Dictionary of Painting and Decorating

A unique reference book that not only defines in simple language the technical terms used in the trade but covers, often in considerable detail, the materials and techniques of the craft, new and old. It also contains much helpful advice on the problems encountered with materials and processes.

Other titles from Griffin's list:

Outlines of Paint Technology	W. M. MORGANS
Structural Surveying	H. E. DESCH
The Storage and Handling of Petroleum Liquids	J. R. HUGHES
Science Data	J. N. FRIEND

Gerard Coleman.

Gerard Coleman

St Gregory's Church, Preston

A colour scheme prepared by Crown Colour Advisory Service

By courtesy of the Church Authorities and Crown Decorative Products Ltd

Painting and Decorating

A. E. Hurst, FRSA, FBID

and

J. H. Goodier, JP, CGIA, FRSA, FBID
Holder of the Queen's Silver Jubilee Medal (awarded for services in Industrial Safety), 1977

Collaborator on Paint Technology sections:

W. M. Morgans, PhD, C. Chem., FRSC
Consultant in Pigments and Paint Technology

NINTH EDITION

CHARLES GRIFFIN & COMPANY LTD
London and High Wycombe

CHARLES GRIFFIN & COMPANY LIMITED

Registered Office:
Charles Griffin House, Crendon Street
High Wycombe, Bucks HP13 6LE
England

First edition (by W. J. Pearce) published 1897
Eighth revised edition 1963
Ninth ,, ,, 1980

ISBN 0 85264 243 1

Set by Preface Ltd, Salisbury, Wilts.
Printed in Great Britain by J. W. Arrowsmith Ltd, Bristol

Contents

MATERIALS

Plates

Foreword to the Ninth Edition

When it became necessary to consider a further revision of this book, the Author knew that his health would not permit him to undertake this onerous task and that he would have to find a successor. But when his old friend, Jim Goodier, agreed to take on the work in spite of his own heavy commitments, the Author was relieved and happy to know that his "baby" could not be in safer hands and would receive the same meticulous care and dedication which produced Mr Goodier's own encyclopedic *Dictionary of Painting and Decorating*.

And so it has proved, for as each section of the revision appeared in typescript the Author has seen his original work pruned, polished, improved, and brought thoroughly up to date. Many chapters have been completely rewritten, with additional sections on Business Management and on Study Methods and Examination Technique which students should find invaluable, for Mr Goodier's experience as Chief Examiner in Painting and Decorating for the City and Guilds of London Institute enables him to write with unique authority on the latter subject.

In the original work the Author consulted the highest authorities in each section of the book so as to ensure maximum accuracy and authenticity. The highly technical sections on paint materials and media, for example, were kindly vetted by the late Noël Heaton, whose *Outlines of Paint Technology* became a classic; in this present revision the Publishers have enlisted the help of Noël Heaton's successor, Dr W. M. Morgans, a recognized expert in the same field.

The Author expresses his indebtedness and gratitude both to Mr Goodier and to Dr Morgans and is proud to have his name linked with theirs in this completely revised new edition.

Darwen, 1980 A. E. H.

Preface to the Ninth Edition

It is a great privilege to have been entrusted with the preparation of this new edition. For many years I have been recommending this book to students as the finest work ever published on the subject. In its original form it was the product of the vast knowledge and experience of the inimitable Walter J. Pearce; when changes in technology made a completely new edition necessary, the book as rewritten by Albert Hurst rose to even greater heights and made a unique contribution to the literature of the craft. Once again materials and techniques have altered to the point where considerable changes in the text are required, but most unfortunately Mr Hurst because of ill-health has been unable to undertake the work. When both he and the publishers asked me to take on the task I was delighted to do so, partly for personal reasons and partly because of the importance of the project.

The personal reasons stretch over almost half a century. Albert Hurst and I were both students at Stockport Technical College under that wonderful craftsman the late Mr Harry Garner. Albert is slightly older than I; when I was beginning my apprenticeship he was already the brilliant student who was winning the highest distinctions in craft education. Our ways diverged. His subsequent career was spent wholly in the service of the Walpamur Co. Ltd,* where he gained an international reputation as a designer and colourist. For my part, after a spell in design and a period as contracts manager in industrial painting, I made my career in technical education. Yet in all that time we have remained close friends, and on many occasions he has given me his unstinted kindness and help. It pleases me greatly that we are now so closely associated, and I like to think that Harry Garner would be pleased to see two of his old students linked together in this work for the benefit of the craft.

The importance of the project lies in the need for a complete and authoritative guide for the rising generation of craftsmen at a time when so many forces are operating against them. Present-day conditions make it very hard for a young person to get an all-round knowledge of the craft. One of the obvious factors is the reduction of the training period. At the time when this book first appeared the apprenticeship period was seven years, and even then it was generally agreed that the newly fledged journeyman still had a lot to learn. Since then the technology of the craft has become enormously more complex, yet the apprenticeship period has been

---

*Now Crown Decorative Products Ltd.

repeatedly shortened. It would now be quite impossible within that period for a young person to gain first-hand experience of more than a fraction of all the skills and techniques of the trade. It is true that there are day release and block release classes to supplement the on-the-job training, but the college lecturing staffs cannot hope to cover the subject adequately in the two or three years that are available to them.

But this is not all. Because of economic stringencies, local education authorities can no longer provide classes when the number of students falls below a certain level, which means that in many areas apprentices are denied the opportunity of attending a course, especially if they wish to pursue their studies beyond craft certificate level. Students wanting post-advanced studies are particularly badly hit. In theory they should travel to the nearest centre where such courses are offered; in practice this is often physically impossible. At one time, as I recall with gratitude, even if there was only one student wanting to take post-advanced studies to a high level the local college could find a way of accommodating such needs. Those days are gone, and even in the big cities there are not always enough students to make certain courses viable. Furthermore, where courses *are* offered, the painting students may be merged for part of the time with numerically larger groups from other dissimilar crafts, and taught by people with neither knowledge of nor experience of a decorator's requirements. It seems a pity that when opportunities for university education are wider than ever before and when all the professional bodies are raising their educational standards, young people who choose to develop their creative skills and become craft apprentices should be deprived of the opportunities for advancement.

One recognizes, of course, that the pattern of craft education is changing and that the Industrial Training Boards are providing excellent schemes of training for craft entrants. By definition, however, such training aims primarily at accelerating the productivity of the students. No-one would criticize the quality of the training schemes, but the members of trade organizations (whether for employers or operatives) are not always interested in craft education; far too often they are satisfied with a minimal standard of efficiency.

The purpose of this book has always been to give the student an insight into *every* aspect of the painting and decorating craft. Of all the crafts there is none with such a wide and diverse range of operations and none which calls for such a blend of abilities. Quite apart from the physical attributes required of anyone who undertakes manual work, a really competent decorator needs a thorough knowledge of many materials, together with the scientific basis of their interaction with various types of surface, a soundly developed sense of design and colour, and the ability to apply these considerations to a host of widely differing skills. Some people would argue that in the early stages of an apprentice's career it is only necessary to give an abridged form of instruction, cutting out any reference to processes and materials which are not in regular daily use on every site; they would say that students need only be given the facts about the commonplace materials and techniques without any explanation of the principles involved. This book approaches the matter from precisely the opposite point of view.

It is Mr Hurst's opinion, and mine too, that the only knowledge which is worth while is that which is based on real understanding. Rote learning, with the assimilation of lists of bare facts, has three drawbacks — that it is boring; that it is easily forgotten; and that it gives the student no basis on which to deal with an unexpected situation. In our view the only way of coping intelligently with any problem that arises is to work it out by a logical consideration of the factors involved. For this reason we have aimed throughout to explain the reasons underlying the use of any material or process. We take the same line about the inclusion of some processes which may not be in daily use by every decorator in the land. If a customer wants a piece of work involving one of these processes, the well-trained decorator should be able to supply it, otherwise the customer will go elsewhere to obtain satisfaction. A thoughtless "take it or leave it" attitude does no good at all either to the craft as a whole or to the individual decorator. The result of paring down craft instruction to include only the basic minimum requirements can only have one end — that of reducing the craft of painting and decorating to the status of a semi-skilled or even an unskilled trade.

A comparison of this edition with previous ones will reveal that several areas of the book have been completely re-written to take account of changing methods and techniques, notably in such matters as spray painting, paperhanging, scaffolding, and general trade practice. In dealing with modern paint materials and media we have had the invaluable assistance of Dr W. M. Morgans, a leading authority on paint technology. On the other hand, there are some areas of study where the main principles never vary. The chapter on Colour, for example, remains substantially as it was in the earlier edition, since in this field Mr Hurst is pre-eminent. Additional chapters have been included to provide some notes on business organization for the craftsman who is contemplating setting up in business on his own account. There is also a chapter on the technique of sitting for examinations, since long experience as an examiner has shown me very clearly that many students have a perfectly sound knowledge of their subject-matter but have never been trained to organize their thoughts in such a way as to do themselves justice.

I would like to thank all those who have made possible the production of this book, especially the Publishers for their wise and courageous decision to proceed with a major updating of the treatise as a whole, while retaining intact a few sections on aspects of the craft that now have a more limited though still definite appeal; thus saving the book from being reduced to a collection of "information sheets" for examination purposes. I am also much indebted to the many manufacturers of materials and equipment who have been generous in providing illustrations and information — in particular to Campbell, Smith & Co. Ltd, Crown Decorative Products Ltd, the DeVilbiss Co. Ltd, Gray-Campling Ltd (concessionaires for J. Wagner GmbH of Friedrichshafen), and SGB Ltd; the courtesy of all these firms is acknowledged individually in the text. Finally I wish to thank my wife who has shouldered the burden of typing and retyping the manuscript.

Newcastle under Lyme, 1980 J. H. G.

1

Introductory

There is probably no other trade or craft so interesting and versatile as that of painting and decorating, nor any which offers greater scope in its higher branches. The painter's field of activity is now wider than ever, embracing as it does every kind of domestic, public, and industrial work, and there must be very few other occupations in which such a variety of materials and processes is employed.

The good all-round craftsman may, and often does, attain a high standard of efficiency in several branches of the craft, but there is usually one particular branch which appeals to him more than the others, and in which he tends to specialize. He may, for instance, specialize in spray painting, paperhanging, or lettering, while some may have a natural talent and a strong preference for preparing perspective drawings and colour schemes, or for the painting of ornament and murals, etc. Those with a scientific turn of mind will be drawn to the technological side and the chemistry of paints, whilst those who excel at mathematics are naturally more suited to the business of estimating and costing. It will at once be apparent that the scope of the craft is so wide that no one can become equally expert in every branch, even if he hopes to do so. Nevertheless, it is very desirable that every student of painting should have, at least, a general working knowledge of the craft as a whole and, if possible, some practical experience in each branch.

We use the phrase "good all-round craftsman" quite deliberately because our purpose in the book is to foster the highest level of craftsmanship. It is important that decorators and students should consider the implications of this, since a great many people would have us believe that in today's world craftsmanship is dead. It may seem that when so many householders do their own decorating, and when so much of the decorative material on sale is specifically designed for the do-it-yourself market, there is very little future for the professional painter; indeed, the general public often believes that the professional has no special skills beyond those that any intelligent amateur can develop. But there are many reasons why these suppositions are misleading. In the first place, the type of operations that can be carried out by the amateur is very limited, just as the range of decorative products that can be successfully applied by the amateur is only a small part of the whole output. The well-trained craftsman can offer the customer a host of decorative effects far beyond the scope of most amateurs' skill. Then, again, most people regard home decorating as a burdensome, time-consuming chore, and are delighted if they can find somebody to relieve them of this irksome task at a reasonable price. It is obvious that intellectuals and professional people such as doctors, lawyers, etc., do not want to spend their leisure time performing manual tasks, and indeed that many of them lack the manual dexterity to produce worthwhile results. Then, too, much amateur work may appear satisfactory when first completed, especially as the householder tends to be blind to its faults in the first rosy flush of pride in achievement, but because of inadequate preparation or incorrect procedure the work may soon become so unsightly as to need renewing. For these reasons the services of a competent craftsman, although initially expensive, produce a quality of work which represents a real saving and economy to the householder in the long run. What is important is that the craftsman should be sufficiently well-trained to offer these benefits to the customer so that the quality of the professional's work is so demonstrably better as to convince the public of its superiority.

Unfortunately some of the present-day schemes of craft training are both ill-conceived and mistaken in their approach. The idea is current in some circles that apprentices should only be taught the rudiments of the craft, sufficient only for the everyday operations of the plainest possible kind of work. If customers discover that by employing a professional decorator they are paying a high price for a standard of work that is no better than what they themselves can achieve, the craft is brought into disrepute. It is also unfortunate that so much of today's craft training is prompted by the erroneous view that painting and decorating is one of the building crafts. In the first place, more than ninety percent of painting and decorating work is concerned not with new construction but with the maintenance and improvement of existing structures; and in the second place the painting work on a new building site is generally of the plainest and least adventurous kind, and is often dismally poor in quality because of price-cutting and because the British public has been conditioned to accept a low standard of work on new property. Many responsible people within the building industry itself imagine that the full extent of the painter's work is to apply emulsion paint on walls and ceilings and to give a quick coat of water-based paint to the woodwork. All these things do the craft a grave dis-service. No wonder the public image of painting and decorating is so unflattering.

It is most important that the decorator should, first, make the public aware of the immense range and variety of treatments that are available, and secondly, should aim to be competent enough to produce these treatments skilfully, tastefully, and at a competitive price. This will bring its own reward in the intense feeling of job-satisfaction that it engenders; there is no incentive for a student to study if the end-product is a dull, repetitive manual job with no degree of skill required, whereas work that demands a high degree of skill and effort is full of interest and fascination. It will also bring considerable financial rewards, and obviously the prospects of prosperity are both highly desirable and attractive. Put at its very lowest level, in a society with a high unemployment rate the competent craftsman has a far better chance of survival than an unskilled worker. More importantly, the decorator who is capable of producing a wide variety of work has no difficulty in finding customers who are looking for something beyond the average run-of-the-mill treatments. And the decorator who combines this knowledge and craft skill with a reputation for a high quality of performance is assured of a continuing and ever-increasing volume of work. It could be argued that Britain's prosperity as an industrial nation was linked with its reputation for the quality of its products, and that our economic decline is coupled to some extent with the decline of these standards. It is a certain fact that in any branch of trade that exists to provide a service for the public, a reputation for good-quality craftsmanship guarantees a full order-book.

It is for these reasons that this book is concerned unashamedly with standards of craftsmanship. This explains too why some of the chapters from earlier editions have been retained, although the skills they present are no longer as widely called for as in the days when they were written. Nobody would suggest, for example, that graining has any great significance in present-day practice. There is still, however, a demand for it to satisfy those customers who for some particular reason occasionally require a piece of woodwork to be imitated in paintwork. In these circumstances a knowledge of the technique is a valuable addition to any decorator's repertoire, and in various parts of the country a small number of specialist grainers find such a constant demand for their services that they are able to command their own price; thus Walter Pearce's original chapter has been considered worthy of inclusion. Similarly, some notes about techniques such as scumbling, glazing, plastic painting, etc., have been included. But clearly, present-day trade practice is far more concerned with developments in spray painting, industrial painting, and the production of modern types of wall covering, and these are the subjects that have been given due prominence.

The training of a painter and decorator consists essentially of two parts — theory and practice. In the first, knowledge is acquired about methods and materials, and in the second, skill is acquired in their application. The acquisition of skill can result only from practical experience, and by seeing and emulating the work of others; constant comparison with work of the highest standard is necessary to avoid becoming too easily satisfied with one's own performance.

Knowledge is gained through persistent inquiry into the why and wherefore of everyday practice and by personal observation and experiment. One needs to guard against taking too much for granted. Theory is often referred to slightingly as "book learning", whereas, in fact, "theory" means any *explanation* whether written or spoken. Theory passed by word of mouth from teacher to class or from craftsman to apprentice, may be misunderstood or misinterpreted if the speaker does not express his thoughts clearly, or if he uses loose or incorrect terminology. For example, the discoloration of certain colours on new lime plaster is sometimes ascribed to their being "burnt" by the lime (when in point of fact there is no burning action), or it is attributed to "acid in the walls" (when alkali is meant) — both statements being grossly misleading. The student must be prepared, therefore, to study independently and to seek information from a reliable source whenever he is in doubt.

It is the duty of the textbook to provide reliable information in a form readily understood. The object of the present work is to present to the decorator and the student apprentice an up-to-date study of painting and decorating technique, based on the best traditional and current practice. Since this is intended primarily as a manual for the practical operative, we shall deal mainly with the practical side of painting, i.e. the preparation of surfaces, the selection and application of paint, wallpaper and other materials, and the production of plain and decorative effects. As far as is possible within the limits of a single volume, however, we shall endeavour to deal briefly with the technical aspect of painters' materials, and also to give some guidance on the artistic side. In the study of colour, the student is often puzzled by the apparently conflicting theories of various authorities and we shall therefore attempt to clarify these points and to give some suggestions for the application of colour.

The Materials section deals with the highly complex matter of modern paint technology, about which a vast quantity of literature has been published. In our section the information must of necessity be condensed and will be confined mainly to those details having the most practical significance to the painter. At no point, however, will accuracy be sacrificed, and within the limits of space the information will be completely authoritative. Those who wish to study the subject of paint technology in greater detail are advised to consult the standard work by W. M. Morgans and the other volumes listed in the bibliography at the end of the book.

In recent years many new materials have been introduced which have revolutionized the techniques of painting, and in the appropriate chapters attention will be given to modern methods of preparation and application. Nevertheless it is still true that the common grounds for painting — namely wood, plaster, cement, iron and steel, non-ferrous metals, etc. — each have their own peculiarities as to absorbency, chemical activity, and corrodibility; and the importance of selecting a preparatory process and a primer suited to the nature of the particular surface cannot be overstressed, since the priming coat is the foundation upon which all subsequent coats must depend. Information about such rapidly developing techniques as blast cleaning, etc., is given considerable prominence in the text. Perhaps nothing in the past has given the painter so much trouble as the treatment of new plaster and cement surfaces, due very often to the customer's insistence on having such surfaces decorated before they were thoroughly dry. We deal at some length, therefore, with the characteristics of different plasters and cements so as to give a clearer understanding of the problems they present. In current practice the increasing use of building boards and sheetings and the introduction of dry building methods and prefabrication have tended to reduce the difficulties associated with wet building (i.e., with traditional bricks and mortar) but have brought with them a fresh group of problems. The scientific investigations at the Building Research Establishment provide valuable data about the behaviour of paints on varying types of surface under different conditions of exposure, and the student is strongly recommended to keep in touch with the reports and bulletins issued by this body from time to time.

Recent legislation has highlighted the appalling record of the construction industry in general, and the service industries such as painting and decorating in particular, with regard to industrial safety and welfare, so whilst introducing modern types of plant and scaffolding we discuss the safety aspects connected with them in some detail.

It is appreciated that many craftsmen nurse the ambition to set up in business on their own account, but that information about business procedures and the techniques of costing and estimating for painting work are not readily available to them; we therefore deal at some length with this important subject, taking the opportunity to point out some of the pitfalls likely to be encountered when starting up in business. It is hoped that this section will enable the aspiring craftsman to recognize the danger signals when they appear and to know what action to take. The notes on the costing of work will be of value to all students, whether or not they intend to run their own business, because no piece of decorative work, however carefully it is performed, can be considered a success unless it is economically viable.

In an ideal world, every job of work would be done as well as the craftsman's knowledge and skill would allow. In our highly competitive world, however, price is a ruling factor and the standard of craftsmanship is governed by the necessity for making every job show a profit. The great danger is that lowered standards may come to be regarded as normal practice, especially by those consistently engaged on work at cut prices. It is therefore the business of the textbook to set forth only those sound principles upon which the best traditional and current practice is based. Questionable methods and "dodges" are foreign to the spirit of fine craftsmanship, which depends not only on the complete understanding and skilful handling of materials, but also on faultless preparatory work.

Our experience in dealing with examination candidates leads us to believe that many students have never been trained to know how to cope with an examination paper. In many cases it is clear that the candidate is probably reasonably good at the subject but is unable to communicate properly. We therefore provide some hints on the much neglected matter of how to embark upon a study programme, how to prepare for an examination, and how to organize one's activity inside the examination room. We hope that this feature will be really useful and that it will prove to be of value to students everywhere.

In most towns of any size there is a college of further education, and in some (but by no means all) of them, painting and decorating courses are available to apprentices and new entrants to the craft. This is a most valuable facility and it is hoped that students will take full advantage of it. Our experience tells us, however, that students are sometimes reluctant to undergo college training, and resent being expected to attend. They regard the college as an unwanted extension of their schooldays, and put up a mental barrier to the learning process. If they could only realize it, the purpose of the college is purely and simply to help them to become proficient and to make a success of their working lives. The members of the teaching staff are practical people, well skilled and qualified in the work, who could have devoted their talents to improving their own position in life instead of helping the younger generation. Eventually most students do realize this, and usually they wish then that they had made better use of their opportunities earlier in the course. It is hoped that students reading these words will recognize this fact and make the most of their chances before it is too late.

Unfortunately, apprentices and young craftsmen in the smaller towns and in rural areas are often too far away from any centre where classes are provided, and consequently they have great difficulty in trying to pursue their studies, and especially in preparing themselves for a qualifying examination. Nevertheless, difficulties can be overcome, and it is possible for people even in remote areas to work out a course of home study if they are sufficiently determined. We hope that this book will be of assistance to them, and indeed to every student.

For the benefit of those who wish to further their studies, or to specialize in some particular branch of the craft, a list of recommended books is included at the end of this volume. Most students will have access to a public library or to the mobile library service provided for people in outlying areas, but they should also try to form their own collection of technical books, to have them permanently available. Incidentally it may be mentioned that *any* non-fiction book, even if not in the possession of the local library, can if necessary be obtained by applying to the librarian. When working from books, students should cultivate the habit of making notes or sketches of anything that interests them, always noting the name of both book and publisher for future reference. And finally, if they will bear in mind the words of Cennini which appear at the beginning of this book, they will be on the way to achieving the ideal combination of theory and practice.

2

Workshop and Stores

The workshop: construction and fitments

Although the decorator's work is carried out on site, at the customer's premises, he obviously needs a base from which to operate and a storeroom in which to house his materials and equipment. The form which these take depends on the size and scope of the firm and upon local conditions. Clearly, the needs of a very small firm employing only three or four men differ greatly from those of the contractor employing several hundred people, but there are certain common factors which apply to every kind of establishment. The ideal workshop for the average firm would include a paintshop, workroom, office and store, with accommodation for vehicles and provision for plant and scaffolding. It is desirable that there should be easy access to the road.

In most cases, brick construction would be the most suitable. Wooden buildings present too great a fire hazard, and although a building of light construction using, say, panels of asbestos-cement sheeting would be less expensive than brick and would moreover be non-flammable, it would be constantly liable to damage by children playing in the vicinity. Asbestos-cement might well, of course, be used for the roofing. The floor should be robust enough to withstand hard wear and should consist of some material that can easily be swilled down and cleaned; a concrete or cement-based flooring is adequate, especially if coated with a suitable floor paint, although a superior finish would be obtained with one of the slip-resistant industrial floorings made with vinyl and aluminium oxide chippings.

Good lighting is essential. For most painting purposes natural daylight is better than any form of artificial light. This means that there must be plenty of windows, and in addition there should be glass or PVC roof-lights, so that advantage can be taken of every moment of natural daylight. Of course there must be artificial lighting as well for the hours of darkness, and this must be plentiful enough to avoid any areas being in deep shadow; there is no point in being penny-pinching in this respect. It is particularly important that stairways should be well lit. The electrical wiring should also provide for a good supply of power points.

Ideally the premises should be heated throughout, but if this is not possible some means must be provided of maintaining the store room at a constant temperature of about 18°C to keep varnishes and enamels in good condition and to prevent stocks of emulsion paint from being spoiled by chilling. Other structural fittings would include toilet facilities, wash-hand basins with hot and cold water, and at least one sink deep enough to be used for filling buckets and washing brushes. Mounted on the wall close to the sinks there should be dispensers containing barrier cream and industrial skin cleaners. Nailbrushes and towels must be provided to comply with the Lead Paint Regulations.

In the paintshop a strong workbench is needed, placed near to a fairly large window with a north aspect, so that colour matching can be done accurately on it. It should be covered with heavy-gauge zinc, making it easy to clean down. A brush keeper should be situated on the bench, and this, together with brushes, will be considered in more detail later. Stout shelving, cupboard and drawer space are required for the storage of paint materials.

Stores and stock-keeping

The type of materials and the quantity of them to be kept in stock will depend largely on local conditions. In present-day practice many decorators carry only a very small stock, and rely on a nearby decorator's merchant to supply most of their requirements as and when needed. In every town there are merchants holding very large stocks, where the decorator can drive in and load his van with only a few minutes' delay, and all the big manufacturing concerns have retail outlets and generally operate a system whereby goods can be delivered to any of their depots within 24 hours of ordering. Many a small firm would see no point in carrying their own stock where such facilities exist. On the other hand, there are times when it is not economical to rely on a merchant's services. If work is being carried out at a remote spot in the heart of the country, deliveries may take place only at weekly intervals and the delivery charges may be high. Then, too, a firm of any size would obtain a substantial discount by buying in bulk and storing the material until it was required, discount that would otherwise be snapped up by the merchant.

Paints and varnishes

It would be reasonable to assume that the average firm would carry a stock of primers and sealers of various kinds, and a fairly large quantity of white undercoating and white gloss paint or enamel; these are materials which are constantly in use, and therefore it would be wasteful and uneconomic to buy them on a piecemeal basis, a few litres at a time. In the same way, any particular colour of paint which is in constant use would generally be stocked in bulk — a decorating firm might, for example, have a contract with a brewery company to paint a number of hotels in the company's standard colour range, in which case it would be known in advance that a fair quantity of these colours would definitely be required. When it comes to stocking emulsion paint, however, the situation is not quite so straightforward. It would be reasonable to buy a bulk quantity of white emulsion and of any colour which is in frequent use, but it would be unwise to lay in too large a stock because of the way that emulsion paint deteriorates if not used within a certain period. The amount to be bought at any one time is therefore regulated by the average turnover of stock.

No such difficulty exists in the case of varnish. Varnish, like good wine, improves with keeping; the writer has had the experience of opening a bottle of what at the time of its purchase had been a very moderately priced varnish and finding that over the course of some years it had matured into a fine full-bodied varnish of excellent quality, producing a quite remarkably good finish. It is a sound investment to buy a large quantity of varnish and to place it in store, to be used gradually as occasion demands. All these materials should be stored on shelves, rather than allowed to stand on the floor where they might become chilled; sound strong shelving is required for the purpose. The stocks of oil paints and enamels should be checked periodically and the tins either inverted at regular intervals to prevent settlement or placed for a few minutes in an agitator (see page 57 (Fig. 4.8)). Varnish bottles should be stored lying on their sides in racks specially made for the purpose.

Other materials which will normally be kept in store include white spirit, knotting and knotting bottles, filler compositions, paint removers, paste powders, containers of ready-mixed adhesives, and sundries such as sponges, chamois leathers, etc. Many firms now buy the white spirit in 5-litre tins; the alternative is to purchase it in large drums or tanks, which are placed on a stillage so that small tins or bottles are easily filled from them. In this case a drip tray is necessary, and a supply of tun-dishes or funnels and measures. Care must be taken to store fillers and other dry materials in bins to protect them from damp.

Other materials

Certain materials will of necessity be stored in the paintshop. These are the materials needed for various kinds of specialist jobs — graining materials, tubes of fine colours for signs and decorative work, various grades of goldsize, etc., together with the tools and appliances for these activities. The policy of the decorator's merchants is to stock only those items which can be guaranteed to have a large sale and rapid turnover. Whereas only a few years ago most of the merchants took a pride in being able to supply whatever a decorator called for, however obscure, present-day sales techniques have altered the whole outlook, and the stocks of small specialist lines have been swept away as being uneconomic. It therefore follows that the decorator is compelled to maintain his own stocks of these things, for if he waited for the local merchant to obtain them when required he would never be able to carry out his commitments.

Racks will be needed for whatever paperhanging materials are kept in stock. There will generally be lining papers, ceiling papers, anaglyptas and ingrains. When it is apparent that a particular pattern is popular and in frequent demand, there is a considerable advantage to be gained by buying in fairly large quantities because of the very attractive discounts that are given on bulk orders. But the careful decorator will keep a watchful eye on the trends, or he may find himself saddled with large quantities of a material which has suddenly and unaccountably fallen out of fashion.

A certain part of the paintshop should be reserved for the stock of buckets, paint cans and similar articles. If these are stacked neatly they occupy little space, but unless there is a definite place for them they tend to become scattered around in untidy heaps. A large pickle-vat or cask containing a strong solution of caustic soda is useful for cleaning paint cans and stripping small metal fittings. It should be remembered, however, that aluminium and zinc are rapidly decomposed by strong alkalis and should never, therefore, be placed in pickle.

There must also be a good supply of dust-sheets. Since so many properties are now carpeted from wall to wall, most decorators have been compelled to adopt the practice of leaving the carpet in position and sheeting the entire floor in addition to covering the furniture, which means that a far bigger stock of dust-sheets is needed than was the case a few years ago. The choice lies between polythene sheets, which are waterproof but are rather easily ripped, and the traditional cotton twill; if the latter are used, arrangements should be made to have them laundered regularly. Clean white dust-sheets, stencilled with the firm's name, give the immediate impression of careful workmanship, and as such they are an excellent advertisement (see Chapter 4).

Inevitably there will be a certain amount of surplus paint returned to the shop from practically every job. Special shelves should be provided for these returns, so that if the opportunity arises for some of them to be used, they can be readily identified. Ideally, of course, the quantities required for a job should be estimated as accurately as possible so as to avoid waste, because the amount of these partly filled tins of paint that can be used up is strictly limited. In some cases there may be an outlet for their use in such things as the painting of rough woodwork or the backs of wallboards before fixing, but it is not advisable to mix them together indiscriminately as the various media and solvents used in their manufacture may be totally incompatible.

Fire precautions

In every paintshop there should be chemical fire extinguishers, situated in prominent and easily accessible positions, as well as buckets of sand for use in case of emergency. On no account should water be used on burning oil or spirit, as it will only make matters worse by spreading the fire. Nor must water ever be used on an electrical fire. It is a wise precaution to invite the Fire Prevention Officer from the local fire brigade to visit the premises and give his advice about the right type of extinguishers and the best points at which to site them; the officer will be glad to do this and the service will be provided free of charge. Arrangements should be made with the firm that supplied the fire extinguishers to come and service them at regular intervals. This will not cost a great deal, and is a very sensible precaution; there are few things so shattering as to discover, at the very moment when an emergency occurs, that the extinguishers have deteriorated and become useless. It is also a very good idea to inform the fire officer if any liquefied petroleum gases, such as Calor Gas containers, are kept on the premises for such purposes as the operation of blowtorches, etc. The bottled gas should always be stored in the same spot, and the fire brigade notified of its location. This means that in the event of fire the brigade knows of the existence of the extra hazard and can take the appropriate measures to deal with it; it also affords some measure of protection to the firemen who might otherwise be faced with an unexpected explosion.

With a reasonable standard of care, the fire hazards in a paintshop are not significantly higher than in any other workplace. Contrary to popular belief, tins of paint are not in themselves a source of danger. In fact, it is impossible to make a tin of fluid paint ignite by applying a light to it, although naturally when a fire has once gained a hold and is sweeping through a building any tins of fluid paint provide additional fuel. But certain hazards do exist, and sometimes they are not particularly obvious. Paint rags, for instance, are a source of potential danger, especially when they have become saturated with graining colour, glaze, or the oily colour used in wiping relief effects. If such rags are left crumpled up, they will generate heat and may ignite through spontaneous combustion. It should therefore be the rule, or better still the unvarying habit, to see that all paint rags are spread out flat on the zinc-topped bench or on the floor before work is finished for the day; the oil will then oxidize freely and harmlessly. When dry they can be disposed of safely along with other waste.

Equipping the workroom

The arrangement of the workroom or painting room will provide for the particular requirements of each business, but it is assumed that signwriting and applied decoration will be included in the firm's regular practice. This room will therefore have to be large enough to accommodate fairly big signboards, for it is much more convenient and comfortable to paint and letter these indoors than *in situ*. A large, solidly constructed zinc-topped bench or table is almost indispensable for supporting work of this kind during the various stages of burning off, rubbing down and painting; but while the sign is being lettered it is better supported upright on tall props made from stout timber drilled with holes about 150 mm apart to hold strong steel bolts or pegs (Fig. 2.1). One or two strong easels will be required for small signs, exhibition panels, and so on.

Glass signs will be lettered flat on the bench, and here again the solid construction of the bench will be an advantage. Stencil and template cutting will also be done on this bench, a large sheet of thick plate glass being kept for the purpose, as well as a piece of thick lino and a steel straight-edge for cutting mounts.

A flat drawing desk or bench by the window will serve to carry an adjustable desk easel supporting an A1 (841 mm by 594 mm) drawing board for setting up scale drawings and colour schemes, a chest of shallow drawers being provided underneath to hold drawing and watercolour papers, also drawings and records of work completed. One large wall should be covered with wallboard (not the hard-pressed type) on which large drawings or cartoons can be prepared and posters painted. A medium-

Fig. 2.1 Sign props
A simple method of accommodating large signs in the workshop

pressed board such as Sundeala takes drawing pins easily and provides an ideal working surface. An alternative method of holding the paper is by the use of self-adhesive masking tape.

The painting of fittings and furniture, and the priming of new work prior to fixing, will all be done in this painting room, and a few strong trestles will be handy to lay the work on. It is useful to have a part of the room glass-partitioned off to form a finishing room where small signs, exhibition panels, and other work requiring a high degree of finish can be varnished or enamelled. This inner room should be made as dust-tight as possible, and its design will be discussed later, in the chapter on Varnishing.

In the average small establishment, office and stores are combined. This is where all unopened stock is kept. Varnish, gloss paints and enamels are stored in an equable temperature — about 18°C — to ensure good condition. It is as well to remember that when varnish or enamel is taken from such a heated room in cold weather for use outdoors or in unheated premises, the material should be given an opportunity to adjust itself to the lowered temperature before use. Valuable stock such as gold leaf and brushes will be kept under lock and key, and a few moth repellent tablets will be placed in the brush cupboard to protect the bristle from damage. Blowlamps and torches, abrasive papers, sponges, chamois leathers, special graining tools, spray guns and spare parts for spray equipment, and similar items will also be in the storekeeper's special care.

Office equipment

The storekeeper needs a fairly tall desk, as he is usually standing when checking out the materials. Very obviously, a careful and accurate record must be kept of all materials and any items of plant or equipment that are sent out to the various jobs, and an equally careful record of any returns.

The old-fashioned method was to enter them by hand into a Day Book, but this is a clumsy and outdated procedure. It is far better to use a modern sophisticated system, with properly designed job cards printed according to the specific requirements of the individual firm. Advice about suitable book-keeping systems is available from numerous sources, and the employers' associations provide useful information to their members, while some of the large paint-manufacturing companies supply booklets about this aspect of running a business as part of their service to customers.

The remaining office equipment will consist of tables, swivel chairs, typewriters, a desk-top photocopier, filing cabinets, telephone, etc. Where possible, a small private office should be included to ensure privacy when business is being discussed with clients or with commercial travellers. In addition, those little differences that crop up occasionally between employer and operative are more likely to be settled amicably by a quiet talk in private than a heated argument in the presence of other members of the staff. A small reference library of textbooks and trade journals can be housed in the office, but every book borrowed should be entered in a loan book with the item signed for by the recipient and the signature cancelled when the book is returned. This is the only way to ensure the safe return of borrowed books.

Storage of plant and fittings

Scaffolding and plant need to be stored under cover to protect them from the weather, but it is desirable that the sides of the scaffold shed should be open or louvred to allow a free circulation of air. Ladders are best stored horizontally on a ladder rack, i.e. a series of strong steel brackets projecting from the walls; ladders are rested on these brackets, not suspended, as this tends to pull the sides away from the rungs. At the foot of each ladder, on the inside face of the stile, the number of rungs should be clearly indicated by means of a boldly painted figure. In this way a ladder of the right length for any particular purpose can be selected immediately without the laborious business of pulling out one ladder after another until a suitable one is found. Step-ladders and small trestles are stored vertically, but long trestles and scaffold tubes are often a problem, especially if they are not in frequent use. It is sometimes possible to lay them flat on overhead tie-beams, or on suitable slung supports from the roof, provided there is an ample safety margin. Planks may be laid flat or on edge, resting on short wood supports to raise them off the floor; laying them on edge makes them more accessible. Lightweight stagings are better laid on edge, and here again it is useful to have the length of the stage indicated at the end by a painted figure; the individual lengths of staging are quite heavy and sorting through several in order to find the particular length required can be a very tiring exercise.

Many decorators nowadays carry a stock of prefabricated aluminium frames with which tower scaffolds can be built; these frames being of uniform size are easily stacked, and can either be laid flat or stored practically upright, tilted slightly to lean against a wall. Cradles are folded down and either laid flat or stored on edge. It is most important that wooden scaffolding equipment, especially ladders, should not be stored close to stoves, boilers or other sources of artificial heat, otherwise they can twist and warp very badly.

Specialized tools and appliances are discussed in their appropriate chapters.

Certain items are best kept separate from the main scaffold store. Small items of scaffold such as bosun's chairs and ladder brackets have a knack of getting pushed out of sight and buried under mounds of heavy equipment, so that when they are next required for use a time-consuming search takes place. Scaffold couplers need collecting together and sorting into piles according to type; they are best kept in strong wire-mesh containers. It is important that operatives should be trained to return the couplers to the storekeeper so that they can be checked off; only too often the couplers are left lying about on jobs and consequently get lost, yet it is obvious that without them the scaffold tubes are useless. The sheave blocks used for suspended scaffolds must also be returned to the storekeeper to be checked and oiled ready for the next time they are needed. Particular care should be given to scaffold ropes, as these are so easily damaged by careless treatment. Ropes that instead of being properly coiled are allowed to become twisted and kinked are greatly weakened; so too are ropes that have been in contact with acid materials. The procedure will vary according to the organization of the firm, but the big painting contractors have a system whereby every rope is labelled and numbered by means of a metal tag, and every rope that is returned from a job is inspected and checked by the storekeeper before being sent out to another site; the result of the inspection is entered on the scaffold register, which complies with the requirements of the Construction Regulations. The ropes are thus used in rotation, which avoids any of them getting undue wear, and they are known to be safe. This is a very sound system, but clearly it can only operate in shops big enough to employ a full-time storeman.

Cellulose materials

Many decorators undertake the lettering of vans and other motor vehicles as a profitable sideline, and some go further and do a certain amount of car spraying. There is no reason why this should not develop into a useful side of the business, but if cellulose materials are to be sprayed it is essential that the provisions of the cellulose paint regulations be strictly observed. Among other things, these regulations demand that the workshop be brick-built, the cellulose stored outside the main workshop, and the light-bulbs be protected by stout glass shields which in turn are guarded by metal cages, so that no breakage can occur which could expose the naked bulb, with its possibility of sparking, to the atmosphere of the workshop. Suitable extraction must be provided, and all electrical switches and equipment must be housed outside the workshop. These stringent regulations are needed because of the highly explosive nature of cellulose. Any decorator intending to engage in this class of work is required to notify H.M. District Inspector of Factories of his intention, and the premises must have been inspected and judged suitable for the purpose before any work is undertaken in them.

3

Plant, Scaffolding, Access Equipment, and Site Safety

The essential thing about painters' plant and scaffolding is that, owing to the need for mobility, it should be as light as possible consistent with strength. The heavy, unwieldy equipment of the builder, designed as it is for greater loads and for comparatively static conditions, is both unnecessary and uneconomic for painters' use. The design, maintenance and use of ladders, cripples, trestles, stagings, and all forms of scaffolding are subject to stringent conditions that are laid down in the appropriate sections of the Construction (Working Places) Regulations.

LADDERS

Ladders are usually constructed from selected Douglas fir or spruce, free from knots, the sides or stiles being made from the two halves of a single pole so that they are of equal strength and resilience. The rungs, staves or rounds, as they are variously called, are of hardwood such as ash, oak, hickory or birch, with some form of metal reinforcement. The old-fashioned reinforcement was a narrow steel tie-bolt inserted under the second rung from the top, the second rung from the bottom and every three or four rungs between. The more modern type is a stout galvanized steel wire sunk into a groove underneath every rung, the wire being tensioned and anchored into the stiles and secured at each side with a ladder button. The purpose of the reinforcement is to support the rungs and hold the stiles firmly together. The holes where the rungs are fitted should have been coated with red lead primer or waterproof adhesive before fixing.

Ladders with half-round sides are known as "pole ladders" or builders' ladders; those with rectangular sides are called "standing ladders". Pole ladders are heavier and more robust than standing ladders, and this type is invariably used when a single ladder of considerable length is required. Standing ladders of more than 5 metres length are reinforced with galvanized iron sunk into grooves along the stiles and inserted under tension. For large-scale work, pole ladders are generally more suitable, but for ordinary small-scale domestic work standing ladders are often preferred because of their comparative lightness.

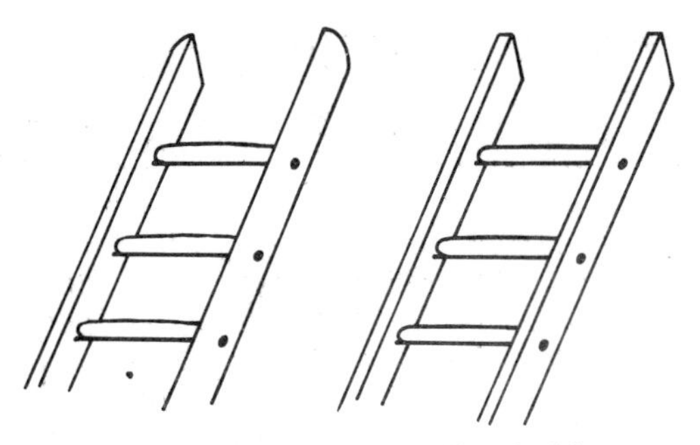

Fig. 3.1 Pole and standing ladders

Extension ladders

These are particularly useful as they can be adjusted to varying heights; they are supplied in two or three sections which can be used separately as well as in combination. Extension ladders are of two kinds, the hook and clip-catch type and the rope-operated catch type; the hook-and-clip type are intended mainly for casual use by the householder and are not really strong enough for the heavy wear and tear of continual everyday use by the painter and decorator. Some extending ladders have steel cable reinforcement inserted along the underside, while the "Eclipse" has a steel rail-track which greatly reduces friction and facilitates the extending operation.

"Splicing" ladders

Where extension ladders are not available, single-section ladders may have to be lashed together or "spliced" to reach the required height. It is sometimes recommended that the ladders be lashed together while they are flat on the ground, but this can make them too heavy to raise, unless a rope is suspended from the roof to assist in the hoisting. The more usual method is to raise both ladders separately, leaning them together against the wall, i.e. one resting upon the other. A splicing rope is first looped round the bottom rung of the under ladder, carried up and over the sixth, seventh or eighth rung (or higher if necessary) of the top ladder; then, with one man pulling at the rope and another man lifting the under ladder from behind, the latter is raised to the required height and secured by taking the rope a few turns round both rungs, now at the same level, and lashing the ladder sides firmly with the remainder of the rope. The top rung of the lower ladder is then tied to the corresponding rung of the upper ladder and the sides are lashed together as before. Splicing ladders, however, is a tricky business which should never be attempted by the inexperienced without the expert supervision of an old hand. Neither should a young apprentice risk hurting himself by trying to raise a heavy ladder by himself, no matter how strong the temptation to show off. In nearly every case, rearing a ladder is essentially a two-man job.

Ladders need some form of protective treatment. They should not be painted, as this adds considerably to the weight and conceals cracks and other defects; in any case the painting of any new timber used for scaffolds, trestles, ladders and step-ladders is forbidden under the Construction Regulations. The best treatment is to coat them with a good quality wood-preservative fluid followed by a coating of clear varnish, with a fresh varnish coating applied at suitable intervals for maintenance purposes.

Metal ladders

Aluminium or aluminium alloy ladders are preferred by some decorators because of their convenience, lightness and strength. They do not rot or decay like timber, and the only protection they require is against strong alkalis (cement, lime, etc.) which attack aluminium. Because of their smoothness and their tendency to slip when used on hard floor surfaces, they are often supplied with rubber non-slip feet; alternatively they can be supplied with interchangeable foot mountings, with spikes for soft ground and rubber suction pads for smooth surfaces. The "Stecalloy" extension ladders made by Stephens & Carter have V-section stiles which permit the upper and lower ladders to engage in a smooth sliding action.

An important point to remember is that metal is a good conductor of electricity, and metal ladders should not be used in any situation where contact with electric current is possible because of the very great danger. It does not follow, however, that wooden ladders are invariably safe and can be used with impunity in such situations; if the stiles of a wooden ladder are reinforced with galvanized wire they too provide a path for electricity, and very great care is needed when they are being used in close proximity to electrical equipment.

Safety in the use of ladders

Ladders are the most commonplace pieces of equipment, and because they are so much a part of the painter's everyday life many people tend to be careless with them; it is the old story of familiarity breeding contempt. Accidents involving ladders are extremely common, and account for a high number of mishaps every year, some of them fatal. Most of these accidents would never happen if people knew the safe methods of working and, more important still, if they made it an invariable habit always to use the safe method.

Ladders should be inspected at regular intervals. Inspection is a two-man job. One man sits astride the ladder and then raises his feet from the ground, and the other man checks the stiles for any defects such as cracks. At the same time the rungs should be inspected by tapping them lightly with a wooden mallet, a dull hollow sound indicating a faulty rung. The wedges should be checked for tightness. Any ladder that is defective should be withdrawn from use at once.

Whether they are stored horizontally or vertically, ladders should always be "at rest" and never left in any position liable to cause twisting or strain. When lying flat they should be supported at the ends and at intermediate points as well. They should never be stored near boilers, radiators or in any other situation where they are exposed to excessive heat. When ladders are being loaded on to lorries or vans, the gantries should be padded and the ladders lashed to prevent them from slipping;

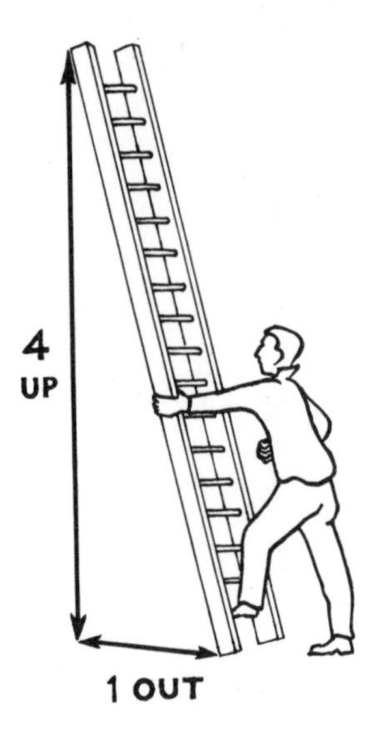

An easy way of making sure that a ladder is inclined at the correct angle . . .

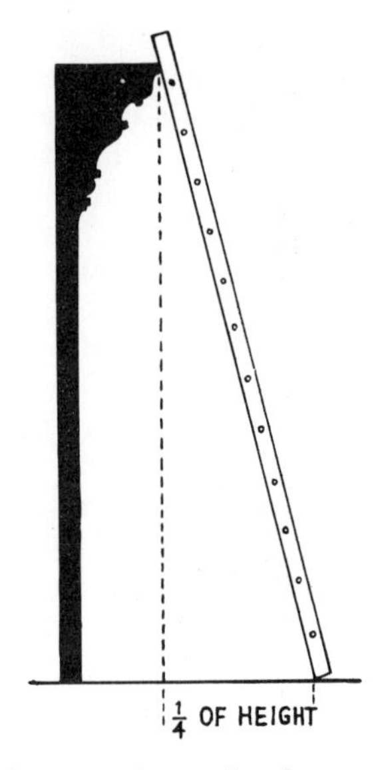

. . . but remember to allow for any projections such as cornices which take the top of the ladder away from the wall . . .

Fig. 3.2 Ladder safety:

failure to do this causes bruising and splintering of the stiles, which allows moisture to enter, thus setting up decay.

When in use, a ladder must always be at the correct *angle of inclination*: if it is too steep it may overbalance, and if it is not steep enough the foot will slide outwards and the ladder will slip downwards when a man's weight is upon it. There is an easy way of knowing if the angle is correct: the vertical height of the ladder is always four times the distance outwards at the foot (Fig. 3.2).

No ladder should ever be used if any of the rungs are missing — and this means not using it if a missing rung has been replaced with a length of wire or string, a piece of metal conduit, or a piece of wood nailed across.

Whilst in use, a ladder must always be securely fixed at the top to prevent it from slipping. This usually means lashing it with a rope to some part of the structure of the building. Where there is no means of doing this the ladder should be securely fixed at its lower end. If for any reason this is not possible (such as for example if the ladder is resting on paving slabs), the ladder must be "footed" by having a man stationed at the foot. But remember that if a ladder once starts to slide the man at the foot has very little chance of arresting the fall. Wherever possible a long ladder should be secured to prevent swaying or sagging.

A ladder must always have a firm and level footing and must never be rested on loose bricks, loose packing, or on tiles and similar brittle materials. On sloping ground, the soil should be dug out to make the feet level.

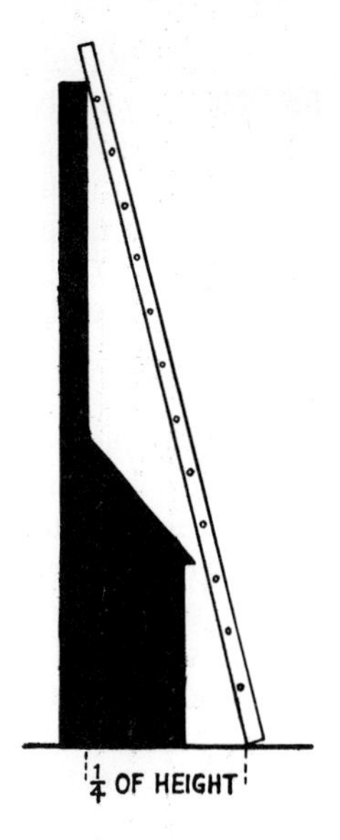

. . . whereas projections at the base, such as outhouses or conservatories, must be reckoned as part of the base-line.

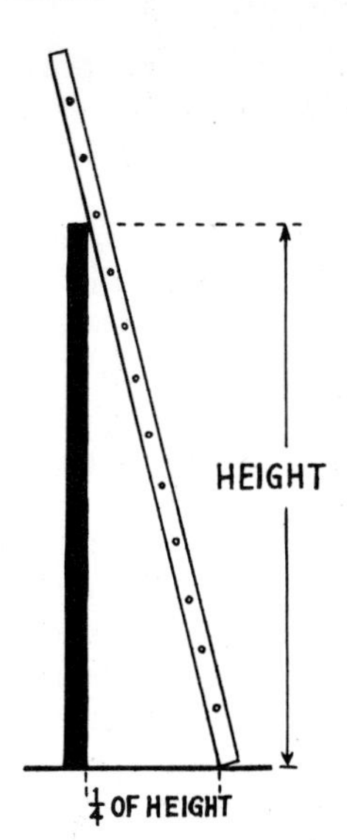

If a ladder extends beyond its point of support, the height is the distance from the ground to the supporting point, NOT the length of the ladder.

angle of inclination

On a sloping pavement an adjustable safety foot should be fitted to the ladder (Fig. 3.3). If it is necessary to use packing under the foot, use a wooden wedge, and keep a man stationed at the foot.

When a ladder is used as the means of access to a working platform, it must by law extend to a height of 1·070 metres above the platform unless there is a suitable handhold in the structure. In all other cases, a ladder must be long enough to extend 1·070 metres above the highest rung upon which the operative's feet will rest. If possible, a ladder should not be used for a vertical distance of more than 9·140 metres (30 ft) unless some form of intermediate landing place is provided, although of course the law recognizes that this not always practicable.

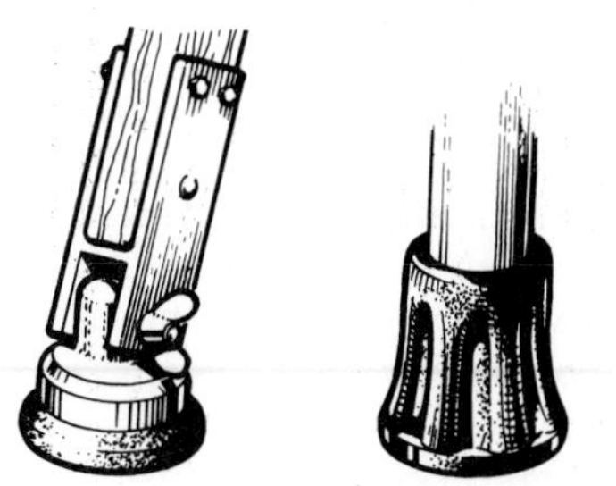
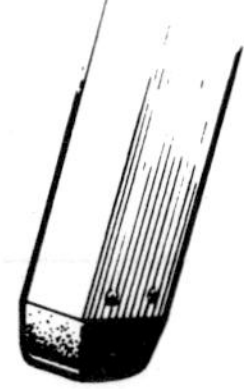

Courtesy: W. C. Youngman Ltd and SGB Ltd

Fig. 3.3 Ladders: anti-slip feet

These are some of the provisions of the Construction Regulations relating to ladders, and the painter should make himself thoroughly conversant with them all. It should be remembered that in the case of any infringement of the Regulations it is not only the employer who is liable to prosecution — the operative may be prosecuted as well. But quite apart from the question of legal penalties, it is only common sense to see that the regulations devised for one's own personal safety are observed at all times.

Roof work

The painter's work often involves climbing over sloping roofs to reach dormer windows, short lengths of guttering, ventilation shafts, etc. The common practice of using an ordinary pole ladder or standing ladder on the roof, with its feet resting in the gutter, is a dangerous one. The guttering itself may be too weak to take the strain, but in addition there is the risk of the gutter brackets snapping off either because, in the case of older property, they have corroded away or, in the case of modern plastic guttering, they are intrinsically brittle. Wherever possible a properly constructed *roof ladder* should be used, with a supporting member which fits over the ridge and anchors the ladder on the other side of the roof (Fig. 3.4). Failing this, a properly made crawling board should be used. Even with the correct equipment in use there is still the possibility of the painter missing his footing or rolling sideways, so it is a wise precaution to wear a safety belt as well.

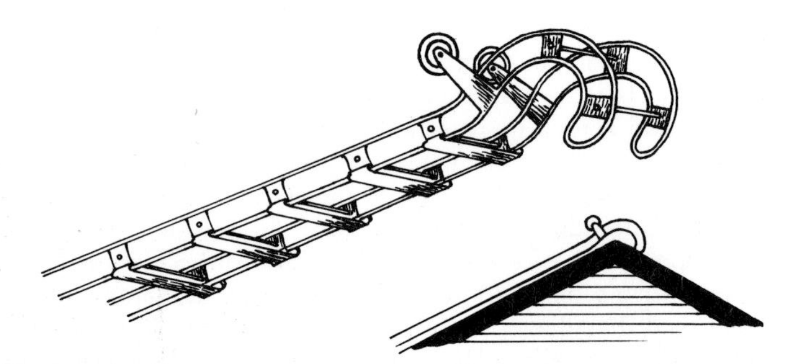

Fig. 3.4 Roof ladder with wheels for use when pushing the ladder into position up a sloping roof

Another very hazardous situation is presented by fragile roofing materials, such as the asbestos-cement sheeting so often used for the roofing of factory buildings. Few people realize how very brittle this material is, and the trouble is that it fractures and splinters without giving any warning. If anybody rests his weight upon it even for a moment it is liable to shatter, and the unfortunate man has no chance of avoiding a precipitate drop through the hole down to the floor below. Some people imagine that they will be safe if they rest their weight on the line of roofing bolts where two sheets overlap, but this gives no protection at all; two sheets are just as brittle as one. When working on fragile roofs it is essential to use properly constructed roof ladders or crawling boards, and the important thing to notice is that two such ladders should be placed side by side so that the operative, when he needs to move one of the ladders to a fresh position, has another ladder to stand upon. Remember always that it is not safe to step on the roofing material even for an instant. As an extra precaution, a safety belt should be worn.

For working on a curved roof surface, such as the exterior of a dome, short ladder sections should be lashed together to conform to the shape of the roof, and they should be firmly anchored.

Ladder brackets or cripples

Ladder brackets or cripples enable the painter to rig up a light scaffold with the aid of two stave ladders and a plank, a method which provides a very convenient way of painting signs on gable-ends, etc. There are several patterns of ladder brackets, both fixed and adjustable, the latter being suitable for use either in front of the ladder or behind it. Preference should be given to those which do not depend on the strength of only one rung of the ladder (Fig. 3.5(A)).

The ladders must of course be secured to prevent slipping, and a length of scaffold tube should be lashed between them at a suitable height to provide a handhold. It is a wise precaution to provide a third ladder standing close to one end of the scaffold (Fig. 3.5(B)), which the painter can use as a means of direct access to the working platform without having to perform the acrobatic feat of clambering over the front face of the bracket.

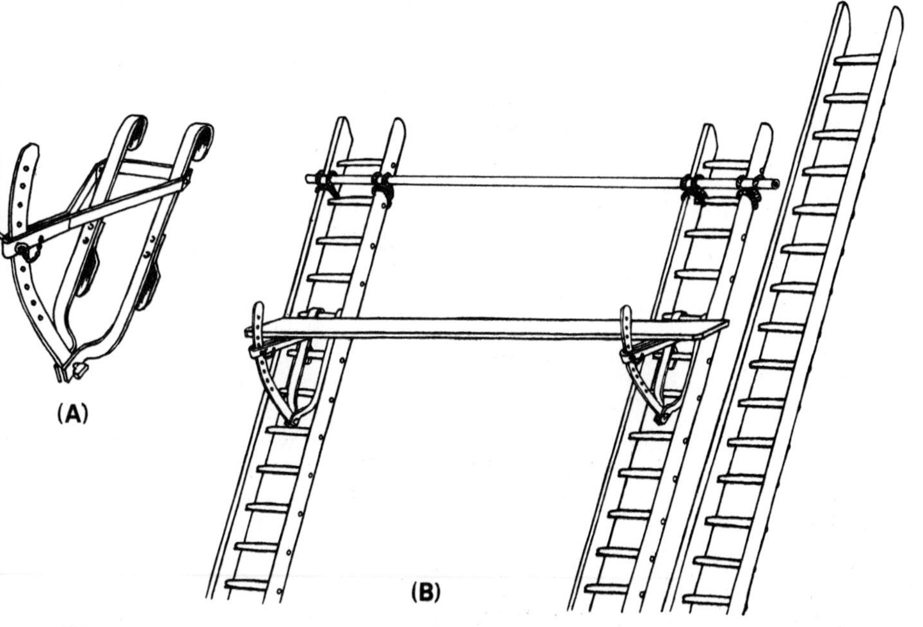

Fig. 3.5 (A) Ladder bracket; (B) bracket scaffold with metal tube lashed across the ladders to form a handhold, and with a third ladder to provide safe means of access

BOARDS, STAGINGS AND TRESTLES

Scaffold boards (generally called "planks")

Planks should be selected for their lightness, toughness and freedom from large knots; they should preferably be of good sound fir or spruce boards of between 225 and 275 mm (9 and 11 in.) width, 3 to 5 metres (10 to 16 ft) length, and of suitable thickness to avoid undue springiness. The corners are best cut off so that there is less likelihood of splitting if the plank is accidentally dropped on one end, and the ends may further be protected by tenoning with a strip of hardwood; this is better than the practice of binding the ends with hoop iron, which can inflict nasty injuries to the operative's hands, rip his clothing, and of course cause untold damage to the flooring in the customer's premises.

Planks, being everyday pieces of equipment, tend to be neglected. They should be tested at regular intervals to detect any flaws that may be developing. But never test a plank by jumping on it: this imposes a sudden and violent strain which can cause damage where none previously existed. The best way is to support the plank at each end, place a heavy weight upon it, and then inspect it so that defects can be detected. Where planks are used as part of a working platform they should, wherever possible, be supported in the middle as well as at the ends, to prevent undue sagging. They should be firmly secured to the transoms to stop them from slipping. On no account must they be allowed to project beyond the end supports to a distance exceeding four times the thickness of the plank, unless they have been effectively secured to prevent them from tipping. A plank which overhangs the support too far is called a "trap end", because an operative walking along the platform may be unable to see the danger that exists, and when he steps on the overhanging part the plank suddenly tips up and sends him shooting downward.

Extending or telescopic scaffold boards are useful in that the length can be adjusted to any given situation; they are an asset in confined spaces and awkward corners, but are very heavy and rather "swingy" in use.

Lightweight stagings

These are working platforms of approximately 450 mm (18 in.) width, constructed with timber stiles, cross-bearers and decking finishing flush with the top edge of the stile so as to provide a flat working surface. They are available in lengths of from 1·8 to 7·3 metres (6 to 24 ft). They are designed to support a load equivalent to the weight of three men with their tools and appliances, approximately to a total weight of 272 kilograms (600 lb). Their great advantage is that they are absolutely safe when supported at each end, without any intermediate support being required; this is because they are so rigid that even when fully loaded they show hardly any deflection. Because of this, they are the ideal form of scaffolding for interior roofing work, especially in factories, where they can be used to span between the roof trusses to provide a running platform for the entire length of the building (Fig. 3.6). The advantages of this are obvious compared with normal planks which, besides being rarely long enough to span between the trusses, are always so springy as to lead to a

Courtesy: SGB Ltd

Fig. 3.6 Lightweight timber stagings

These provide a stable platform for interior roofing work

considerable amount of swaying and sagging. Anyone accustomed to roof work knows only too well the dangers and difficulties of working with two planks side by side; the two planks are always at different levels and are both moving upwards and downwards as the weight is transferred from one to the other, making it very easy for the operative to miss his footing and trip up. Lightweight stagings provide a stable platform which makes for safe, confident working. To comply with the Construction Regulations a handhold must be provided for the operatives, but no guard rails or toeboards are required.

The only difficulty arises when the stagings are being hauled into position: the term "lightweight" is then found to be rather misleading. Admittedly the staging weighs less than the equivalent quantity of scaffold planks. For example, the weight of a staging 6 metres long by 450 mm wide is certainly less than the total weight of four planks each 3 metres long by 225 mm wide; but, of course, the four planks would be handled separately, whereas the staging has to be lifted as a complete unit, which is quite a strenuous task.

Steps

The correct name for the type of steps used by the painter is "swing-back steps". They are generally known in the trade (quite wrongly) as "step-ladders": step-ladders are really single-section ladders with rectangular sides and flat treads, used chiefly by householders and shopkeepers, whereas the painter's steps have a hinged frame which swings back to make the ladder self-supporting.

Steps should be protected with a clear varnish coating, and should be regularly inspected; check the ropes for fraying, the hinges for rusting, and the screws for loosening. Make sure the steps are always fully opened and extended whilst in use; never stand on the back frame, and never put bricks on top of the steps to give extra height to a plank.

Trestles

Folding trestles are made with two frames hinged together, each fitted with cross-bearers suitable for supporting a working platform; to relieve the hinges of strain when the trestles are being carried in the closed position, cheek blocks are fitted to the inside faces of both stiles on one of the

Courtesy: SGB Ltd

Fig. 3.7 Adjustable steel trestles, providing a convenient working platform for medium heights

frames. As with steps, they should be given a clear protective coating and should be regularly inspected. A trestle scaffold is formed with two pairs of trestles supporting a plank or two planks, or better still a lightweight staging.

The provisions of the Construction Regulations relating to trestle scaffolds are quite stringent. Such scaffolds must not be used if a person would be liable to fall from the working platform a distance of more than 4·570 metres (15 ft). It is forbidden to put two working platforms at different heights between any two pairs of trestles. A trestle scaffold must not be erected on any other sort of scaffold platform unless the platform is wide enough to allow a clear space around the trestles for the transport of materials, and unless the trestles are braced and firmly secured to the platform to prevent movement.

Adjustable splitheads provide an easily erected and rigid platform for medium heights, and this is a great convenience where spray plant is being used or where quick manipulation over a large area — as in blending operations — is required. For the purposes of the Construction Regulations such a platform is classed as a trestle scaffold and is subject to the same restrictions. More widely used, however, are the adjustable steel trestles shown in Fig. 3.7.

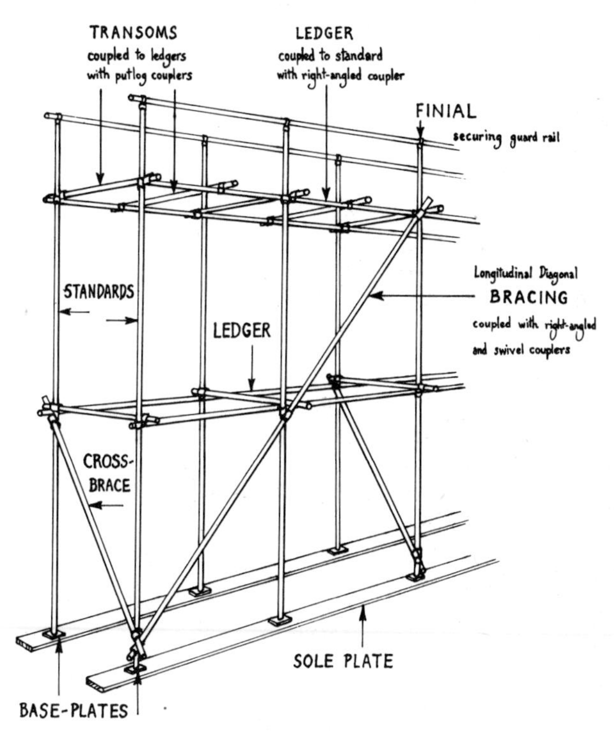

Fig. 3.8 Scaffold terms

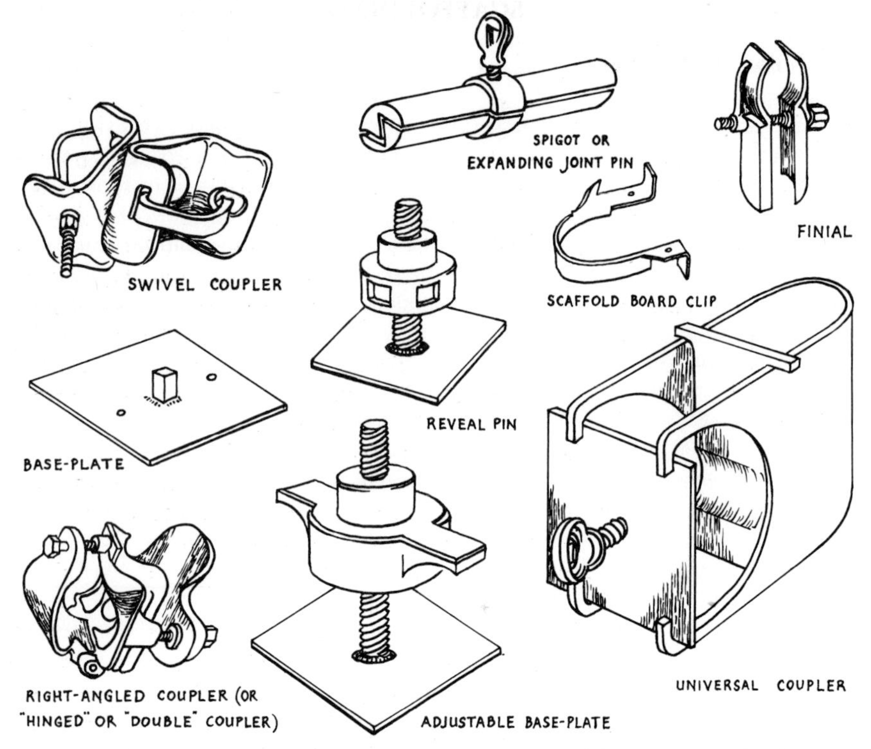

Fig. 3.9 Some of the couplers and fittings used with tubular scaffolding

Right-angled Coupler Also known as a hinged coupler or double coupler; it is the load-bearing coupler used to connect standards and ledgers at the principal node points. It should be capable of supporting a load of 1270 kilograms (25 cwt) applied to the horizontal tube without slip or distortion. If it is desired to increase the safe load, an extra coupler can be placed directly beneath the one used to couple the vertical and horizontal tubes.

Universal Coupler A patented type of load-bearing coupler of excellent design, used in any situation where a right-angled coupler is appropriate.

Swivel Coupler A coupler designed for joining tubes at any angle other than a right-angle, and intended for use in forming diagonal bracings.

Putlog Coupler (not illustrated) A non-loadbearing coupler intended solely for the purpose of fastening a putlog or transom to a ledger to prevent rotation of the tubes, or any tendency for the putlog to become displaced or to slip. Because of its ease of application, many people use it where a load-bearing coupler should be employed; this is an extremely dangerous practice.

Sleeve Coupler (not illustrated) An external coupler for joining two tubes end to end.

Spigot or Expanding Joint Pin An internal connection for joining two tubes together end-to-end; it is fitted with a screw by which it is expanded to grip the inside of the tube firmly. It is not intended for use when the joint is subject to bending stresses or axial tensions.

Base-plate A square flat plate with 150-mm (6-in.) sides and with a centre shank 50 mm long over which the tube fits loosely; it is intended to distribute the load from a standard or from a raker. There are two small holes in the plate through which spikes can be driven to secure the fitting to the sole plate.

SCAFFOLDING

Tubular scaffolding

For all practical purposes wooden scaffold poles and the various forms of prefabricated scaffold frames made from timber are now obsolete, having been superseded by metal units. Some of the large painting contractors, especially those that specialize in industrial work or church decoration, carry large stocks of scaffolding materials, and employ a number of trained men who are competent to erect and maintain scaffolds as and when they are required. But for the average decorator it would not be economic to lock up part of the capital by purchasing the stocks of scaffolding needed for really big jobs such as public halls, etc. The more usual practice is to hire it when occasion demands from one of the various specialist scaffolding firms, branches of which exist in all the main centres of population, and indeed this is a very sensible practice.

Specialist plant-hire firms provide an excellent service and carry a big enough stock to meet any contingency; they will inspect any prospective work for which the decorator has received an inquiry, will suggest the best methods of scaffolding, and will give an estimate of the cost. They also employ a highly expert team of men who can design a scaffold to meet the needs of any particularly awkward or difficult job, and if the decorator is successful in gaining the contract they will deal with the whole business of erecting and maintaining the scaffold on a hire charge basis, thus relieving the decorator of a great deal of worry and leaving him free to concentrate on his own particular part of the work. It must be emphasized, however, that every employer of labour has an obligation by law to assure himself that the working conditions of his men are perfectly safe, just as every employee has a personal duty to comply with the Construction Regulations. Nobody can evade this responsibility. It is for this reason that the notes on the next few pages are provided. There is no intention of suggesting that the painter should also be an expert scaffolder, but it is most important that everybody engaged in the trade should have the knowledge that will enable him to appreciate whether or not the correct methods have been employed and to recognize any faults that may exist.

Tubular scaffolding (Fig. 3.8) consists of either steel or aluminium alloy; the diameter of the tubes is the same in either case, but steel is con-

---

Adjustable Base-plate A base-plate with a screw-jack stem, used to compensate for rising or uneven ground.

Reveal Pin A screw-jack fitting used for tightening a reveal tie between two opposing surfaces, e.g. a window opening.

Finial A fitting designed to hold a horizontal tube directly above the vertical tubes to form a barrier rail or guard rail. Two types of finial are available — a fixed type for making a right-angled joint and a swivel type for making a joint at an angle other than a right-angle.

Scaffold Board Clip A clip by which scaffold boards can be attached vertically or horizontally to transoms, putlogs or standards and which will prevent the boards from being displaced by any normal movement on the scaffold. The holes allow of nailing to the boards.

siderably stronger than aluminium and for this reason is usually preferred. There are occasions, however, when aluminium is the better choice — on very lofty structures, for instance, where the very much lighter weight of the alloy tubing is a major factor, and on buildings of historic importance where any rust stains on the stonework or fabric of the structures would be disastrous. On the other hand, on any job which involves cleaning down with chemical detergents or where the scaffold is in contact with new cement or concrete, it is better to use steel scaffolding because aluminium alloy is softened by alkaline action. Where an aluminium alloy scaffold is used, the upright members or "standards" must be spaced more closely together to compensate for the lesser strength of the tubes. It is important never to mix the two types in any one scaffold, because the differing rates of expansion of the two metals could lead in certain conditions to the scaffold buckling.

The various members of the scaffold are connected together with couplers, which are generally made of sherardised steel, and the correct choice of couplers is important (see Fig. 3.9).

SCAFFOLD FOR EXTERIOR WORK

(1) Putlog scaffold

This is essentially the kind of scaffold used by construction workers when a new building is in course of erection (Fig. 3.10). It extends upwards in stages, being increased in height to keep pace with the rise of the brickwork. It would not be used to scaffold an existing building and would never, therefore, be erected by painters, but of course the situation often occurs when the painter is required to treat the exterior surfaces of a new property before the builders' scaffold is dismantled, so he needs to know how it is made.

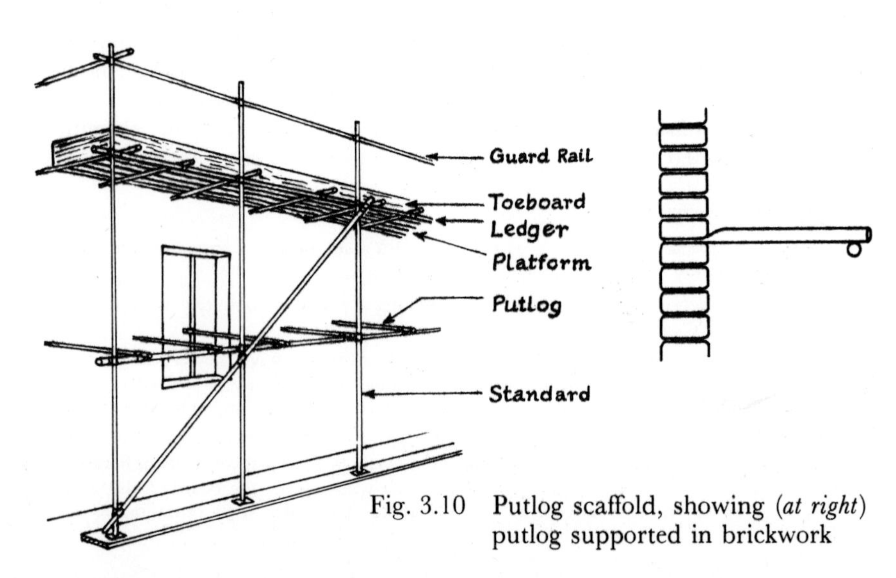

Fig. 3.10 Putlog scaffold, showing (*at right*) putlog supported in brickwork

The word "putlog" is the name given to a short scaffold tube with one end flattened to enable it to fit into the joints between brickwork (Fig. 3.10). A putlog scaffold consists of one row of standards parallel with the wall and about one metre away from it, the standards being joined by ledgers which support one end of the putlogs while the flattened ends are inserted in the brickwork joints. The putlogs support a working platform, and the ledgers, putlogs and platforms are spaced about 1·5 metres (5 ft) apart, a fresh lift being provided whenever the height of the wall reaches the level where the bricklayer can no longer work conveniently. Diagonal tubes should be used to brace the scaffold and prevent distortion. The working platforms must have guard rails and toeboards.

(2) Independent scaffold

This is the kind of scaffold used for existing buildings and, as such, is of great importance to the painter. It is called independent because it does not depend on the building for support. It consists of two rows of uprights or standards, each connected with horizontal ledgers, upon which short scaffold tubes called *transoms* are placed; the transoms support the working platforms, which must have guard rails and toeboards fitted (Fig. 3.11). The first row of standards is sited as close to the wall as possible, after allowing for projections such as sills and cornices, and the outer row is placed about one metre out from the wall. Adequate diagonal bracings must be provided to keep the scaffold rigid.

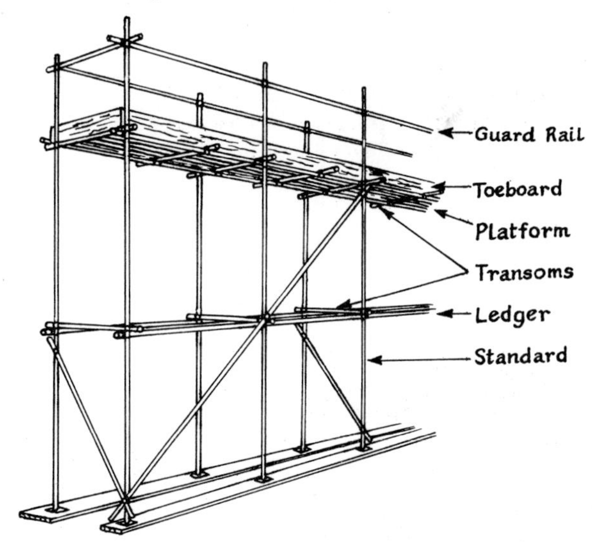

Fig. 3.11 Independent scaffold

It is most important that an independent scaffold should be "tied-in" to the building at frequent intervals to prevent it from overturning. The best way of tying-in is by transoms passed through the window openings and secured to tubes inside the building bearing upon the inside wall, as

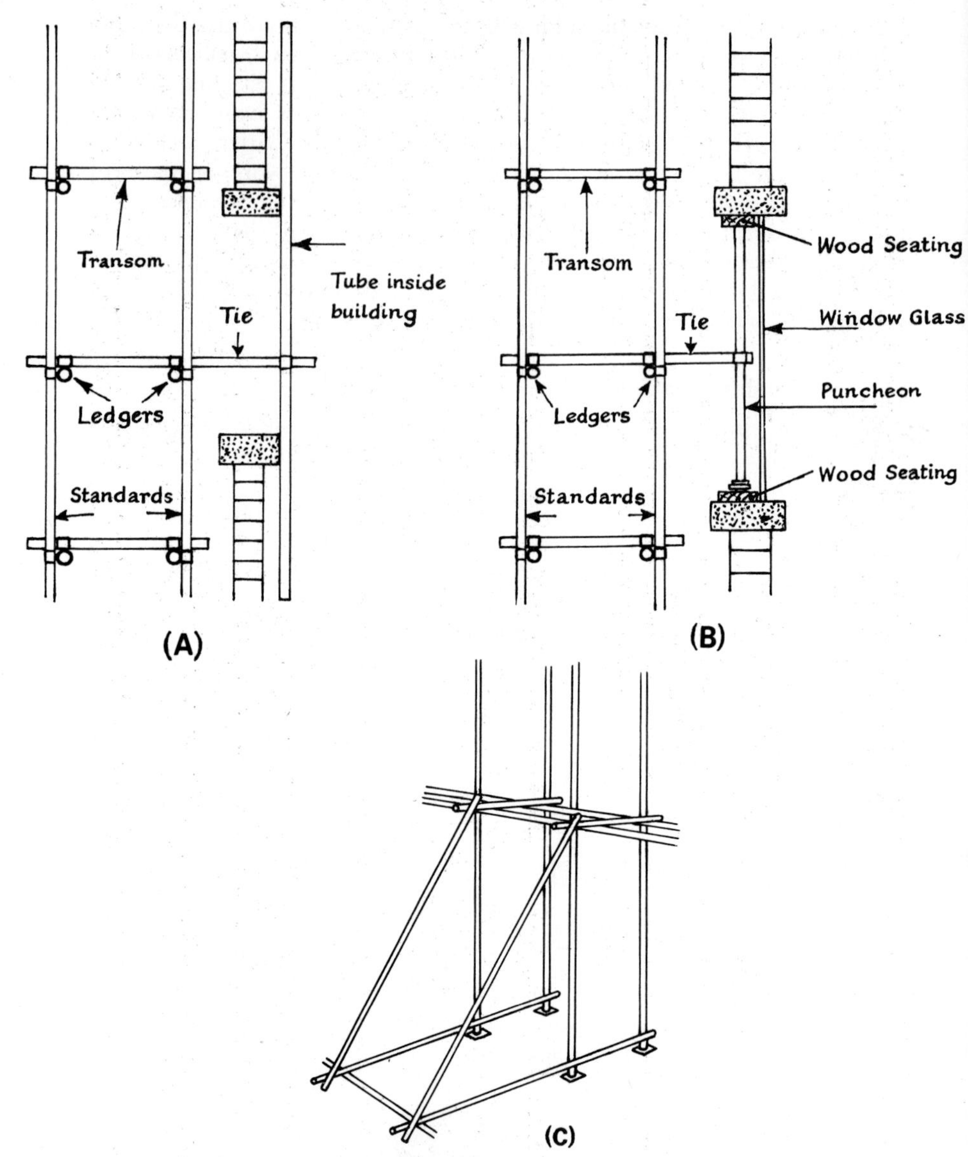

Fig. 3.12 Support of independent scaffold

(A) and (B) show two methods of "tying-in"; (C) method of support by the use of raking struts where tying-in is not feasible

shown in Fig. 3.12(A). Where the type of window fitting does not permit this, transoms are coupled to short upright or horizontal tubes called "puncheons", which are firmly secured in the window openings by means of reveal pins operating on the screw-jack principle (Fig. 3.12(B) and 3.9), though this latter method is not so strong as tying to the inside of the building, and therefore a greater number of ties are needed to compensate.

Where there are not enough window openings to provide adequate ties, an independent scaffold needs supporting with raking struts diagonally from the ground (Fig. 3.12(C)).

(3) Jib, cantilever, or trussed scaffold

This is another kind of scaffold which would never be erected specifically for painting and decorating work but might already be in position and available for use by the painter before it is dismantled. It is used in towns and cities, especially for shop work on busy thoroughfares where the flow of pedestrian traffic along the pavement must not be obstructed. The scaffold is made by cantilevering tubes out of window openings, the outer ledgers being supported by raking struts brought up from window sills or other projections below (Fig. 3.13). Very often a length of corrugated metal sheeting is fastened to the raking struts to form a barrier guard, which protects people walking along the pavement from the danger of being struck by falling objects.

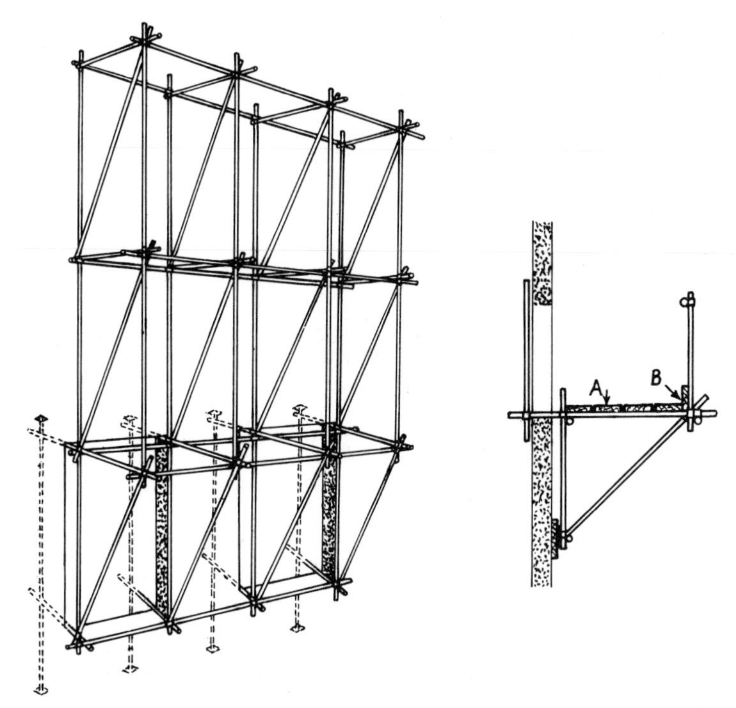

Fig. 3.13 Jib or cantilever scaffold, showing method of anchorage within building

At right: a method of tying-in behind window. A: working platform; B: toeboard

Since there are no vertical members or standards resting directly on the ground it is essential for the scaffold to be securely tied in to the building. On no account must the frames be supported by dogs, spikes, or other fixings inserted into the stone or brickwork joints.

SCAFFOLD FOR INDOOR WORK

(1) Birdcage scaffold

This is a type of scaffold particularly suited to ceiling work, especially in large halls, churches, and similar work. The reason for this is that on large flat or slightly curved ceiling areas, especially those that are not broken up into smaller panels by way of beams, etc., a birdcage scaffold provides the only means of painting the continuous width that enables the edges to be kept alive. The scaffold is made by spacing rows of standards at approximately equal distances apart along both the length and the breadth of the building; ledgers are spaced at vertical intervals of about 2 metres, and the top ledgers support a close-boarded working platform some 2 metres below the ceiling. Sometimes narrow working platforms are supported on the intermediate ledgers around the sides of the hall so that the walls can be painted from the same scaffold. The standards should not be more than 3 metres apart — less in the case of aluminium alloy tubes (Fig. 3.14).

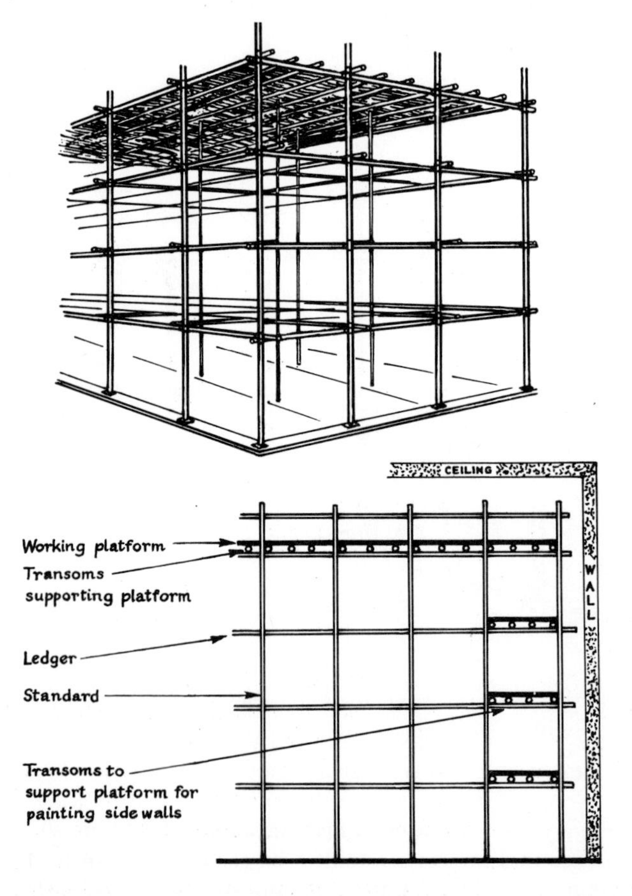

Fig. 3.14 Birdcage scaffold; (*below*) sectional diagram

(2) Slung scaffold

This is a type of scaffold which is suspended by wire bonds from the underside of ceilings or roofs in large structures, leaving the ground space clear and unobstructed (Fig. 3.15). It is used in cinemas and theatres, where the proprietors appreciate the fact that a working scaffold can be in position without any loss of income due to interruptions in the showings

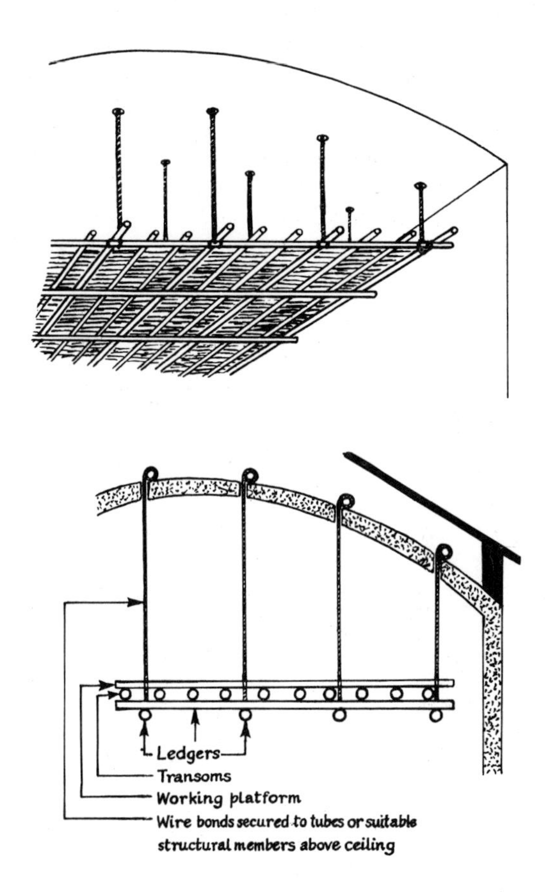

Fig. 3.15 Slung scaffold; (*below*) sectional diagram

and performances; another useful factor from the decorator's point of view is that building up a tubular scaffold on a sloping floor with close-packed rows of seating is a difficult operation. The ceilings of theatres and cinemas are usually furnished with rows of small circular holes through which the bonds can be lowered, the holes when not in use being covered with a hinged flap so that they are inconspicuous. Slung scaffolds are also used in painting the roofwork of railway stations, because it is clearly impossible to erect a tubular scaffold from the ground without obstructing the rail track.

The wire bonds and ledgers are spaced about 3 metres apart and the transoms about 1 metre apart; the whole framework is then covered with a close-boarded working platform. Care is needed to see that the bonds are secured round scaffold poles rather than twisted round sharp-edged roof members, because if a wire rope is kinked or distorted its strength is seriously diminished. There must be an adequate number of suspension points — it is not sufficient to rely on one at each corner, because if one of them fails the platform will tilt. The suspension bonds must be vertical, not slanted.

(3) The mobile tower

A square- or rectangular-based tower made with tubular scaffold and mounted on castor wheels is most useful in painting and decorating work (Fig. 3.16). Because it can be moved around freely from place to place, it

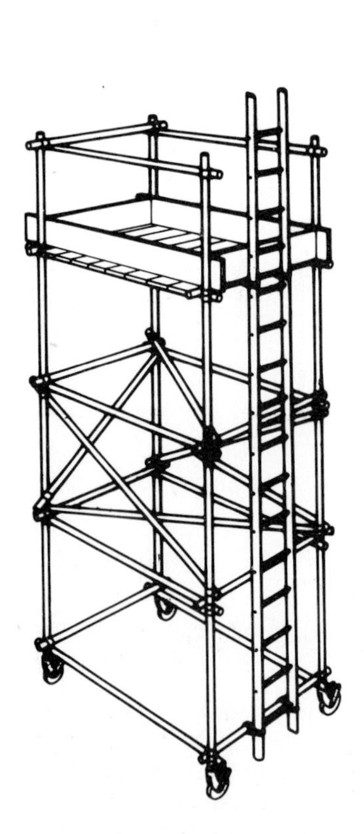

Fig. 3.16 Mobile scaffold tower

The working platform has a guard rail and toeboards, and the wheels are fitted with a locking device

Courtesy: SGB Ltd

Fig. 3.17 Prefabricated alloy access tower

Access to the working platform is by a hinged trapdoor

enables large areas of roof and ceiling work to be painted from a comparatively small working platform, using only a small quantity of tubing. It occupies very little floor space, whereas most other forms of scaffold obstruct the whole floor area; and because it is so easily and quickly erected it makes for a considerable saving in labour costs. For these reasons, and because of its convenience, it is very widely used by painters and decorators. Its usefulness on ceiling work is limited, of course, to those areas which are divided up in such a way that no problems arise in keeping the edges of the paintwork alive.

Unfortunately, accidents involving the use of mobile towers are very common. Most of these mishaps occur due to the tower tilting or overturning. There is a simple rule to prevent a tower from becoming top-heavy; the height of the tower should never exceed 3 times the length of the base-line, and in the case of a rectangular tower this means the length of the shorter side. If for some reason this height must be exceeded, the base must be stabilized with counterweights. Other essential rules are:

(a) A tower should always be moved by pressure applied at the base.
(b) A tower should never be moved if anyone is on the working platform.
(c) People using the tower must not attempt to move it themselves by pulling on the beams or roof trusses.
(d) The castor wheels must be firmly secured to the standards, and locked to prevent accidental movement while the tower is in use.
(e) When the floor is covered with protective sheets, it must be ascertained that there are no concealed pitfalls before a tower is moved; accidents have been caused by placing a tower with one corner over some unsuspected hole such as an inspection pit in a garage.

A tubular tower needs to be braced along the horizontal plane as well as on the vertical face of the framework.

Prefabricated frames

All the major companies which manufacture scaffolding equipment now produce prefabricated frames which can be slotted together to form a self-locking tower, which offers great advantages over tubular scaffolding, in both outdoor and indoor work, for many of the operations in painting and decorating (Fig. 3.17). The frames are light and easy to handle, being made of lightweight steel; the work of erection is extremely simple and rapid, and there are no loose parts to get lost or damaged. Almost any length and shape of scaffold can be formed by joining the towers together with ledgers or tie-bars. Because of their convenience, and the considerable saving in labour costs as compared with the price of erecting tubular scaffold, these self-locking units are very popular with decorating firms. They can be bought outright or hired, and many decorators prefer to buy them, starting with a small number of units and adding to them as their needs increase.

It should be noted that when prefabricated frames are used in place of any of the foregoing forms of tubular steel scaffold, they are subject to the same safety rules as the piece of scaffold they replace. For instance, self-locking frames are often used to form an independent scaffold for outdoor use, and when this is done it is essential that the frames be adequately tied-in to the inside of the building. Similarly, mobile towers for indoor use are often made with prefabricated frames, and exactly the same rules apply as those given for towers made with scaffold tubes.

ACCESS EQUIPMENT

Cradles

These are used a great deal in painting and decorating, especially in the large towns and cities, notably the textile towns of Lancashire and Yorkshire. Very often they provide the easiest and most practical method of access to the elevation of a building. They are particularly useful over busy thoroughfares (enabling a building to be painted without obstructing the pavement), on lofty structures where the work would not be accessible from the ground, and on buildings which rise above a river-bed, a canal, or some similar feature which would preclude the use of a scaffold built from the ground.

Hand-operated cradles depend on the scientific principle of the pulley, whereby a mechanical advantage is obtained relative to the number of pulley wheels employed. Thus, in a system where the rope passes over two pulley wheels a force equivalent to half the load is required to move the cradle; where the rope passes over four wheels the required force is one-quarter of the load, and so on (Fig. 3.18); the limiting factor is that for every wheel introduced into a system the rope needs to be extended by a length equivalent to the height of the building.

The cradle or "boat" itself is usually made of wood, with iron stirrups which fold down flat when the cradle is not in use, but there are more modern alternatives made either from aluminium alloy or glass fibre. The fall ropes are traditionally made of vegetable fibre such as hemp or manilla, but in situations where they are exposed to chemical fumes or solutions, man-made fibres such as polypropylene are to be preferred. The sheave blocks are units consisting of single, double or treble pulley wheels, attached to a swan-neck hook.

The main types are as follows:

(a) *Fixed cradle, or standing cradle*

Here there are two outriggers fixed in one position, the top sheave blocks being attached directly to them (Fig. 3.19 and 3.20). The cradle is thus capable of upward and downward movement only; to extend the work sideways in either direction involves dismantling the rig and re-erecting it in a new position.

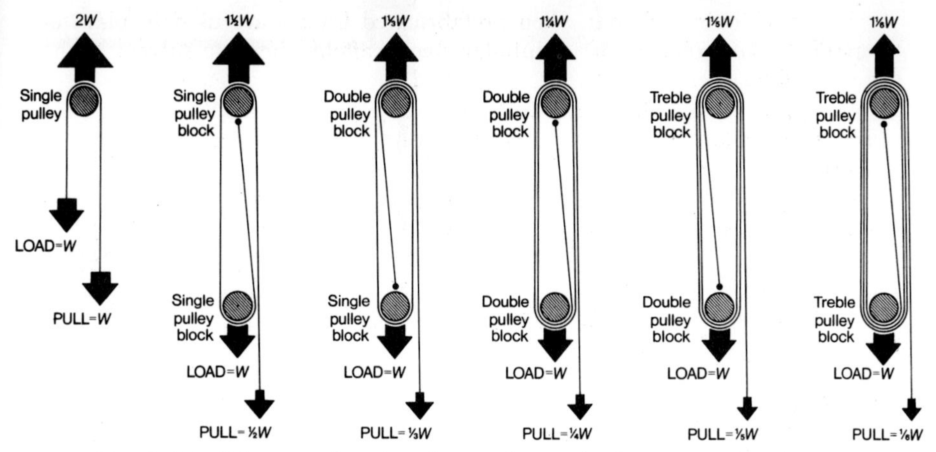

Fig. 3.18 Diagram showing the mechanical advantage of various pulley systems

The diagram shows only the relative mechanical advantage of the various systems; it does not take into account the frictional losses, nor the dead weight of the ropes and sheave blocks.

System 1 offers no mechanical advantage. It is not used in cradle rigs; its use is restricted to the rigging of gin wheels for the hoisting of materials and equipment on the site.

System 2 is widely used for cradle work, and one of the largest firms supplying and hiring scaffold equipment uses this system and no other. It has the advantage of keeping the length of rope to a minimum.

Systems 3 and 4 are preferred by some scaffold firms, and are popular with decorators because they reduce the amount of effort required by the operatives.

System 5 is also quite widely used for cradle work, and is the method solely employed by one of the biggest firms of industrial painting contractors in the North of England. Its value lies in reducing to a minimum the amount of effort by the operatives; the drawback is that extremely long ropes are needed, and these tend to become tangled if they are not kept neatly coiled.

All the systems illustrated, including System 6, are used for general lifting purposes; the selection of a particular system is dictated by specific circumstances of the work in hand.

When the roof of the building is flat, the rigging is comparatively simple, consisting of outriggers some 7 metres long overhanging the roof edge by about half a metre, with counterweights at the extreme end sufficient to give a safety factor of 3 (Fig. 3.21). In the case of a sloping roof, or in any situation where long outriggers are not feasible, a framework is designed suitable to the particular conditions of the site, and fitted with the appropriate counterweights to give a safety factor of 3.

Fig. 3.22 illustrates a typical use of a fixed cradle.

(b) *Travelling cradle*

Here a track consisting of an aluminium alloy joist is suspended from the

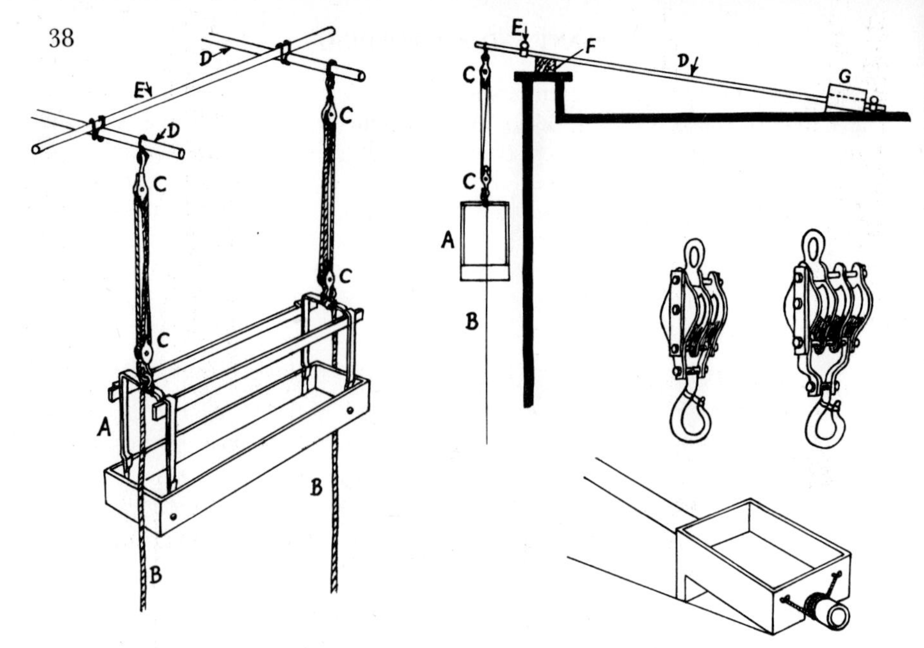

Fig. 3.19 (*left*) The fixed cradle (sometimes called the "standing cradle"), permitting only vertical movement

Fig. 3.20 (*right*) Diagram to show the rigging of a fixed cradle on a flat roof with parapet

Key to both the above

A Cradle (or "boat")
B Fall rope
C Single sheave blocks (pulley blocks)
D Outrigger
E Lateral tube to stabilize the outriggers
F Wood block seating (to prevent possibility of outrigger rolling on edge of parapet)
G Counterweights in box attached to outrigger

Mid right: Double and treble pulley blocks as used with larger cradles

Bottom right: Specially constructed box to fit over the outriggers and in which the counterweights are placed

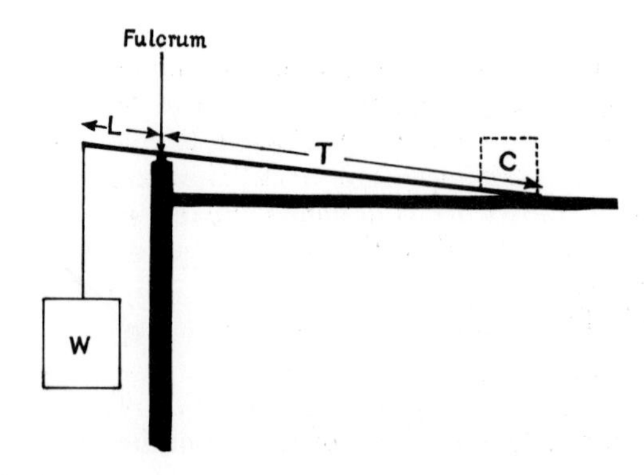

Fig. 3.21 Calculating the amount of counterweighting required for rigging a cradle

W = Weight of cradle, associated equipment and the operatives
L = Length of outrigger from fulcrum to point of suspension (overhang)
T = Length of outrigger on the roof side of the fulcrum (tail length)
C = Position of counterweight

Calculate what weight would be required at C to make C × T equal to W × L. Multiply this weight by 3 to get the statutory safety factor of 3.

outriggers, and the cradle itself is suspended from jockeys which run along the flange of the track (Fig. 3.23). At each end of the track there is a guy block and quadrant over which a traversing line is passed and which also acts as a safety stop. The cradle can thus be moved in an upward, downward or sideways direction. The complete elevation of a building can be painted by using separate sections of track, the total length being equal to that of the building, and each length having its own cradle rigged. Alternatively the sections of track can be linked together with connectors to give a continuous run for one cradle.

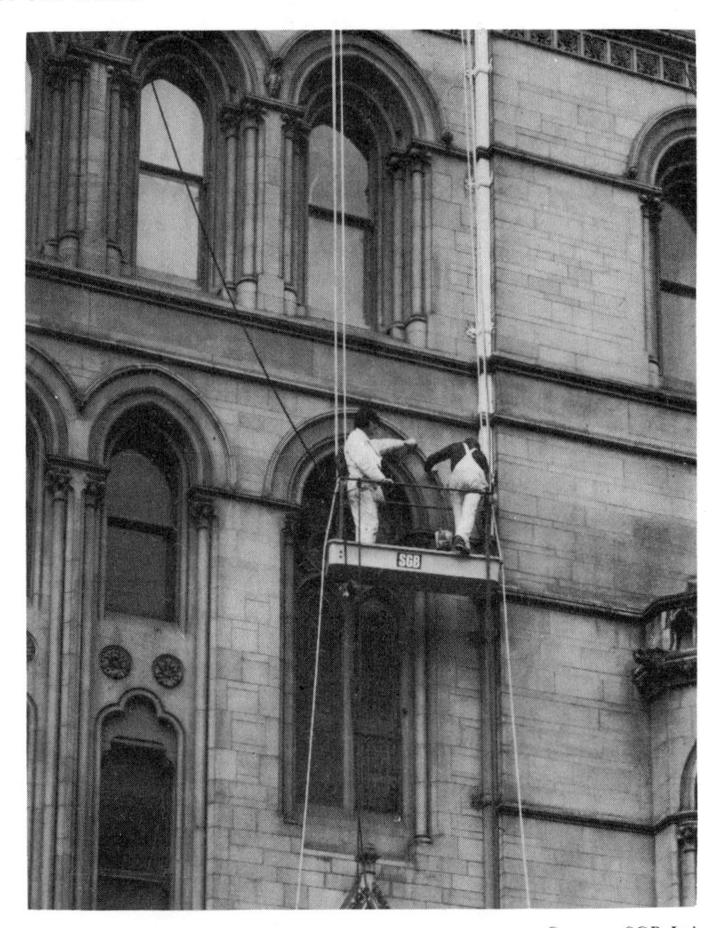

Courtesy: SGB Ltd

Fig. 3.22 Rope-operated painter's cradle

(c) *Mobile tower cradle*

Where there is a smooth flat roof with a parapet, a cradle can be rigged from outriggers extended from a mobile tower, provided a sufficient quantity of counterweight is used (Fig. 3.24). The cradle itself is fixed and can give only vertical movement, but the mobility of the tower provides sideways movement as well.

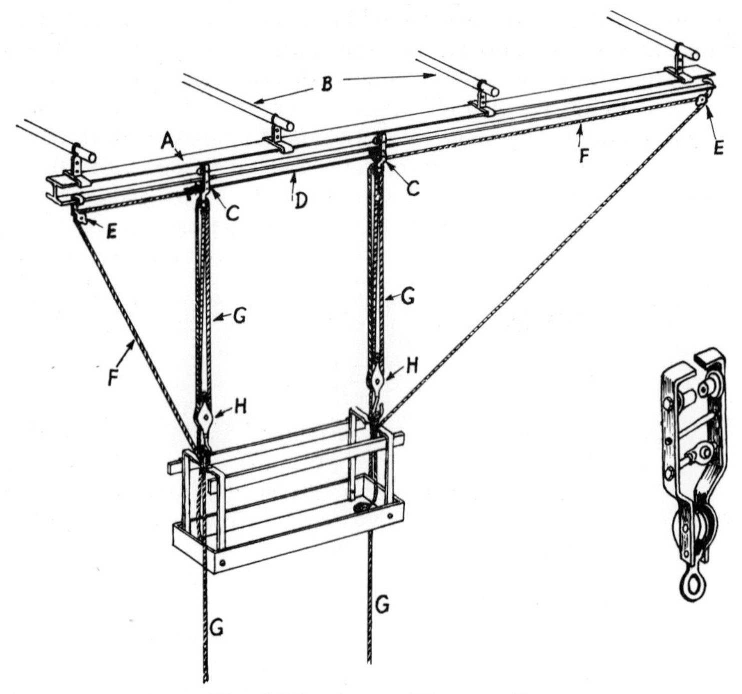

Fig. 3.23 A travelling cradle

A	Aluminium alloy joist	E	Guy blocks
B	Outriggers	F	Traversing lines
C	Jockeys or joist roller blocks	G	Fall ropes
D	Headwire or connecting line	H	Sheave blocks

Inset: Jockey or joist roller block

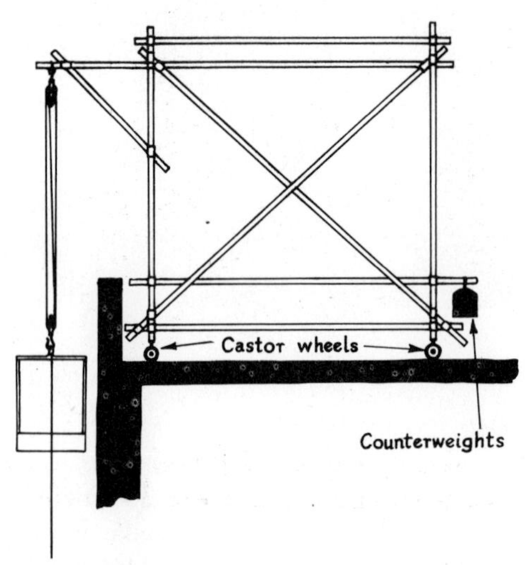

Fig. 3.24 Cradle suspended from a mobile tower on a flat roof with parapet

The ropes used in cradle work should be inspected by a competent person every time they are put into use; the ropes must be of the correct diameter to fit snugly into the grooves in the pulley wheels. The pulleys must be well lubricated, the fall rope tied off securely on to the stirrup and around the swan neck, and the swan-neck hook should be fitted with a locking device to prevent the gear from slipping off.

Makeshift materials such as rocks, bricks or sandbags should not be used as ballast for counterweights. Outriggers supporting the track of a travelling cradle should not be more than 3 metres apart. Where two separate cradles are in use, no attempt should be made to bridge the gap between them with planks.

(d) *Winch-operated cradles*

On very lofty buildings where the amount of rope needed would be excessive and inconvenient, cradles are often suspended by wire ropes wound on to a hand-operated winch which is mounted on the cradle decking (Fig. 3.25). The erection of such cradles is always carried out by specialist firms. The wire rope must always be long enough to allow two full turns of rope to be taken round the drum when the cradle is in its lowest position.

Courtesy: SGB Ltd

Fig. 3.25 Winch-operated painter's cradle

(e) *Permanent installations*

On modern multi-storey buildings, electrically operated cradles of a permanent nature are often installed during the construction of the premises, providing access to all parts for maintenance purposes.

Bosun's chairs

These are of more importance to painters than to workers in any other trade. Very often they provide the only feasible means of access to some isolated piece of work, especially in the field of industrial painting. They consist of a seating for one person, suspended by sheave blocks, the operative lowering himself as required (Fig. 3.26). Under the Construction Regulations the use of bosun's chairs is permitted where the work is of short duration and where other forms of scaffold would be impracticable.

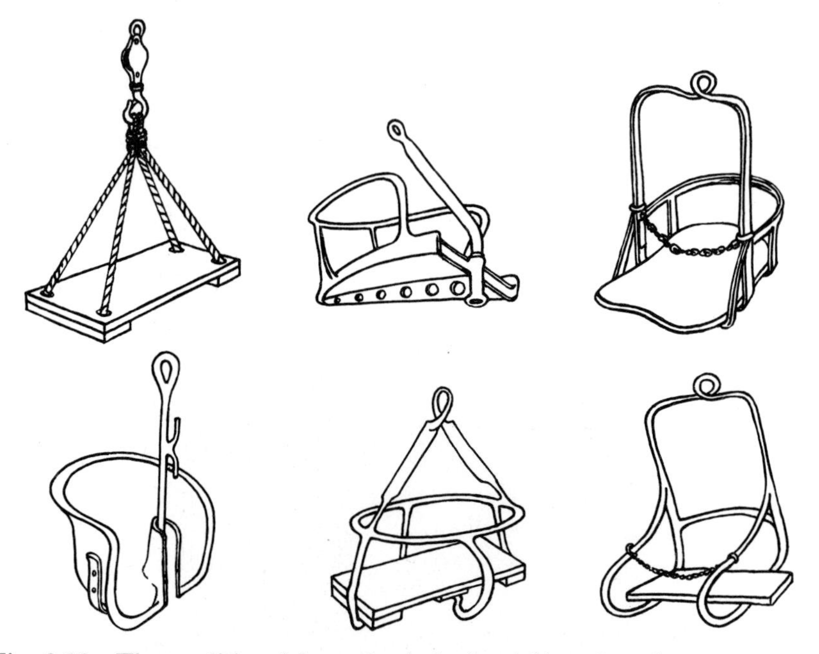

Fig. 3.26 The traditional bosun's chair (*top left*) and various patterns of modern safety chairs

A man working on a bosun's chair is very vulnerable. He is beyond the reach of effective help from anyone else if anything goes wrong, and the work is usually so lofty that a mishap is bound to have serious results. It is important that the anchorage of the chair be firm and secure, and that steps be taken to prevent the chair from twisting and spinning whilst in use. The operative should be provided with a guard rope or strap so that he cannot accidentally slip out of the chair.

Safety chairs approved by the British Standards Institution include such features as a backrest, a forward member arranged in such a way that the operative has one leg on either side of it, and a continuous metal hoop or safety belt to prevent the operative from falling out; the suspension point is connected in such a way that the chair is stable under all conditions. In spite of these desirable features, industrial painters often prefer the old-fashioned type of chair, largely because when it is suspended high above the ground it is so much easier to swing the body into it than into the enclosed guard of the safety chair.

Skips and baskets

In Lancashire, the homely skip is used for painting the windows of tall mills and warehouses, this being a large wicker basket fitted with wheels which roll down the wall at each side of the windows, and suspended from above by sheave blocks. Under the Construction Regulations the skip must be at least 920 mm (3 ft) deep and carried by two strong bands of metal which are continued round the sides and bottom of the basket.

POWERED ACCESS EQUIPMENT

It sometimes happens that a piece of work the actual painting of which is a matter of short duration is situated in a difficult and inaccessible posi-

Courtesy: SGB Ltd

Fig. 3.27 The Simon hydraulic platform

An appliance providing a convenient means of access to difficult positions without the use of scaffolding

tion, and the cost of erecting a conventional scaffold would be considerable and quite out of proportion to the labour cost entailed in the work. In such a case, one answer to the problem might be the use of a hydraulic or electrically operated platform. A well-known example of this type of equipment is the hydraulic platform manufactured by Simon Engineering and suppplied by Scaffolding (Great Britain) Ltd; Fig. 3.27 illustrates one of the larger models. This consists of a platform on the end of a crane-like boom which can be raised to a vertical working height of 28 metres (92 ft) and can swing around to any point within a 28-metre diameter, with a load of up to 135 kilograms (300 lb). The boom is mounted on a vehicle turntable which can be manoeuvred into any desired position, being controlled either from the turntable or by the men on the platform. For overhead work in busy streets, underneath bridges, on awkward gable walls, tall masts etc., it provides a convenient way of reaching the work safely and quickly. The Simon platform is available in a range of sizes having a vertical working height of from 7·3 to 31·4 metres (24 to 103 ft).

SGB Ltd also produce several kinds of electrically powered access equipment, including the recently introduced Power-Tower (Fig. 3.28).

Courtesy: SGB Ltd

Fig. 3.28 The Power-Tower, electrically operated

Two sizes are supplied, the M1400, having a vertical working height of 14·5 metres (47 ft 6 in.) and capable of carrying a working load of 300 kg (660 lb), and the M2200, reaching a height of 22 metres (72 ft) free-standing or 60 metres (197 ft) with special wall ties, and designed for a maximum working load of 300 kg (660 lb). This equipment can be towed by ordinary light commercial vehicles and is rapidly assembled on the site. The operator controls the movement from the working platform, and 220/240 volt current used to operate the motor is reduced to 24 volts for all control circuits. As with the hydraulic platform, this piece of equipment would have few applications to ordinary painting and decorating work, where the convenience as opposed to normal ladder work would be heavily outweighed by the purchase price or hire charge. Its value lies in the painting of large numbers of isolated units spaced out over a considerable distance, and as such it is used by corporation works departments, etc., for various types of maintenance painting.

Courtesy: John Rusling Ltd

Fig. 3.29 The "Flying Carpet" access platform

Fig. 3.29 illustrates another type of access platform known rather fancifully as the Flying Carpet, supplied by John Rusling Ltd. This also is electrically operated, being powered by a 36-volt DC battery. The platform is raised or lowered by an expanding frame, the whole equipment being mounted on a mobile platform-chassis. The operator has complete control from the working platform, with a single lever to control ascent, descent, forward and reverse movement, steering and locking. Because of the capacity of the working platform, this equipment is eminently suited to certain classes of industrial painting sites.

SITE SAFETY

Safety nets and protection sheets

For work of an extremely hazardous nature it may be necessary, for the protection of the operatives, to supplement the scaffolding and access equipment with a second line of defence in the form of man-catching sheets or nets suspended below the work area. In the event of a mishap a man who falls from the staging is caught and saved by the net. For a long time sheets and nets of this kind have been used in industrial painting operations. Their importance has been recognized, and they have been used successfully on some recent civil engineering contracts. An outstanding example was their use in the construction of the Forth Road Bridge in the 1960's, a contract which set new standards of accident prevention in a field which has always been subject to very high risks.

Until recently, the man-catching sheets used by painting contractors usually consisted of heavy canvas reinforced with stout webbing, but these unfortunately have not always given the measure of safety that was expected of them; when the sheets were suspended the tension was uneven, and some parts retained a certain amount of play whilst other parts were tightly stretched. If a man fell on to one of the taut areas the tendency was for the sheet to burst apart, causing him to shoot right through without his fall being arrested. Present-day man-catching nets (Fig. 3.30) are designed to prevent this. They are made with a mesh of man-made fibre ropes, which are immensely strong and unaffected by

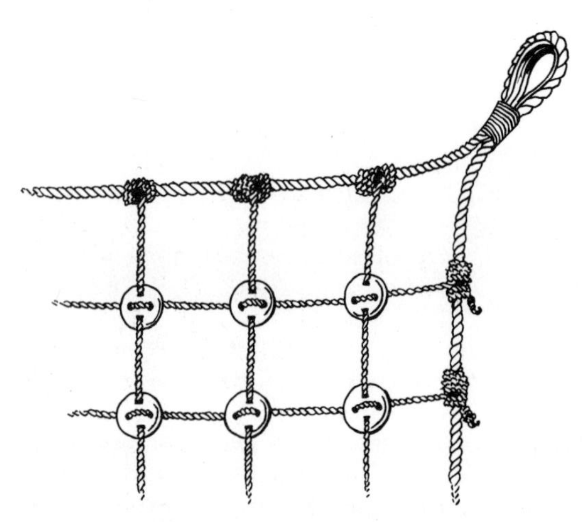

Fig. 3.30 A corner of a modern man-catching net

chemical pollution or exposure to severe weather conditions. Furthermore, instead of the ropes being tied and knotted, at each point of intersection there is a plastic disc with four holes in it, the ropes sliding with comparative freedom through the holes; because of this, the entire net area is uniformly slack — sufficiently so to absorb the shock of any weight that falls into it. These modern nets are completely reliable.

Safety belts

Obviously there are many occasions in painting and decorating where an operative working in hazardous conditions needs some extra form of protection but where, for economic reasons and because of the short duration of the work, the expense of providing man-catching nets is not justified. It is in these cases that the safety belt gives a useful second line of defence for an individual painter. Safety belts are not used nearly as often as they should be, and if only they were more widely used a number of serious accidents would be prevented. They should be provided, for instance, when men are working on fragile roofing materials, or on tall steel chimneys, beneath factory roofs where there is a lot of dust and smoke, and in enclosed spaces where there is a risk of being overcome by fumes.

In its simplest form a safety belt consists of a leather or canvas belt worn round the waist and secured by a length of rope which a second operative pays out or draws in as required. This is better than nothing, but it presents its own hazards; if the operative misses his footing and drops only two or three metres the sudden jerk when his fall is arrested may cause internal injuries of a serious nature. A safety harness made on the same principle as a parachute harness is far more satisfactory. In some situations it is better that the lifeline should be secured to some rigid part of the structure rather than held by another operative, and in these cases an inertia reel attachment gives the most complete protection, allowing freedom of movement in normal use but coming into operation when there is a sudden pull. Various forms of harness made with chemically-resistant synthetic fibre webbing are available.

Safety helmets

For most painting work a safety helmet is not necessary, and indeed a helmet would be an encumbrance to anyone decorating small rooms in domestic property. On building sites, however, and on scaffold work, the situation is quite different. On construction work of any size, most contractors make the wearing of safety helmets compulsory, and this is a very good rule based on sound common sense. The risk of being struck by falling objects is high, and the human skull is unfortunately very fragile.

Since safety helmets came into common use, a great many fatal accidents have been prevented, and it would be sheer folly to take a chance instead of wearing adequate protection on any site where such risks existed. The main point to notice is that each man should have his own helmet, adjusted to his own particular fitting. A helmet that does not fit properly may cause a severe headache, with the result that the operative prefers not to wear it and takes the first opportunity to discard it.

Safety footwear

In many industrial operations the use of boots and shoes with protective steel toecaps is advisable because of the risk of the operative's toes being trapped and crushed under heavy weights. Such a risk is a very remote one in painting and decorating, so although pressure is sometimes exerted to make painters wear them, there is little point in using safety shoes of this kind. What *is* important is the risk of stepping on upturned nails, etc., which is a frequent source of accidents, particularly on building sites. The risk is considerably lessened if reasonable standards of site tidiness are insisted upon, but while the painter could hardly be expected to wear the heavy wellington boots with steel reinforced soles that are used by construction workers, he would be well advised to make sure that the ordinary shoes he wears are in good condition, with good stout soles kept in good repair. Light plimsolls and sports shoes are totally unsuitable for work on a building site. A different situation exists when the painter is working on an indoor scaffold, where there is much advantage to be gained by his having a pair of plimsolls to change into, because the thin soles enable him to "feel" the position of the planks so that he doesn't need to be constantly looking downwards.

Safety and inspection of scaffolding

It is important that the painter should be able to recognize whether or not a scaffold has been properly constructed and maintained, even though some other person has been entrusted with the actual construction. He must therefore be aware of the fundamental principles involved. Among the points he must learn to observe closely are the following:

(1) That all vertical members of a scaffold are plumbed and properly upright, and all horizontal members level.

(2) That every scaffold rising from the ground has a firm base and secure foundations; adjustable baseplates should be used if the ground is uneven.

(3) That the joints between separate lengths of tubular scaffold are "staggered"; a continuous line of joints is a line of weakness along which a scaffold may fall apart.

(4) Any rectangular framework of scaffold tubes is liable to distort; it needs bracing diagonally to make it rigid.

(5) The correct couplers must be used; in particular the main couplers joining standards and ledgers must be right-angled couplers of the load-bearing variety; putlog couplers and swivel couplers are not suitable.

(6) Every working platform must have properly fitted guard-rails and toeboards, the correct distance apart.

The inspection of scaffold is a highly important matter. It must be inspected by a competent person at stated intervals, and the results of the inspection entered in the appropriate register. The times when it is to be inspected are

(1) After erection, within the 7 days immediately before it is put into use.

(2) Every 7 days whilst in use.

(3) Whenever it has been substantially added to, altered, or partially dismantled.

(4) At any time when it has been exposed to severe weather conditions which may have affected its strength.

No scaffold may be used when partly erected or dismantled.

The recommended types of materials used in scaffolding equipment and the methods of construction are to be found in the appropriate British Standards, published by the British Standards Institution, 2 Park Street, London, W1A 2BS.

The complete regulations governing every aspect of the use of scaffolding under the terms of the Factories Acts are to be found in the Construction (Working Places) Regulations, 1966 (Statutory Instrument No. 94), obtainable from Her Majesty's Stationery Office. These should be read and studied in conjunction with the Construction (Health and Welfare) Regulations.

4

Tools and Appliances

TOOLS

A journeyman painter is expected to provide his own tool-kit, but fortunately his requirements are modest and are easily contained in a leather grip or tool bag. The man who is keen on his job will take the trouble to equip himself with all the tools he is likely to require and will not depend on borrowing this or that from his workmates, a practice which should be strongly discouraged. Each man should be independent in this matter, so that when he is sent to a job on his own, he is not held up for lack of tools. Needless to say, the good workman will always endeavour to keep his tools clean and in good condition.

One or two broad knives or scrapers will be required for stripping wallpaper and burning-off, etc., and a broad knife, fairly stiff in the blade but with a flexible end, is desirable for filling good-class work (Fig. 4.1); this should not be used for stripping wallpaper because the edge would be quickly damaged, nor for burning-off as the heat of the lamp would soon make it too stiff for filling purposes. A 30-mm chisel knife will serve both for stirring paint and for burning-off narrow, flat members or fillets, while shavehooks with large and small heads will be needed for dealing with

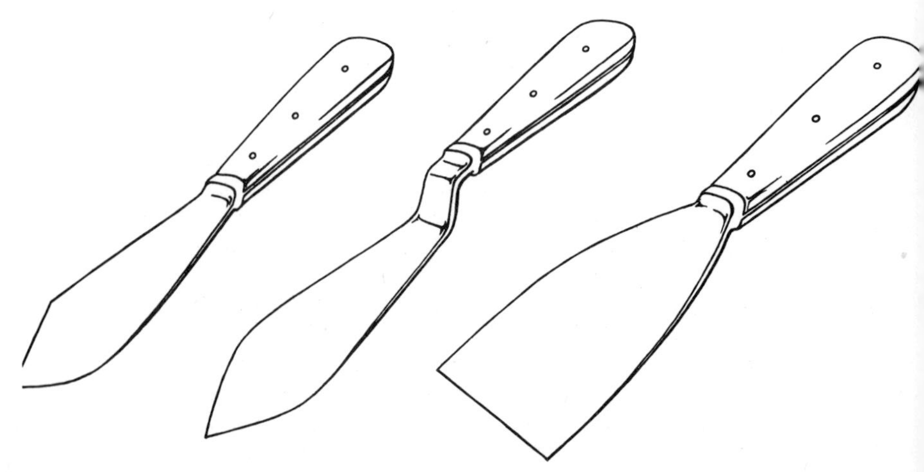

Fig. 4.1 Stopping or putty knife, trowel stopping knife, and broad chisel knife

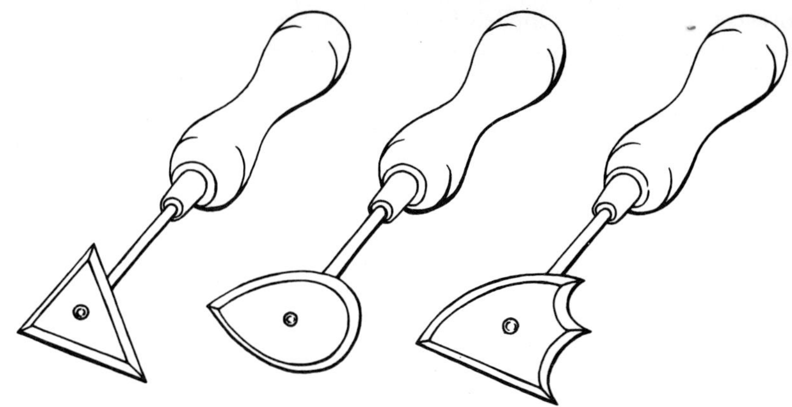

Fig. 4.2 Shave-hooks: triangular, pear-shaped, and combination type with head shaped for mouldings

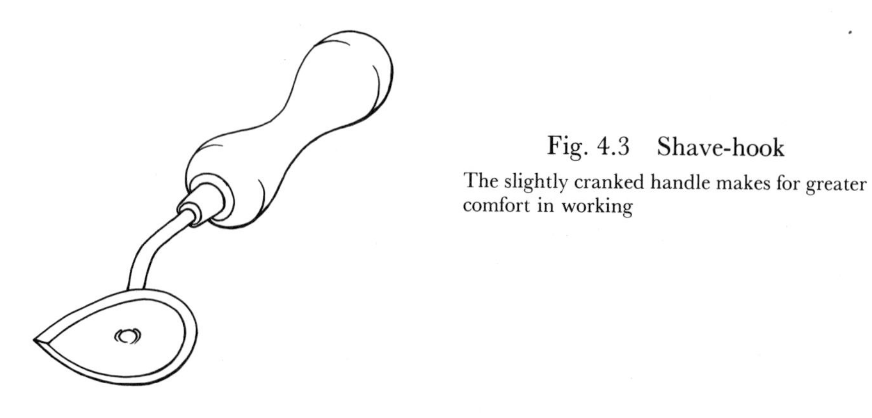

Fig. 4.3 Shave-hook
The slightly cranked handle makes for greater comfort in working

various mouldings (Fig. 4.2), those with the stem slightly cranked being perhaps the most convenient in use (Fig. 4.3).

Putty knives should be stiff enough to press stopping well home. A small trowel for making good defective plaster and one or two small plasterer's steel tools for repairs to cornices and other fine pointing should be in the kit. A plumber's hacking knife, a small tack hammer, screwdriver and pliers should also be included. The paperhanger will require a boxwood metric rule, plumb bob and chalk line, scissors and trimming knife. One of the patent pocket trimmers such as the Ridgely, Champion, or Morgan-Lee will be useful for any paper that is not supplied in ready-trimmed form (see Chapter 30).

The signwriter usually has his own special kit, a wood or metal compactum housing tube colours, media and thinners, dippers, brushes, palette and mahlstick. The grainer, too, has his own particular set of tools which comprise steel and rubber combs, fitches, hog-hair mottlers, overgrainers, badger or hog-hair softener, and possibly a set of Berries graining tools or other means of incising wood grains (see Chapter 27).

Dust brushes (or jamb dusters) and paperhanger's "putting on" brushes are usually supplied by the employer, but many craftsmen prefer to use their own. The employer, however, supplies all other brushes and appliances, usually handing a set of paint brushes to each man on joining the firm and holding him responsible for their safe keeping and good condition during his stay.

Paste-boards

Paste-boards for paperhangers may be supported in the centre by an X-shaped collapsible trestle or by a close-up trestle at each end; the latter is the more rigid, and is a necessity when knife and straightedge are used for trimming. The folding type is a light but rigid table which folds neatly into a compact box when not in use (Fig. 4.4) and was designed originally for use with the Ridgely patent straightedge and trimmer. Many of the

Courtesy Eclipse Rail-Track Ladder Co. Ltd

Fig. 4.4 Paperhanger's folding table

paste-boards on sale are much too narrow, being just the same width (533 mm) (21 in.) as the standard width of English wallpaper. If wallpaper is to be pasted properly without risk of paste being transferred to the face, the board needs to be at least 610 mm (24 in.) wide; this allows the paperhanger to detach the top piece of paper from the pile on the board, move it over to the front edge and then bring it over to the back edge for pasting each side, leaving the remaining pieces in the middle. In fact, both 610-mm (24 in.) and 750-mm (30 in.) boards will be needed to accommodate the various widths of paper in general use, but for extra wide materials such as the Canadian "Salubra" or some of the PVC-coated fabrics two boards can be placed side by side, unless such materials are being hung so frequently as to warrant having specially wide boards made.

Dust-sheets and protective sheeting

Dust-sheets or drop-sheets form a most necessary part of the painter's outfit and a good stock should be kept. Heavy twill sheets of various sizes, hemmed and eyeletted, will be required and they should be stencilled at frequent intervals with the decorator's name and address, both as an advertisement and as a safeguard against pilfering. Sheets should be laundered fairly regularly, otherwise they soon become choked with dust and, if sent on a job in such a condition, are liable to give a very bad impression, whereas clean sheets are always a good advertisement. Heavy sailcloth or sun-blind material is very useful for protecting polished floors from damage, and "Sisalkraft" paper also has many protective uses, being waterproof and practically untearable. Transparent polythene sheeting also has many uses both for indoor and outdoor protection (see Fig. 4.5). It is supplied in rolls up to 4 metres wide (and wider, to order) in light, medium, heavy and double heavyweight, and apart from its use as a waterproof dust-cover, it can be affixed to light timber framework or to scaffolding to form a transparent screen to enable outside work to proceed in inclement weather.

Fig. 4.5 External painting under cover of Visqueen polythene sheeting

Sundries

A glance through any decorators' merchant's catalogue is sufficient to show that the painter is well catered for in such necessities as buckets, paint cans and strainers, etc., as well as in many useful gadgets. Some paint manufacturers supply emulsion paints and distempers in free buckets and these are quite useful additions to the painter's stock; the ears or clips used for fastening the lids down are a source of danger, however, and should always be removed before use, for they are liable to catch in the clothing and thereby get upset or dragged off scaffolding. Paint cans are supplied with either a beaded edge or a plain edge, and the latter is to be preferred as being cleaner in use; they should be of galvanized iron or other rustproof finish and of fairly heavy gauge so that they withstand hard wear and keep their shape well. Strainers are best made of zinc which is easily cleaned and is non-rusting, an important point when they are kept in water. Loose gauzes in brass or copper are usually supplied in 30, 40 and 60 mesh for emulsion paint, oil paint and enamel respectively.

APPLIANCES

The painter and decorator has in the past been notoriously conservative and has tended to regard innovations with some caution. When he is finally convinced, however, that a new method or new appliance is a sound practical proposition, he usually becomes enthusiastic. This is perhaps as it should be, for the man who refuses to avail himself of any equipment of proved merit is simply handicapping himself. Nowadays the notion tends to gain ground that if a job can be done better, more speedily and with less fatigue by using a machine, well then, use the machine! There are some who see the end of all craftmanship if such a principle is accepted, but this, surely, is taking a narrow view. The man who wields the spray-gun with expert precision and dexterity, who realizes its full possibilities (and limitations) with various materials, is just as much a craftsman as the man who uses the brush.

No piece of equipment is entirely foolproof, but if there were no idiosyncrasies to master, no snags to overcome, there would be no need for the craftsman. The wise man, therefore, will choose the best tool or appliance for the job and will persevere until he becomes expert in using it. Items of equipment such as spray plant, wallpaper trimming machines and a variety of others will be dealt with in their appropriate chapters.

Stripping machines

The removal of wallpaper presents little or no difficulty when only a single layer of paper is involved. It is not uncommon, however, for the decorator to be confronted with walls covered with nine or ten papers, or several layers of varnished paper, the removal of which generally proves an expensive, disheartening and unpleasant job. Anything which lightens or facilitates such a task is a boon and a blessing.

The "Lightning" wallpaper stripper (Fig. 4.6), marketed by Sandersons of Berners Street, London, is indeed such a boon for in addition to speeding up the removal of plain and varnished papers, it can be used for stripping difficult materials such as Anaglypta, Lincrusta, and similar wall-coverings. It is designed to generate steam at a low pressure which is conveyed by a hose to a perforated metal concentrator. When the concentrator is held against the papered walls for a few seconds, the steam penetrates the surface and softens the paste, enabling the paper to be removed easily with a wall scraper. The surface of varnished or emulsion-painted papers should be well scored or scratched to assist penetration of steam; in stubborn cases a strong solution of sugar soap

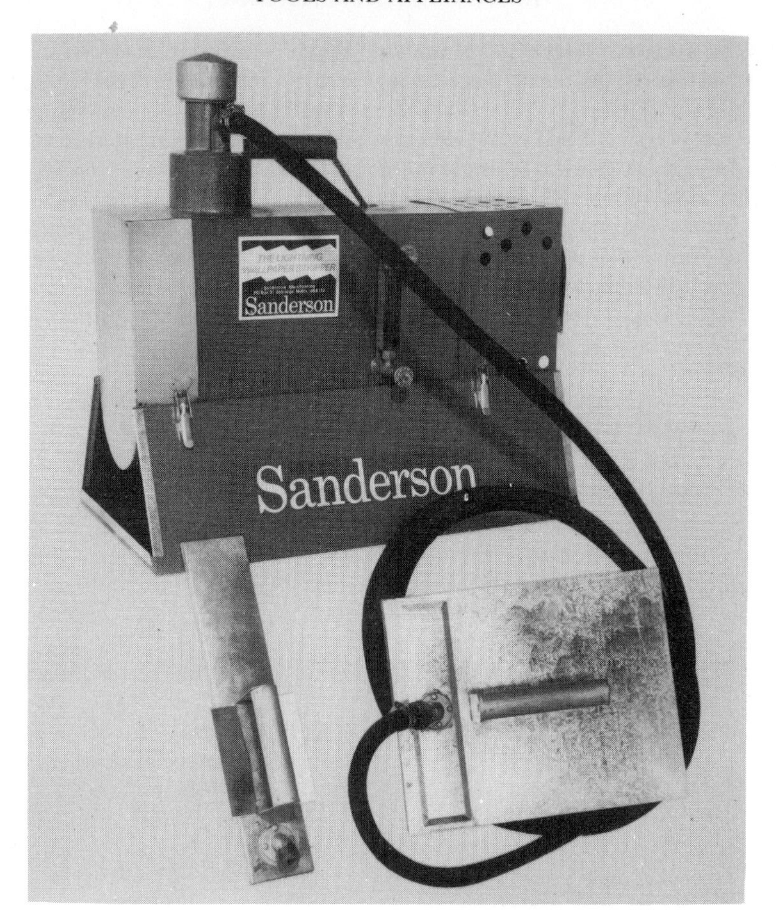

Courtesy: Arthur Sanderson & Sons Ltd

Fig. 4.6 The "Lightning" Wallpaper Stripper, showing two sizes of steam concentrator

may be applied to soften up a varnish coating before using the stripper. Distemper and water paint can also be quickly softened with the stripper, then finally removed with a sponge and water.

The Stripper comprises a 15-litre ($3\frac{1}{4}$ gallon) water tank with enclosed burner, which operates on Calor Gas. The use of a 14·5 kg (32 lb) container is recommended, with high-pressure regulator generating steam at $\frac{1}{2}$–$\frac{3}{4}$ bar (7–10 psi) at full pressure. Steam is generated within 10 minutes. A smaller concentrator, shown in the illustration, is supplied for narrow or awkward places, and a two-way attachment enables both concentrators to be used simultaneously if need be.

The stripping machines on sale today, mostly designed to operate from an LPG cylinder, may not be fitted with a pump, but of course a great many machines of the older type are still in use and giving satisfactory service.

There is also an electrically powered steam wallpaper stripper supplied by Damproof Engineering Services, of Marple, Cheshire. This equipment comprises a 9-litre (2 gallon) stainless steel tank and concentrator, with connecting hose, 3-kW heater element, and 4½ metre (15 ft) mains lead.

An important point about steam stripping is the sterilizing effect of the concentrated steam which leaves the surface free from bacteria and germs. Also, since the most obstinate papers can be removed without hard jabbing with the stripping knife, there is less damage to plaster and therefore less making good. It should be remembered, however, that if the work is done carelessly and the concentrator held in one position for too long, the surface of the plaster may be damaged and the skimming lifted and buckled.

High-pressure washing and steam cleaning

The cleaning down of machinery, vehicle chassis and other greasy surfaces prior to repainting can be a costly process owing to the time required to obtain a surface clean enough to receive paint. Moreover, when normal oil solvents are used there is always a danger of merely spreading the grease over the surface instead of completely removing it.

The old steam jenny has been superseded by more sophisticated high-performance machines suitable for all industrial cleaning tasks, including the degreasing and de-waxing of vehicles, cleaning and sterilizing agricultural machinery, farm buildings, etc. The Wickham "Cyclone" steam cleaner (Fig. 4.7) is a high-pressure hot- or cold-water washer and steam cleaner, operated either electrically or by petrol engine. The high-pressure steam jet works at a pressure of 16 bars (250 psi), incorporating a detergent or sterilizing agent where required; it emulsifies

Courtesy: Wickham Industrial Equipment Ltd

Fig. 4.7 The Wickham "Cyclone" high-pressure water washer and steam cleaner

A machine suitable for preparing buildings for repainting

any soluble oils, grease, tar and dirt which is finally flushed off with dry steam, leaving the surface thoroughly clean and ready for painting.

Such appliances are practically indispensable to large firms engaged in the painting of factories and plant, farm buildings, road and rail vehicles, etc., but where they are required only occasionally they can be hired from small specialist firms such as Steam Clean Services, Longton, Preston.

Paint agitators

Where large stocks of ready-mixed paints are regularly held, as is often the case with some of the larger painting contractors, the number of man-hours expended in stirring paint which has settled may be considerable. There are various types of paint shaker or conditioner available such as the "Quickway" machine (Fig. 4.8), which will accommodate tins from one litre to five litre size and which, by rapid agitation, will rejuvenate old stock in a few minutes. By going through the stock periodically, or by placing each container on the machine for a few

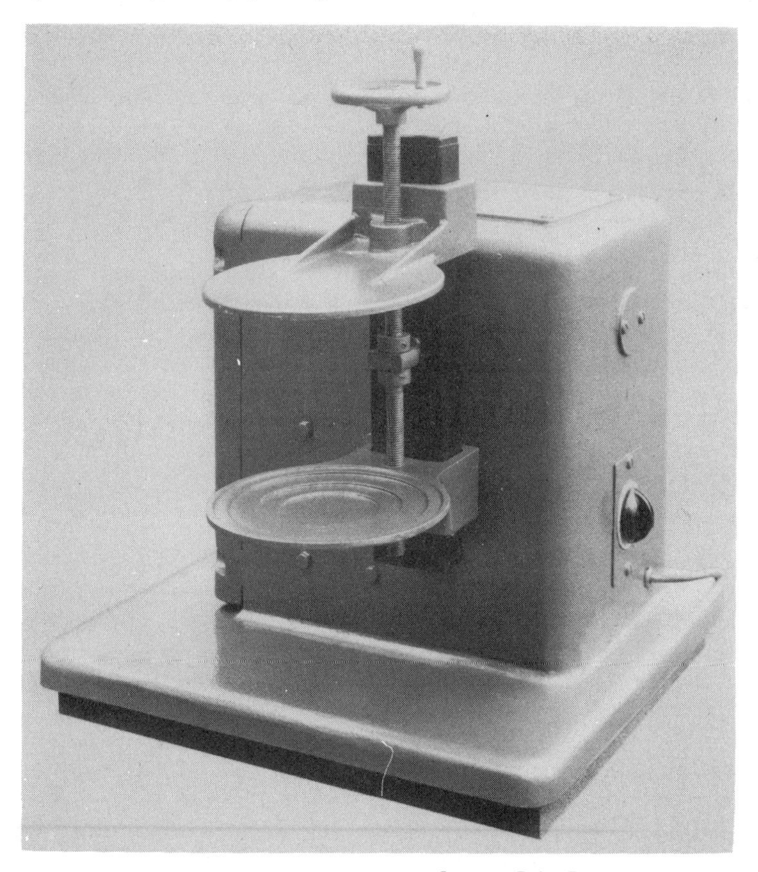

Courtesy: Baker Perkins Holdings Ltd

Fig. 4.8 The "Quickway" paint-shaking machine for restoring settled paint to original consistency

minutes before sending it out from the paintshop, the material is kept in good working condition and much time is saved. A paint-agitating machine is also essential for use in conjunction with a paint dispensing machine. A paint agitator is embodied in certain types of spray painting equipment, especially those for applying heavy pigment paints (see Chapter 6).

Electric dryers

Although these machines were originally designed for use in hairdressing, there are many occasions where an electric dryer can be used to great advantage by the painter and decorator for interior work where a power point is available. Warm or cold air can be blown on to a damp or wet surface to hasten drying after washing down, etc., when it is necessary to apply a coat of paint the same day; mouldings and quirks which hold the water and normally take a long time to dry need not delay the job if treated in this way. Similarly, doors grained or glazed in water often dry slowly in damp or cold weather. A few minutes treatment with the dryer will ensure the surface being thoroughly dry when the varnish is applied.

Signwriters often have much trouble when lettering shop windows in cold weather, but steaming and condensation can be quickly eliminated with hot dry air. Glass gilding can also be expedited in the same way, and the sign man packs an electric dryer as a normal part of his kit.

These dryers are sometimes used for drying trial tints in water paint and distemper, but it should be remembered that when samples are force-dried in this way, the colour may be slightly different from that obtained by normal air drying.

In the drawing office and studio, the writer has found the electric dryer indispensable for speeding up the drying of water-colour paper after stretching it and also for drying water-colour washes in perspective renderings, etc.; on damp, humid days a wash of colour may easily take half an hour or longer to dry normally and the electric dryer therefore effects a considerable saving in time.

Flame-cleaning equipment

The removal of rust and scale from iron and steel surfaces has always been one of the painter's most arduous tasks; moreover, no matter how conscientiously the business of chipping, scaling and wire-brushing is carried out, rust often appears on the surface in a surprisingly short time, or continues to spread beneath the paint film until the latter is ruptured and finally pushed off by fresh scaling.

The introduction of the flame-cleaning process provided a most effective practical method for removing rust and scale *from existing structures*; but, what is even more important to the painter, the process leaves the surface in an ideal condition to receive paint, even when working under adverse circumstances. The process was originally developed by BOC Ltd and consists briefly of "scrubbing" the corroded surface with an intensely hot flame; this dehydrates the surface and reduces the rust and scale to a loose black powder (black oxide of iron) which is easily removed with the wire-brush. The surface remains warm for 30 to 45 minutes after the flame has passed over it and thus gives ample time for the priming paint to be applied before cooling; it is essential that priming should be done while the surface is still warm and dry, since the value of dehydrating it is completely lost once the surface has cooled and moisture has re-condensed upon it.

The significance of flame-cleaning is dealt with under "Priming of Iron

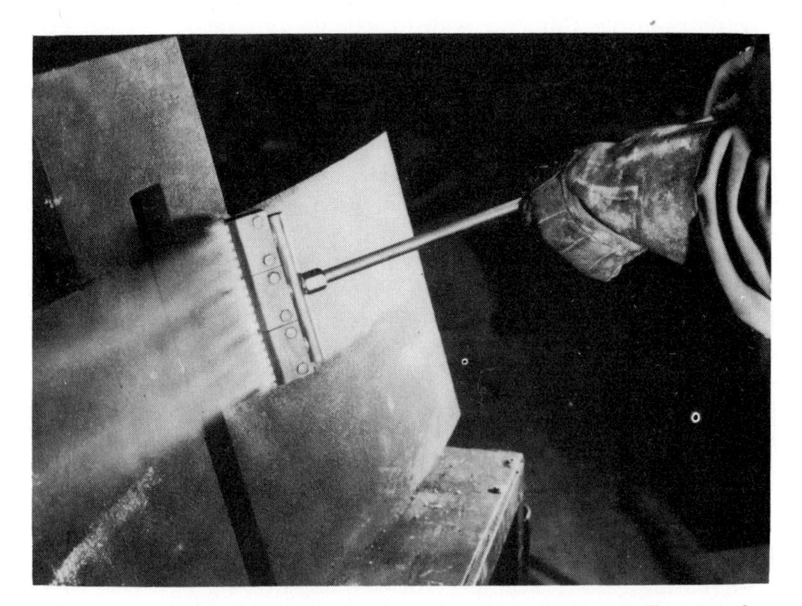

Fig. 4.9 Flame cleaning a casting, using a 150-mm flat nozzle

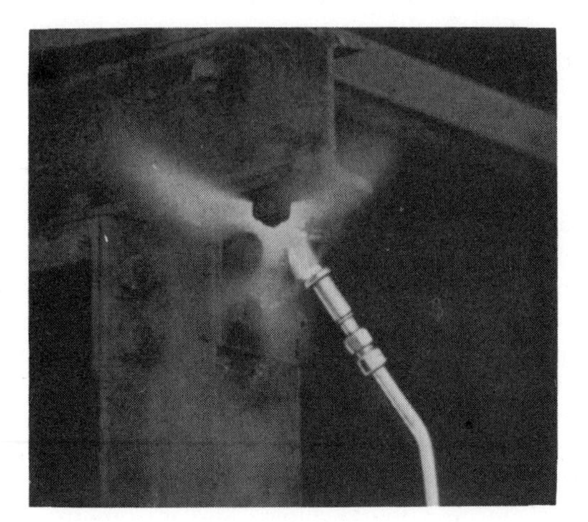

Fig. 4.10 Flame cleaning bolt heads, using a round nozzle

Courtesy: BOC Ltd

and Steel" in Chapter 19. Until recently it was invariably an oxy-acetylene flame that was used, the equipment consisting of oxygen and acetylene gas in separate high-pressure cylinders connected by suitable independent pipes to the torch itself; in present-day practice increasing use is being made of propane used in a special mixer and similar LPG equipment to provide the flame. The torch is very similar to any oxy-acetylene or LPG torch used in other applications of these gases, and the main difference lies in the type of burners employed. These are usually in multiples 50 mm wide, each 50 mm piece having 14 small flames along its length (Fig. 4.9). The multiple burners can be made up in widths of 50 mm, 100 mm, 150 mm and up to 300 mm, while in addition there is a round nozzle whose purpose is to deal with round objects such as rivet heads, etc. (Fig. 4.10). No great skill is required in actually operating the flame torch, and the method can be quickly learned after a short demonstration; the work should nevertheless be done under expert supervision so that no essential part of the process is missed and to ensure that the priming is applied at the right time.

The process is of interest mainly to the big industrial painting contractors who carry out large projects such as the initial painting and maintenance of bridges, dock installations, ships and industrial plant, and it would obviously not be an economic proposition for a small firm to acquire equipment for which there might only be an occasional need. The process is widely used in America, while in this country British Rail have adopted it to a certain extent for bridge painting, and difficult structures like tall radio and television masts have been successfully treated *in situ*. In 1961 the whole of the steel structure of the new Runcorn–Widnes bridge was prepared by flame-cleaning, and since then the process has been increasingly used on important civil engineering schemes. In spite of the rapidly expanding use of other preparatory processes such as blast-cleaning, there are still many occasions when flame-cleaning provides the best and most feasible method of preparing structural steelwork, especially in damp, foggy or frosty weather conditions when work would otherwise be brought to a standstill.

Blast cleaning

Blast cleaning is probably the most effective of all the methods of preparing steelwork, provided it is done correctly; it produces a clean, slightly roughened surface which offers good adhesion to paint. Though regarded mainly as a factory process, it is extensively used on site work where conditions are suitable. The term "blast cleaning" covers the range of operations known variously by such names as abrasive blasting, abrasive cleaning, grit-blasting, and shot-blasting. The process consists of propelling various kinds of abrasive towards a surface at very high speed by means of compressed air, the force of the impact and the hardness of the abrasive particles serving to dislodge and remove the rust and scale. The blast-cleaning process can also be applied to the cleaning of stonework, concrete, and metal alloys, and is sometimes further used for decorative purposes by blasting motifs and patterns into the surface of brickwork, stone, concrete, etc. The efficiency of the cleaning operation depends on the rate of working and the coarseness of the grit, and this aspect of the matter is dealt with more fully under "Blast Cleaning of Iron and Steel" (Chapter 19).

Blast cleaning is usually carried out on the site by an open blasting process. The equipment consists of a hopper into which the abrasive particles are loaded and from which they are passed into a pressurized pot. From here they are fed through a metering valve into the stream of compressed air, being forced through the hose to the nozzle. Obviously the impact of the abrasive is increased by raising the air pressure, and pressures in the order of 6·5 bars (100 psi) at the nozzle are generally employed. Air consumption is high, and a large compressor unit is essential.

The design of the nozzle has an important effect on the efficiency of the process; the type with a venturi tube or narrowed throating inside the cylinder increases the velocity of the abrasive, and produces a more regular cleaning pattern than a straight cylindrical nozzle. The inner casing of the nozzle is lined with tungsten carbide or a special alloy resistant to the harsh abrasive action. The reinforced rubber hose contains an ingredient which minimizes the build-up of static electricity, and external hose couplings are used so as not to decrease the internal diameter and thus impede the flow of the abrasive.

There are obvious difficulties about the use of blast-cleaning equipment on erected steelwork (particularly when a painting contract is being carried out in factory premises where normal production has to be maintained with as little interruption as possible) because of the danger to other operatives on the site, and also because the wastage of shot, unless carefully controlled, would be prohibitive. There is, however, a type of blasting equipment, shown in Fig. 4.11 and 4.12, which overcomes these difficulties. It incorporates a vacuum device which reclaims the shot immediately after impact, passes it back into the plant, cleans it by removing all the dirt and debris, and re-circulates it — all in one operation. There is a shielding device on the blasting head which conforms readily to the contours of the steel members and which prevents the shot from straying beyond the confines of the steelwork, so that the equipment is perfectly safe for site work. The wastage of shot is reduced to the level of less than 0·45 kg (1 lb) in a full day's working, and the application of paint can follow immediately after the cleaning, with the painter working in close proximity to the blasting operation.

The pioneers in the development of this equipment were Vacu-Blast Ltd. The equipment is now widely available and varies considerably in size and capacity. Fig. 4.13 illustrates some of the models produced by Hodge Clemco Ltd, of Handsworth, Sheffield. The smallest of the units illustrated weighs 36 kg (80 lb) and holds 25·4 kg (56 lb) of expendable abrasive, while the largest weighs 295 kg (650 lb) and holds 362 kg (800 lb) of abrasive. Obviously a small machine has its limitations, especially with regard to the size of nozzle that can be used. The most popular machine for general use is one weighing 181 kg (400 lb) which can readily be moved by one person and has a capacity of 272 kg (600 lb). Generally speaking, the greater the diameter of the machine in relation to

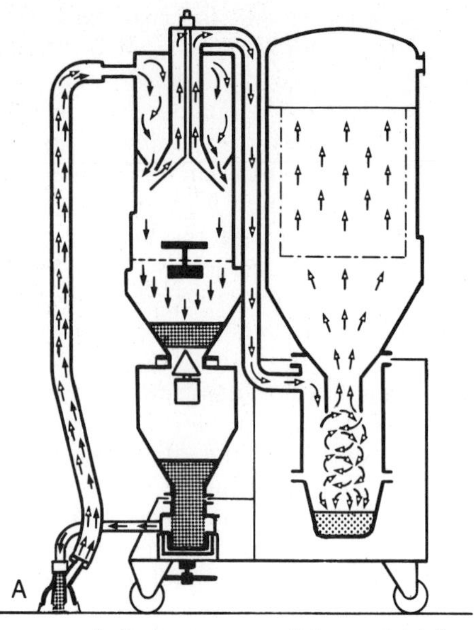

Fig. 4.11 Vacu-Blast closed-circuit cleaning equipment

Diagram showing, at left, pressure feed hopper, gun head, and reclaimer section for abrasive; and, at right, method of returning debris via vacuum hose to dust extraction cyclone

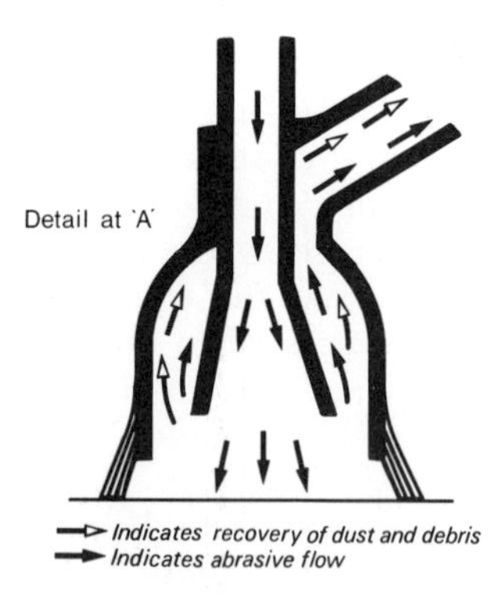

Fig. 4.12 Detail of gun and vacuum pick-up

Courtesy: Vacu-Blast Ltd

its height, the more stable it will be and the more easily it can be loaded with abrasive, but the taller narrower types are useful where the operation is to be carried out in a restricted working area.

Of particular interest for small-scale blast-cleaning in situations where the abrasive and dust need to be fully contained (as, for example, where other workers are in the immediate vicinity) is the portable hand-operated

Courtesy: Hodge Clemco Ltd

Fig. 4.13 Range of Hodge Clemco blast-cleaning machines

machine supplied by Hodge Clemco Ltd and illustrated in Fig. 4.14. This unit, which recovers by suction and re-cycles its own abrasive, weighs only 3 kg plus the weight of the abrasive, up to a further 2 kg. Any re-usable abrasive such as steel shot or grit, chilled iron, aluminium oxide, silicon carbides, glass beads, etc., can be used. The blast pattern is about 32 mm wide.

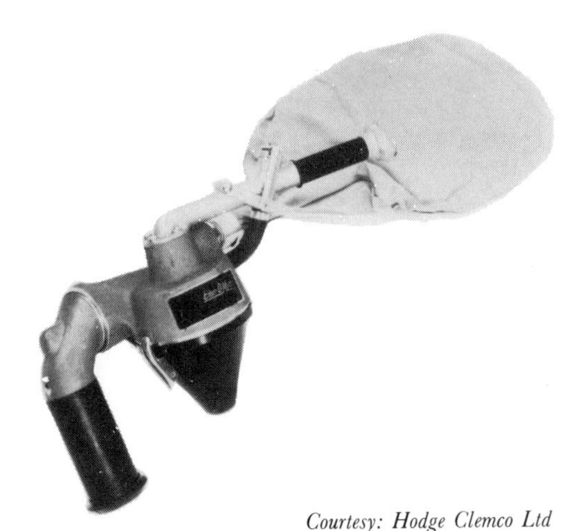

Courtesy: Hodge Clemco Ltd

Fig. 4.14 The "Educt-O-Matic" lightweight blast-cleaning machine

A hand-operated unit for small-scale dust-free cleaning

Wet blast-cleaning

Blast-cleaning equipment is also available in which water or steam is combined with the abrasive in order to minimize the amount of dust produced. This method, though effective, is generally slower than the normal dry process, but great improvements in cleaning rates have recently been made with this technique, and the firm of Hodge Clemco Ltd has developed the versatile "Kleenblast" water-injection unit, a system which gives a choice of either wet or dry blasting, the changeover being controlled by the operator (Fig. 4.15). The water, with the addition of an approved rust inhibitor if required, is drawn from either mains or static supply direct to the pump. It is then carried by high-pressure hose to the nozzle holder and is injected immediately behind the nozzle, thus leaving the blast hose completely free of atomized water. The water not

Courtesy: Hodge Clemco Ltd

Fig. 4.15 Cleaning the steelwork of a motorway bridge using the "Kleenblast" water-injection unit

only suppresses dust but also serves to remove ingrained salts and chemical residues from the surface. Virtually any degree of surface preparation can be achieved by use of the correct abrasive (type and size), nozzle pressures, water additives, and blast cleaning techniques. By using the pressure regulator to reduce the nozzle pressure, and decreasing the angle of incidence of the abrasive stream to the work surface, a much more sensitive cleaning can be achieved, using the same basic equipment. It is even possible, where circumstances indicate, to remove only the top coat of a paint system, leaving well adhering sub-coatings on the substrate.

A simple flow-control valve at the nozzle enables the operator to change at will from wet to dry blasting and vice versa, and also permits him to regulate the rate of injection of water, irrespective of the length of the hose. An additional safety factor is a "deadman" control which prevents the possibility of any accidental discharge of abrasive. The system can be used with any of the normal range of Clemco blast machines and with a wider choice of abrasive sizes than any other comparable wet-blasting equipment.

Total air consumption of the blast machine and Kleenblast unit using a 12 mm (½-inch) nozzle is 338 cfm (160 litres/sec) at 100 psi (7 bars). Using the same equipment, however, it is possible to reduce consumption to 137 cfm (64 litres/sec) at 30 psi (2 bars) — pressure figures being measured at the nozzle.

Pneumatic tools

Rust and scale can also be removed by mechanical chipping hammers and other pneumatic tools driven by compressed air. Large-scale equipment of this type suitable for heavy industrial work is supplied by Armstrong Whitworth & Co. (Pneumatic Tools) Ltd, of Manchester.

A type of pneumatic equipment suitable for use where small quantities of steelwork are to be treated, and in a price range that puts it within the reach of ordinary painting and decorating firms, is the Jason de-scaling pistol (Fig. 4.16). This consists of a group of hardened steel needles,

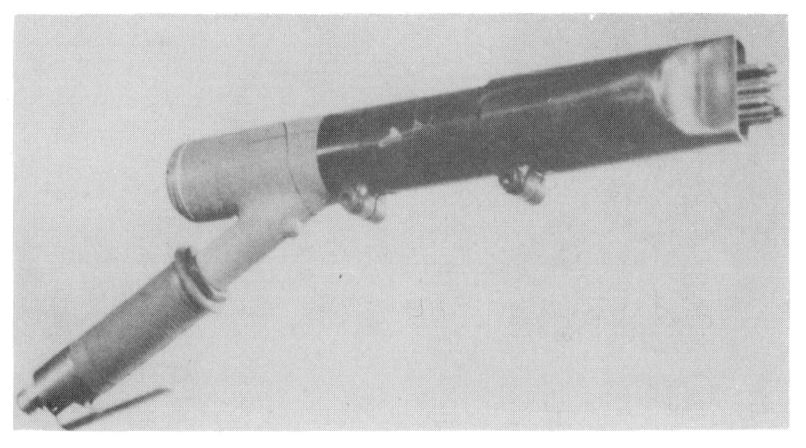

Fig. 4.16 De-scaling pistol

propelled and retracted from an open valve by a spring-loaded piston. The vibrating needles adjust themselves to an uneven surface or contour and will clean out angles, pockets, rivet heads, and areas of rust-pitting down to the parent metal, though the process is not economic for larger areas. There are three types of needle — round, pointed, and chisel-ended — and the tool will also operate larger chisels for chipping purposes. The action is trigger-controlled and air flows continuously over the working surface, blowing off the loosened rust and scale.

BURNING-OFF EQUIPMENT

For removing paint, there is a choice of equipment in the form of several types of blowlamp and blowtorch, as well as electrical devices.

Blowlamps

Petrol and paraffin blowlamps (Fig. 4.17) each have their own adherents, but petrol is probably the more popular since it generally requires less cleaning and maintenance and is lighter in weight. On the other hand, of course, in today's economic conditions petrol is an expensive fuel, and it may be that cheaper alternatives are preferred. The paraffin lamp is fitted with a pump which is used to maintain pressure, but petrol burning-off lamps do not require a pump, for once the lamp is warmed up, pressure is maintained automatically.

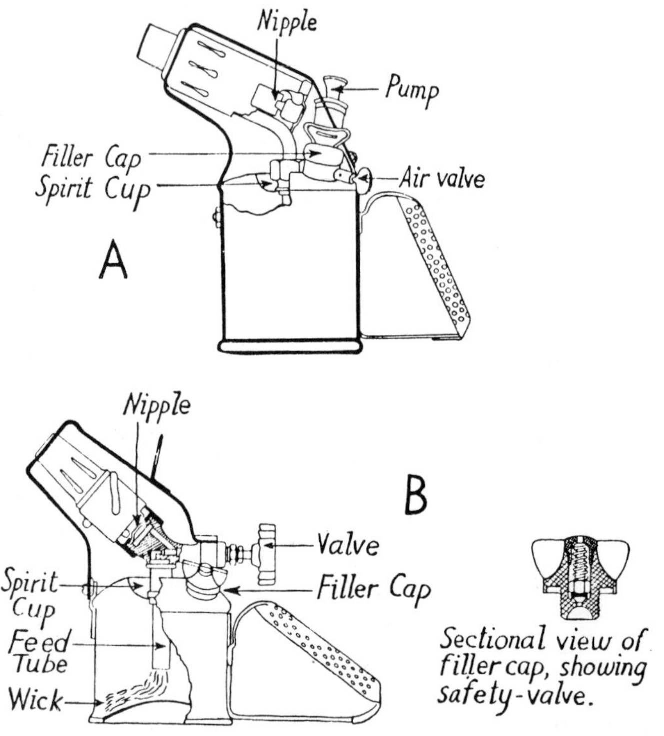

Fig. 4.17 Primus blowlamps — sectional views
(A) Paraffin, and (B) petrol types

The *petrol lamp* is extremely simple in operation. It should be filled through a funnel fitted with a fine gauze filter; when filled, pour out a little of the petrol to allow for expansion when the spirit is vaporized. Screw on the filler cap tightly, then fill the spirit cup with petrol and light it, first making sure that the valve is closed. A cotton wick inside the lamp draws the petrol up to the heated nozzle where it becomes vaporized. When the petrol in the spirit cup is almost burnt out, open the valve; the lamp should then light and burn with a blue flame which gets stronger as the body of the lamp warms up. If the valve is opened too soon (i.e. before the petrol becomes vaporized) petrol will spurt out of the nozzle and, if indoors, may cause a fire (for this reason blowlamps are best lit outside, sheltered from the wind). If the petrol in the cup is allowed almost to burn out before opening the valve, however, the petrol inside the lamp will be properly vaporized and will burn correctly.

When working outdoors in cold weather or in a wind, it is necessary to use an asbestos-lined metal windshield to keep the body of the lamp warm, otherwise the lamp will cool off and the flame will weaken or blow out. Every lamp is fitted with a safety device, either in the form of a spring-loaded valve or a metal plug in soft solder; if the lamp becomes overheated the solder melts, releasing the plug, and so prevents an explosion. The best types are self-pricking, but others have a pricker supplied to clear the nipple when this clogs up with carbon. Lamps should be cleaned periodically and the filler-cap washer should be renewed when necessary to prevent leakage and loss of pressure.

Paraffin lamps tend to clog more frequently owing to the nipple becoming charred up with carbon deposit. The lamp is warmed up in much the same way as the petrol type, except that a few strokes of the pump are required both at the start and occasionally during operation to obtain and maintain the required pressure. Paraffin lamps are economical in use, but there is a tendency to leave a greasy deposit on the work if a low-grade paraffin is used. It should be noted, however, that in some branches of the armed services paraffin blowlamps are issued as part of the standard kit of tools, and under the strict schedule laid down for the maintenance of equipment they are kept in first-class condition by regular servicing. When properly maintained in this way they are surprisingly efficient and provide a hot, powerful flame with no residual deposits.

Another alternative is the *methylated spirit blowlamp*, which is highly efficient, does not clog up, burns with a very hot flame, and leaves no residue whatsoever. Of the three types this is probably the cleanest type of lamp, but although widely used in some of the metal-working industries it has never been widely adopted by painters and decorators, in spite of efforts to popularize it.

Blowlamps are perfectly safe if used with normal care, but it is as well to have an old wet sack handy to smother the lamp in case of emergency. When lighting the lamp always remove the fuel container to a safe distance and replace the screw cap tightly.

Blowtorches

In present-day practice, considerably more use is made of blowtorches operated from a fuel consisting of liquefied petroleum gas (LPG), i.e. butane or propane; "Calor Gas" being a trade name for butane. This type of equipment offers distinct advantages as compared with the blowlamp. One very important feature is that the torch lights instantly and is ready for use at full power without any delay, whereas with the blowlamp there is always some time lost in filling the lamp and waiting for it to warm up before it can be used. This compensates to some extent for the fact that the fuel is more expensive, because while a blowlamp will run for about 2½ to 3 hours from one filling, there can be as much as 20 minutes lost time with every filling; furthermore the flame produced by a gas torch is hotter, and this makes for quicker working. Another important factor is that there is no loss of pressure whilst work is in progress, and far less chance of the flame being affected by the wind; one of the most frustrating things about using a blowlamp in a strong wind is the continual blowing out of the flame and the cooling down of the lamp casing, resulting in loss of pressure.

With the gas torch, there is much greater control over the size of flame, which can be adjusted from a narrow pencil shape for burning off narrow window glazing bars to a wide fan-shape for broad surfaces. Then again, there is no residue left on the surface to affect the adhesion of subsequent coats of paint. The torch itself is lighter in weight than a blowlamp, and when ladder work is involved this makes for considerably less fatigue. Various nozzles are available, with a range of jets that provide differing flame-shapes and varying degrees of heat. The main types of blowtorch are as follows:

Torches operated from large bottled gas cylinders

In this type, the fuel is contained in a large cylinder which is connected to the burner or torch by a long flexible hose (Fig. 4.18). The cylinders

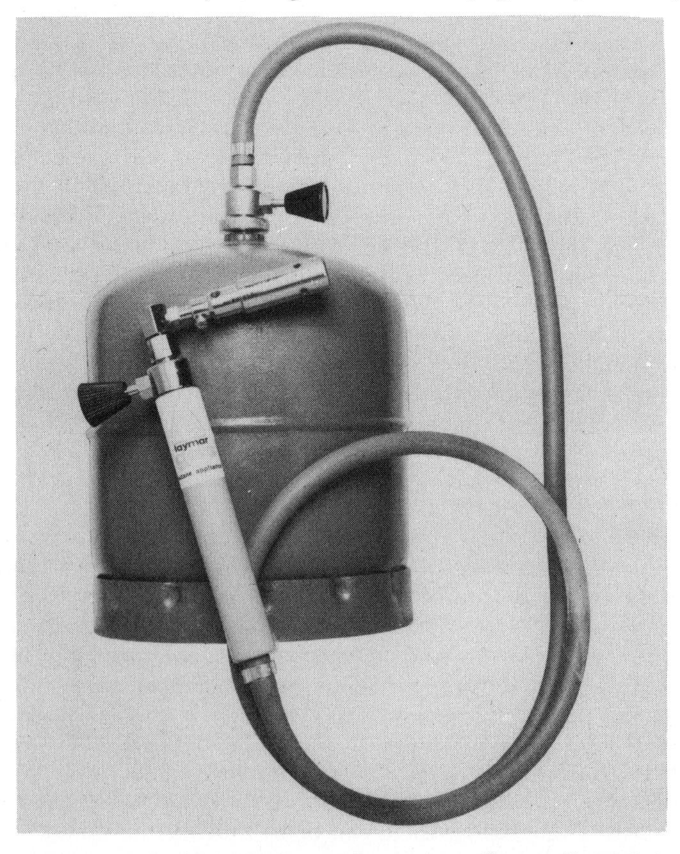

Courtesy: Taymar Ltd

Fig. 4.18 LPG blowtorch with large capacity cylinder

are stocked by service depots covering most parts of the country; they are usually delivered to the site by the stockist and the discharged cylinders collected for refilling, but they can of course be collected from the stockist by the decorator where this is more convenient. Various sizes of cylinder are available, ranging from those containing 4·5 kg (10 lb) of gas to the larger ones containing 14·5 kg (32 lb). Two torches can be operated from one cylinder simultaneously. A special advantage of this type is that the painter has only the handpiece to hold and this is very light in weight — a much easier thing than holding a blowlamp all day; in awkward places this is an even greater advantage. A drawback is that the equipment is not completely portable, since it cannot be used on scaffold work beyond the extent of the hose.

Fully portable blowtorches

These consist of a small cylinder which is screwed underneath the torch. According to the type of nozzle used, this provides enough fuel for from 2 to 4 hours burning; it then needs refilling. This type of torch has the advantage that it can be carried into any position at any height, although in small confined spaces it is less manoeuvrable than the torch fed by a hose; another slight disadvantage is that if the torch is tilted acutely it has a tendency either to flare up or to extinguish itself.

Disposable cartridge torches

These are a more recent development, and consist of a torch head into which a disposable butane gas cartridge is screwed. They are very handy indeed, and are very light in weight. They can be carried in the tool-kit without any trouble and are most useful for those occasions when unexpectedly a small amount of burning-off is found to be necessary. The burning time, however, is rather short, and if this type of torch were to be used continuously it would prove to be more expensive than the use of a refillable cylinder. Fig. 4.19 shows the standard Taymar disposable cartridge torch, together with the wallet set including some of the special-purpose nozzles that can be used with all Taymar torches. For more difficult jobs requiring much heat the larger model shown in Fig. 4.20 may be used.

Acetylene blowtorches

The apparatus for these consists of a metal cylinder of compressed acetylene gas which is fed to the torch through a rubber hose. The torch is a light handpiece with adjustments for mixing the gas with air; the size of the flame can be regulated with great precision. This type of equipment was the pioneer in the field of bottled gas torches, though in recent years it has declined in favour of the butane and propane types.

Electric paint strippers

These consist of a long handpiece at one end of which is a small rectangular head containing an electric element, round which there is a guard to prevent the element from coming into direct contact with the painted surface; from the other end of the handpiece there is a length of electric cable which is attached to the plug. For practical purposes the

Fig. 4.19 Disposable cartridge blowtorch
A flat attachment for paint removal is included in the wallet set

Fig. 4.20 Craftsman blowtorch with engineers' burner

Courtesy: Taymar Ltd

electric stripper is not a particularly efficient tool. The great drawback is that it cannot be used with the usual burning-off technique; with a blowlamp or blowtorch the burner is held in one hand and moved in a slow continuous movement across the surface, while the stripping knife in the other hand moves in unison with it, the knife following immediately upon the flame; but with an electric stripper the burner head has to remain in one position until a rectangle of paint has been softened and then scraped off before the next small area is heated. This makes the operation painfully slow. There may be a slight advantage in that the tool can be used in situations where a naked flame might cause damage to the surface, but in fact it would usually be better to use the solvent method of paint removal in a situation such as this.

Mains gas torches

A blowtorch may be used with the ordinary domestic coal gas or natural gas supply where this is available. The method can be quite convenient in the paintshop, but has little application to site work. The torch should be checked by a competent person to make sure that it is suitable for use with natural gas.

Hot air stripping

A more recently developed method of stripping old paint consists of softening the film by means of a stream of hot air directed upon it, a process which offers certain distinct advantages. Clearly, since neither flame nor chemical substances are involved, it is much safer than many traditional methods. This entirely new idea originated in Holland, in response to a

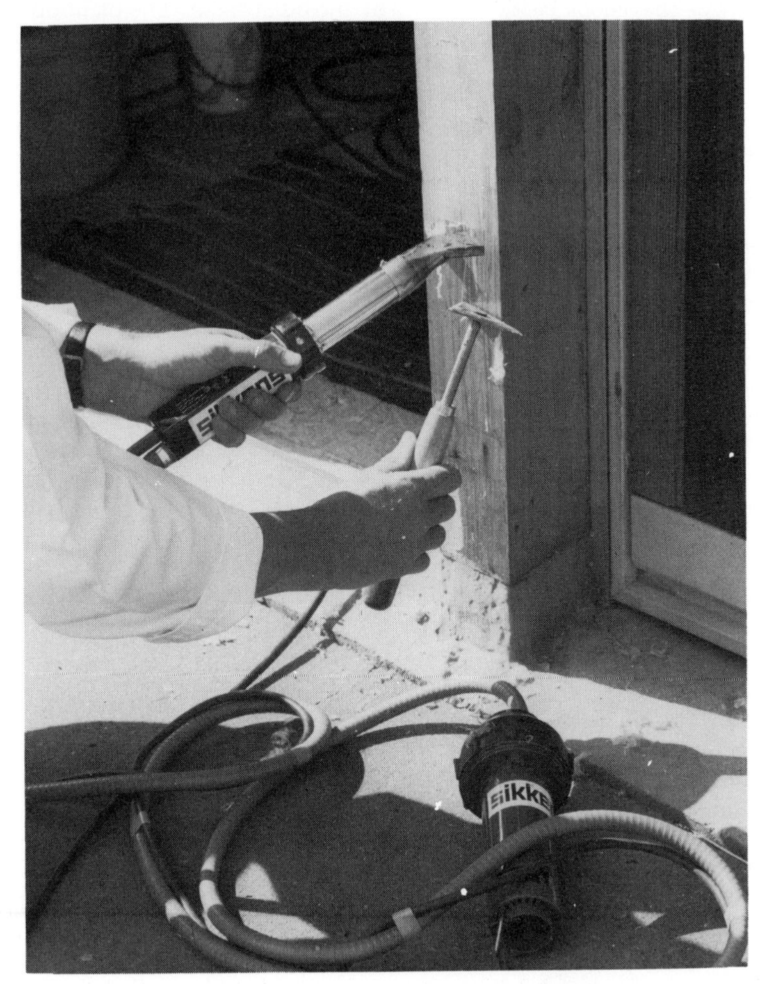

Courtesy: Sikkens (UK) Ltd, Didcot

Fig. 4.21 The Sikkens hot air paint stripper

problem concerning the removal of large quantities of old paint in a situation where the use of a naked flame was not acceptable, at the Royal Palace of Noordeinde in The Hague. The equipment is now available in Britain.

The plant (Fig. 4.21) comprises a blower, heater and handpiece with the associated tubing, cable and electrical connections. The blower, which weighs only 1·2 kg, propels air along a flexible tube at a rate of 400 litres per minute to the heater situated in the handpiece. The heater raises the temperature of the air immediately before it is directed to the paint surface; a control on the handle enables the air temperature to be varied anywhere between 20° and 600°C. The power cable leading to the heater is attached to the air pipe, and the handle is insulated so as to remain cool. The handpiece is very light in weight and is easily held in one hand while the scraper or shavehook is manipulated by the other hand. The plant operates normally on a voltage of 220–240, but a 110-volt model is also available. The consumption rating of the heater is 1300 watts and that of the blower 60 watts.

It is not suggested that this equipment will supersede burning off, but it has obvious attractions for use in high-risk areas such as where flammable materials are stored, and for those occasions when it is necessary to avoid charring or scorching a wooden surface whilst stripping paint, because the subsequent finish is to be varnish or some other clear material. In these situations hot air stripping offers an economical alternative to the use of liquid paint removers, being far quicker and less messy.

Sanding machines

One of the most important factors in any successful painting operation is the preparation of the surface, and attention is given to this in the descriptions of the various decorative processes in later chapters of the book. Traditionally, a great deal of the preparatory work consists of using abrasives by manual rubbing down to smooth the surfaces and slightly etch the preceding coatings. There are, however, numerous mechanical aids which in certain cases can considerably lighten the work and effect a substantial saving in the cost. A wide range of sanders, driven by electrical or compressed-air equipment, is used in the industrial finishing and furnishing trades, but the mechanical hand tools of greatest interest to the painter include rotary disc sanders, belt sanders, and orbital sanders.

The *rotary disc sander* consists of a flexible rubber or composition head to which is fitted the type of abrasive paper appropriate to the particular purpose. It can be used for preparing new woodwork or previously painted surfaces, for the removal of rust from steelwork, or for re-surfacing floors. It can also be used for polishing when fitted with a lambswool mop, or used for removing rough surfaces from welds when fitted with a carborundum grinding wheel, and it has the advantage of being suitable for use on curved surfaces. One great disadvantage is that it is a difficult tool to control, and if used inexpertly it can easily cause damage to the surface. This fact needs to be recognized. It would be foolish to regard a disc sander as a general handyman's tool that can be given out indiscriminately to any of the operatives employed by the firm; the cost of building up a good surface again can heavily outweigh any advantage gained in the speed of rubbing down.

The *belt sander* operates with a continuous belt of abrasive paper with a straight-line action. It has very little application to general painting and

Fig. 4.22 Preparing final finish with a straight-line sander

Fig. 4.23 Range of high-speed pneumatic sanders
Courtesy: Sundstrand Sanders Ltd

decorating work, but can be useful where, for instance, a long run of new woodwork needs to be shop-primed before delivery to the site; and a heavy-duty belt sander is a great asset on a job where a considerable amount of flooring is to be renovated ready for a stained or clear lacquered finish.

The *orbital sander* is the one of greatest interest for general painting and decorating work, and it can be used to prepare woodwork, metal, or plaster surfaces or to rub down previously painted work. It is much easier to control than a rotary disc sander, and being only about 1·5 kg in weight (just over 3 lb) it is light enough to be used for long periods without causing undue strain to the operative. It consists of a rectangular platform pad to which the abrasive paper is attached, the platform moving with a small circular or orbital motion, the movement being controlled by a lever in the palm of the hand. It is worth noticing that the abrasive used with a mechanical sander can be a grade or two coarser than that used for manual rubbing down.

While an orbital sander is suitable for most types of work, it is not entirely satisfactory for woodwork that is to be finished with stain or clear lacquer, where it is essential for the sanding to run parallel with the grain, because the cross-scratching of the grain caused by the orbital sander will show through the finish. To avoid swirls or pressure marks on fine surfaces a straight-line sander is used (Fig. 4.22).

A most important point to remember when choosing a sander, especially an orbital sander, is that although the equipment can be powered by either electricity or compressed air, most of the preparatory work in painting and decorating consists of wet rubbing-down; for this reason the compressed-air type is much to be preferred because of the safety hazard presented by using electrical tools with water.

Fig. 4.23 shows three types of single-pad orbital compressed-air sander in the Sundstrand high-speed range, together with the larger twin-pad straight-line models. Several types of pads are available, including felt, cellular and sponge rubber, and tubular rubber, the last-named being specially adapted to curved surfaces.

5

Brushes and Rollers

There are various methods by which paint can be applied to a surface. The three most important methods from the painter's and decorator's point of view are the brush, roller, and spray. Part of the skill of the painter consists of knowing which of the three is best suited for the particular job in hand. Of the three methods the oldest and the traditional tool of the painter is the brush, and in spite of all modern developments it is still of the utmost importance. Obviously there are many occasions on which it would be uneconomic to use the brush, but on the other hand there are a great many operations for which the brush is still the best choice, and a number of things which can only be done by brush. Paint rollers are dealt with in a later section, and spray painting in a separate chapter (Chapter 6).

BRUSHES

The materials which go to the making of painters' brushes are obtained at present from two main sources — animal hair and vegetable fibre. Hog-hair or bristle is by far the most important brush material, and it is important to remember that the word "bristle" properly refers only to hog-hair. British brush manufacturers have agreed on strict trade definitions, and a brush stamped "Pure Bristle" is guaranteed to contain only hog-hair; if the bristle is mixed with other animal hair, it is sometimes described as "all hair".

The only suitable bristle for paint and distemper brushes comes from the semi-wild pig or boar found in Poland, the U.S.S.R., India and China: a lean, narrow-backed animal (Fig. 5.1) which has a good covering of hair and bristles, particularly along the spine where they are longest and stiffest. The hair of the domestic pig is generally too fine and short for paint brushes. The best bristles came from the U.S.S.R., but after the Revolution and the intensive Soviet development of agriculture, the wild pig practically disappeared owing to selective crossing with domestic strains to produce a fatter animal for food. Hence the growing world scarcity of best-quality long bristle. The supply position is unlikely to improve since it is not a commercial proposition to breed pigs for their bristle alone; it takes three years for a pig to produce about 2 lb (0.9 kg) of bristles of suitable length and quality, whereas the improved domestic strains yield their maximum pork value in twelve months.

India and China were for a long time rather slow in adopting this policy of domesticating the wild pig, but now these countries too are becoming com-

Courtesy: Briton Chadwick Ltd

Fig. 5.1 Wild boar: source of the best bristle

mercially developed, so that the supply of bristle is continually shrinking; for many years the price has been soaring, and it is clear that scarcity in the face of ever-growing demand will result in a serious situation for the painter unless an alternative source or an adequate substitute for bristle can be found. In America, optimistic claims are being made for certain types of domestic bristle, but these do not make up for the lack of long bristles. Scientific research has produced nylon, rayon and other synthetic filaments which are now well established as paint-brush materials, but for various reasons (as explained later) they are not completely satisfactory as an alternative to pure bristle. No doubt further research will eventually produce improvements and possibly alternative materials.

Good craftsmen have always taken great care and pride in the use of their brushes, but it is now more than ever incumbent on everyone to see that pure bristle brushes suffer no ill-usage, either through neglect or ignorance. There is much to be said for giving each man a set of brushes and holding him responsible for them whilst in his care, but no brush should be handed out until the employer has satisfied himself that the user fully understands the necessity for looking after these precious tools and, what is even more important, *knows* how to care for them. Moreover, it should be seen to that no bristle brush (except a worndown scrub) is used on rough work such as washing down, limewashing, etc.; mixed hair or fibre brushes are suitable and quite good enough for this sort of work. Every effort should be made to conserve our dwindling supplies of bristle by using it only where a good finish cannot be obtained by other means.

The main characteristics of bristle and its two chief adulterants, horsehair and vegetable fibre, are briefly as follows.

Bristle The distinguishing feature about bristle is that it tapers from root to tip where it is split into two or more fine strands; this split tip is known as the "flag". The effect of the flag is to provide the brush with a very soft tip which is invaluable in laying off. Bristle also has minute serrations along its length, pointing towards the tip. These serrations cannot be seen except under the microscope but if a bristle is rubbed between finger and thumb the serrations propel it always in one direction; they have the effect of helping to anchor the bristle in its binding.

The value of bristle is determined by its length and stiffness. The lengths used in paint and distemper brushes range from 50 mm to about 175 mm. In colour bristle may be white, yellow, grey or black. Siberian white or "lily" bristle has the best qualities and commands the highest price. Indian grey or black bristle tends to be rather coarse, whilst black China is stiffish but fine, and is generally used in varnish brushes. At one time colour was a fairly reliable guide as to quality or origin, but since bristles may be bleached white or dyed black, this is no longer so. When burnt, bristle and other animal hair singes and shrivels away, leaving a characteristic smell, whereas vegetable fibre burns with a bright flame like wood and leaves ash; this provides a convenient test in case of doubt.

Horsehair is obtained from manes and tails. It is bundled and cut into suitable lengths called "drafts", the best drafts being produced in the U.K., South America, Australia, and China. Horsehair has little or no spring and no taper, nor has it any natural flag, but it is now being artificially "flagged" by the brushmaker. Its principal use is in mixing with bristle as an adulterant, but it is also mixed with fibre in jamb dusters and cheap retail paint brushes.

Other animal hair Central European *badger hair* is used for softeners and gilders' tips; it is sometimes adulterated with goat hair, or with civet hair from a species of wild cat found in Africa and India. Red and black *sable hair* is obtained from the Siberian sable, the marten, and other members of the weasel family; it is used in signwriters' pencils and artists' best-quality brushes, while *ox hair* is used in second-quality brushes, being stronger but coarser than sable.

The so-called "camel hair" is obtained generally from squirrel tails, though sometimes pony hair is used. There is much speculation as to the origin of the term "camel hair", some holding that it is a corruption of "Kemls' hair" from the name of a Dutchman who first used it about 1860, while others state that "camel hair" was used to camouflage the origin of Japanese pony hair when this first came on the market. There is evidence, however, that the ancient Egyptians used pencils made of hairs taken out of camel wool, so that the term, whilst being a misnomer today, has, in fact, a legitimate historical origin. Camel hair (for so we must still call it) is extremely soft and has little or no spring; it is used in French polishers' mops, in "dabbers" for pressing gold leaf into mouldings and quirks, etc., in glass-gilding size brushes, and in artists' pencils. Some signwriters prefer camel hair for glass work on the bench as its lack of spring keeps it on the glass and makes it more manageable for this work than the livelier sable.

Fibre There are many kinds of vegetable fibre, split cane and coarse tropical grasses used in brushmaking today, but in painters' brushes the fibre most commonly used is Mexican "istle" or "tampico" from the agave plant (sometimes called the American aloe). Cleaned, combed, bleached or dyed as required, and drafted into suitable lengths, it is used as an adulterant with bristle or horsehair to reduce the cost of distemper brushes, or used alone in cheap grade limewashing and washing-down brushes. Being practically unaffected by alkalis, it is suitable for applying caustic paint removers, silicate of soda solutions, etc.

Synthetic fibres The search for bristle substitutes led to the experimental use of artificial silk fibres such as nylon and rayon, and although the results are far from perfect as yet, they are definitely encouraging. These synthetic fibres, or *filaments* as they are technically called, are exceptionally hard-wearing, so much so that a nylon brush on test showed only 3 mm wear after painting 4000 sq. metres. The chief difficulty appears to be that the filament is smooth and non-absorbent, which means that it carries less paint than bristle owing to the tendency of the paint to run off; this is less noticeable with a viscous medium such as varnish or enamel than with ordinary oil paints. Paint-carrying properties are improved when the filament is roughened, and once the brush is properly "broken-in" the nylon seems to acquire an absorbent quality which holds the paint better. The breaking-in period requires patience, however, and an extended period on a moderately rough surface appears to be required. In America, experiments in artificial roughening have been tried with some success. Tapered nylon has also been produced successfully with artificial flag, but of course when the flag is worn down it does not renew itself as the flag on natural bristle does. A rayon filament which has been produced with taper and flag is said to wear equal to bristle, but it is handicapped by being unsuitable for use in water and certain solvents. This indicates some of the difficulties which the scientist is up against in his quest for the ideal bristle substitute. Another possible line of development may lie in doping vegetable fibre in order to toughen and strengthen it.

Some of the man-made filaments used in brushmaking are known by their trade names, such as Orel, Perlon, Tynex, etc., and are listed in the manufacturers' catalogues as such.

Brush manufacture

The story of brushmaking, like the story of painting, probably dates from prehistoric times, for some of the wall paintings found in caves suggest that a form of brush may have been used by stone-age man. In Egypt, brushes made of vegetable fibre and split palm leaf have been discovered — still clogged with paint; they probably date from the 14th century B.C., but "pencils" of even earlier date have also been found. Pencils made of bristles were in common use by the Greeks and Romans.

In modern brushmaking, many processes such as mixing and grading the hair, shaping the ferrule, etc., are done by elaborate machinery, but the actual building up of the brush filling and placing it in the ferrule is done by hand — one might say, by *sleight* of hand, for the dexterity with which these craftsmen (and girls) handle bristle is nothing short of miraculous. The bristle or hair mixture is arranged in layers of carefully graduated lengths, the slightly short middle being surrounded or "cased" by longer bristles which are set to curve inwards to form a slight taper from heel to tip. The filling is then inserted in the metal ferrule and secured with a special cement (Fig. 5.2).

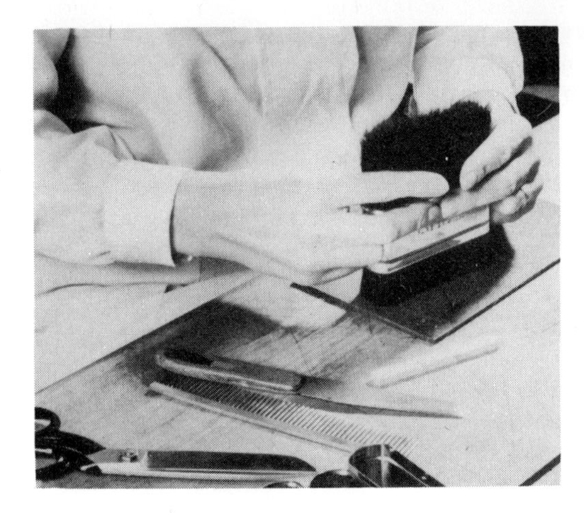

Fig. 5.2 (A) Handmaking high-class varnish brush (Acorn 27)

Fig. 5.2 (B) Inserting wedges into Acorn 27 varnish brush

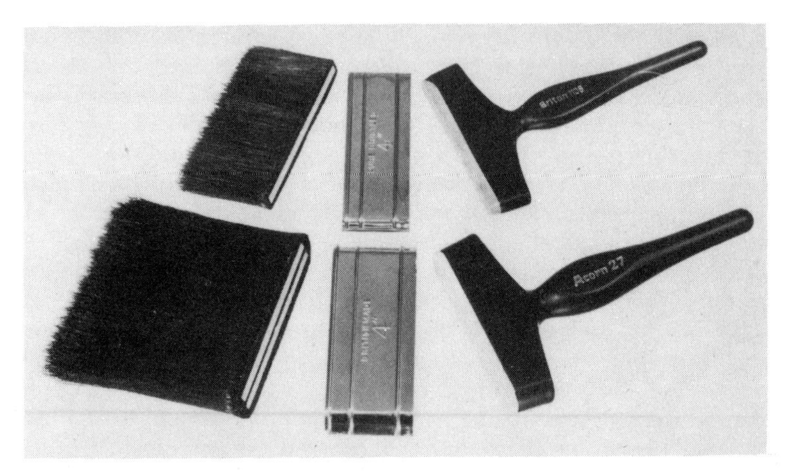

Fig. 5.2 (C) Difference between inexpensive brush and high-quality brush — note length out, thickness, etc.

Courtesy: Briton Chadwick Ltd

The old brush settings — rosin, pitch, glue, etc., have been superseded by vulcanized rubber and synthetic cements which are insoluble in water, oil, white spirit, or any other normal paint solvent in which they are liable to be used. The obvious advantage is that the bristle is held firmly in its binding in all circumstances. Only turks' heads, dusters (Fig. 5.3) and similar brushes which are not likely to require cleaning in spirit solvents are now set in pitch. The small tufts of hair in signwriters' pencils and artists' brushes are set in metal ferrules, or in quills which have been expanded by soaking in hot water and which, when cool, contract and firmly grip the hair (see Chapter 34).

There are of course many qualities of brush on sale, and as with every other commodity the cheapest is not always the most economical. When buying brushes it is sensible to look beyond the initial price and examine the

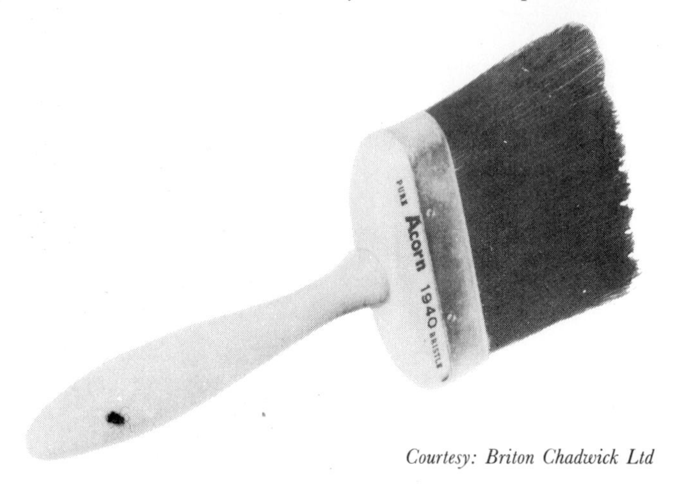

Courtesy: Briton Chadwick Ltd

Fig. 5.3 Painter's dust brush or "jamb duster"

brush itself. The quantity of bristle in the brush is no indication of quality. A brush can be packed with bristle to the point of flabbiness and offered at a cheaper price than a very much slimmer brush, simply because a blend of cheaper, inferior bristles has been used. In fact, it is a mistake for a brush to be too full, for reasons that are apparent when it is put into use: it tends to clog up, and its working life is considerably shortened. Brushes which have not been properly blended with the right selection of bristles will "twist" and "finger" when in use. Length of bristle — though usually a sign of good quality — can be similarly misleading; in a cheap brush the bristle may appear longer because it has not been set deeply enough in the cement. Another point to look out for is the way in which the brush-pins are inserted. The nails or pins that go through the ferrule ought to be "clenched" or turned over on the reverse side of the ferrule; by this means they lock together the bristle, ferrule and setting as a complete unit. As the bristle fills with liquid and the strain of expansion is exerted on the brush, the stress is placed first on the strength of the setting, then on the ferrule, and finally the pressure is on the brush-pins; if these are not strong enough they will break or loosen. In a cheap brush the pins, instead of being clenched, are simply driven in on either side of the ferrule, and there is no locking action. When the brush is in use they are liable to be forced out, so that the brush head becomes loose and eventually falls off.

Fig. 5.4 is an illustration from the original version of *Painting and Decorating* by Walter J. Pearce and is included for historic interest as showing how the old-fashioned pound brush was bridled with string before use. These brushes are no longer manufactured in this country due to lack of demand; they are, however, still popular in France, Norway, and certain other Continental

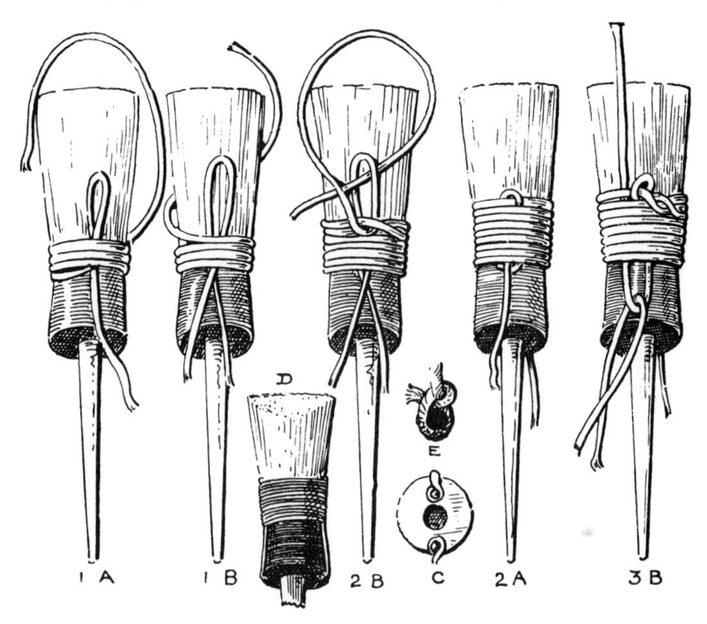

Fig. 5.4 Bridling a brush — string-bound type

At 1 and 2, B shows the reverse side of the brush to that seen in A. D shows the completed brush, with the ends of the cord tacked into the stock (as at C).

The "length out" (i.e. the amount of usable bristle projecting from the ferrule) was far greater in the case of the old-style pound brush than in a modern paint brush. Every so often, as the brush wore down, a few turns of the bridling were removed, thus enabling the effective length of bristle to be maintained without variation.

countries, though the method of bridling is different. Bridling is now incorporated at the time of manufacture and takes the form of a moulded plastic tube in sections, the tube being removed section by section as the brush wears down; or, alternatively, as a prefabricated wire or other type of binding with a similar facility.

Fig. 5.5 and 5.6 show modern types of wire-bound brushes.

CARE OF BRUSHES

All new brushes are liable to shed a few bristles since they invariably contain a few "shorts" and it is impossible to secure every hair; they generally contain also a considerable amount of dust. A new brush, therefore,

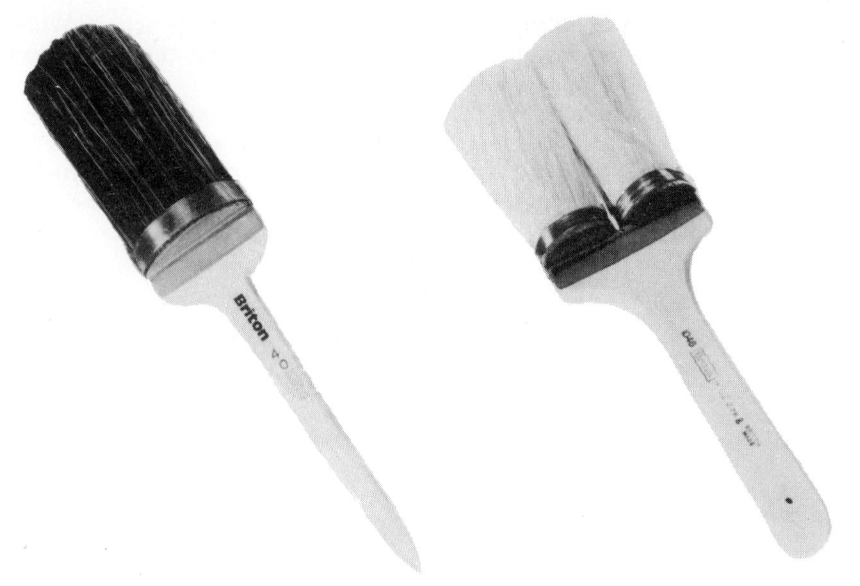

Fig. 5.5 Copper-bound ground or pound brush

Fig. 5.6 Two-knot fibre washdown brush

Courtesy: Briton Chadwick Ltd

should always be regarded as a dirty brush which requires cleaning before putting into paint or varnish. A large amount of dust can be ejected by pressing the bristle back and releasing it sharply or by twirling it rapidly between the hands; it should then be thoroughly rinsed in white spirit and again flirted out. It is then ready for its "breaking-in" period, when it may be used for priming woodwork or wall surfaces, etc., until all loose shorts are eliminated and the brush becomes fit for finishing work.

Some painters imagine that before a new brush is put into use the correct procedure is to soak it in water overnight. This is a piece of folk-lore, dating back to the days when paint brushes were bridled with string which had to be tightened before use. With modern brushes this is unnecessary and in fact could be detrimental, because the cement in which the brush is set could absorb moisture and swell, putting a strain on the ferrule which might cause it to burst.

At the end of the day's work, the usual trade practice is to keep paint brushes in water to prevent the paint drying in them, but whilst this may be effective in keeping them soft from day to day, it is wrong in principle and cannot be defended by any sound argument. Before putting such brushes in paint, they are usually "rubbed out" on the paint chest lid, or on any rough surface which might be handy, in order to expel the water, but no matter how well the brush is rubbed out, one can never be sure that all water has been expelled. In point of fact, there is a danger of rubbing the water *in* rather than out, for in the friction set up some of the water may become temporarily "emulsified" with the paint in the brush, and water introduced into a paint film in this way may become a serious source of weakness — all the more serious because the cause is unsuspected.

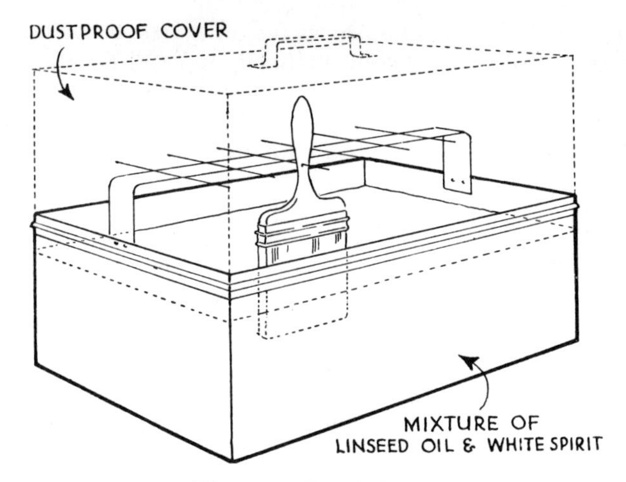

Fig. 5.7 Brush-keeper
Method of suspending brushes under dustproof cover

The safer way is to keep paint brushes in something compatible with oil paint, and this is most conveniently met in a mixture of equal parts raw linseed oil and white spirit. Do not keep them in pure turps for, apart from being expensive, it tends to harden bristle. If brushes are kept in a covered brush-keeper similar to the one illustrated (Fig. 5.7), the spirit will not evaporate to any great extent and dust is excluded. The liquid may be poured out into a can whilst the brush keeper is cleaned out from time to time and the same liquid either replaced, or a fresh lot used as is found necessary. Brushes should be *suspended* and not allowed to rest on the tips. Varnish brushes (Fig. 5.8) may be similarly suspended in linseed oil and white spirit;

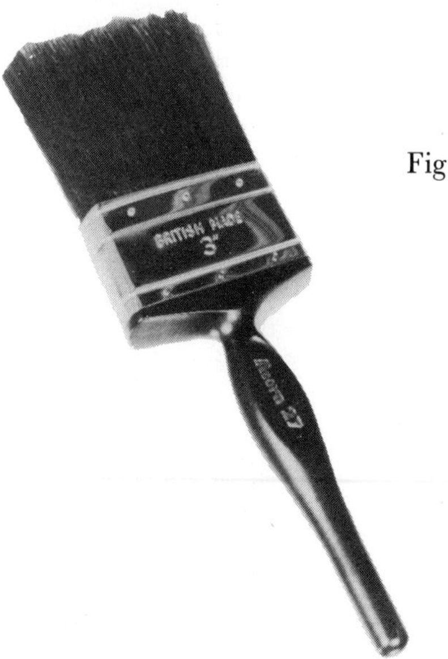

Fig. 5.8 Flat metal-bound varnish brush
Courtesy: Briton Chadwick Ltd

before putting back in varnish, they should, of course, be well rinsed out in white spirit, twirled well between the hands to expel the spirit, and then carefully worked into the varnish before commencing work.

Of course, if the brush is to be taken out of use and put into storage for any length of time, it should be well rinsed out in solvent, washed thoroughly clean in soap and water, and then allowed to dry.

Brush cleaners and renovators

Trade practice is undergoing a considerable change in this respect. Some of the well-known paint-manufacturing firms now produce a brush cleanser and renovator which represents a quite different approach to the care of brushes. The cleanser is a liquid intended for the cleaning and maintenance of brushes in constant use and for bringing back into use brushes that have been neglected and allowed to become hard. It is suitable for brushes used in ordinary paint materials but not for those used in two-pack materials.

For normal day-to-day use the brushes are suspended in the liquid overnight and the next morning are lifted out, rinsed under a cold water tap and dried; after an overnight soaking they are clean. Brushes which have gone hard may require soaking in the liquid for several days. Obviously the system has many advantages; time is saved because the brushes clean themselves whilst they are in overnight storage, and remain soft because they are cleaned right through to the stock. The liquid can be used repeatedly over a long period, even when soiled, and there is very little loss by evaporation. The cleanser should not be put into a plastic container. It is said not to damage clothing, but any splashes on the skin or on paintwork should be washed off as soon as possible.

One of the problems posed by modern quick-setting paints is the difficulty of keeping the brushes in good condition during actual working hours. Paint on the ferrule hardens quickly, and gradually the brush itself stiffens up whilst in use, which can lead to twisting of the bristles. The ferrule should be wiped frequently with a rag moistened in solvent, and at regular intervals the brush should be rinsed in solvent to free it from hardening paint.

There is another way in which trade practice has completely changed. The old-time painter always carried three brushes, even when embarking on the painting of some quite small area of work; he would have felt himself badly equipped unless he was carrying his pound tool, seconds tool and sash tool — or (in terms of Imperial measures) his 3-inch, 2-inch and 1-inch brush, or metrically, 75-mm, 50-mm and 25-mm. Nowadays the general practice is to carry one brush only and make it serve for every purpose, even for cutting-in the window putties or cutting the skirting board to the floor line. This most emphatically is not a change for the better. Cutting-in should be done with a small brush, namely the cutting tool specifically made for the purpose, which is generally a 25-mm brush. By this means cutting-in can be done without twisting the brush. The present-day practice of using a 75-mm brush for all purposes and twisting it over on its side in order to do a piece of cutting-in is lazy and wasteful. Brushes which are repeatedly twisted in this manner develop the fault known as "fingering" — an expressive phrase coined by brush manufacturers to denote a brush which has separated into two or more round clumps. Once a brush has developed this fault it is practically impossible to redeem it.

Distemper brushes (Fig. 5.9) need washing out immediately after use; so, too, do *flat wall-brushes* (Fig. 5.10) that have been used in emulsion paint. They should be rinsed out first in cold water to get rid of the bulk of the colour, then washed in warm (not hot) water and soap, rinsed well with clean water, shaken out, and hung up to dry. What to do with them during a mealtime break is always a problem. If they are left on top of the bucket, the quick-drying emulsion paint sets hard and they will be hopelessly stiff when work is resumed. If they are washed out, when work is resumed the paint will run through the wet stock and down the handle. Leaving them standing in a bucket of emulsion is a dirty practice which causes them to be overloaded with paint. The best solution is to wipe out all surplus paint on the rim of the bucket and then wrap them in a polythene bag so as to exclude the air.

Courtesy: Briton Chadwick Ltd

Fig. 5.9 Metal-bound flat distemper brush

When oil-bound water paint has been used, a brush sometimes becomes greasy and sticky while it is being washed out. When this happens, rinse it well in white spirit or paraffin and wash in lukewarm water and soap as before. This stickiness is the result of oil being released from the emulsion, which is sometimes caused by the continual friction of the brush on wall surfaces. Washing in very hot water is almost certain to cause greasiness. If a sticky brush is allowed to dry, it will become hard and unusable; when this happens, spirit paint remover or similar solvents (non-caustic) should be used to soften it. Since all animal hair is destroyed by strong acids and alkalis, brushes containing bristle and horsehair should not be used in these types of liquid. Freshly slaked lime, caustic paint removers, and strong solutions of soda, potash, or sugar soap, etc., are all damaging to bristle, and only fibre brushes should therefore be used in these solutions.

Brushes should not be put away whilst damp or stored in damp conditions or they may become mildewed; store cupboards should be ventilated to

Courtesy: Briton Chadwick Ltd.

Fig. 5.10 Flat wall brush

provide a good circulation of air. Moths lay their eggs in bristle and when the grubs or maggots hatch out they feed on the flag and so destroy the most valuable part. Camphor, paraffin and other preventatives cannot altogether be relied upon, and the only real safeguard is to inspect regularly and shake out any eggs before they hatch out.

Stipplers (Chapter 28) should be rinsed in white spirit immediately after use in oil paint, then carefully washed in lukewarm water and soap, holding the stippler with the bristles *down* to avoid wetting the wood back which may warp badly if it gets soaked (many stipplers are now made with an aluminium plate to obviate this danger of warping). Rinse in clean water, beat with a dry cloth to separate the bristles, and hang up to dry, bristles downwards. *Stencil brushes* (Chapter 33) should be well washed, dried thoroughly and bound with paper secured by a rubber band, to preserve their shape. *Signwriters' pencils* (Chapter 34) should be rinsed in white spirit and greased with vaseline to preserve their shape, or if used only occasionally, washed out in lukewarm water and soap, rinsed, dried, and greased. When fitting sticks to quill pencils, first soak the quill (but not the hair) in hot water to swell it, then slide on to the stick which has been carefully shaped to fit; when cool, it will be firmly held. Fitting it cold might easily split the quill.

Rubber stipplers should be cleaned immediately after use in oil paint; first remove the colour with a rag moistened with white spirit (but do not soak in the spirit) then wash out immediately with warm water and soap. On no account should oil paint be allowed to dry on them, for if solvent is used to soften the paint it will soften the rubber also. During mealtimes stand the rubber stippler in water up to (but not touching) the wood back; this will keep the paint soft during the recess.

Pressure-fed brushes

These are brushes which are connected by a long flexible hose to a pressure-feed tank of the type used in spray painting. The flow of paint is controlled partly by the air pressure at the tank and partly by a flow-control valve on the brush. It is clear that this device offers several potentially attractive features, in that there is a constant flow of paint to the brush, there is no paint can to be carried around and therefore the operative has one hand free all the time, and the work proceeds without the continual interruption caused by having to dip the brush into a can; all these factors contribute to a much faster rate of working. The less pleasing features are that because the flow is continuous there is a tendency for the paint to be applied in too thick a coating, and that it is difficult to lay off the paint to eliminate brushmarks.

In spite of its possibilities, the pressure-fed brush has never become accepted; it is by no means a new idea, but it has never caught on with the decorator. In recent years, however, it has made some advance in the field of industrial painting, where on large surface areas speed of application is a decisive factor, and on the painting of steelwork where a thicker paint film is often desirable.

The brushes are usually supplied in either 75 mm or 100 mm width; the brush heads are detachable so that they can be replaced when necessary, and they are available with fillings of either nylon or bristle.

PAINT ROLLERS

A great deal of painting work is now carried out by roller, and this in itself has prevented the chronic shortage of bristle from becoming such a disaster as to cripple the activities of the trade. But it would be wrong to consider the roller merely as a second-rate substitute for the brush. On the contrary, in present-day practice the roller is an important method of paint application in its own right.

From an economic point of view the roller can make a substantial contribution to cutting down painting costs. If a comparison is made over two identical areas of plain wall or ceiling surface the roller is about three times as quick as the brush on smooth plaster, while on a rough-textured surface such as pebbledash it may well prove to be five or six times as fast. This of course is an over-simplification; it never happens that a job consists of plain surfaces with no breaks, obstructions or items needing to be cut in. Nevertheless there can be a considerable saving of time, and this is not the only advantage. In many cases roller application is easier and less tiring than the brush, and there are occasions when there is a distinct saving in the amount of scaffold required, and in the time spent in shifting steps and planks around a room. The roller is therefore accepted nowadays as a legitimate part of trade practice, yet there are still some conservatively minded people who regard it with suspicion, and occasionally one finds a customer who refuses to allow a decorator to use the roller on a given piece of work. This is a very short-sighted view. Present-day rollers are highly sophisticated pieces of equipment, strongly made, well finished, supplied with a variety of fittings suitable for all kinds of materials or surfaces, and far removed from the flimsy contraptions produced for the cheap amateur market.

Basically a roller consists of a wooden cylinder or strong metal drum or frame, revolving on a spindle attached to the handle either at one side — as in the case of the small sizes — or on both sides, and with a roller head consisting of material or fabric which picks up paint from the container and spreads it evenly on the surface to be painted (Fig. 5.11). The three kinds of covering commonly used for the roller head are (a) medium-length lambs-

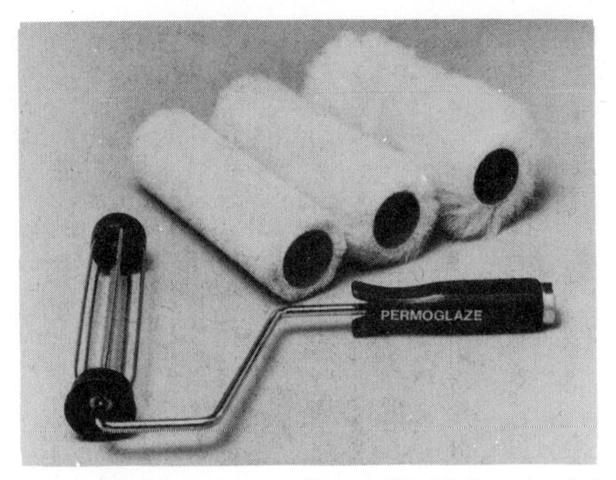

Courtesy: Blundell-Permoglaze Ltd

Fig. 5.11 Master roller frame and interchangeable heads

The short pile is used for smooth surfaces, the medium pile for general purpose work, and the long pile for very rough areas such as pebbledash and roughcast

wool, (b) short-pile fabric such as mohair, woven nylon, etc., and (c) sponge rubber (foamed latex). The current trend, however, is towards abandoning natural lambswool and sponge rubber in favour of specially developed water-resistant man-made fibre fabric which is available in a variety of pile lengths to suit many differing forms of application. The reasons for this are (i) the greater consistency of a man-made fibre compared with a natural product which varies in length and density and tends in the course of use to matt down and present an uneven surface, and (ii) that such a fabric has less tendency to skid around on a surface than foam rubber. Fabric covers are mounted spirally to avoid "join-patterns" in the finished paintwork. Sponge rubber covers are jointless, being moulded in one piece.

Paint rollers are supplied in a variety of widths, the usual sizes in the U.K. being 50 mm, 75 mm, 175 mm, 225 mm, 250 mm, and 350 mm, though greater widths are available on the Continent and in America. Some rollers are contoured for special purposes such as the painting of corrugated iron and asbestos sheeting, and there are concave rollers for pipes, shaped to fit varying diameters. Many modern rollers are heavier than the equivalent size of brushes, but by virtue of the balance of the tool this does not mean that they are more tiring in use.

Extension handles of lengths varying between 450 mm (18 in.) and 1250 mm (48 in.) are available and are quickly fitted when required; these are extremely useful where steps, planks and other scaffolding would otherwise be needed (Fig. 5.12). They are especially useful on ceiling work.

The paint for roller application is contained either in a tray or, on large-scale work, in a bucket. The paint trays are pressed into a series of ribs which enable the roller to be loaded evenly with paint. The buckets are made of galvanized iron and incorporate a meshed grid which allows the roller to take up an evenly distributed quantity of paint and removes any excess; the position of the grid can be varied to suit operational needs.

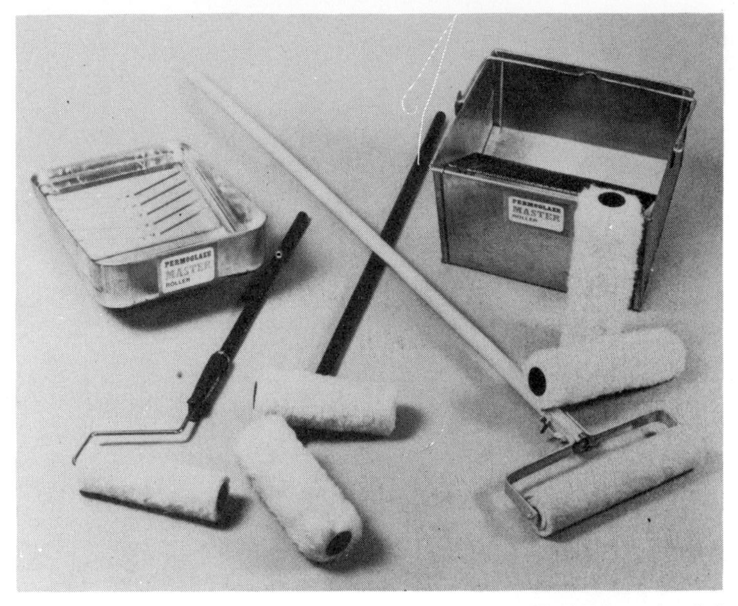

Courtesy: Blundell-Permoglaze Ltd

Fig. 5.12 Roller equipment showing extension handles, bucket with grid for removal of excess paint, and paint tray

Pressure-fed rollers

On large-scale work, especially industrial painting work, there is a considerable advantage to be gained by using pressure-fed rollers, which do away with the necessity for trays and buckets. The paint is conveyed through a long flexible hose to the hollow handle of a fabric-covered perforated roller; it is forced along the hose by compressed air, a moderately low pressure being sufficient, and the flow of paint is regulated by trigger action controlled by the operator. By means of this equipment a continuous supply of paint is provided, and the roller head does not need to be lifted away from the surface, which obviously makes for economical working. If the roller is attached to an extension handle, a considerable area of wall and ceiling surface can be painted without the use of scaffold, which again makes for a saving of time.

Pressure-fed rollers of this type have been available for several years, but a more recent development is the pressure-fed roller operated from an airless spray unit. In this equipment the roller is mounted on an extension handle, which can of course be lengthened with additional sections as required; the paint flows through the hollow stem of the handle, and when the trigger is operated the paint is forced through a series of flexible rubber nozzles which are attached to the hollow roller frame; these distribute the paint on to the roller head. The value of this device is the surprisingly high rate of coverage which is achieved; on broad areas the rate of application can be as high as 25 square metres per hour.

Fig. 5.13 shows the Wagner automatic roller, a lightweight system for

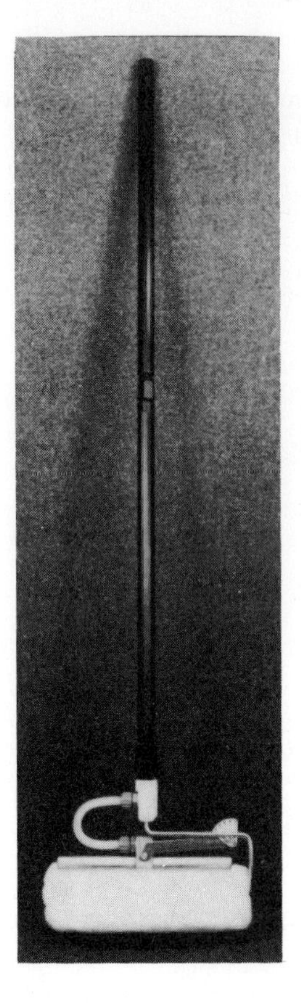

Fig. 5.13 Wagner automatic paint roller with 50-cm tube extensions
Courtesy: Gray-Campling Ltd

attaching to an airless spray gun; it uses standard lambswool rollers, and with some conversion parts alternative to the tube extensions illustrated it can be made into a pole gun with swivel joint, for spraying ceilings and places difficult of access. Tests have shown that with this system output can be increased up to three times as compared with traditional roller methods.

Care and cleaning of rollers

Just as new brushes need to be broken in when new, to work out the loose bristles, a new roller needs treating to prevent loose fibre particles from marring the finish; the method used is to wash the roller head vigorously in the solvent appropriate to the material to be used, either water or white spirit, the loose fibres being removed with a cleaning tool.

Rollers are cleaned by first removing as much as possible of the surplus paint; they are then rinsed in the appropriate solvent (water or white spirit), washed in warm soapy water, and finally rinsed and placed to dry. Nowadays rollers are usually supplied with a key fitting which squeezes out the surplus paint to speed up the time spent in cleaning. During meal breaks or other temporary stoppages, rollers should be placed in a polythene bag and tightly secured to exclude the air. Soaking in water is not to be recommended.

Cost-effectiveness

Broadly speaking, paint rollers become economically attractive where there are sufficient areas of plain work to counterbalance the amount of cutting in, so that the overall speed is higher than brush application. On very large work a point is reached when the spray becomes more economical; this occurs where the areas of plain work are large enough to counterbalance the amount of masking and sheeting required, plus the time spent in setting up and dismantling the spray. There is thus a band of medium-sized operations in which the roller is the most "cost-effective" of the three application methods. Sometimes, however, the roller may be preferred to the spray even when the work it produces is not as quick; a typical example might be in premises where delicate machinery is installed which could be spoiled by spray mist, and in this circumstance roller application would be safe to use, whereas the spray would be risky. Certain operations, of course, cannot be carried out by rollers, the obvious example being cutting-in. It is generally agreed that priming paints for wood or steel can be better applied by brush than by roller.

One of the objections sometimes raised about roller-applied paint is the "orange peel" appearance, which is particularly noticeable with gloss paints on smooth surfaces. To remove it, it may be necessary to feather off the paint with a brush after rolling. Thixotropic paints should be stirred up well before roller application.

Other manual methods of paint application

For the sake of completeness two other forms of paint application which cannot be classified elsewhere should be mentioned: these are painting mittens and paint pads.

Painting mittens were originally produced to make the painting of metal railings, pipes, etc., somewhat easier. They are made of lambswool lined with polythene, and are worn like a glove. The operative dips his mittened hand into the paint, withdraws it, and then grips the railing, moving his hand backwards and forwards to spread the paint. Hand-painting of railings is always a dismal job, so perhaps it is unfair to be too critical of this method of applying paint.

Paint pads consist of a flat rectangular plastic holder, attached to a handle; a detachable head consisting of mohair or foam rubber is fastened to the holder, and this is what serves to spread paint over the surface. The pads are available in various sizes, the smallest being intended for cutting-in window bars and the largest having an outside measurement of 150 × 100 mm. Some makes have wheels at the side of the plastic holder to facilitate cutting-in. The pads are intended for the application of all normal decorative paints, emulsion paints, etc., to flat surfaces. Their significance is not such that any further description is necessary.

6

Spraying and Spray Equipment

The third of the methods of paint application open to the painter and decorator is spray painting, and since there is such a wide range of spray equipment available this aspect of the subject merits a separate chapter. Spray painting is a tremendously important part of present-day practice, partly for economic reasons due to the speed of application, partly because certain paint materials can only be applied effectively by spray, and partly because there are some finished effects that cannot be achieved by any other means.

Advantages of spray painting

Undoubtedly the most important factor commercially is the speed of application. It has already been stated that if a comparison is made over identical areas of smooth ceiling or wall surfaces the roller can be up to three times as quick as the brush; over a similar area, spray application may be in the order of ten times as quick in the case of high-pressure air-propelled equipment, and very much faster still in the case of airless spraying. There are of course other factors to take into account. Before the actual spraying begins, the equipment has to be assembled, and when the spraying is finished the equipment has to be cleaned out and dismantled, which is all unproductive work. Then again, the surfaces which are not to be coated with paint need to be protected, and this means masking up the windows, etc., which is a lengthy process, and it may also involve the sheeting up of quite considerable areas, all of which consumes time and therefore costs money. In cases where the individual wall areas are small, where several different colours are to be used, and where a great deal of cutting-in is required, the overall speed of paint application is greatly reduced, so that in much painting and decorating work the spray offers no advantage at all. But where the areas of plain work are large enough to counterbalance the time spent in masking up and in assembling and dismantling the plant, the point is reached when the spray is the most economical form of application; and in general terms it follows that the bigger the work area the greater is the economy in labour costs. On many kinds of work, particularly in the industrial field, the saving in labour cost is so high that any other form of application would be unthinkable. To any firm whose range of operations includes a significant amount of industrial work, the high initial cost of the plant is soon recouped and spray application is an essential part of everyday trade practice (Fig. 6.1).

Courtesy: Gray-Campling Ltd

Fig. 6.1 Factory redecoration by the use of Wagner airless spray equipment mounted on a cradle

2000 sq. metres were painted by a two-man team in two days

In the case of *rough-textured surfaces* the speed advantage swings heavily in favour of the spray. Applying paint to roughcast or pebbledashed areas or to rough unplastered brickwork is a laborious and wearisome business by brush, but it presents no difficulty to the spray, every kind of surface is sprayed with equal simplicity and with saving of cost (Fig. 6.2). It is obvious that reversible materials such as cellulose, which soften up under the solvent action of each new coating, can only be applied by spray, but there are other materials more commonly used by the decorator which can only be applied efficiently with a spray gun. In multicolour paints, for example, the flecks of separate colour would merge together if brushed or rolled, and certain kinds of anti-condensation paint can only be applied in regular even coatings when the spray is used. Certain metallic finishes, flock finishes, spattering, and Lymnato are other examples of materials which depend upon the spray for their application. Another point to be considered is that a sprayed coating is of more uniform thickness and generally presents a more even appearance than a brush-applied coating because of the absence of brushmarks.

Limitations

Clearly there are many advantages in spray application; it is also clear that there are limitations. Apart from those occasions when some kind of paint is specified that can only be properly applied by spray, the decision about what method of application to use in any given circumstance is usually dictated by economics. On small domestic work there is not usually much room for doubt. When the work areas are very small, there would be no point in going to the trouble of masking and sheeting and setting up spray plant if the whole job could be carried out in less time by some other means. When there is much cutting-in, or where several different colours are to be used on various wall surfaces, the spray offers no advantage. Sometimes, of course, other factors come into play. For instance, it is generally advisable for priming paints to be applied by brush, because the friction set up by the brush strokes forces the paint

Courtesy: Binks-Bullows Ltd

Fig. 6.2 "Safari" airless spray equipment in use for exterior renovation

into more intimate contact with the surface. In the case of lead paints, indoor spraying is forbidden by the Lead Paint Regulations. Exterior spraying might not be a feasible proposition in windy weather, especially on an exposed site. The spraying of cellulose is governed by stringent regulations, and any decorator thinking of installing spray plant in his workshop should make himself acquainted with them.

The principles of spraying

Spray painting depends basically on the phenomenon known as "atomization", in which a stream of liquid is subjected to some force which stretches it until its surface tension can no longer take the strain, so that it breaks up into very fine particles or droplets. Liquid paint is fed into a spray gun, and the gun is designed to provide a means both of atomizing the paint and conveying the resultant fine particles towards the surface which is to be painted; when the particles reach the surface they merge together again or coalesce into a coherent film.

There are various methods by which a liquid can be made to atomize. Much spray equipment at present in use depends on a system of *air atomization* whereby a controlled trickle of paint is fed into a forced airstream, a vacuum area breaking up the liquid flow. Another method consists of releasing a fluid which is held under great pressure and changing its directional flow at the point of release, causing it to spread over a larger surface area than its surface tension can take, and thus breaking it up; this method is employed in what is called *airless spraying*. A further method is to use centrifugal force to stretch a liquid over a wider area than its surface tension can maintain, thus breaking it into particles; this method is employed in *electrostatic spraying*.

The way the paint is atomized is linked with the way in which it reaches the surface. Air-atomized paint is propelled in a blast of air which, when it strikes a hard surface, bounces back from it. What is required of course is that the paint particles should stay on the surface and coalesce to form a film, but here the tendency is for some of the particles to bounce off as well, causing what is known as "spray-mist" or "spray-fog". There may also be a tendency, when paint is being sprayed by this method into an internal angle (such as the corner of a room, the joint between wall and ceiling surfaces, and so on), for a cushion of air to build up in the corner, thus deflecting the paint particles and preventing them from getting into the angle. By contrast, when paint is atomized by one of the other methods these difficulties do not arise. Of course there are other ways of overcoming the difficulties; given paint of the right viscosity and the correct solvent balance, given the right operating pressure, and provided the spray gun is held at the correct distance from the wall, spray-fog can be substantially eliminated; in the same way, a proper spraying technique enables corners to be sprayed successfully. It is also true that other spraying methods can present their own particular problems. The important point is that if the painter and decorator is to get the best results from spraying, he needs to appreciate the capabilities and limitations of each particular method.

Terminology

For many years the main types of spray equipment in general use were those based upon air atomization, and much of the terminology that is used in spraying was brought into being before some of the other developments took place. This means that the terms commonly used are now liable to cause confusion. For instance, we use the term "high pressure" to refer to certain specific types of equipment based on air atomization, although the pressures used are quite moderate; whereas the pressures involved in airless spraying are very considerably higher. Even the name of "airless spraying" is itself rather misleading; it strongly implies that no air-pressure is used in the process, whereas of course a great deal of the airless spray plant in use at the present time incorporates a compressed-air supply to the pump in order to develop the very high fluid pressures which are a fundamental feature of the process. What the student needs to understand is that the term "airless spraying" merely means that the paint particles are not carried or conveyed from the gun to the work surface by a stream of air.

The point is that the technology of paint spraying has grown up on a piecemeal basis, and the terms which were used quite logically at one stage of development carry a misleading inference at some later stage. Manufacturers in various countries developing new types of equipment tend to introduce their own technical terms and phrases, which do not always correspond with what other manufacturers are doing, and which sometimes conflict with terms previously used in a different connotation. It is a pity that there is no over-riding authority to examine the whole subject critically and work out from scratch a completely new and accurate code of technical terms which everybody, manufacturers and operatives alike, would accept. Much confusion would be avoided, for example, if the spray systems in which paint is conveyed to a surface by a current of air could be clearly established under some name such as "pneumatic spraying" or "air-propelled spraying", allowing the ambiguous term "high-pressure spraying" to be dropped. But this of course is a matter that bedevils the whole subject of paint technology, and at present there is no means of enforcing any universal terminology.

In the circumstances, therefore, we have thought it best in this chapter to group the various kinds of equipment together in their appropriate order according to the operating principles that are involved, and in each case a simple explanation is provided of the way these principles affect the manner in which the equipment is used *from the point of view of the painter and decorator*. This will differ in certain respects from the descriptions given in some previous textbooks and in some of the earlier trade literature published by the various spray-plant manufacturers, but it is suggested that a student reading our description carefully will be able to relate it to other publications.

Motive power: electric versus petrol/diesel

Any piece of mechanical equipment needs a source of power to drive it, and mobile spray plant of the types used by the painter and decorator is driven either by electric motor (Fig. 6.3) or petrol engine (Fig. 6.4). (There are also, of course, air-compressing plants driven by diesel engine for large-scale construction work and civil engineering sites.) By far the greater proportion are electrically driven — not surprisingly because electric motors are much more convenient. When plugged in and switched on they come into operation without delay; they are of robust construction, and given proper care they provide many years of trouble-free service. Petrol and diesel engines obviously need a periodic pause for filling, and in the course of use they require far more attention in the way of care and maintenance. They are also inclined to be temperamental if not placed on an absolutely level surface, or if filled from fuel tins that are not kept scrupulously clean. Much time can be wasted and energy consumed in trying to coax a reluctant engine to start, especially if the operative does not happen to be mechanically minded. On

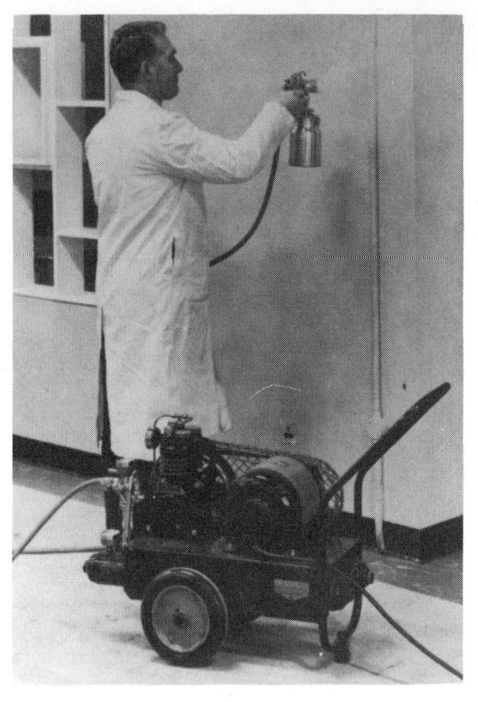

Fig. 6.3 (*left*) Painting with a mobile electric-powered spraying plant
Fig. 6.4 (*right*) Petrol-driven portable air-compressor for one operator

Courtesy: DeVilbiss Co. Ltd

the other hand, of course, most young painters run their own transport and are familiar with the intricacies of the internal combustion engine, so perhaps this is not such a problem as it was to people of an earlier generation.

A petrol engine is a necessity for work on buildings where for some reason electric power is not available, or in rural districts where the current characteristics do not conform to the normal pattern. For work on outdoor installations on sites of a scattered nature a petrol-driven plant is invaluable, such as, for example, on a contract for painting all the small wayside installations along a stretch of railway line between two stations several miles apart. The writer once had the strange experience of having to bring in a petrol-driven engine on the site of an electric power station that was in course of construction, just for the purpose of spraying the transformers, which were situated a short distance from the main structure, because no power could be made available at that particular point.

For most purposes, however, an electrically driven plant is infinitely preferable; it is a necessity in situations where fire hazards preclude the use of petrol. On many industrial sites, if a petrol engine is used it has to be located outside the building, and this means feeding the air lines in through window openings; as spraying proceeds there are continual irritating delays while the lines are disconnected, the plant shifted along the outside wall from one position to another, and the lines rethreaded through another opening.

TYPES OF PLANT

(A) THE AIR-PROPELLED RANGE

(1) Air-volume spraying

This is a simple type of equipment and is so called because it uses a considerable volume of air which is drawn along continuously all the while the motor is switched on, though very little air-pressure is generated. This type of plant does *not* incorporate a compressor, or anything that would cause air to be compressed or could store compressed air; some confusion exists on this point, and one often hears the machine wrongly described as a "compressor". The flow of air is produced by means of an enclosed fan or vaned wheel, like a vacuum cleaner in reverse, and the air is blown along the line, reaching a very moderate pressure of about 1½ bars* (20 psi), though a slightly higher pressure can be reached with a fan of larger size and speed.

In air-volume plant a "bleeder-type" gun is used, which means that there is no air valve and the air flows through it continuously, emerging at the nozzle; the paint feed is of the gravity type, the cup being mounted on top of the gun (Fig. 6.5). When the trigger is pulled a trickle of paint is released which descends into the airstream, where it is broken up and conveyed to the work surface. Paint cannot flow until the trigger is operated.

Because the pressure is so low the paint is not so finely atomized as with a compressed-air plant, and the degree of finish is not so fine. The rate of application is very slow, and of course a gravity feed gun cannot be tilted to enable ceiling work to be painted. Because of the low pressure the incidence of spray-fog and rebound is very slight, but against this must be placed the fact that because the paint particles are comparatively large, any spray-fog that does form is far wetter than that which usually emerges from a spray gun and is consequently much harder to remove and clean up. For commercial purposes, air-volume plant has been completely superseded by more modern equipment, but for decorative spraying it is extremely useful; the continuous flow of air tends to hold templates and stencils close to the wall, and the simplicity and light weight of the unit make it very handy, especially where scaffold work is entailed. On some models, as with Fig. 6.6, spatter work can be produced by opening a cock at the base of the gun to release some of the air before it reaches the nozzle.

(2) Compressed-air plant, usually termed "high-pressure spraying"

At present the greater part of the spray equipment in general use in this country is still probably of this type.

* 1 bar = 10^5 N/m^2 = 10^5 pascals = 14½ lb/in^2.

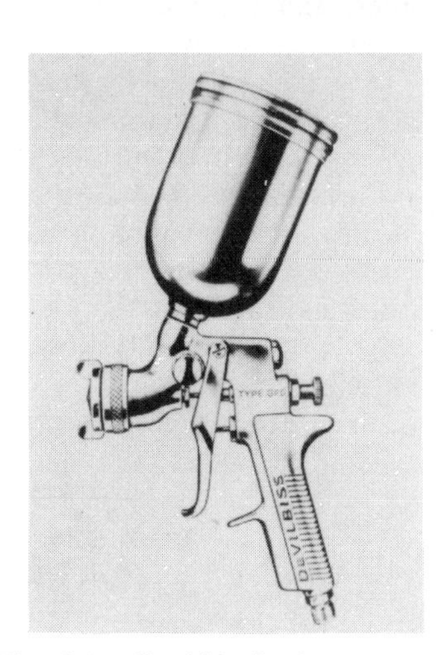

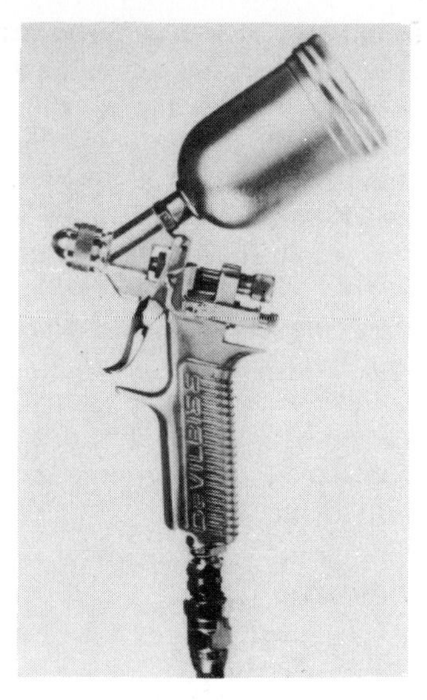

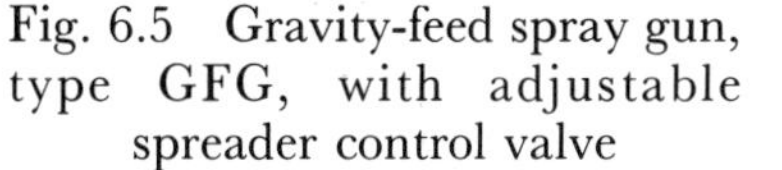
Fig. 6.5 Gravity-feed spray gun, type GFG, with adjustable spreader control valve

Fig. 6.6 Gravity feed gun, type MPG, with 1-oz (28 grams) paint cup
Suitable for small, intricate stencil work

Courtesy: DeVilbiss Co. Ltd

The essential feature of the equipment is the *compressor*, which can be defined as a mechanism designed to draw in air, compress it, and supply it continuously as required at a predetermined maximum pressure and in sufficient volume to operate efficiently a given number of tools (in this case spray guns) of a stated capacity. Compressors of many sizes and types are made for a number of purposes, some of them capable of driving several powerful tools. The limiting factor in the case of a decorator's compressor is that it should be readily transportable from one site to another and easily manoeuvred by hand, so that it can be moved around freely on the site. This implies the use of a compact unit consisting of a fairly small compressor and all its associated equipment, mounted together on a wheeled chassis with a suitable handle.

The complete portable compressor unit consists of the following parts: (i) the motor or engine that drives the compressor; (ii) the compressor itself, with filtering devices to clean the air; (iii) an air storage tank or air receiver to retain the air when it is compressed; and (iv) an air transformer, with (v) a reducing valve and (vi) a pressure gauge. The *compressor* takes in air at normal atmospheric pressure and compresses it; usually the air passes first through an air intake filter to prevent the entry of dust. The *air receiver* is a cylinder in which the compressed air is stored until it is required. While it is stored it is being cooled down, and this is necessary because the air temperature rises while it is being compressed, but it is desirable that it should have returned to the atmospheric temperature before it is withdrawn; in addition, the cooling allows any moisture vapour to condense out, so that it can be separated. When maximum pressure has been reached in the air receiver an automatic unloader comes into operation, allowing the motor which drives the compressor to run idle; as the guns are brought into play and air is withdrawn, at a certain point the motor cuts in again. This device enables air to be drawn off at a steady, even pressure, with the pulsations of the compressor completely damped out. The *air transformer* filters out any dirt or oil or moisture from the compressed air so that what reaches the gun is absolutely clean. The air transformer incorporates the *reducing valve* which regulates the air pressure from the compressor and brings it to the pressure required for spraying. This pressure is indicated on the *pressure gauge*, and the required pressure is selected by the operator who rotates a circular control valve until the gauge reaches the required readings.

Fig. 6.7 shows in diagrammatic form the essential features of compressed-air spray plant.

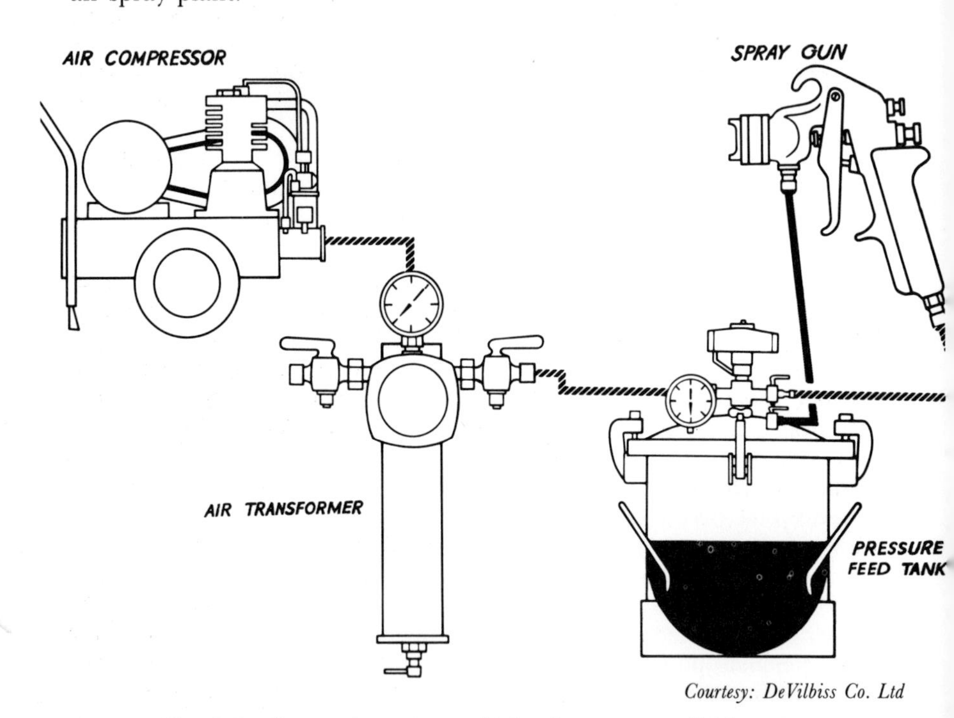

Courtesy: DeVilbiss Co. Ltd

Fig. 6.7 Set-up for spray painting by pressure-feed system

Choice of compressor: optimum size

There are many different kinds of compressor on sale, varying considerably in size, and the range is such that it is possible for any paint user to choose one that is ideally suited to his own particular needs. It must be remembered that spraying is not confined to painters and decorators; on the contrary, there are a great many kinds of industrial activity (the furniture industry and car finishing are obvious examples) in which spraying is an essential part, and in many of these the range of materials sprayed and the conditions under which they are sprayed are far more diverse than our own rather limited class of work. The variety of equipment is so wide that the decorator who is considering buying some item would be well advised to discuss the matter carefully with the suppliers, because when faced with a bewildering choice he may easily select a type that is not very well suited to his needs.

One of the commonest mistakes a decorator can make, if not properly advised, is to choose a piece of equipment which is not big enough to do what he expects of it. This can happen for one of two reasons. In the first place, a decorator can be confused about the output of a particular kind of compressor, by not realizing that the *displacement* of a compressor is often quite considerably more than its *delivery*. The displacement is the amount of air taken in per minute by the compressor; the delivery is the actual output of compressed air per minute. No piece of mechanical equipment is 100 percent efficient, and the delivery can be as low as 70 percent of the displacement. It is usual for compressors to be rated by their displacement, and in the same way the guns are rated by their volumetric consumption, but of course the consumption of the guns is related to the *output* of the compressor and not its displacement. Consequently the decorator who considers what kind of materials he will normally be spraying and looks in the trade literature for the type of guns recommended for them can be misled into choosing a compressor which is not big enough to give satisfactory service.

The other trap into which a decorator can fall is that of not allowing for the expansion of his business. It may well happen that when he first buys the plant he has no intention of using anything but normal decorative paints, emulsions, etc., but at some later stage, having realized the economic advantages, he undertakes a contract calling for the spraying of much heavier materials, only to find that his plant is inadequate.

There are several firms supplying spray equipment; each firm works to its own specifications, but while there are obviously individual points of difference and design between them, it can be clearly stated that there is little to choose between them since they are all equally reliable.

From the compressor an air line conveys the compressed air either directly to a spray gun or, more often, to a pressure container. It will be as well to consider these two possibilities separately, although in either case the sort of gun used is one in which the air is fed in through the base of the handle and the paint is fed in beneath the gun at a position just in front of the trigger. In each case the gun is of the "non-bleeder" type, which means that neither air nor paint can flow until the trigger is pulled. As a matter of fact the gun is delicately adjusted so that when the trigger is slightly depressed, air alone is released, but when it is further depressed paint is released to be ejected by the airstream.

Suction-feed system

When air is fed directly to the gun, the gun used can be of the suction-feed type (in which the air cap is designed to create a vacuum which syphons the paint upwards from the container), but it is usually of a later and improved pattern which operates equally well with either suction or pressure feed. The container is called a "suction-feed cup" and has either a screw top or a clamp top which is attached to the fluid inlet underneath the gun and is secured by a locking nut. The cup holds slightly more than a litre of paint, and this type of operation is suitable for light- or medium-weight materials such as lacquers, synthetic gloss paints, bronzes, latex, etc. (see Fig. 6.8).

Fig. 6.8 Suction-feed spray gun
Courtesy: DeVilbiss Co. Ltd

The suction-feed gun is useful where only small quantities of such materials are required, because it is easily assembled and quickly cleaned out after use. It is also handy in situations where there are to be numerous changes of colour — such as, for instance, on show or exhibition work — because the changeover can be made rapidly. But there are also obvious disadvantages. For commercial work the capacity of the cup is hopelessly inadequate, and on a job of any size there would be constant stoppages to recharge the cup. The cup itself with its contents is heavy, which makes the gun cumbersome in use; and the extent to which the gun can be tilted is limited.

Pressure-feed system

In normal trade practice a pressure-feed system is used, which means that instead of the air line from the compressor being taken to a spray gun it is taken instead to a pressure-feed tank, otherwise known as a "pressure container" or "pressure pot". This consists of a sturdy steel cylindrical tank, galvanized both inside and outside, and fitted with a clamp-on lid with a gasket to prevent air leakage, inlet and outlet taps and valves, pressure gauges, and a safety valve. The tank is usually supplied with a hand-operated agitator to keep the paint properly stirred. The tank capacity ranges from 4·5 litres to 270 litres, but since mobility is an important factor on site work the size generally used by the decorator is one containing 9 litres; anything larger would be too cumbersome, especially if scaffold work was involved (see Fig. 6.9).

Leading out from the container there are two lines or hoses, an air line

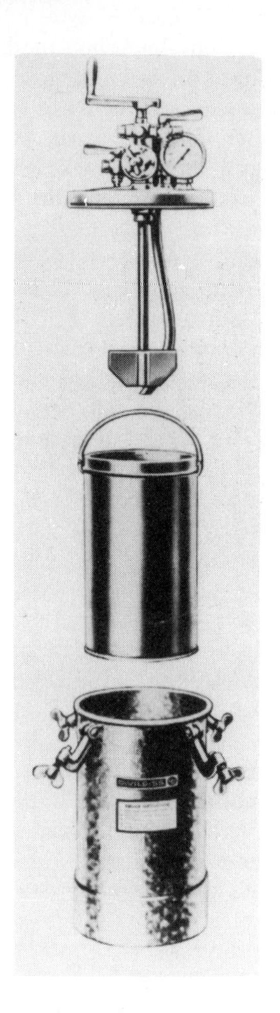

Fig. 6.9 Pressure-feed paint tank with insert container and hand-operated agitator

Courtesy: DeVilbiss Co. Ltd

and a fluid line; the air line is connected to the base of the gun handle, and the fluid line is taken to the fluid inlet in front of the gun trigger. To avoid any confusion between the two, the hose connections are of differing sizes and the lines themselves are colour coded, the air hose being *red* or orange and the fluid hose *black*. The principle of the pressure-feed system is the application of air pressure on the paint in the tank to force it through the fluid hose to the spray gun. The most useful type of pressure container for the decorator is the "double-regulator" type which provides independent control over the air pressure to the spray gun and the fluid pressure in the tank; there are simpler varieties such as the equalized pressure tank and the single-regulator type, but these do not give nearly so accurate a measure of control.

The pressure-feed system offers a number of advantages, and in fact large-scale spray painting would be unthinkable without it. A very large area of work can be painted before it becomes necessary to refill the tank, thus saving a lot of time, and the paint can be fed to the gun in greater volume than by other methods, especially when heavy materials are in use. Equally important from the operator's point of view are the ease and convenience of the method. There is no heavy and cumbersome paint container fastened to the gun; all the operator has to carry is the gun in one hand and the two lines in the other. This makes the application so much easier, because the gun can be turned around freely to any angle, horizontal or vertical.

Special tanks are available for the spraying of cement, asphalt, sound-deadening materials, and other products for which standard tanks are not suitable.

A very useful addition to the pressure container is the simple device called an *insert container*, shown in Fig. 6.9; this is a thin light-metal vessel placed inside the tank to hold the paint, instead of the paint being poured directly into the tank; it makes the work of cleaning out the equipment much easier and quicker, and reduces the time spent when a refill of paint is required, because a spare insert can be filled ready for a quick changeover.

There are various pieces of equipment in which the principle of pressure feed can be adapted to some particular type of operation. For large-scale spraying, *fluid pumps* driven by compressed air are available for feeding paint to the spray gun directly from the drum in which the paint is supplied by the manufacturers, giving a rapid rate of flow and offering a considerable saving of time. At the other end of the scale, *remote cup outfits* are available for the spraying of small areas; these combine all the advantages of standard pressure-feed equipment with the portability of small containers. The outfit consists of a small container with a capacity of just over two litres, fitted with controls which balance the air and fluid flow, and connected to the gun by short lengths of air and fluid hose. This device is useful to the decorator who wants to produce the full range of spray effects on a small scale for showroom purposes, or again for the preparation of samples, or in situations where small areas are to be sprayed with differing individual effects.

Fig. 6.10 shows the remote cup device connected to the versatile "Tuffy" lightweight spray unit; the remote cup can be clipped to the operator's belt or slung from the shoulder, giving freedom of movement over large areas or when working from ladders. Larger pressure-feed tanks can be used with the "Tuffy" for the bigger jobs.

Extension handles are available which can be fitted to a spray gun quite easily, enabling the operator to reach lofty or inaccessible areas of work without the inconvenience of providing scaffolding (Fig. 6.11).

Spray guns for pressure-feed systems

The gun, being the instrument which is actually used to apply the paint, is clearly of the utmost importance; it should be selected with care because the success of the whole operation depends very largely on its efficiency. The modern spray gun (Fig. 6.12) is a beautifully made piece of precision engineering. It is forged from a specially compounded aluminium alloy which has the strength of steel and yet is extremely light in weight. The many types available from the various manufacturers are

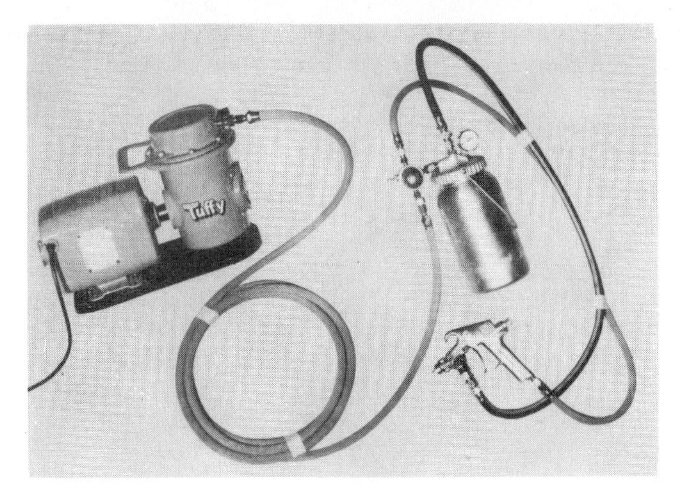

Fig. 6.10 The "Tuffy" pressure-feed unit with remote cup

An outfit suitable for light decorative spray work

Fig. 6.11 Spraying with an extension handle

The equipment in use here is the "Greyhound" portable airless spray unit (see Fig. 6.22)

Courtesy: DeVilbiss Co. Ltd

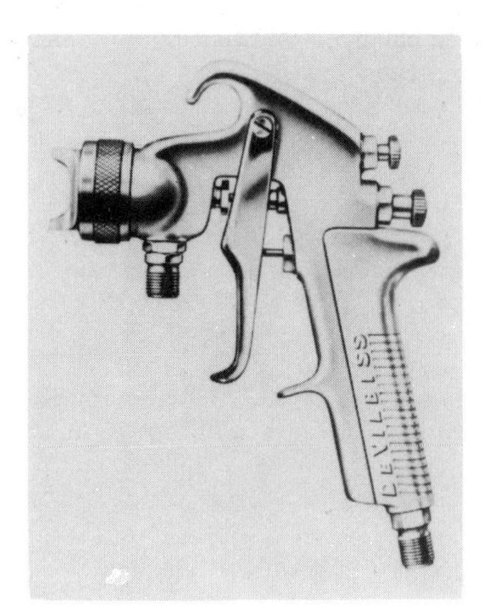

Fig. 6.12 Typical pressure-feed spray gun

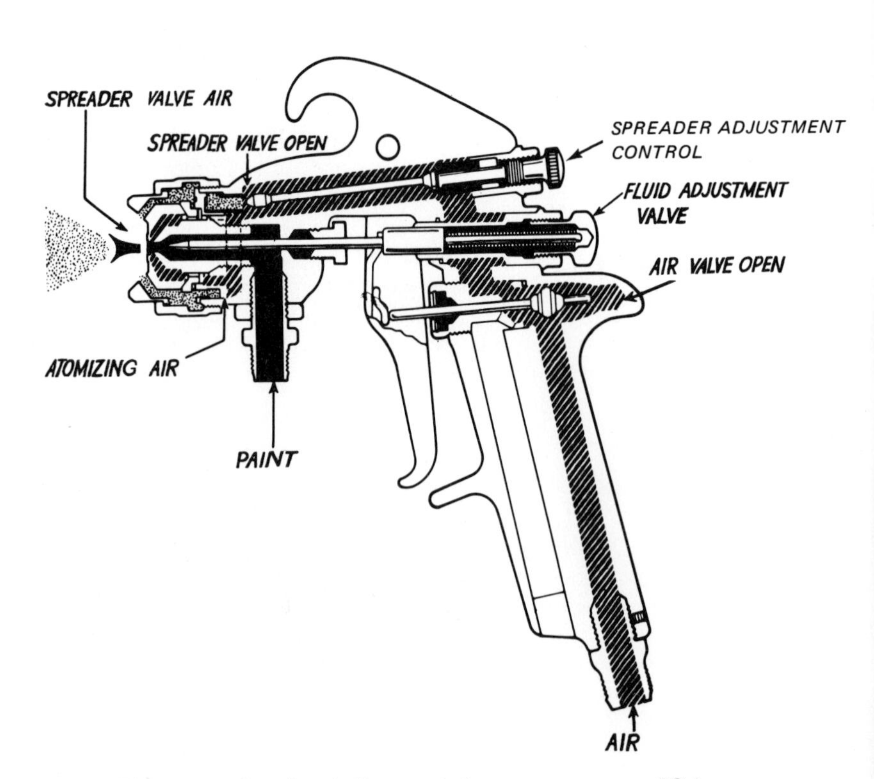

Fig. 6.13 Sectional diagram of spray gun, type JGA

Courtesy: DeVilbiss Co. Ltd

all carefully designed to give a perfectly balanced tool that handles easily and can be used for long periods without causing fatigue.

The parts of the gun which, together with the body and handle, make up the complete unit, consist of the *fluid tip* which meters and directs the paint into the airstream, the *fluid needle* which stops and starts the flow of the paint, and the *air cap* which atomizes the paint and forms the spray pattern. These parts are not assembled at random; they have to be matched up correctly, taking into account the type of paint in use, the amount of air available, and the nature of the surface; advice and information about the correct choice for any given situation can be obtained from the literature produced by the manufacturers. The operation of these various parts is controlled by the *air valve* which is opened and closed by the action of the trigger to start and stop the flow of compressed air through the gun, the *spreader adjustment valve* which controls the air flow and thus the width of the spray pattern, and the *fluid adjustment screw* which, by restricting the travel of the fluid needle, determines the amount of paint which can flow (see Fig. 6.13).

The contractor engaged in large-scale painting operations will generally use high-speed guns of large capacity with a wide range of air cap and fluid tip combinations, so that any kind of material can be thoroughly atomized and properly applied to produce a good finish. Stainless steel inserts for the fluid passages can be obtained to provide extra resistance when corrosive or abrasive materials are in use; and for the spraying of certain specially abrasive substances it may be necessary to use nitro-alloy fluid tips and needle tips.

The decorator's requirements are somewhat different from those of the industrial painting contractor in that he uses a wide range of materials and in many cases needs several changes of colour in fairly quick succession. These demands are catered for with a gun which has a sprayhead barrel complete with fluid inlet nipple, air cap, fluid tip and fluid needle, all of which can be removed as a single unit from the gun body. The detachable sprayhead speeds up the business of changing colours or materials and makes the cleaning easier; and there are special sprayheads for plastic spraying, mottling, and Lymnato effects (see Fig. 6.14 and 6.41).

The twin-headed spray gun

This gun (Fig. 6.15) has been developed for the spraying of two-pack catalyst materials which have too short a pot life if mixed before use. The two fluids are fed simultaneously to the spray gun from two separate pressure-feed tanks; they mix together in the spray pattern after leaving the gun and before reaching the surface. The fluids must be blended in precisely the correct proportions, and this demands accurate setting of the air and fluid pressures according to the manufacturer's recommendations.

Special guns

Special guns are made for a variety of uses such as the spraying of powders, cork granules, flock, plumbago and other dry materials and the application of sound-deadening or insulating materials, asbestos mixtures, and similar heavy and fibrous mixtures.

Fig. 6.14 Spray gun (type MBC) with detachable spray head for quick change of spraying material or method

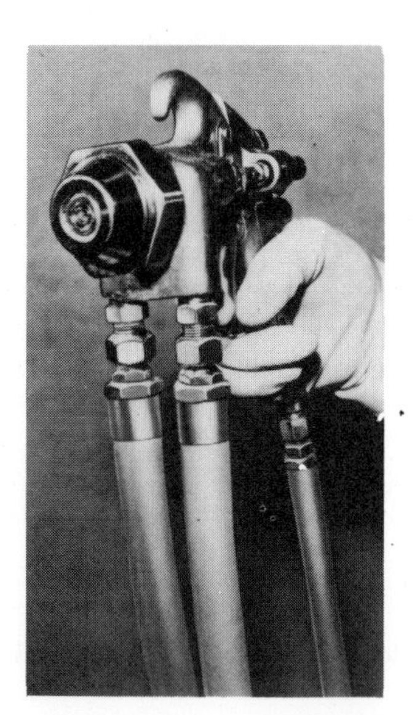

Fig. 6.15 Twin-headed gun (type JGC 505) for spraying two-pack catalyst fluids

Courtesy: DeVilbiss Co. Ltd

Paint application by air-propelled spraying

Successful spraying, like any other process, depends partly upon the selection of the correct equipment and partly upon the skill of the operative. Assuming that the paint has been properly mixed, the plant set up correctly, and the fluid and atomizing air pressures adjusted according to the particular material in hand, the spraying technique consists of following certain basic procedures.

In the case of suction-feed and pressure-feed guns, the initial light pressure on the trigger releases air alone, and paint does not flow till a firm pressure is applied to bring the trigger further back. This flow of air produced by the light trigger pressure enables the operative to dust the surface before paint is applied. This point is apt to be overlooked; we have often noticed that a conscientious painter who would automatically dust off the surface before applying paint with the brush seems to forget that a dust-free surface is just as important in spray application, to ensure the adhesion of the paint.

When applying paint, the gun should be held at right-angles to the surface at all times, at a distance of between 150 and 200 mm (6 to 8 in.); as a rough guide, a hand span with the thumb and fingers fully extended between the air cap and the wall indicates approximately the right distance. Each stroke is made by moving the gun at a constant speed across the surface of the work and at the same distance from the surface (see Fig. 6.16). At the beginning of each stroke the trigger is firmly depressed to its fullest extent and at the end of each stroke the trigger is released — beginners forget this and continue to squirt paint at the wall between strokes, which leads to a heavy build-up of paint, sagging and running wherever each stroke has ended. Each stroke should overlap the previous one by about 50 percent, and this is achieved by aiming the gun at the outside edge of the previous stroke. A good operative keeps his wrists flexible and trains himself to spray equally well with either hand, so that he is able to spray a larger area from one position.

When spraying into the corner of a room, or indeed into any internal angle, it should be remembered that an air cushion will be formed which prevents the paint particles from reaching properly into the corner; the technique is to turn the gun on its side and make a stroke at right-angles to the direction of the rest of the work along the face of each adjoining surface. When spraying narrow structural members, the width of the spray pattern should be adjusted to minimize the extent of the overspray.

Defects in air-propelled spraying

Failure to develop the technique just described leads to many of the faults commonly seen in sprayed work. "Arcing the gun" means waving the gun around in a semicircle instead of keeping it parallel to the wall (Fig. 6.16); this produces a heavy build-up of paint in a narrow band in the centre of each stroke, the band opening out wider at each side, with the thickness of the paint decreasing; it also leads to a great deal of overspray and spray-fog. Tilting the gun leads to an uneven spray pattern and a considerable amount of spray-fog. Incorrect triggering leads to a patchy finish with areas of thick sagging paint at intervals. Holding the gun too close to the work causes the surface to be overloaded with paint and reduces the area covered in each stroke. Holding the gun too far away means that much of the solvent evaporates from the atomized paint, causing it to fall in a coarse, dry, dusty coating; it also leads to excessive spray-fog, and because the paint particles cannot flow together properly it produces an "orange-peel" effect. Varying the speed at which the gun hand moves also causes an uneven coating.

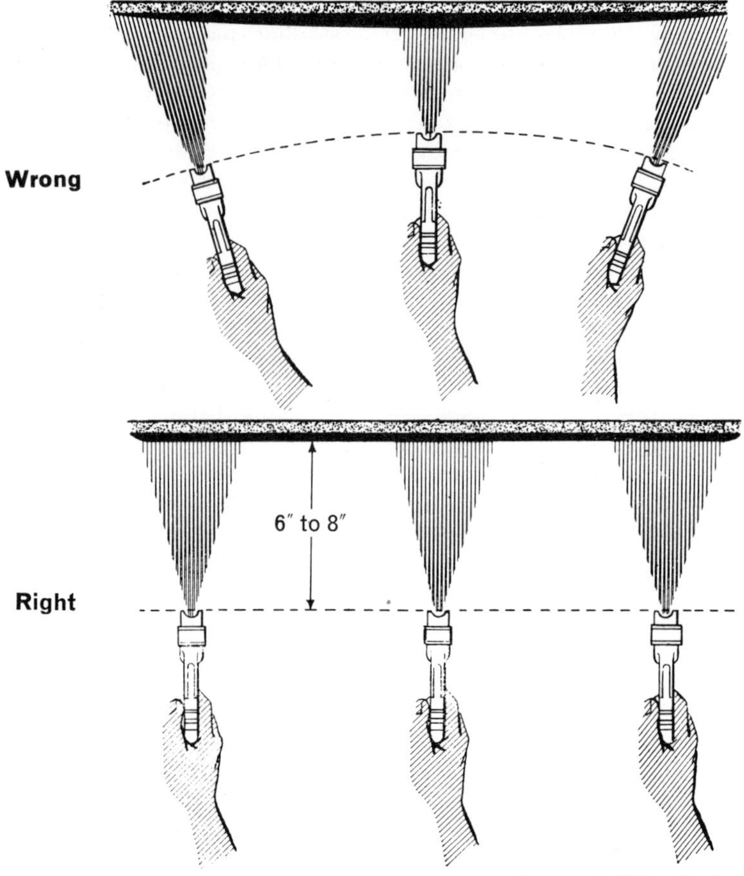

Courtesy: DeVilbiss Co. Ltd

Fig. 6.16 (*above*) "arcing the gun", which produces a coating of uneven thickness; (*below*) correct method of holding the gun to obtain an even coating

Other defects in spraying may be caused by failure to adjust the gun controls or the pressure settings correctly, by faults caused through incorrect operation of the plant, or by failure to maintain the plant in good condition. Excessive spray-fog is very often caused by using too high an atomizing air pressure or by over-thinning of the paint. Incidentally it used to be recommended that the paint for spraying should be thinned down to a lower consistency than that required for brush application, but with modern paints no such adjustment is necessary, and indeed in most cases it is unwise to add thinners.

Effects of faults in equipment

Spray equipment which has been subjected to heavy usage may develop slight faults, and the various moving parts may become worn; this in turn causes defects to appear in the sprayed finish, but usually such defects are more likely to be due to inadequate cleaning of the equipment after use. Jerky or fluttering spray may be due to a loose fluid-needle packing nut, or a cracked fluid tube in the case of a suction cup, but the more probable causes are either a coupling nut on the fluid line not having been properly tightened, or a loose fluid tip in the gun, or some obstruction in the fluid passageway. Similarly the spray pattern may be distorted; Fig. 6.17 shows the normal spray pattern produced by a correctly adjusted gun, but if the spray pattern takes onc of the forms illustrated in Fig. 6.18 there is a fault that needs to be traced. It may be that the atomizing pressure is too low

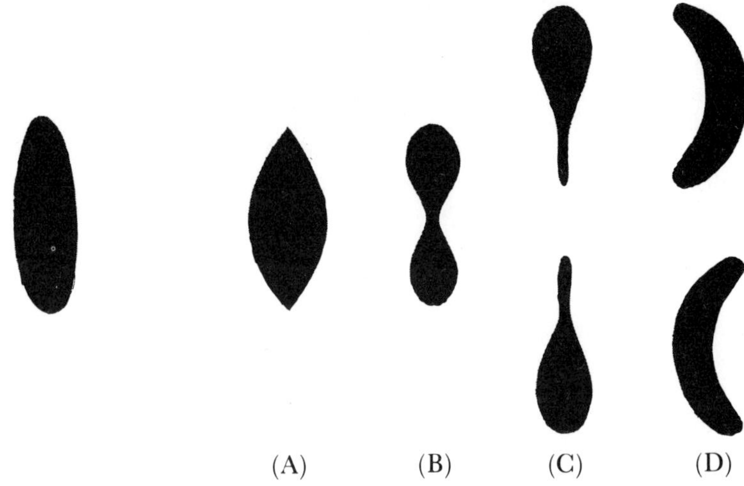

Fig. 6.17 Normal spray pattern

Fig. 6.18 Irregular patterns caused by dirt or defects in the nozzle

Courtesy: DeVilbiss Co. Ltd

(A), or that the atomizing pressure and fluid flow are not correctly balanced (B), but a more likely cause is the presence of dirt on the fluid tip (C), or the right- or left-side horn holes in the air cap being partially clogged with dried paint (D). Paint leakage from the front of the gun may be due to a worn or damaged fluid tip, fluid needle, or fluid needle spring, but it may also be due to dried paint or dirt on the fluid tip. The very common fault of "orange-peeling" is generally caused by an incorrect atomizing air pressure — too high or too low — or by holding the gun too close or too far away from the surface, or it may simply be due to the paint not having been thoroughly stirred.

Cleaning the spray gun and equipment

Spray guns, fluid lines and containers must be kept scrupulously clean, otherwise clogging of the spray nozzles will occur, causing delays in the progress of the work and thus adding to the general cost. The proper time to clean the equipment is *immediately* after use, when the material can easily be washed out. The method of cleaning is as follows.

Empty and drain the paint container or fluid tank, wash it out with the appropriate solvent, empty it, and wipe it out till it is clean. In the case of a cup container this is straightforward, but with a pressure-feed system there are certain additional points to notice: the air supply to the tank is turned off, the relief valve on the lid is opened to release the pressure, the four retaining clamps are unscrewed (taking care to loosen them alternately rather than in rotation), the lid is lifted slightly, the air turned on again, and with a piece of rag held over the spray-gun air cap the gun trigger is pulled so that the air is blown back through the fluid tank to empty the paint contained in it back into the tank. The cup or tank is now partially refilled with cleaning fluid and this is sprayed through the gun; occasionally a rag is held over the air cap so that the cleaning fluid is directed through the fluid passages to cleanse them thoroughly. This is continued until the fluid emerging from the gun is seen to be clean and uncoloured. The air cap of the gun is removed, soaked in the cleaning fluid, and scrubbed with a stiff brush. If the horn holes are clogged, the dried paint should be dislodged with a matchstick, toothpick or other soft implement, but *never* with a metal implement because this would cause irreparable damage to the cap.

While the cap is off, make sure that the outside of the fluid tip is clear of paint. Reassemble the gun, wiping the outside clean with a cloth soaked in cleaning fluid, and lightly lubricate the moving parts; the fluid needle spring is lubricated with vaseline or grease, though this only requires occasional attention. The cleaning fluid should obviously be the one appropriate to the material which has been sprayed — white spirit for oil paints, water for emulsion paints or oil-bound water paints (warm water may be used, but hot water might upset the emulsion and cause a sticky coating to be deposited in the nozzle and inner channels), and cellulose thinners for cellulose materials.

If a spray gun has been neglected and has become choked with dried paint, it should be dismantled, and the various parts steeped in cellulose thinners for half an hour before cleaning and reassembling. Caustic soda solutions or "pickle" should *never* be used as this will seriously damage the aluminium alloy with which the gun is constructed. Similarly, the gun should never be immersed in cleaning fluid as this would destroy the lubricant in the fluid needle and air-valve packings. Some operatives adopt the dirty practice of keeping the gun connected to the hoses and leaving it dangling in a bucket of solvent overnight so that it is ready for immediate use the next day; this practice cannot be too strongly condemned.

Fluid hoses should be thoroughly cleaned out each day after use because they become hopelessly choked if material is allowed to set in them. Certain cement paints and plastic paints containing calcium sulphate commence to set up fairly quickly, and when using such materials it is advisable to flush out the container and hoses with water before meal-breaks or at any time when spraying is interrupted.

Further information on cleaning, lubrication, and the avoidance of spray gun troubles may be found in Maintenance Charts supplied by the manufacturers.

(3) Hot spray

This process is widely used in vehicle painting and in the industrial finishing trades, though seldom used in painting and decorating and not to a great extent in everyday industrial painting. In spite of this it offers distinct advantages in the spraying of certain types of material, and for this reason some knowledge of the underlying principles is desirable.

The actual spraying techniques are no different from those already described; similar spray guns are used, and there is no difference in the way that the atomized paint travels from the gun to the surface. The difference lies in the fact that the paint is raised to a high temperature before application and is ejected from the spray gun in its hot state, so that the viscosity of the paint is considerably reduced. It is a well-known fact that a rise in temperature leads to a reduction in the viscosity of a drying oil or paint medium; the youngest apprentice knows that a gloss paint applied on a cold winter day is stiffer and exerts a greater pull on the wrist than the same brand of paint applied on a hot day or in a heated room. Hot-spray plant extends this principle by raising the temperature of the paint far above atmospheric temperatures — often to 70°C, and sometimes not far below 100°C, the boiling point of water. If a low-boiling, fast-evaporating solvent such as toluol or xylol is added to the paint before heating, the viscosity of the paint when heated becomes so low that the material can be fully atomized by air pressures as low as 1½ to 2 bars (20–30 psi). This means that there is practically no spray-fog or overspray, which makes for a great saving of paint. Because of the heat, the low-boiling solvents evaporate immediately they leave the gun, so that the paint reaching the surface has become highly viscous and contains a very high proportion of pigment; this means that a considerable film thickness is deposited in one application. At the low pressures employed, the high-boiling solvents in the paint are not evaporated before they reach the surface, so they enable the paint to flow out into a smooth coating that is free from "orange-peel" effect.

The fact that hot-spraying gives complete viscosity control irrespective of variable atmospheric conditions is obviously of great importance in industrial finishing, where mass-produced articles are to be coated to critical standards of specification. But with the types of paint material used by the decorator, which are normally made in a form that is equally suitable for brush or spray application, there is less to gain, and whatever advantages hot spray offers would rarely justify the cost of providing the equipment; and of course there is no advantage at all in using hot spray for water-thinned materials such as emulsion paints and water paints. In the case of the industrial painting contractor, however, where extra-heavy materials are to be applied, especially on large-scale work, there are occasions when hot spraying can prove to be the most satisfactory of the various available methods.

Hot-spray equipment is supplied in various forms. The hot-water circulating system can be used for both stationary plant (for workshop use) or portable plant; the latter type of plant is of course used for site work. Water in a closed tank is heated by a thermostatically controlled electric element and is forced by an electric pump round a closed circuit through a heat-exchanger and back to the tank. The paint is brought from a pressure-feed tank through capillary tubes in the heat-exchanger where its temperature is raised, and it is then fed to the spray gun through a fluid hose that is flanked on either side by a hot-water hose, the three hoses being encased in a jacket to retain the heat. A subsidiary flow of hot water circulates through these flanking hoses so that the paint reaches the gun at the required temperature. The portable heating unit (Fig. 6.19) is designed to connect up with any conventional spray outfit, using the existing pressure-feed tank, compressor and gun. This equipment is ideal for normal site painting operations.

Fig. 6.19 Portable paint-heating appliance

Courtesy: DeVilbiss Co. Ltd

Another method makes use of a heat-exchanger with heated compressed air to raise the paint temperature, but this can be applied only to a stationary plant for workshop use. A third method consists of a heater unit in which both fluid and air are heated by a thermostatically controlled electric element. An additional method which can be used when only small quantities of paint are required consists of a controlled heating water chamber in which four suction- or pressure-feed cups, each of one litre capacity, are accommodated. The cups, which can each contain a different colour, are fitted to an attachment connected directly to the spray gun.

(B) EQUIPMENT IN WHICH THE PAINT IS *NOT* PROPELLED BY AIR

(1) Airless spray and its advantages

There are fundamental differences between airless spraying and the air-propelled systems we have already discussed. Although it was introduced so much more recently, airless spraying has developed very rapidly indeed, and it is now a highly important factor in both industrial and decorative painting. The reasons why it has been so widely adopted are largely economic — the extremely high speed of paint application, together with an almost complete absence of spray-fog, which results in a considerable saving of paint and enables spraying to be carried out in enclosed areas without interfering with other operations that may be in progress in the vicinity.

The great difference between this and the previous systems lies in the way in which the paint is atomized and conveyed from the spray gun to the work surface. Instead of a forced stream of compressed air being used for this purpose, perfect atomization of the paint is achieved by first exerting a very high pressure on the paint itself and then releasing the pressurized paint through a tiny orifice in the spray cap. As it emerges it immediately explodes into a cloud of minute particles which travel with diminishing impetus across the space between the gun and the wall or ceiling surface that is being sprayed; the slowing down is due to the resistance of the air, the paint reaching the surface and settling down upon it gently. There is no moving airstream to fling the paint particles on the surface and then bounce off again; as a result, there is no spray-fog. Furthermore, the paint can penetrate fully into corners, sharp angles or crevices, because there is no build-up of compressed air to prevent it from entering such areas. Again, because it is not mixed with air the paint does not lose its volatile solvents by evaporation whilst travelling outwards from the gun, but retains them so as to form a good wet coating with the full flow properties with which the paint was compounded. All these advantages are achieved with plant that is simple to operate.

Equipment for airless spraying

The equipment, as first introduced and still widely used, consists of a paint container fitted with an agitator, a high-pressure air-motor-driven fluid pump, a paint strainer, a length of fluid hose, and an airless spray gun, together with a supply of compressed air to the pump motor; an appliance of this type is shown in Fig. 6.20. Pressurization of the pump is produced by the pump unit which has an air-to-fluid ratio varying between 24 to 1 and 50 to 1 according to the type of plant.

There are several manufacturers of airless spray equipment, each marketing its own range of plant, and between them they produce pumps with air/fluid ratios of 24, 26, 28, 30, 32, 36, 40, 44, 48 and 50 to 1. Put in simple terms this means that the air input pressure is converted and multiplied into a fluid output pressure which is between 24 and 50 times as great; thus, with an air input pressure of 100 lb per sq. in. (psi) (7 bars approx.) a 24:1 pump ratio will deliver a fluid output pressure of 2400 psi (168 bars approx.) and a 48:1 pump will develop a fluid output pressure of 4800 psi (336 bars approx.); with an air input of 60 psi (4 bars) a 40:1 pump ratio will produce 2400 psi (168 bars); and so on. An output of 2400 psi is adequate for most of the paint materials normally used by the decorator.

High-pressure plants are of particular interest to the industrial

Fig. 6.20 Lightweight airless spray unit with 5-gallon (22·7 litre) paint container

Courtesy: DeVilbiss Co. Ltd

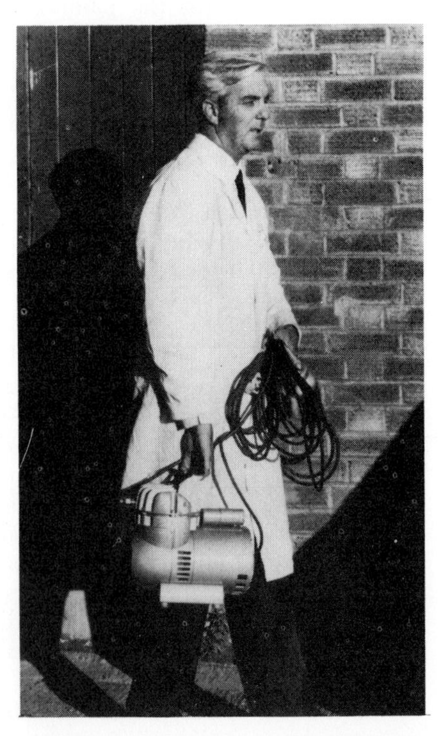

Fig. 6.21 The compact "Spray-Bee" airless unit, fitted with a quart capacity cup (1·1 litres)

Courtesy: Binks-Bullows Ltd

contractor called upon to spray heavy-viscosity materials such as zinc-rich primers, coal-tar epoxies, anti-condensation paints, heavy-bodied latex, etc., all of which can be properly atomized provided a high enough pressure is employed. It should be noted, too, that a certain amount of pressure is dissipated in forcing the paint along the fluid line, and that when the work involves an unusually long fluid line a considerable pressure drop takes place; for this reason it is often advisable for a contractor with an expanding business to select a plant with a higher pump ratio than is needed for his immediate requirements.

Elimination of air motor 

Although a great deal of the equipment still in use at the present time consists of the original air-motor/reciprocating fluid-pump type described in the preceding paragraph, various improvements and modifications have been developed which are incorporated into later models. The underlying principle remains exactly the same, but one important innovation is for a pump operated by hydraulic fluid to be substituted for the air motor, the hydraulic fluid itself being circulated by an electric pump. With this

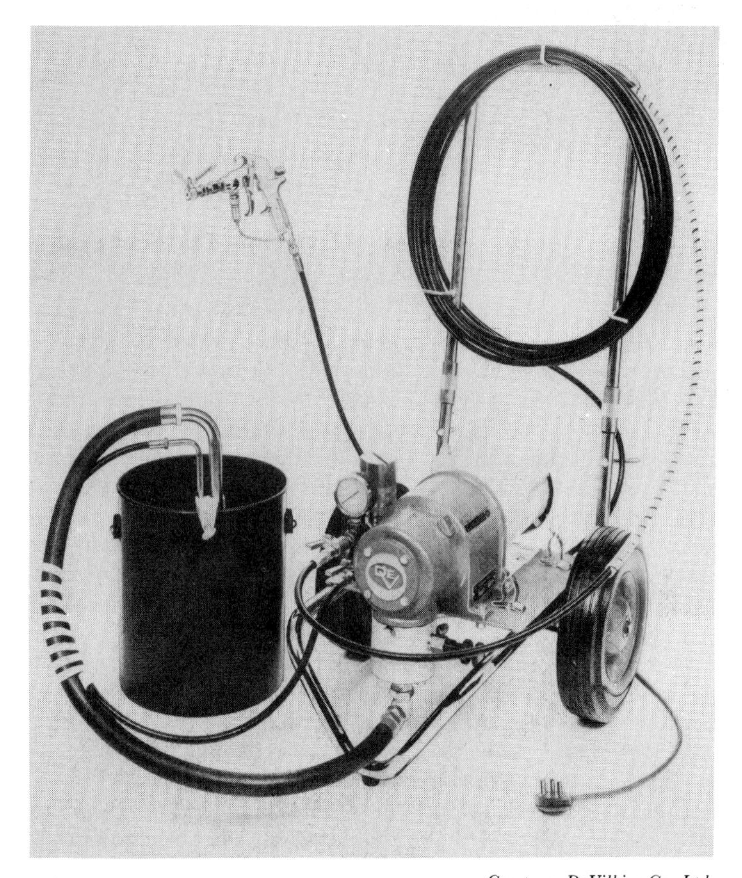

Courtesy: DeVilbiss Co. Ltd

Fig. 6.22 The "Greyhound" portable plant will spray 3·79 litres of paint per minute at a hydraulic pressure of up to 206 bars (3000 psi)

A suitable plant for maintenance painting in places such as hospitals and hotels, where quietness is essential

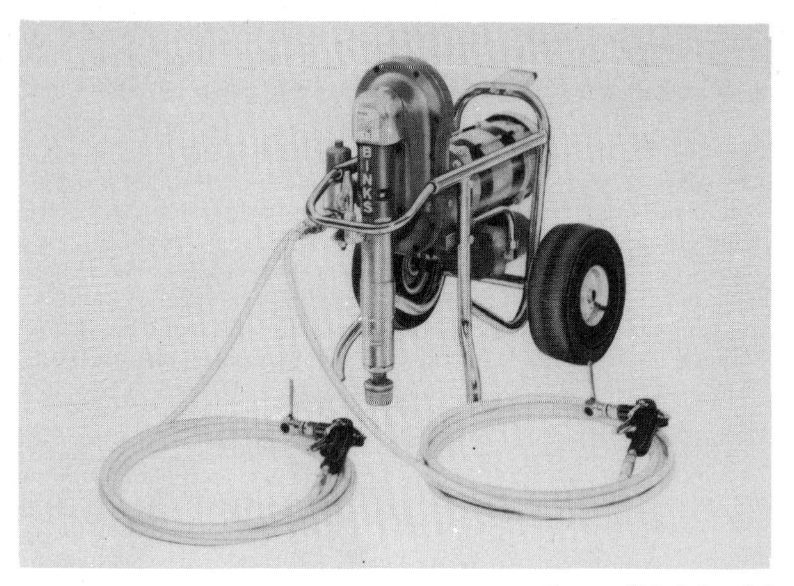

Courtesy: Binks-Bullows Ltd

Fig. 6.23 The "Hornet" airless plant will spray 4 litres of paint per minute

modification there is no longer any need for an external compressor unit. A further improvement is for the fluid pump to be driven by an electric motor and reduction gearing, which provides a more direct mechanical type of drive. A more recent development is the airless spray unit which incorporates a diaphragm-operated pump driven by an electric motor, in which hydraulic fluid is on one side of the diaphragm and paint on the other. Unlike the original fluid pump, in this equipment there are no seals to be worn down, and consequently both the initial cost and the maintenance costs are lower. In practice the most significant difference between this equipment and the older type of unit is in the pump capacity and speed of operation; the original pumps had a higher capacity but operated at a lower number of strokes per minute, whereas the diaphragm pump has a lower capacity which is compensated by a higher operating speed. Because of this, the unit sustains its high fluid pressure with a very much lower degree of pulsation, which is a decided advantage when heavy-bodied materials are being sprayed.

Many different patterns of portable plant are available, some designed to be lifted and carried and others mounted on a wheeled chassis. Some are extremely compact and can be manoeuvred through doorways, lifted up and down stairways, and transported to the site in the boot of a car. Generally the high-powered plants are mounted on a pneumatic-tyred trolley and are equipped with take-off points for more than one operator.

The *fluid line* needs to be capable of withstanding extremely high fluid pressures. It is usually made from a specially strengthened type of nylon surrounded with terylene braid and a plastic outer cover, although in

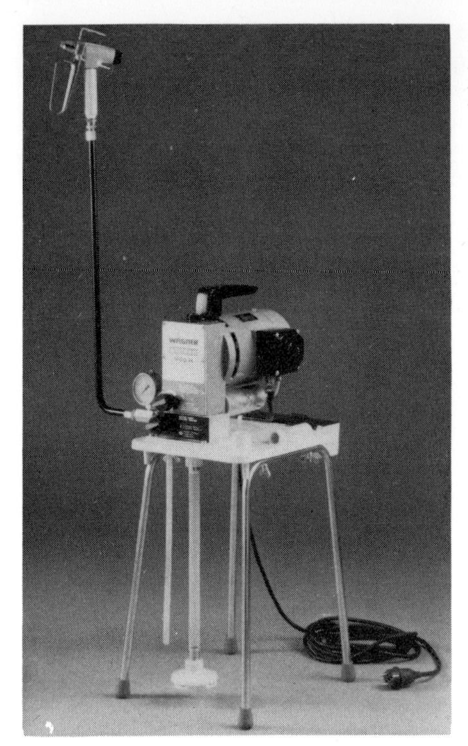

Fig. 6.24 The 1400H has an output of 2·6 litres per minute (0·57 gpm)

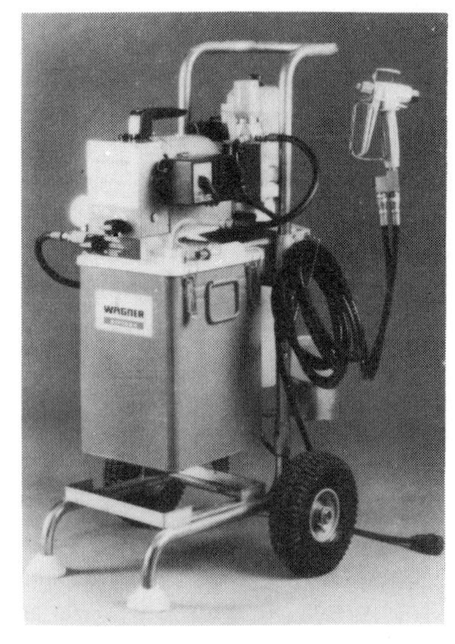

Fig. 6.25 The 2600H has a flow rate of 4·3 litres/min (0·95 gpm)

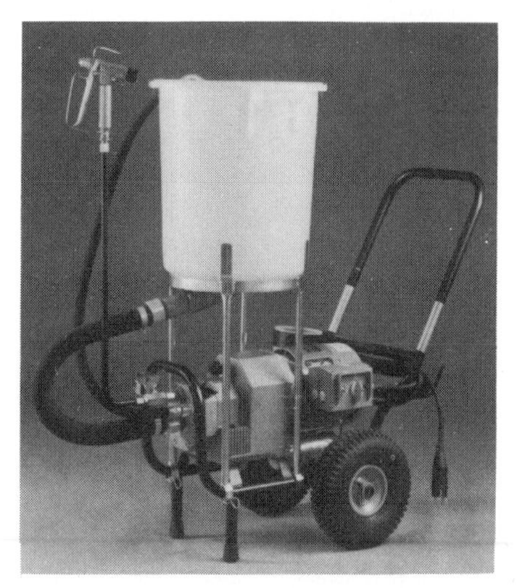

Fig. 6.26 The 6500H is a high-performance unit with a capacity of up to 6 litres/min (1·3 gpm)

APPLIANCES FROM THE WAGNER ELECTRIC AIRLESS RANGE, OPERATING AT 240–250 bars (3410–3555 psi) MAXIMUM HYDRAULIC PRESSURE

Courtesy: Gray-Campling Ltd

some cases a flexible stainless steel hose with a solvent-resisting lining is used. A static conductor wire is fitted to eliminate any build-up of static electricity. Generally the hose will have been tested by the manufacturer at a pressure three times as high as the normal maximum operating pressure. If the hose becomes damaged it is advisable to fit a new one rather than to attempt running repairs or to try to refit the connections which might give way under pressure.

A surge vessel is fitted between the fluid pump and the spray gun to provide back pressure in the fluid system, thus reducing the pulsation effect caused by the time-lag between strokes.

In Fig. 6.21 to 6.26 inclusive are shown various models of portable electric-motor-driven airless spray plant embodying the diaphragm-type compressor, most models being capable of operating more than one spray gun. Fig. 6.27 shows the Clemco range of syphon cart airless units which will spray most of the paints and primers used by industrial contractors, including high-viscosity materials; in these units the air compressor — either mobile or static — and paint pump are separate.

The airless spray gun

The airless spray gun (Fig. 6.29) is very different in construction from the guns used in air-propelled spraying. It has to be immensely strong to withstand the high hydraulic pressures involved, and must also be able to withstand the abrasive action of the paint and pigment particles passing through it at very high velocities. The nozzle tip is made of tungsten carbide, which, compared with alternative materials, is the most resistant to abrasion. Even so, it must be expected that the tip will eventually become worn and will need replacing; the actual life of the tip depends on a great number of variable factors but is chiefly dependent on the composition of the materials in use, silicates being especially erosive, though even the titanium oxide used in so many decorative paints has a markedly abrasive effect. The orifice through which the paint is released is generally elliptical in shape. It is necessary, of course, that the paint should not be ejected at random, but should be channelled in such a way as to direct it towards the work surface. This is done by means of an angled slot or deflector groove immediately beyond the orifice, so that the atomized paint stream is forced into a predetermined fan pattern (Fig. 6.30).

It is fairly obvious, however, that the viscosity of the paint material will have a pronounced effect upon its ability to flow through a small aperture, and upon the rate of flow. The viscosity of the material also affects the spray angle; and another factor in the distribution of the paint is the relative coarseness or fineness of the pigment particles and the proportion of solids in its composition. The whole matter of balancing the viscosity and composition of a material with its rate of flow is highly complex, and the selection of the most suitable spray angle and orifice size is too intricate to be dealt with by random adjustments carried out on a trial-and-error basis. This is the reason why an airless spray gun differs from the guns used in air-propelled systems, in that there is no means of varying the width of the spray pattern. Only one adjustment is necessary

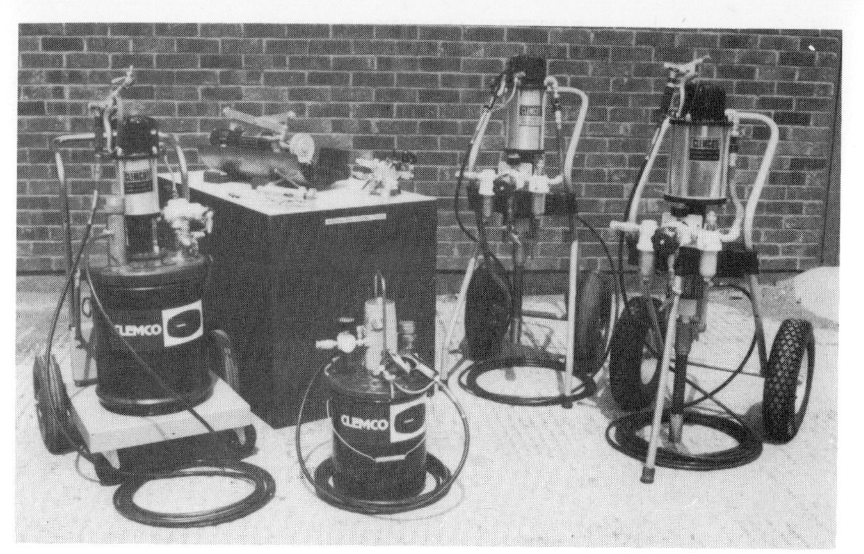

Fig. 6.27 Clemco portable airless units, with internal pipe-coating tools

The largest spray unit shown is fitted with a 150-mm air motor 44:1 ratio pump and has a flow rate, measured on water, of up to 6½ litres/min (1·43 gpm)

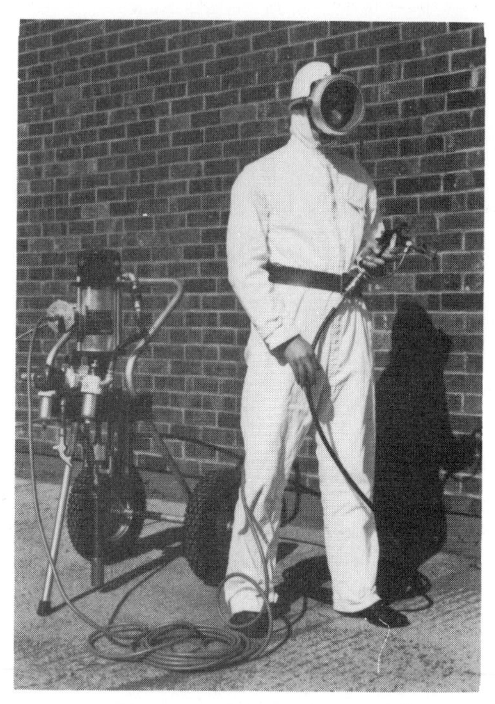

Fig. 6.28 Clemco airless spray equipment: operator wearing an air curtain mask and other protective clothing

Courtesy: Hodge Clemco Ltd

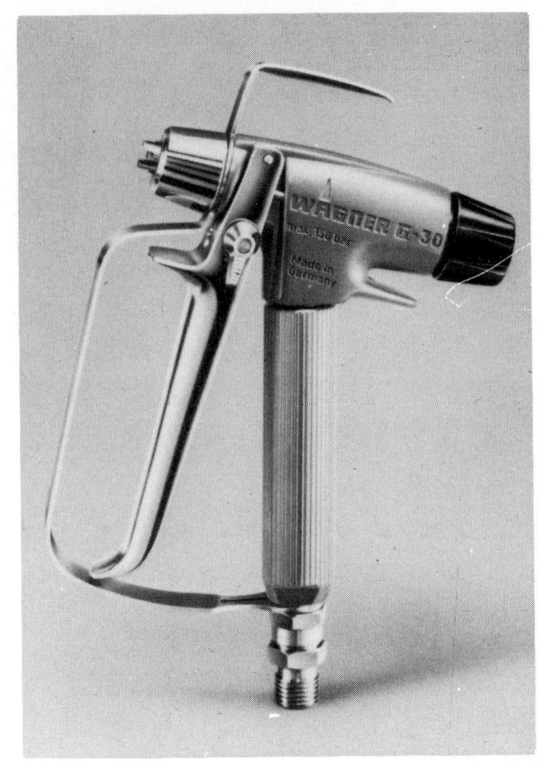

Courtesy: Gray-Campling Ltd

Fig. 6.29 Typical high-capacity airless spray gun (Wagner G 30) suitable for spraying both hard gloss and emulsion paints

The safety catch near top of trigger prevents inadvertent opening of the gun. (Here seen in the spraying position, it is turned back for the safety position, and forward for continuous spraying)

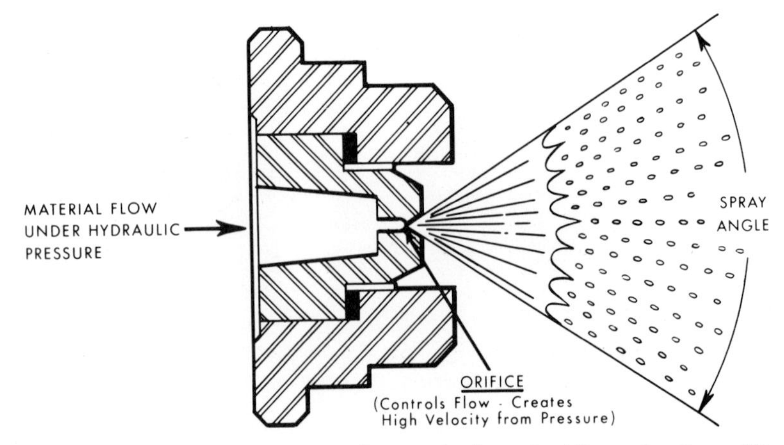

Courtesy: Aro International Corporation, Bryan, Ohio

Fig. 6.30 Cross-section of airless spray tip

to control the flow of paint to the gun, and there is no critical balancing of air and fluid pressure to obtain perfect atomization.

Spray caps

The size of the spray pattern is not altered by fluid pressure adjustment but is controlled by the type of cap fitted and by the viscosity of the material. A wide range of spray caps is available to deal with thin, medium, and extra heavy materials and giving a choice of spray-pattern widths to suit the particular job in hand. Spray cap angles vary from 5° for a very narrow pattern to 95° for a wide pattern. The manufacturers all provide tables indicating the spray cap range and giving complete information about the selection of the appropriate cap in any situation.

An unusual and interesting development of airless spraying technique has been evolved to cope with the enormously difficult problem of providing paint protection to the underwater structures of the North Sea oil rigs. Here a modified form of airless spray plant is used to apply paint under water at a rate of between 30 and 40 square metres per hour. The paints used are generally (though not invariably) based on epoxy resins, with the addition of non-reactive diluents such as pitch or coal tar to increase the water resistance. The application of the paint commences below water and proceeds upwards, through the level where the water meets the air, and eventually onwards to where the structure is completely clear of the water.

Hot airless spray

Portable hot airless spray units are available which combine the advantages of normal airless spraying with the reduction of viscosity obtained by the hot-spray process. This is of great importance in industrial painting as it enables heavy material of very high viscosity to be sprayed even in the coldest weather.

Static electricity

Static electricity, caused by friction between a relatively conductive material and non-conductive material and which may cause sparking that could ignite combustible vapours, is generated in airless spray plant by the high rate of paint flow through the hose and gun. Although in the U.S.A. there are no safety regulations governing this factor, in most European countries it is required that the equipment be earthed to prevent any risk. All equipment supplied by reputable manufacturers in the U.K. is protected in this manner.

Airless spraying technique

The technique of airless spraying differs in some respects from other forms of spraying. The gun should be held further away from the work and moved more quickly; the distance of the gun nozzle from the work may vary from 300 mm to 450 mm (12 to 18 in.), although for most decorative work 300 mm is usually recommended. The normal spray pattern tends to be more elongated than that produced by air-atomized spray, and the edges of the pattern are more sharply defined in that there is less feathering of the outline. The strokes should therefore be overlapped by 50 percent to produce an even coating without sagging. The speed of the stroke is very much faster than with other forms of spraying. The stroke movement should begin before the trigger is pulled and should end after the trigger is released, so that the operative part of the stroke is at a constant speed. The trigger needs to be depressed firmly and completely every time, and released abruptly and completely; it is not possible to feather the stroke by pulling the trigger part-way as in air-atomized spraying. It is of course essential that the gun be held at right angles to the work and kept at the same distance from it at all times, and that arcing the gun should be avoided (page 109). Because of the high pressures involved, proper precautions are required in handling the gun, and the operator should avoid aiming it at any part of the body.

Cleaning of airless spray equipment

It is most important that paint should never be allowed to dry or harden in the pump, hose, or gun. The equipment must be thoroughly cleaned immediately after use with the solvent appropriate to the material which has been sprayed, and the manufacturer's detailed instructions for cleaning must be rigorously followed. This is especially important to the decorator in view of the rapidly increasing use of water-thinned materials in ordinary trade practice; airless spray is particularly well adapted to the spraying of emulsion paints and similar quick-drying materials, and for this reason it is being used more and more by decorators in comparatively small room areas where until recently spray application would never have been considered. There are many small decorating firms now adopting the process who are not so keenly aware of the need for strict controls as are the industrial painting contractors to whom it has long been a commonplace.

There is one significant difference in the actual operation of airless spray equipment as compared with the pressure-feed system. On no account should a cloth be held over the spray nozzle and the trigger depressed to clear the gun: with the very high pressures involved this would be disastrous and could easily result in a serious injury such as the

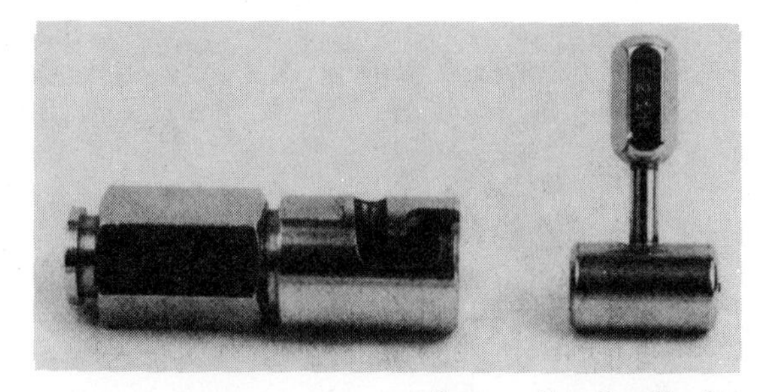

Courtesy: Gray-Campling Ltd

Fig. 6.31 Quick-change twist tip and holder for airless nozzle cleaning

loss of one or two fingers. All the manufacturers produce an adaptor which reverses the flow through the spray tip and flushes out the gun. The adaptor is also used if the spray cap becomes choked whilst in use; the obstruction is easily removed by reversing the flow, but it would become more firmly embedded if blown from behind. The operator should take care not to pull the trigger whilst turning the handle of the adaptor. A typical attachment is shown in Fig. 6.31; however, an adaptor is now incorporated in the gun assembly in some models.

Faults in airless spraying

The normal spray pattern appears as in the sketch (Fig. 6.32). If trails appear on the surface (Fig. 6.33), either the pressure is too low, the material too viscous, or the orifice too small, indicating that the next higher nozzle-group is required. A distorted spray pattern indicates dirt or dried paint on the spray cap, partial blockage of the strainer screen, too low a fluid pressure, or an excessively worn spray cap; it is difficult to detect by visual inspection when the cap is becoming worn, and the distortion may occur suddenly and without warning. If the spray pattern is too wide, the spray angle is too large; if it is too narrow the spray angle is too small. If too much material is sprayed, the nozzle group is too high; if there is not sufficient material the nozzle group is too low. A thick skin on the work indicates too heavy an application of paint, an operating fault.

Fig. 6.32 Normal airless spray pattern

Fig. 6.33 Defective airless spray pattern

Courtesy: DeVilbiss Co. Ltd

Usually there is practically no air-fog with airless spraying; should any occur it indicates too high an air-input pressure, the wrong choice of nozzle, or paint which is too thin. If the gun continues to spray when the trigger is released the fluid pressure is too high and is holding the needle open. Paint dripping from the spray cap when the trigger is released indicates a worn tip or seating.

By far the commonest fault which occurs on site work is the stoppage caused by blocked tips and filters. Many operators place too much reliance on the filters incorporated in the equipment; in spite of the presence of these filters it is most important that all paint should be strained before use.

(2) Electrostatic spraying

In this process an entirely different principle is used to convey the atomized paint from the gun to the work surface, the principle being that when bodies are electrically charged, those with a similar charge repel one another whilst those of opposing charge are attracted. Electrostatic spraying consists of imparting a strong electrical charge either of positive or negative polarity to the paint, and making sure that the object to be sprayed is well earthed. The atomized particles of paint are dispersed because of their mutually repellent charge, but they are vigorously attracted to the object which is earthed. Because the paint particles are drawn to the object by electrical attraction they all reach the surface and there is no overspray or spray-fog and no wastage of paint. Furthermore, the atomized cloud of paint particles will travel to the sides and right round to the back of the object in order to get to the earth. This means that, provided the object is slender enough, there is no need to twist and move the gun around in order to reach the sides and the back because the paint will automatically go around and form a coating of uniform thickness all over; this phenomenon is called the "wrap-round" effect (see Plate 1).

The principle has been known for a long while, and for many years highly sophisticated electrostatic equipment has been extensively used in the industrial finishing trades; obviously it is of most value for metal objects with high electrical conductivity, but it is also used in the spraying of wood furniture, where it has been found that the moisture content of the wood is sufficiently conductive to provide an adequate earth.

The painting and decorating trade has been slow to take advantage of the process. Clearly there would be few occasions in normal decorating practice where its usefulness would justify the cost of purchasing the equipment, but there are certainly some operations on which a sufficiently large firm or maintenance department would find the method economically viable. Examples are the painting of metal railings or of angle irons in factory roofs, where instead of a team of men carrying out a tedious task slowly and with a good deal of paint wastage and much splashing over the skin and clothing, electrostatic spraying would enable one man to work speedily, in perfect cleanliness, without leaving any parts unpainted, and with economy in use of paint. Some manufacturers of spray equipment are now marketing a mobile electrostatic spray outfit comprising an AC-powered generator and all the associated equipment, which can be mounted in a light van; such outfits are being used by public undertakings and by the maintenance departments of civic authorities for the painting of motorway fenders, railings, lamp-posts, pylons, bridge work and other installations. The process also finds manifold uses in re-spray and general painting work on site (Fig. 6.34); and there is electrostatic equipment designed for decorative purposes such as the spraying of nylon and rayon fibres, a treatment known as "flocking".

Typical electrostatic spray guns are shown in Fig. 6.35 and 6.36, while Fig. 6.37 shows the Wagner "Spray-o-Round" conversion set which can be used for electrostatic spraying in conjunction with standard airless equipment, in whatever way powered.

PLATE 1

Painting the base of a radar unit by electrostatic spray

Courtesy: J. Wagner Gmbh and Gray-Campling Ltd

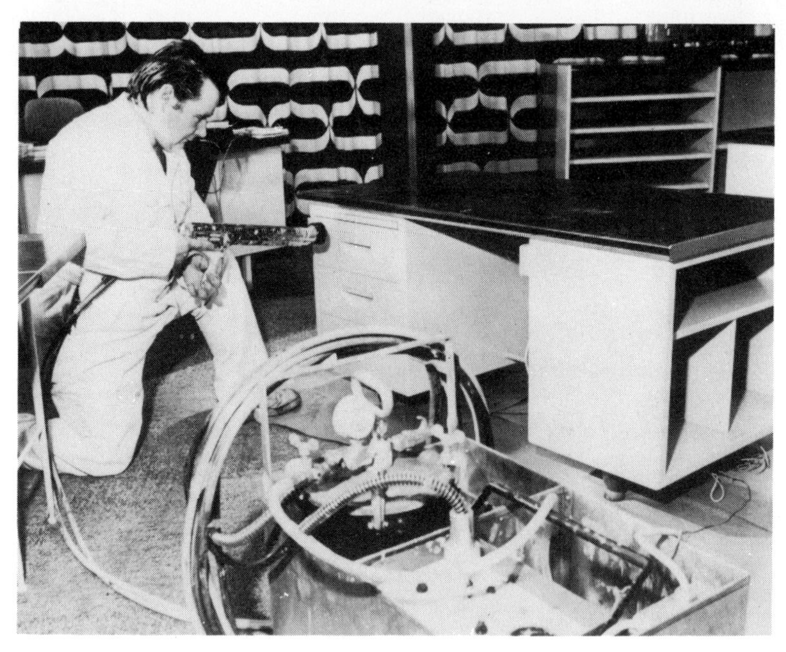

Fig. 6.34 Operator using an electrostatic hand gun to paint steel office furniture *in situ*

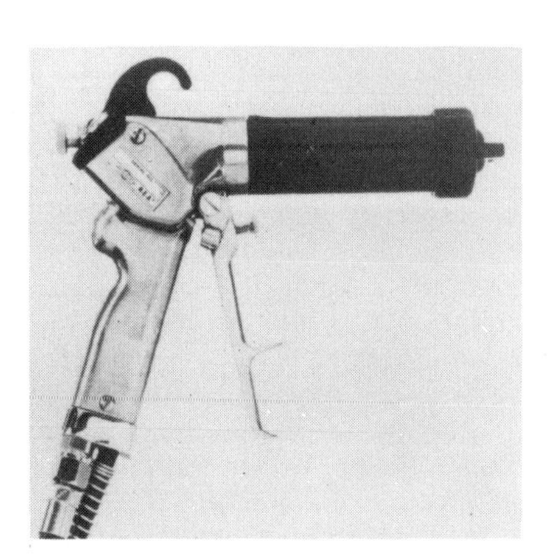

Fig. 6.35 Electrostatic spray gun using air atomization

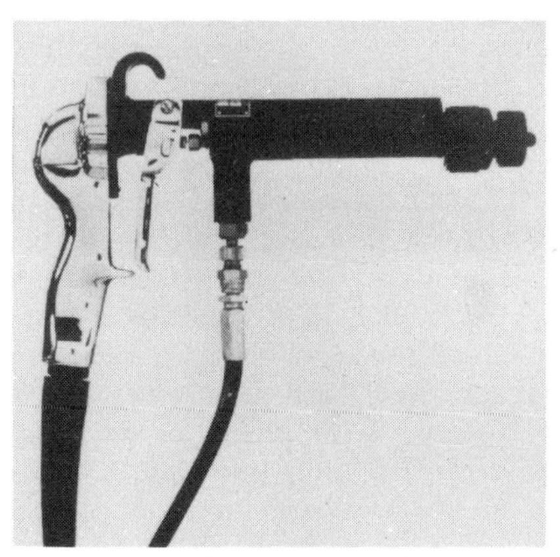

Fig. 6.36 Electrostatic spray gun designed for use with airless equipment

Courtesy: Ransburg-Peabody Ltd, Weybridge

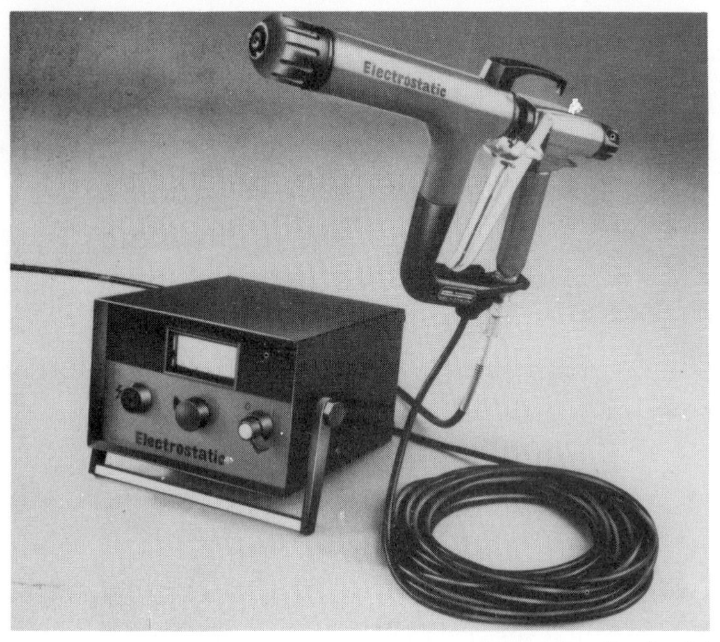

Courtesy: Gray-Campling Ltd

Fig. 6.37 Electrostatic conversion set

Masking in sprayed paint application

With the exception of electrostatic spray, it is impossible to limit sprayed paint to a precise area or to produce a sharply defined edge to the work; there is also the difficulty that with most spraying systems, and especially with air-atomized spraying, there is always a certain amount of spray-fog which settles indiscriminately on every surface and every object in the vicinity. Before commencing operations, therefore, the parts which are *not* to be sprayed must be masked or protected in some way. Floors may be protected with newspaper or dust-sheets, but the difficulty here is that the sheets are disturbed by movement of the equipment or scaffold; a better plan is to use wide sheets of sisalcraft paper, secured in place with small pieces of masking tape; sisalcraft is sturdy enough to be reclaimed and used many times over. An alternative is to use polythene dustsheets, secured with pieces of masking tape. Lining paper or odd lengths of discarded wallpaper can be fixed with masking tape to woodwork, dadoes, and other parts needing protection. Self-adhesive masking tape should generally be used, because ordinary gumstrip tends to damage the surface and also requires soaking before it can be removed; gumstrip is convenient when spraying water-thinned materials, however, as these naturally damp the gummed paper so that it can easily be removed before it dries. Gumstrip has the advantage of being much cheaper than masking tape, although the latter may with care sometimes be re-used. Masking tape, on the other hand, is not rigid but can be manipulated to conform to curved shapes or around the contours of any projecting obstructions (Fig. 6.38

Fig. 6.38 Broad band of paper, edged with adhesive masking tape, used on a concrete column to produce a dado effect

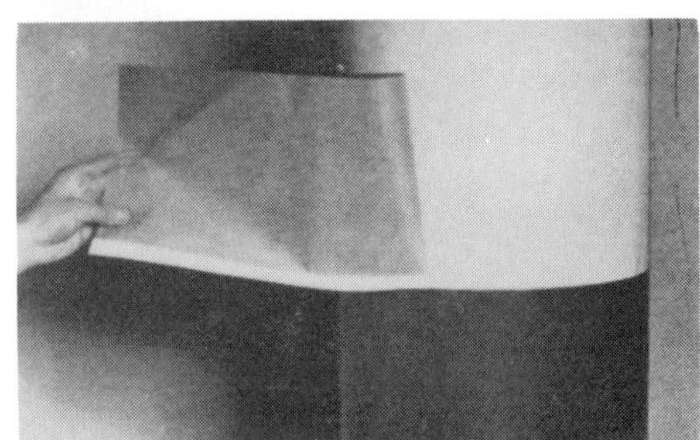

Fig. 6.39 The clean edge obtained when masking tape is removed

Courtesy: Durex Abrasives Ltd

and 6.39). Masking-up can take a great deal of time and may easily cancel out the time saved by spraying, but in some cases close-masking is not necessary; for instance, when spraying walls with water-thinned paint it is sometimes quicker to sponge down the architraves and skirting board than laboriously to mask them.

Various proprietary masking pastes and solutions are available, some of which may be applied by brush or spray to the areas to be protected; when the exposed surface has been sprayed and allowed to dry, the paste is sponged off with warm water. One of the most useful methods of masking, and the one which is by far the cheapest in the long run, is a metal shield which any sheet-metal worker can produce from some lightweight metal such as aluminium. The shield should be about 2 metres in length and should have a handle firmly riveted or welded on. The operator holds the shield in one hand whilst spraying with the other, and can easily mask off a ceiling from a wall surface, moving the shield along as required, or mask off a window reveal. At ground level the shield can be laid along the top of the skirting board and moved along as required. The only proviso is that such a shield should not be used when airless spraying is in progress, as the operator might bring his hand or arm too close to the spray gun for safety.

DECORATIVE SPRAYING

Spray equipment is produced for practically every purpose connected with the application of paints, colours and fluids, and for fine art work and small-scale decorative work several models are available; Fig. 6.40 shows one of these, which operates like an artist's airbrush. The airbrush is much too fine for anything larger than, say, a perspective drawing, and the writer has found the small MPG model shown in Fig. 6.6, *above*, also made by the DeVilbiss Company, to be excellent for detailed wall decoration and poster work; it is, in fact, the ideal link between the ultra-fine airbrush and the normal spray gun.

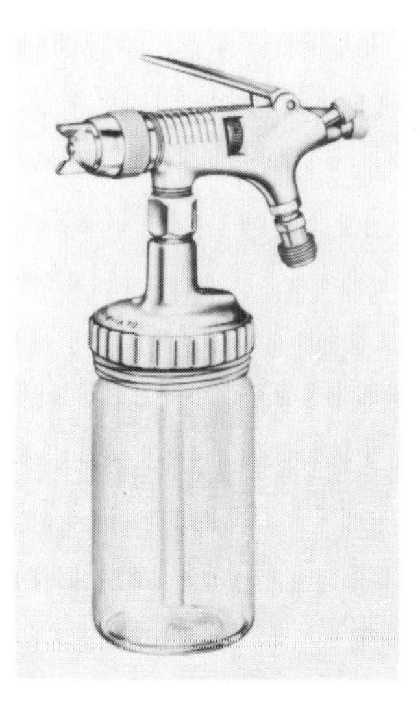

Fig. 6.40 The DeVilbiss type EGA suction-feed spray gun

Useful for detail decorative and sign-writing work

Courtesy: DeVilbiss Co. Ltd

Mural work and the production of decorative motifs is a fascinating exercise, and makes a highly individual contribution to the completion of a well-executed scheme of decoration. It involves working with stencils and templates cut from stout cartridge paper or detail paper, and the speed and adaptability of the spray gun makes the work very rapid compared with hand methods. The air-volume equipment is particularly suitable since the constant flow of air from the nozzle holds the template snugly to the surface and so helps to prevent blurring due to colour being blown under the edges. Colour and gradations can be gauged to a nicety, but it is as well to keep the spraying on the light side, for it will be found to appear much darker when the template is removed. Flat or eggshell cellulose is an excellent spraying material to use on a ground of water-thinned paint. It may also be used on hard oil-painted grounds, provided the cellulose is lightly misted on, for if it were sprayed heavily the solvents would soften the ground. Cellulose applied as a mist coat dries almost immediately, and multi-plate effects are quickly built up. The tendency to over-elaboration must be avoided because the facility of the spray method, once mastered, can easily tempt the decorator to overdo things and so commit errors of good taste — but this is bound up with the question of art training and is outside the scope of this chapter.

The spray gun is the ideal instrument for the production of blended or "shaded" effects, either in conjunction with stencils and templates or in the treatment of small decorative panels where a graduated background is required; such panels may be blended from the base upwards or, alternatively, from the edges of the panel, softening in towards the centre. Blending is made easier by using the mist cap of the decorator's combination sprayhead (Fig. 6.41 and Fig. 6.14, *above*). Ceiling and wall

Courtesy: DeVilbiss Co. Ltd

Fig. 6.41 Interchangeable sprayheads for producing decorative effects

panels of manageable size can be made to look very fine if the colour is well chosen (a warm buff, for instance, blended into cream looks far better than a fierce orange similarly blended, and the former is much easier to execute since there is less contrast with the ground colour). It is often assumed — probably because of the undoubted ease with which blended effects can be produced with the spray gun — that large unbroken ceiling and wall surfaces may be blended with equal facility; we would point out, however, that this is a greatly mistaken notion. We very much doubt whether the most skilful spray operator could compete with the best efforts of the orthodox brush and stipple method as outlined in "Blending" (see Chapter 28).

"Shot" effects

These are produced by side-spraying or misting relief textures with two or more colours. For this purpose the gun is held at an angle to the wall so as to spray one side of the relief with one colour, another side with another colour, and possibly introduce further colours from below or above. The use of metallics makes this particularly effective.

Soft mottled effects in polychrome are quickly produced by mist-spraying patches of various tints in an irregular or haphazard manner, and the appearance may be further enhanced by over-spattering. Careful choice of colour and restrained handling of the gun are both necessary if a bizarre effect is to be avoided, but contrasting colours can be used if they are reduced with white or cream to very pale tints of equal tone-value (which means that one must not be darker than another). A great advantage of spraywork is that if any one colour in a mottled polychrome treatment stands out too insistently it can easily be corrected, either by over-spattering or by misting with the ground colour.

Spatter

For this purpose paint is ejected from the gun in such a way as to fall on the surface in thick blobs which remain separate and do not flow together. This is achieved by using an air pressure so low that the paint is not atomized; the gun is used with a circular motion, the trigger pressed continuously. The wall surface is first painted by brush or spray to produce a solid colour, and when this is dry, one, two or three colours differing from the ground colour are spattered on to it, the darkest colour being spattered first. A final spatter of metallic paint enhances the effect. A special plastic sprayhead can be used to facilitate the process (Fig. 6.39).

Multi-coloured spraying

This is a method of obtaining something akin to a spatter effect in one operation. Multi-coloured paints specially made for the purpose contain globules of colour which remain separate in the dried film, being insulated from each other by an aqueous colloidal solution which prevents them from merging. The advantage is that only one application is required, but the coating is so thick that the coverage of the paint is very low, and the time spent in applying it is not markedly less than the time taken to produce a spattered finish. It should be noted that multi-coloured paint cannot be applied by airless spray as the high atomizing pressure breaks up the globules to produce a single mixed colour instead of allowing the colours to remain separate.

Sprayed plastic

Here a thick plastic paint composition is applied in a heavy coating with the special sprayhead, using a low air-pressure on the gun, so as to produce a relief texture. Afterwards the texture may be spattered with contrasting colours or metallic paints.

Lymnato, otherwise known as "veiling"

This is a decorative effect made up of sharply defined lines of colour intertwined at random on a suitably coloured ground. Fig. 6.42 shows a

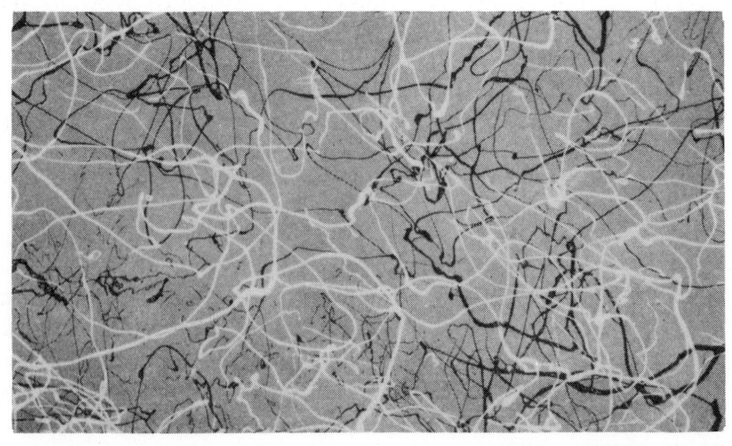

Courtesy: DeVilbiss Co. Ltd

Fig. 6.42 A specimen of Lymnato

specimen of Lymnato with lines of two colours on a ground of intermediate colour. The paint for this effect needs to be extremely viscous, and a cellulose leather paint, which is available in several colourways, gives excellent results. Ordinary paints are less satisfactory as they tend to sag. The air pressure on the gun is kept low, in the order of $1\frac{1}{2}$ to 2 bars (23–28 psi). Variation in the thickness of the lines is produced by altering the air pressure or by adjusting the fluid-needle adjustment screw.

Marbling

An effect vaguely reminiscent of marble can be produced with the help of a marbling screen or stencil of the type offered by decorators' merchants, though an inventive operative can experiment with home-made screens formed with teased-out steel wool or stiffened hemp. The method is to spray thin cellulose or spirit lacquer of the desired colour through the screen on to a ground of a different colour, the whole surface being then sprayed over with clear varnish. Some old-time craftsmen, notably Ernest Sanderson (formerly of Bradford), achieved remarkable imitations both of marble and of graining by the spray process, but they of course were able to grain or marble by orthodox methods besides being proficient with the spray gun; without such knowledge the attempt would be doomed to failure.

Stencilling is an extension of the normal brush-stencilling technique carried out with the spray gun; an air-volume gun is the most satisfactory for the purpose because the airflow holds the stencil plate flat to the wall.

Flock spraying

Flock spraying can be used for decorative purposes, since it produces a very pleasing and attractive surface texture; it can also be used for thermal or acoustic insulation or for anti-condensation purposes. Basically the process consists of coating the surface with an adhesive which, when tacky, is sprayed with powdered rayon, rayon fibres, or for a really outstanding effect with nylon fibres; the resultant finish gives the appearance of velvet or suede. The dry fibrous material can be sprayed by a normal air-propelled spray system, using a special flocking gun and container; or it can be sprayed by electrostatic equipment. Other dry materials, such as granulated cork for anti-condensation purposes, may be sprayed by the same methods.

Masking for lettering

The lettering or decoration of cellulose surfaces (for example, the lettering of commercial vehicles) is often done by first applying a sheet of stout tracing paper or thin lead foil to the surface, using vaseline for adhesive; the mask is smoothed out free of all wrinkles and the design or lettering is then pounced on and cut out with an exceedingly sharp stencil knife (see Fig. 33.2), using only sufficient pressure to cut the paper or foil without damaging the cellulose ground. The exposed cut-out shapes are thoroughly cleaned with white spirit to remove all trace of the vaseline, and the lettering or design is then sprayed in the requisite colour or metal (i.e. gold, bronze, aluminium, etc.). When this is quite hard, the mask is peeled off, the surface is thoroughly cleaned with white spirit and finally finished in one or more coats of clear cellulose lacquer. In some cases the lettering only is clear-lacquered before the mask is removed.

Courtesy: Gray-Campling Ltd

Fig. 6.43 Airless spray is speedy: the gable end wall of this hotel was painted in two hours by an operator using the Wagner 7000H electric airless unit, which sprays 8·3 litres of paint per minute

7

Brief History of Paint

From prehistoric times onward, man has used paint in one form or another, either for symbolic or for decorative purposes, such as the adornment of the human body — both living and dead — and the decoration of the home, the temple and the burial chamber. As far as we can discover, the use of paint as a means of *preservation* is comparatively "recent" — recent, that is, in relation to its long history.

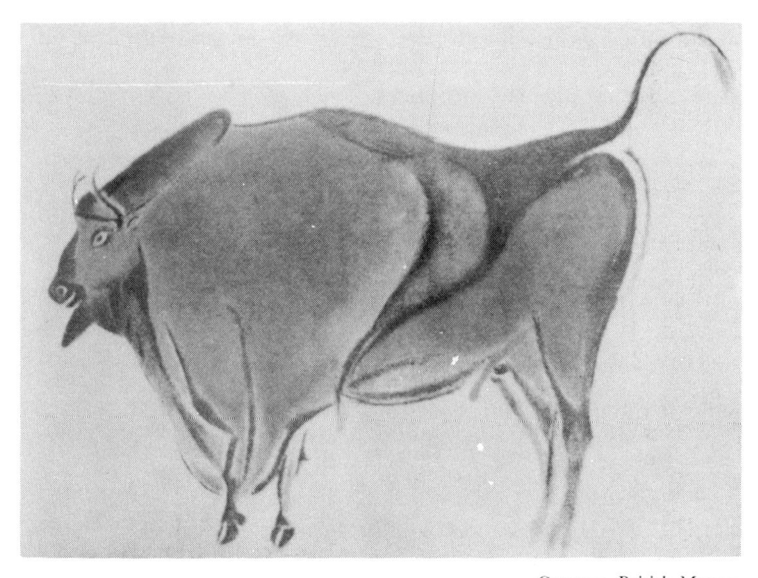

Courtesy: British Museum

Fig. 7.1 An example of prehistoric cave painting from Altamira, Spain

Executed in natural earth pigments mixed with water or animal fat. Age unknown, but probably painted not less than 35,000 years ago

Interior decoration may be said to have commenced when palaeolithic or early stone-age man began to adorn the walls of his cave dwelling with representations of deer, bison and other familiar animals which he hunted, as well as with childlike portrayals of the human form. Some of these drawings or paintings — the earliest known mural decorations — have endured to this day after 30,000 years or more. Notable examples have been discovered in France, Spain and Italy, those at Altamira in Northern Spain being especially of a high order and fine state of preservation. The pigments used were natural coloured earths, chiefly red, yellow, black, brown and white, according to the geological character of the district. We can only speculate on how these rock paintings were done, but it seems probable that the earliest cave-man artists collected moist clay, or made a puddle of coloured earth and water, and applied it to the walls with their hands. Such a method could give permanent results only in a dry climate and in a sheltered position, for the pigment in this unbound condition could not possibly have withstood the effects of wind and rain. The vigorous drawing and direct rendering of later work (as in the Altamira bison shown in Fig. 7.1) suggest that brushes were used, and it is also thought that the pigments may have been mixed with animal fat.

If neolithic man in Europe practised any form of mural art no examples have yet been discovered. His change of abode, however, from rock caves to impermanent mud huts and pile dwellings would account for their non-existence, and we do not meet wall painting again until we find it in Egypt about 4000 B.C. (Fig. 7.2). By this time the ancient Egyptians had discovered how to bind or fix their pigments so that they did not rub off. Among the earliest binders were sappy exudations from certain plants and trees, such as water-soluble gum arabic from the acacia. Egg, glue, milk and even honey

Courtesy: British Museum

Fig. 7.2 A good example of early Egyptian wall painting

The ancient Egyptians worked in tempera (distemper) and bound their pigments with plant juices, egg, honey, milk, animal glue, etc.

were also used to bind these early distemper (or tempera) paintings, and to the earth colours previously mentioned the Egyptians added the copper pigments malachite green and Egyptian blue (the first artificial pigment), also the mineral *orpiment*, also known as arsenic yellow or "King's yellow".

Fresco painting was probably first practised in pre-classical times by the early Cretans and some very fine examples of this art were discovered about 1900 at Cnossos in the island of Crete (palace of Minos, 2000 to 1500 B.C.) (Fig 7.3). Early Roman frescoes showing Greek influence were unearthed at

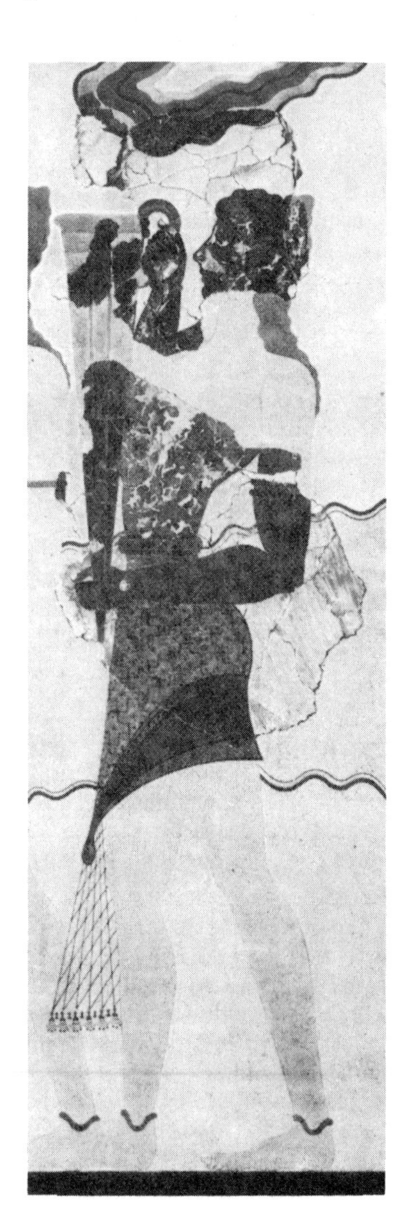

Fig. 7.3 Portion of wall fresco from the Palace of Minos at Cnossos, Crete (about 2000 B.C.)

Probably one of the earliest examples of fresco painting in which the colours were mixed with lime water and painted direct on wet plaster

Courtesy: British Museum

Pompeii and Herculaneum in the 18th century. *Buon* or true fresco, which flourished in Italy until the 15th–16th centuries, was painted direct on wet lime plaster, using lime-proof pigments mixed with lime water; the carbonating action of the atmosphere rendered the lime insoluble in water, and so fixed or bound the pigments. This method is still practised in parts of southern Europe, but although very permanent in a dry climate, it is totally unsuitable for damp industrial atmospheres. Another early method was the "encaustic" or wax painting practised by the Greeks and Romans in which they employed melted beeswax, using heated metal tools to blend the colours together.

The introduction of oil paint

The use of paint prior to 4000 B.C. seems to have been purely decorative or symbolic, and there appears to be no evidence of its application as a preservative. Hitherto, building materials had been mainly of stone, sun-baked clay bricks, and mud thickly smeared over interlaced twigs (wattle and daub), but as wood and iron came to be worked and fashioned the need for some form of finish or preservative arose.

The earliest varnishes were spirit varnishes, i.e. natural semi-liquid pine tree balsams or soft oleo-resins. They were probably warmed and applied by spreading with the hands; the ancient Egyptians are believed to have finished their distempered mummy cases in this way, and it is interesting to note that in some parts of China this primitive method of applying varnish is still practised. Incidentally, it is highly probable that the Chinese, about whose early civilisation we know so little, were the first to use varnishes and lacquers (Fig. 7.4).

It is possible that the added richness and depth of colour which resulted when varnish was applied over tempera paintings led to the actual grinding of pigment in a varnish medium, so producing the first "gloss paints". There is also evidence that beeswax and some form of tar or natural pitch was mixed with pigments for the decoration and preservation of ships on the Nile.

The first use of oil as a paint medium is obscure, but we learn from early manuscripts that vegetable drying oils (walnut and poppy) were known in Roman times and that lead and umber had been found to assist the drying of these oils as early as the 2nd century A.D. Linseed oil is mentioned in a 6th century manuscript; it was mixed with natural gums and balsams and used as a picture varnish. Professor Laurie in his book *Materials of the Painter's Craft* quotes a reference to preservative painting, from an 11th century manuscript, to the effect that cinnabar (natural vermilion), ground in linseed oil, was painted on door panels, exposed to the sun to dry and afterwards varnished. Efficient driers were then unknown and paint dried very slowly, taking several days for each coat. Another early chronicle tells of "fires being lit to dry the oil paintings on the walls". Accounts at Ely and Westminster in the 13th and 14th centuries show that oil was used in the early decoration of English cathedrals.

Distillation of spirits was practised by alchemists as far back as the 2nd century A.D., but it is not until the 16th century that direct reference is made in manuscripts to the use of turpentine and petroleum spirits for dissolving resins. Wax painting, in a thin smooth film (probably thinned with turpen-

Courtesy: British Museum

Fig. 7.4 Early Chinese lacquer toilet box of the Han dynasty (206 B.C.–A.D. 220)

The Chinese were probably the first to use lacquer, a natural resinous exudation from the so-called varnish tree (*Rhus vernicifera*). Their magnificent lacquer and gilt screens and cabinets have never been surpassed

tine), was carried out in the 15th century. It seems certain that oil painting was fairly well established in the northern countries such as Germany, Flanders and Britain (where the climate demanded some form of protective painting) long before it became popular in Italy and the Mediterranean. In fact, it was only after the Italians heard of the wonderful technique of the Flemish brothers Van Eyck in the 15th century that oil painting came to be practised on a wide scale in Italy.

The introduction of oil paint was revolutionary and widened the scope of paint tremendously. Whereas frescoes had to be painted quickly over plaster laid freshly each day, oil paintings could be done more leisurely on a dry ground, with ample time for careful modelling and fine detail. This fact alone had a great influence on the development of art. Moreover, linseed oil gave to paint valuable preservative and weather-resisting properties which extended its use to all surfaces liable to corrosion and decay. White lead, for instance, was known to the ancients in Egypt, Greece and Rome (where, incidentally, it was used as a *face* powder!); but not until the introduction of linseed oil, with which it combines so well, could its great preservative value have been fully realized or exploited. For very many years it remained the painter's standard and yardstick by which other paints were judged, but on account of its reactivity (toxicity and tendency to darken in urban atmospheres) and the introduction of synthetic media it has now been almost completely replaced by titanium dioxide.

Throughout the centuries the use of oil paint slowly developed as new materials were discovered and old materials gradually improved. Paint-making was essentially an individual craft, the painter selecting and grinding his own colours with loving care, preparing his oils and varnishes from his own secret recipes. In the monasteries, the preparation of colours and oils was invariably accompanied by the recitation of Paternosters, as a means both of timing the operations and of warding off evil spirits. Many of the medieval painters died with their secrets undivulged. One can only speculate, for instance, on the methods of the Van Eycks who were reputed to have "invented" oil painting; we do not know for certain whether they did in fact paint in oil or whether they worked in tempera and finished with oil varnish; their work can be seen today at the National Gallery — perfect, after more than five hundred years!

It was not until the late 18th century or thereabouts that artificial or chemical colours began to replace or add to the traditional minerals, earths, and natural dyestuffs; artificial ultramarine took the place of the semi-precious *lapis lazuli*, Prussian blue was introduced, and in 1856 Perkin produced the first artificial dyestuff from coal tar, a direct ancestor of our modern synthetic lakes and pigment dyes. Linseed oil remained the chief paint and varnish medium and turpentine the chief diluent for a number of years, but with the advent of tung oil and, later, the alkyd resins, the use of linseed oil and linseed oil varnishes as paint media has declined, whilst white spirit has replaced turpentine as the chief diluent.

Nowadays the formulation and manufacture of paints is based on scientific principles and the behaviour of pigments and media is more clearly understood. Paints of every conceivable hue are now produced in high gloss, semi-gloss, eggshell, and matt finishes and in media designed for their specific purpose. Fresh discoveries and improvements are constantly being made. Outstanding developments in modern times have been the introduction of a wide range of synthetic polymers into paint manufacture, and these materials have already brought about some revolutionary changes in painting technique such as electrostatic and airless spray (see Chapter 6), electrodeposition, infra-red drying, electron beam curing, etc.

Regarding the future of paint, the increasing use of glass, plastics, non-corrodible stay-bright metals, and other permanent building materials, has been viewed with misgiving by some members of the painting and decorating industry as being a threat to their very livelihood. It should be borne in mind, however, that permanent finishes, no matter how attractive at first, are apt to pall, especially if one has to live with them day in day out for a few years, and it is here that the decorator is in a strong position in being able to transform, by the least expensive and quickest means, the entire colour and texture of any existing structure. This indisputable fact, coupled with the inherent human craving for change in fashion, assures the decorator that he will be just as indispensable in the future as in the past.

8

Decorative and Building Paints

Modern paint technology is a highly specialized science which demands an advanced knowledge of chemistry and physics, and every paint firm of repute is obliged to employ a staff of chemists and technicians in order to cope with the complex problems of manufacture. In addition, the services of the Paint Research Association, Waldegrave Road, Teddington are at the disposal of the industry.

It will be obvious, then, that the painter and decorator cannot be expected to learn the deeper technicalities of all the materials he uses, nor is it necessary for him to do so. It is very desirable, however, that he should have a sound knowledge of their general behaviour on various surfaces; for besides the peculiar chemical and physical characteristics of paints and varnishes, building materials also may be chemically active, as in the case of new cements and lime plasters; moreover, the atmosphere itself may be chemically polluted as in industrial areas, or strongly saline as in coastal regions. Ignorance of the fundamental principles involved in dealing with such conditions is far too common in the trade and is probably responsible for most complaints regarding premature paint failures.

The present chapter deals not so much with the complex chemical make-up of paint materials, but rather with their general physical characteristics and behaviour in certain circumstances. It will be necessary, however, to consider the elementary chemistry of certain pigments and media in relation to their reactions with each other and also to note the various conditions to which they may be subjected, but it is felt that if the rudiments of paint chemistry are dealt with in simple terms, and then only when absolutely necessary, this will be of most service to the great majority of craftsmen and apprentices. Advanced student apprentices and others wishing to study the subject more fully are recommended to the list of suggested books for further study given in Appendix 6.

THE MAIN CONSTITUENTS

Air-drying decorative and building paints consist essentially of four main constituents, namely pigment, binder or film former, drier, and thinner. The pigment is the solid portion which provides colour and opacity (hiding power). It also contributes to durability and, in primers for metal, to resistance to corrosion. The binder or film former is the resin or synthetic polymer which carries the pigment in suspension and which, when exposed to the air in a thin layer, dries to a tough skin, so binding the pigment particles firmly to the surface in a cohesive film. The function of the drier is to accelerate the normal drying of the resin, whilst the thinner is used to convert the paint to a suitable consistency for application. The mixture of binder and thinner is often referred to as the "medium" or "vehicle". Each of these constituents will be considered separately in subsequent chapters.

GENERAL CHARACTERISTICS OF PIGMENTS

Apart from the variation in colour, pigments differ greatly in their chemical and physical properties, and it may be useful for purposes of comparison to outline here some of their more important characteristics before proceeding to consider pigments individually.

Generally the majority of modern pigments are chemically stable, that is to say, they do not react with other pigments or with the media in which they are dispersed, and they are resistant to attack by atmospheric impurities. The exceptions are zinc oxide, which tends to react with small amounts of organic acids in the medium, leading to embrittlement of the film, and lead pigments which are discoloured by reaction with sulphur compounds in the atmosphere. However, as will appear later, the use of these pigments in decorative paints is rapidly decreasing.

The toxicity of pigments and, indeed, of paint components generally is now watched carefully, and for this reason the amount of lead pigments such as white lead and lead chromes used in paints has decreased considerably in recent years. Red lead continues to be used in primers for steelwork, but its use is subject to the restrictions of the Lead Paint (Protection against Poisoning) Act and other Statutes. The topic of toxicity will be considered in some detail later.

The physical properties of pigments may vary according to their origin or method of manufacture. Some are hard and crystalline, while others are soft and amorphous. However, this classification is not rigid and the degree of softness or hardness can vary considerably.

Particle size

All pigments used in paints have been ground to a very fine state by the suppliers, and no further reduction in size takes place in paint manufacture. The particle size of pigments is expressed in micrometres (written μm); 1 micrometre = 10^{-6} metres or 10^{-3} millimetres. For conversion purposes it is useful to remember that 10^{-3} inch (one "thou") is equal to 25·4 micrometres. The particles in a particular pigment are not all of the same size but vary over a range. For example, the particles of titanium dioxide range from 0·25 to 0·35 micrometres in diameter, but some extenders such as barytes can range up to 20 micrometres. Carbon black, on the other hand, is extremely fine, the particle size ranging from 10 to 80 millimicrons (mμ). Particle size distributions in pigments are usually expressed by means of a histogram (Fig. 8.1) in which the percentage of particles, by weight or number, is plotted against particle size.

The particle size of a coloured pigment influences the tinting strength or staining power, which is the ability of a coloured pigment to tint or stain a

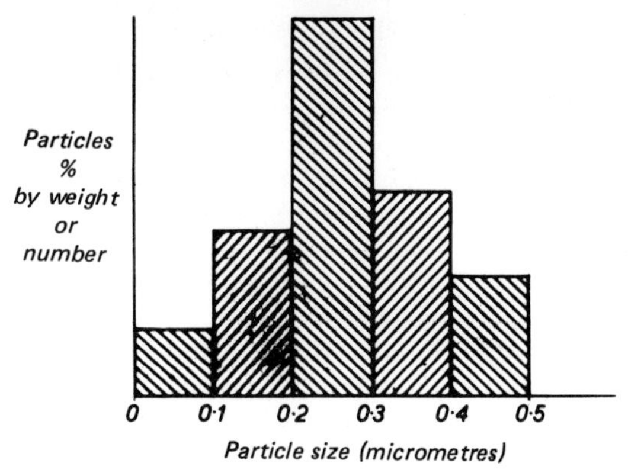

Fig. 8.1 Histogram illustrating distribution of particle size in a pigment

given mass of white to a desired shade. The finer the particle size of a pigment, the greater is the tinting strength.

For a given weight of pigment, the total surface area of the particles will increase as the particle size decreases, and therefore, in order to wet the particles thoroughly, more paint medium will be required for fine than for coarse pigments. A guide to the amount required is indicated by the Oil Absorption Value which is defined as the weight of linseed oil (usually acid refined) required to convert 100 parts by weight of pigment to a smooth paste. This is determined by the method set out in British Standard 3483.

Particle size also influences the rate at which a pigment settles in a paint, the larger particles settling more rapidly than the finer ones. However, the rate of settling is influenced to a greater degree by the specific gravity of the pigment as compared with the specific gravity of the paint medium used (the media average about 0·9). Thus red lead (S.G. 8·8) settles much more rapidly than titanium oxide (S.G. 4·0). The rate of settlement also depends on the shape of the particles, those with a compact shape settling more rapidly than those which are angular or elongated. For this reason the extender asbestine, which has a fibrous texture, is often used to check the rate of settlement of denser pigments and to prevent them from forming a compact mass in the bottom of the container.

In addition to indicating the tendency of a pigment to settle, specific gravity figures are used by the paint formulator to calculate the volume contributed by the pigment to the liquid paint and also the volume of a dry paint film occupied by the pigment. The latter has an important bearing on the performance and durability of the paint.

The differences in weight per litre between gloss paints of different colours is partly accounted for by the differences in specific gravity of the pigments used (the other important factor is opacity). For example, a black gloss based on the very opaque carbon black (S.G. 1·8) will contain about 50 grams of pigment per litre, and 1 litre will weigh about 0·95 kilograms, whereas a white gloss based on titanium dioxide (S.G. 4·0) will contain about 350 grams of pigment per litre and the weight of a litre will be about 1·2 kilograms.

Opacity

Pigments differ widely in their ability to hide or obliterate a surface when they are embodied in a paint. This property, known as opacity, hiding power or obliterating power, depends on the amount of light which is reflected by the particles of pigment and this, in turn, depends on the difference between the refractive index of the pigment and that of the medium. When this is wide — as, for example with titanium dioxide — the paint possesses good opacity and hides the surface well. With some organic pigments the difference is small, resulting in paints of poor opacity, and the underlying surface "grins through". With extenders such as barytes or china clay the difference is extremely small, and these materials are practically transparent in paint media. For this reason extenders make no contribution to the opacity (or colour) of a paint but are used to modify some of the physical properties such as flow or gloss.

The opacity of a pigment is measured by the Contrast Ratio method after it has been made up into a paint at a definite concentration. The paint is applied to a colourless transparent sheet, e.g. cellulose acetate, at a known film thickness and the reflectance measured when the dry film is placed on white and on black substrates. The difference gives the amount of light passing through the film and being absorbed by the black substrate. The Contrast Ratio (or opacity) is the reflectance over black expressed as a percentage of the reflectance over white.

Lightfastness

It is essential that coloured pigments do not fade or change colour when exposed to sunlight. In other words, they must possess good lightfastness. Inorganic pigments, which are compounds of metals, are generally good in this respect, but there is wide variation between the various types of organic pigments. Some of these show good lightfastness both in the pure state and when mixed with large quantities of white ("reduced"), but others are satisfactory when pure but fade rapidly when reduced. Lightfastness is assessed by reference to the Blue Wool Scale and is measured by dispersing the pigment in a paint medium and applying the dispersion to a strip of metal or card. When dry, the strip, together with a standard for comparison, is exposed to ultra-violet light in a fugitometer. The method is described in British Standard 1006. Figures for the lightfastness of organic pigments are available from the manufacturers.

Corrosion inhibition

Certain pigments are known to be active inhibitors of corrosion and so are used in primers for steel and other metals. Among these are red lead, zinc chrome, zinc phosphate, calcium plumbate and the metallic pigments zinc and lead. These pigments are little used other than in primers. When pigment particles are in the form of flat plates they tend to become orientated in the paint film with their planes parallel to the surface, a behaviour known as "leafing". The particles overlap and this increases the resistance of the film to moisture penetration and so reduces corrosion. Such pigments are described as "laminar", typical examples being aluminium and micaceous iron oxide.

The colour of pigments and particularly that of pigment mixtures is, of course, of paramount importance. This aspect, together with the nature of colour and its application in decoration is discussed in detail in Chapter 36.

Origin of pigments and extenders

Broadly speaking, there are two main classes of pigments and extenders: (a) those occurring as natural mineral deposits, and (b) manufactured types. The natural minerals include oxides of iron, graphite, and mineral extenders.

Table 8.1 Classification of the principal pigments and extenders according to origin

Colour	Natural	Chemically prepared
Whites	Barytes, Whiting, China clay, Asbestine, Silica, Talc } Extenders	Titanium dioxide Zinc oxide Zinc phosphate Antimony oxide Calcium plumbate (buff colour) Lithopone White lead Blanc fixe, Fumed silica } Extenders
Yellows	Ochres Siennas	Cadmium yellows Zinc and lead chromes Yellow oxide of iron (Ferrite yellow) Organic yellows
Reds & browns	Oxides of iron Umbers	Synthetic iron oxides Cadmium reds Red lead Organic reds
Blues & greens	Ultramarine (Lapis lazuli)	Phthalocyanine blue Prussian blue Ultramarine blue Phthalocyanine green Chrome green Chromium oxide Guignet's green (hydrated chromium oxide)
Blacks	Black oxide of iron Micaceous oxide of iron Graphite	Carbon black Vegetable (lamp) black Black oxide of iron Graphite
Metallics	–	Aluminium Lead Zinc Bronzes (copper alloys)

The manufactured products cover a wide range of inorganic and organic compounds as well as some extenders. Table 8.1 classifics the principal pigments and extenders into the two groups — natural and chemically prepared.

PROCESSING OF PIGMENTS

Before proceeding to consider the physical characteristics of the principal pigments individually (see Chapters 9 and 10), a few words in general on methods of processing may well be interposed here in order to avoid repetition later.

Grinding and air-flotation

With mineral pigments the large pieces of ore are washed to remove earthy matter and are then crushed to a coarse powder. This is then ground to the size required by the paint manufacturer. There are several types of mill employed for this purpose, but the finest pigments are produced by the hammer mill or the fluid energy mill (the Micronizer). The latter is an unusual mill in that it contains no moving parts and size reduction of the pigment is achieved by attrition between the particles. This is brought about by compressed air or high-pressure steam in a circular chamber of special design. In both this and the hammer mill the finely ground pigment is removed continuously by a stream of air and then graded, a process known as air-flotation.* This process is much faster than grinding followed by levigation, which is described in the following section.

Grinding and levigation

This is the simplest and oldest process for producing and grading powders and is now used mainly for certain types of extender, notably china clay and whiting, which are produced in large quantities. The materials are soft and can either be quarried, crushed and levigated or, as is the case with china clay, high-pressure jets of water are used to remove the clay and the resulting slurry is then levigated. The process of levigation consists in allowing the pigment/water mixture to flow into a series of settling tanks of increasing size and lower elevation (Fig. 8.2). When the first tank is full, the water overflows

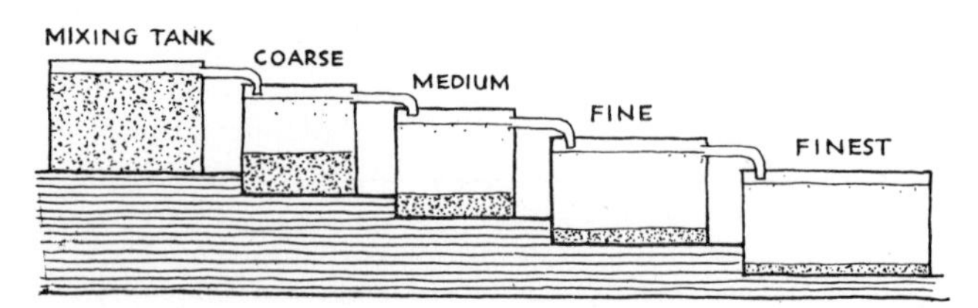

Fig. 8.2 Levigation

The pigment is automatically graded into fine and coarse particles in a series of settling tanks

*For a more detailed description of these mills see *Outlines of Paint Technology* by W. M. Morgans Vol. 1 (Charles Griffin & Co. Ltd).

into the next, carrying with it the lighter particles, and this is repeated until the last and largest tank is reached. The increasing size of the tanks slows down the rate of flow, giving the pigment more time to settle, thus automatically grading the material and ensuring that only the finest particles reach the last tank. When the pigment has settled, the upper liquid is run off and the graded "sludge" is filtered and dried at a low temperature.

Precipitation

This consists in mixing together appropriate solutions which react to form an insoluble compound (the "precipitate") which separates as a fine solid. This is kept in suspension by stirring, and the suspension is then pumped into a filter press where most of the water is removed. The wet pigment is washed in the press with further quantities of water and then transferred to drying ovens. It is finally ground. Pigments produced in this way include Prussian blue, lead and zinc chromes, and cadmium colours. In some cases precipitation is followed by calcination (*below*).

Calcination

In this process pigments are prepared by roasting or calcining the raw materials at high temperatures in suitable types of furnace; red lead and ultramarine blue are produced in this way. It is also used to develop the properties of some pigments produced by precipitation, e.g. cadmium colours. The preparation of Burnt Sienna and Burnt Umber from the respective "Raw" types is also carried out by calcination.

Vapour-phase oxidation

The metals zinc and antimony can be volatilized by heating to a sufficiently high temperature, and if the metal vapour is mixed with air then oxides of the metals are formed. In the case of zinc, the oxide is drawn away by the stream of air and passed through a series of chambers. The coarser particles settle first, and the finer particles settle progressively in the more distant chambers. In this way the zinc oxide is graded into various sizes (an example of air-flotation).

9

Natural Pigments and Extenders

NATURAL PIGMENTS

The majority of the minerals which are processed for use as pigments are oxides or hydroxides of iron, usually mixed with varying amounts of siliceous matter or clay. They range in colour from fairly bright yellow ochres to the dark umbers and black oxides, and the range is further extended by calcining some of the members. This reddens and deepens the colour and gives greater transparency. These natural oxide pigments are among the most permanent colours known and, as mentioned previously, have endured in some cases for thousands of years. They are usually processed by grinding followed by air-flotation, and the so-called "Micronized" grades are widely used.

Ochres

Ochres consist of mixtures of hydrated oxide of iron with mineral silicates (clays), and both the colour and composition vary considerably from one source to another. They are widely distributed, and good-quality grades occur in France and in South Africa. On account of the low iron oxide content, ochres possess poor opacity and their main use is in stainers for paint. However, for this purpose they are being supplanted gradually by the cleaner and brighter manufactured material (ferrite yellow).

Siennas

The raw siennas are brownish yellow in colour and contain a small quantity (about 1 percent) of oxide of manganese. Like the ochres, they possess poor opacity and vary in composition, the iron oxide content varying from 40 to 70 percent. They occur principally in Italy and Sicily. Calcination of the raw sienna produces the orange-red burnt sienna. Both forms are supplied in micronized grades and are used as stainers.

Umbers

Raw umber is dark brown in colour and contains up to 15 percent of oxide of manganese, which accounts for its pronounced drying influence on oxidizable media. The capacity and comparison vary, the iron oxide content ranging from 30 to 50 percent. The best grade (Turkey umber) originally came from Cyprus but is found in many parts of the world. Calcination of raw umber gives the rich reddish-brown burnt umber. Like the siennas, umbers are supplied in micronized form and are used as paint stainers.

Red oxides

Deposits of high-grade iron oxide occur in many parts of the world, but English deposits are little used as pigments. Good quality ore is imported from Spain and the ground material contains 80 to 90 percent of iron oxide. Spanish oxide is a bright yellowish red and has been popular with paint manufacturers for many years. A blue shade of red oxide is obtained from the Near East and is known as Gulf Red. The natural red oxides are very stable and lightfast. They are used widely in metal primers and gloss paints.

Micaceous oxide of iron

The name of this pigment is derived from the fact that the particles are in the form of flat plates which glisten in a similar way to mica. It is a greyish or greenish black, and when the pigment is used in paint the leafing action of the particles increases the resistance of the film to penetration by moisture. Micaceous oxides of iron are used in paints for structural steelwork.

Graphite

Deposits of natural graphite (blacklead or plumbago) occur in Ceylon and Germany. Graphite is one of the crystalline forms of carbon, and the quality of the natural material is assessed on the carbon content, the remainder being siliceous matter. A pure form of graphite is manufactured by calcining a high-carbon grade of coal (e.g. anthracite) in electric furnaces.

Graphite is not a true black, but rather a very dark grey with a slight metallic lustre. It is laminar in structure and the leafing properties improve the moisture resistance of paints. Graphite paints possess exceptional spreading power due to the "slippery" or lubricating property of the particles, which enables the paint to be brushed out to a very thin film. When in direct contact with steel in the presence of moisture, graphite will stimulate corrosion and it is therefore most unwise to apply graphite paints to bare steel. However, a mixture of graphite and red lead is an effective pigment for primers for steelwork. Graphite is used in undercoats and finishes for structural steelwork, but as these paints often show poor intercoat adhesion, a proportion of a crystalline extender such as silica or barytes is incorporated into the undercoat to give "tooth" and so improve the adhesion of the finish.

EXTENDERS

These are, in general, chemically inert materials, some of which are processed mineral whites while others are manufactured. Although varying in chemical composition, they have one property in common, namely refractive index. In all cases this is very close to the average for paint media, and consequently extenders are virtually transparent in paint media and make no contribution to the opacity or colour of the paint. However, they are opaque in water and can be used as pigments in certain types of water paints.

Extenders are sometimes used to cheapen paints, but more often they are employed to modify or control some of the physical properties such as flow, gloss, prevention of hard settlement, etc. The following are the most commonly used extenders.

Whiting (calcium carbonate) is prepared from natural chalk which comprises the basic structure of south-east and north-east England. Chalk is a soft material and is readily crushed to powder. It is then washed, dried and bagged. The purer levigated grades are known as Paris White. Whiting shows an alkaline reaction in water and is readily soluble in dilute mineral and acetic acids; it should not, therefore, be used in paints which are likely to encounter acidic conditions. Glazing putty for wooden sashes is made from whiting ground in raw linseed oil. Other uses of whiting are in stoppers, fillers, and some types of primer. It is also used to a limited extent in latex emulsion paints.

Barytes (barium sulphate) occurs extensively in many parts of the world as the mineral "heavy spar", but the quality and colour vary considerably. The heavy ore is crushed, treated with acid to remove impurities such as iron oxide, washed, dried, ground, and finally graded. The finest grades are produced by micronizing followed by air-flotation.

Barytes is a hard, crystalline material, opaque in water and practically transparent in paint media. Chemically it is very inert. It is used in stoppers, fillers, and also in undercoats where it is thought to give physical reinforcement to the film. A variety of barium sulphate produced by precipitation is known as "blanc fixe". This has a fine texture and, like barytes, is practically transparent in paint media. It also has good opacity in water and is used by commercial artists under the name "Process white".

China clay, also known as "kaolin", is a hydrated aluminium silicate and large deposits occur in the St Austell area of Cornwall. It is formed by the decomposition of felspar which, in turn, is one of the components of granite. Granite consists of a matrix of crystals of quartz, felspar and mica. Breakdown leads eventually to china clay mixed with fairly coarse silica (quartz) and mica. The china clay is quarried by jets of high-pressure water and forms a slurry from which the coarse quartz and any unchanged granite settle out. The slurry is then levigated to separate silica and mica and the various grades of china clay. The latter are filtered off, dried, and powdered.

China clay is a fine white crystalline powder, practically transparent in paint media. It forms a plastic mass when mixed with water. It is used in limited quantities in undercoats as a flatting, suspending and thickening agent. Excessive quantities will impair the flow properties of the paint and will result in a soft or "cheesy" film. Calcination of china clay destroys its crystalline character and gives an amorphous product with an oil absorption higher than the original material. The calcined clay has a wider application as a flatting and thickening agent.

Asbestine The source of asbestine is a fibrous mineral occurring in the United States and elsewhere. It is powdered and then subjected to air-flotation but, on account of its fibrous nature, precautions must be taken to avoid inhaling the dry powder; the powder has an adverse effect on the lungs similar to that produced by asbestos dust. The fibrous texture renders it a useful addition to certain paints containing heavy pigments, where it checks hard settlement. It also increases the mechanical strength of certain types of paint film.

Talc, also known as "French chalk", is a hydrated magnesium silicate and occurs as a fairly soft mineral which breaks down on grinding into compact rhombohedral crystals. It possesses the normal properties of extenders and is supplied in micronized form as well as from normal grinding. The micronized form possesses good flatting and anti-settling properties and does not affect flow properties to the same extent as do certain other extenders.

Silica The paint industry uses natural silica (in the form of ground quartz) as well as chemically produced material. Natural silica is very hard and its use is mainly confined to fillers and stoppers, where the hardness is an asset in the rubbing-down operation. The more finely ground natural silica finds some use in undercoats to give "tooth" to the film and this, it is claimed, improves the adhesion of the finishing coat. Manufactured silicas are produced in an extremely fine particle size by the controlled decomposition of compounds of silicon. They are characterized by low specific gravity, high bulking value, extremely fine particle size, and very high oil absorption. They are transparent in paint media and give colourless gels at relatively low concentrations. The addition of small quantities exerts a profound effect on the gloss and consistency of paints.

10

Chemically Prepared Pigments

The listing of pigments under the headings "Natural" and "Chemically Prepared" is done merely to make clear the origins of the various pigments. The demand for brighter colours and clean pastel shades has stimulated the use of chemically prepared pigments, and many of the duller natural products are now used only in primers or stainers. Another advantage of the chemically prepared pigments is their uniformity, and many modern pigments, particularly in the organic types, possess outstanding tinting strength and excellent lightfastness.

WHITE PIGMENTS

A very high proportion of the paint made today is either white or pastel shade, i.e. white tinted with a small quantity of stainer. The white pigment used is almost invariably titanium dioxide which has replaced white lead and zinc oxide in gloss paints and undercoats and has also replaced lithopone in water-based paints. However, small quantities of white lead, antimony oxide, zinc oxide and lithopone continue to be used for specific purposes. Calcium plumbate (which is actually a buff colour due to impurities) and zinc phosphate are more recent additions to the range, but their use is confined, in the main, to primers for metals.

Titanium dioxide

The two principal sources of titanium dioxide are the ores *ilmenite* and *rutile*. Ilmenite is a black ferrous titanate occurring chiefly in Norway and India, while rutile is a natural titanium dioxide which is found in various places.

The manufacture of titanium dioxide from the ores involves complex chemical reactions and there are two methods in use. In the older sulphate process, the ilmenite is dissolved in sulphuric acid and the iron compounds are then removed. The solution of titanium sulphate is hydrolysed to precipitate titanium hydroxide and this is filtered, dried, and then calcined to form titanium dioxide. By adjustment of the conditions of manufacture, one or other of two forms of titanium dioxide can be produced: these are known as "anatase" and "rutile" grades and differ in their crystal structure. This difference is reflected in their properties. The anatase grade is a cold bluish-white and if used in exterior finishes it causes them to chalk very rapidly on exposure. Chalking is a condition of the paint surface arising from breakdown of the organic binder, resulting in loose pigment or "chalk" on the surface. The rutile type is a slightly creamier white but gives finishes which are resistant to chalking on exterior exposure.

The more recent chloride process of manufacture converts the titanium compounds into the volatile titanium chloride which is then converted into a rutile grade of titanium dioxide. The pigment produced by this process is a little superior to that from the sulphate process in colour, opacity, and tint resistance.

The predominant position of titanium dioxide is a consequence of its superior properties. It is resistant to heat and has greater opacity and tint resistance than any other white. It is also non-toxic and non-reactive toward paint media. However, in highly acid media some grades may show reactivity. This arises from the fact that the titanium dioxide particles have been coated with substances to improve their dispersibility and general film properties.

The rutile grade is used in decorative gloss finishes, undercoats, emulsion paints and in marine and industrial finishes. For interior finishes where good whiteness is important, as in domestic and hospital equipment, the anatase grade is frequently employed.

Antimony oxide ("Timonox")

This is obtained from the ore *stibnite* which is subjected to a roasting process whereby the metal is first produced and then volatilizes. The metal vapour combines with oxygen from the air to give antimony oxide vapour which is carried by the airstream into condensing chambers, where the solid pigment is deposited on cooling.

Antimony oxide is non-reactive toward paint media and was at one time used to counteract film embrittlement in zinc oxide paints. When exposed to sunlight, paints based on antimony oxide do not chalk, and a mixture of this pigment with anatase titanium dioxide was used in exterior enamels for some years. The introduction of the chalk-resistant rutile titanium dioxide eliminated the need for antimony oxide in such paints, and the main use of the pigment today is in fire-retarding paints. These are described in Chapter 17.

White lead

White lead is a basic carbonate of lead and has been known for many centuries. A natural carbonate of lead was used by the ancient Greeks and Romans, and it is believed that the ancient Egyptians produced a form of white lead by treating metallic lead with vinegar. The action of acetic acid, water and carbon dioxide on metallic lead was the basis of our methods of manufacture for many years. The processes were very slow and have been replaced by a precipitation method in which a solution of basic lead acetate is treated with carbon dioxide. The basic lead carbonate precipitates, and by adjusting the conditions three grades of white lead can be produced. These are known as high stain-, medium stain-, and low stain-resistant white leads.

White lead is a reactive pigment and reacts with oils and varnishes to produce lead "soaps". These contribute to the elasticity of the paint film and to the reputation of white lead paints for good exterior durability. Although the protective qualities of white lead oil paints were excellent, the flow and gloss retention were inferior to gloss finishes based on alkyds which also showed excellent durability. The introduction of these finishes based on alkyds and rutile titanium dioxide resulted in a decline in the use of white lead in finishing paints. It continued to be used in undercoats and in primers for exterior woodwork (see p. 241), but the highly toxic nature of the pigment has resulted in severe restrictions on its use. The high reactivity of white lead renders it readily ingested into the human system and it is contained in few modern paints. It continues to be used by artists under the name "Flake white".

Zinc oxide

This pigment has been used in oil and varnish type paints for very many years and is known to artists as "Chinese white". With the introduction of titanium dioxide and the use of alkyds and other synthetic resins the use of zinc oxide in paints has declined and today the amount used for this purpose is small. A certain amount is used in cellulose nitrate lacquers, and limited amounts of colloidal (extremely fine) zinc oxide are thought to assist gloss retention in some types of finish and are used for this purpose.

Zinc phosphate

Zinc phosphate is a relatively recent addition to the range of white pigments. It is not used in finishes but possesses pronounced corrosion-resisting properties. This, together with its non-toxic nature and compatibility with a wide range of media, accounts for the fact that it is displacing red lead in primers for steelwork.

Lithopone

Lithopone was the major pigment used in oil-bound distempers. These have now been replaced by polymer emulsion paints pigmented with titanium dioxide, and although a small amount of lithopone is incorporated into some emulsion paints the amount used by the paint industry is extremely small.

Calcium plumbate

This pigment is produced by heating calcium hydroxide (slaked lime) and litharge (lead oxide) to a high temperature. It is a light buff colour, but if pure calcium hydroxide is used the colour is white.

Calcium plumbate shows a slightly alkaline reaction and disperses readily in various types of media such as linseed oils, stand oils, long oil, alkyds, and chlorinated rubber. Primers containing calcium plumbate are used extensively for the protection of steelwork against corrosion; they also adhere satisfactorily to new galvanized iron, a surface to which conventional primers show poor adhesion.

YELLOW, ORANGE AND RED INORGANIC PIGMENTS

Lead chromes

Lead chromes range in colour from a pale greenish yellow to orange and scarlet and are classified as primrose, lemon, middle, orange, and scarlet (or molybdate) chromes. They all contain lead chromate, but some members (primrose and lemon) also contain lead sulphate, whilst the scarlet chrome also contains lead molybdate as well as a little lead sulphate. They are manufactured by precipitation from solutions of lead acetate or lead nitrate by addition of solutions of either sodium chromate or mixtures of sodium chromate and sodium sulphate, or (for scarlet chromes) mixtures of sodium chromate, sodium molybdate and sodium sulphate.

The lead chromes are all bright pigments with good opacity and high tinting strength. Their general durability is good, though their usage in paints has decreased in recent years due to a general agreement to eliminate toxic pigments from decorative, toy and some other paints. They continue to be used in many types of industrial paints either as chromes or in admixture with Prussian blue in the form of chrome greens.

Zinc chromes

There are two zinc chromate pigments in use in the paint industry: they are known as zinc chrome (or zinc potassium chromate) and zinc tetroxychromate.

Zinc chrome is manufactured from zinc oxide by treatment with potassium dichromate and chromic acid. The process produces a pigment free from any by-products (such as chloride or sulphate) which would impair its efficiency when used in primers for metals. It is a greenish-yellow pigment with poor opacity. The main use of zinc chrome is in primers for steelwork and light alloys, and the effectiveness is associated with the small amount of soluble chromate which is part of its constitution.

Zinc tetroxychromate is manufactured by treating zinc oxide with chromic acid. It also is a greenish-yellow pigment with poor opacity, but it contains less soluble chromate than does zinc potassium chromate. The main use of zinc tetroxychromate is in etch primers (Chapter 17) but it is also used in conventional primers for metals.

Cadmium colours

This term covers a number of colours ranging from a pale primrose to lemon, orange, scarlet, crimson, and maroon. They are all produced by precipitation from cadmium sulphate solution. The paler shades consist of cadmium sulphide and are made by passing hydrogen sulphide gas through, or adding sodium sulphide solution to, the cadmium sulphate solution. For the deeper shades a mixture of sodium sulphide and sodium selenide is used to give a mixed precipitate containing cadmium sulphide and cadmium selenide. In all cases the precipitate is filtered, dried, and calcined.

Cadmium colours are bright with good lightfastness and outstanding heat resistance. It is for the latter property that they are most highly valued, and their main use is in heat-resisting paints. In common with other compounds of cadmium they are very expensive.

Titanium nickel yellow

This pigment is prepared by addition of salts of nickel and antimony to titanium hydroxide before calcination. It has a dull yellow colour but gives clean, bright tints on reduction with white. The resistance to heat is excellent and the principal use of this pigment is in heat-resisting paints.

Chemically prepared iron oxides

These range in colour from yellow, through red and purple-brown to black. They are all prepared from the same raw material, ferrous sulphate, which is obtained as a by-product in the galvanizing, tinplate and titanium dioxide industries.

Yellow oxide of iron (Ferrite yellow) is prepared by addition of a solution of alkali, e.g. sodium hydroxide, to a solution of ferrous sulphate to give a precipitate of hydrated ferrous hydroxide. This is then carefully oxidized in a current of hot air. The yellow oxide is a soft pigment, cleaner and brighter than ochre. It also possesses much greater tinting strength and so is widely used as a stainer. The lightfastness is of a high order and it is non-reactive toward paint media in general.

Synthetic red oxide of iron is produced by a modification of the precipitation process described for Yellow oxide. Sodium carbonate is used in place of sodium hydroxide, and by adjustment of the temperature and duration of the air-blowing, yellow-shade or blue-shade red oxides are obtained. These oxides are brighter and softer than the natural types but show the same permanence and chemical inertness. Like the natural oxides they are used in primers for steelwork, but for this use the synthetic materials must be checked carefully for freedom from chloride or sulphate ions. These can be present if the precipitated oxide has not been washed effectively and they can promote corrosion. Mixtures of synthetic iron oxide and zinc chrome are used in both air drying and stoving primers for steelwork. Red oxide is also used in finishing coats, and these show good durability as a result of the photochemical property of the pigment. The breakdown of paint films on exposure to sunlight is largely due to the action of ultra-violet rays on the organic binder. Red oxide of iron has the property of absorbing the ultra-violet radiations and so protects the binder.

Deeper and bluer iron oxides are produced by calcination of ferrous sulphate crystals (green copperas) and are known as Turkey reds, Indian reds and Purple browns. They tend to become progressively bluer and harder in this order.

Red lead

This is an orange-red pigment characterized by high specific gravity and low oil absorption. It is manufactured by oxidizing litharge (lead monoxide) by heating in a stream of air. The oxidation is never complete, a certain amount of litharge remaining unchanged. It is the amount of litharge present which determines the end use of red lead, as litharge reacts very readily with linseed oil and varnishes to form cements. For use in paints the litharge content is kept below 5 percent and this type is known as "non-setting" red lead as it will not set to a cement with linseed oil. Red leads with higher litharge contents are used for jointing and similar purposes.

The main use of red lead is in primers for steelwork, where the protective qualities are second to none. It works best in linseed oil, but as such primers dry and harden very slowly the linseed oil is often mixed with a proportion of alkyd resin. The protective action of red lead primers is associated with the presence of lead "soaps" produced by reaction between the pigment and fatty acids in the oil. Red lead primers based on other media such as chlorinated rubber and epoxy ester are also known.

Red lead is a highly toxic pigment and therefore, in spite of its efficiency, it is being replaced gradually by primers based on the non-toxic zinc chrome and zinc phosphate.

BLUE PIGMENTS

The three most important blue pigments in use today are Phthalocyanine, Prussian and Ultramarine blues; of these, phthalocyanine is by far the most widely used.

Phthalocyanine blue

The molecular structure of this pigment is very complex and the molecule contains an atom of copper (Fig. 10.1).

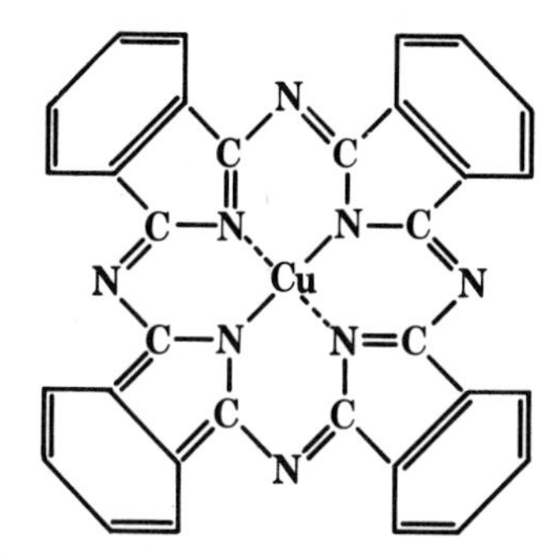

Fig. 10.1 Structure of phthalocyanine blue

Both green-shade and red-shade phthalocyanine blues are available and, in common with other organic pigments of complex structure, they are characterized by outstanding chemical and physical stability. Their tinting strength is very high and they are lightfast at all concentrations. They are insoluble in all paint media and withstand heat up to 200°C. Phthalocyanine blues can be regarded as the "universal" blues and have replaced Prussian blues in all but a few applications. When mixed with the organic arylamide yellows they produce a range of bright lightfast greens which are free from the inherent defects of chrome greens. These mixed greens are now used in all decorative green paints and many industrial types.

Prussian blue

This is an iron compound produced by precipitation using solutions of ferrous sulphate and potassium hexacyanoferrate (II) followed by oxidation. The precipitate is formed in a very fine state of division and, in order to make it as coarse as possible to help filtration, boiling solutions are used. Even so, the precipitate is difficult to filter and to wash free from soluble salts produced as by-products. Traces of soluble salts contribute to the hard nature of the dry pigment, making it difficult to disperse in paint media. Prussian blue is produced in "non-bronze" and "bronze" types. The latter gives a bronzy finish and is used in printing inks, the non-bronze type being preferred for paints.

Prussian blue is a bright pigment with high tinting strength, good lightfastness but poor opacity. Resistance to acids is good, but it is readily attacked by alkali and so cannot be used in paints for application to alkaline surfaces. In admixture with lead chromes, Prussian blue forms the well-known range of chrome or Brunswick greens (*below*). However, since the introduction of the phthalocyanine blues the use of Prussian blue in paints has declined considerably and very limited quantities are now used. The terms "Chinese" and "Milori" blue are sometimes applied to the best qualities of the pigment.

Ultramarine blue

The natural material, a mineral known as "lapis lazuli", has been used as a pigment for many centuries. It was used by the early scribes in their manuscripts and by medieval artists in fresco and tempera painting. The use of lapis lazuli as a paint pigment was discontinued many years ago, but small quantities continue to be used in artists' colours. The ultramarine blue in use today is a synthetic product made by calcining a mixture of sulphur, soda ash, siliceous material such as china clay, and carbonaceous matter such as coal. It is available in a range of tones which probably correspond to variations in composition. Ultramarine is one of the reddest blues available, and the colour is not easily matched with other pigments. As a consequence of the method of manufacture, the heat resistance is excellent. The pigment is resistant to alkali but is very sensitive to acids, and this may account for its poor reputation for lightfastness. Pale tints on ultramarine blue "fade" on exterior exposure, and this is probably due to attack by atmospheric acidity.

The pigment is not easily dispersed in paint media and has an unfortunate tendency to settle out in the can and to form a dense sediment. The amount used by the paint industry is relatively small, but this pigment is useful in coatings which must be resistant to lime and other alkalis.

GREEN PIGMENTS

In many of the bright green paints produced today the pigment is a mixture of phthalocyanine blue and organic (arylamide) yellow. Such a mixture, apart from being clean and bright, shows good lightfastness and satisfies the modern demand for pigments that are free from lead. Other green pigments in use are the following.

Phthalocyanine greens

These pigments have the same basic constitution as phthalocyanine blue (p. 157) but differ from the blue in containing either chlorine or both chlorine and bromine atoms in the molecule. Chlorine alone gives bluish greens, but a yellowish green is produced by the presence of both chlorine and bromine atoms. Phthalocyanine greens possess the same outstanding stability, lightfastness, heat resistance and tinting strength as phthalocyanine blue. They are not cheap pigments and are used in good-quality gloss paints.

Chrome greens

These are sometimes known as Brunswick greens and comprise a range of greens varying in colour from fairly light yellowish-greens to deep blue-greens. They consist of mixtures of lead chromes and varying amounts of Prussian blue. As mixtures they embody the properties of the two components. They are toxic and are attacked by alkali and so cannot be used in paints exposed to alkaline conditions. The opacity is good, but the wet paint films often show a deepening in colour before they are dry. This is caused by the Prussian blue rising to the surface and is known as "flooding". "Floating" or "flotation" is due to a similar cause, but in this case the surface shows a mottled appearance. The use of chrome greens in paints has decreased owing to their toxicity and tendency to flooding or floating. In decorative and many other types of paint they are being replaced by the brighter and non-toxic mixtures of phthalocyanine blue and organic yellow mentioned above.

Chromium oxide

This pigment is a sage-green colour and is dull and "chalky" compared with the chrome greens. The opacity is very high as also is the resistance to light, heat, acids and alkalis. On account of its dull colour it is little used in paints but finds wide use in colouring concrete, roof tiles and plastics.

BLACK PIGMENTS

Much of the black pigment used in paint manufacture consists of carbon either in a fairly pure state (carbon and vegetable blacks) or in admixture with inorganic matter (bone or drop blacks). Other blacks in fairly wide use are graphite (p. 149), black oxides of iron, and micaceous oxide of iron (p. 149).

Carbon blacks

These blacks are produced by burning petroleum gases or oils in a limited supply of air to give "soots". A range of blacks is available, differing in intensity of "blackness". The intensity appears to be related to particle size, which is of a very low order (10 to 80 millimicrons). As a result of this fine particle size, carbon blacks possess very high oil absorption and tinting strength. Black surfaces absorb all the incident light, and this, together with the very fine particle size, means that very little pigment is required in black paints to give complete opacity. A 5-litre can of black gloss paint will weigh only about 4·75 kilograms, which is roughly the same density as the paint medium.

Carbon blacks are completely fast to light and chemicals and when reduced with white produce greys with a brownish tint. For tinting purposes the vegetable black (*below*) is preferred. Carbon blacks can be safely used in most types of paint, but black air-drying finishes can become very slow-drying on long storage as a result of adsorption of part of the drier metals by the carbon black. A further addition of liquid driers is required to restore the drying properties.

Vegetable (lamp) black

As the name suggests, vegetable blacks were originally made by burning certain vegetable oils in a limited supply of air. This pigment is now made by burning creosote and waste oils in furnaces of special design. The term "lampblack" was formerly used for the poorer grades of vegetable black, but today both terms are used for the one product.

Vegetable black is grey and dull in comparison with carbon black. It is not employed as the principal pigment in black paints, but it possesses high tinting strength and, when reduced with white, gives clean bluish tints. The principal use of vegetable black is therefore as a stainer.

Bone (or drop) black

This pigment is produced by calcination of degreased bones. The quality is assessed on the carbon content which can vary from 5 to 20 percent, the balance consisting mainly of calcium phosphate. The carbon content determines the colour and tinting strength. Bone black mixes more readily with oils and water than does carbon black, but it is similar to carbon black in that it tends to adsorb drier metals from the paint medium. The addition of liquid driers will correct the fault, and the paint will then dry in the normal time. Bone black is used in fillers and some types of undercoat.

Precipated black oxide of iron

The manufacture of this pigment is on similar lines to that of the yellow and red oxides except that the conditions of oxidation are altered to give the black type. It has a soft texture and moderate tinting strength. It is used in fillers, stoppers, and some black undercoats.

METALLIC PIGMENTS

Aluminium

Finely divided aluminium powder appeared at the beginning of the century and its use in paint has steadily increased. It is a very dusty material and, wherever possible, manufacturers of paint prefer to use the paste form in which the powder has been damped with a solvent.

Metallic aluminium pigment is an interesting material as the particles are in the form of small flakes with a bright metallic lustre and they possess the property of "leafing" in paint media. The flakes become arranged with their flat sides parallel to the surface of the paint film, and they lie immediately under the surface. The flakes overlap and in this way they increase the resistance of the film to penetration by moisture. This type of aluminium pigment is known as the "leafing" grade, but there is also a "non-leafing" grade the flakes of which become distributed at random throughout the paint film. The leafing grade is used in both decorative and heat-resisting aluminium paints. By the use of silicone resins as binders it is possible to produce heat-resisting aluminium paints which will withstand temperatures up to 500°C. The non-leafing grades are used in wood primers and in some industrial coatings such as hammer and polychromatic finishes.

Lead

Metallic lead pigment is manufactured in a fine state of division, but the dry powder has a very great affinity for oxygen. It is therefore always supplied to the paint manufacturer in the form of paste. This can be in water, plasticizer, linseed oil or other medium. If the paste in water is thoroughly mixed with an organic paint medium, the metallic lead migrates to the paint medium and the water separates. This process, known as "flushing", enables the paint manufacturer to use media of his own choice. The main use of the pigment is in corrosion-inhibiting primers for steelwork where it can be used in a variety of media. The manufacture and use of metallic lead primers are, of course, subject to the Lead Paint Regulations (see Chapter 19 and Appendix 2).

Zinc

Metallic zinc pigment is marketed as a blue-grey powder under the name "zinc dust" and is manufactured by vaporizing spelter (commercial metallic zinc) and conveying the vapour into large condensers. The conditions are such that the metal condenses as a fine powder. The main use of zinc dust is in primers for the protection of steelwork. These primers contain between 92 and 95 percent of metallic zinc in the dry film and are known as "zinc rich", but there is another variety of metallic zinc primer which contains only 80 percent of pigment in the dry film. The small amount of binder present in zinc rich primers is usually of the non-oxidizing type and the following types are in use — polystyrene, chlorinated and isomerized rubber, epoxies, acrylics, and vinyls. Zinc dust is also used in "inorganic zinc silicate" primers in which the medium is a silicon ester. These primers will be considered in more detail in Chapter 19.

Bronzes

This term covers a range of copper or copper alloy pigments some of which are akin to the brasses. The paler and brighter shades are marketed under the name "gold powder". The particles are in the form of flat plates, similar to aluminium, and possess leafing properties. However, the specific gravity of the bronzes is higher than that of aluminium and consequently the leafing is less pronounced. The bronzes are very sensitive to small amounts of acidity in paint media and also tarnish rapidly on exterior exposure. They are used for interior decoration, usually dispersed in solutions of polymers such as polyvinyl acetate or polystyrene.

LUMINOUS PIGMENTS

This term covers two types of pigment which differ in the way they react to illumination.

Fluorescent pigments

In these pigments the luminosity exists only as long as the paint is exposed to the exciting radiation and disappears on its removal. They are activated by daylight and are used in high-visibility paints. A widely used type contains a fluorescent dye dissolved in a hard resin which is then powdered and used as a pigment. The paints are translucent and so are applied over a flat ground coat. The performance of the paints can be improved by a coat of protective lacquer, but the exterior durability is limited to about 12 to 18 months.

Phosphorescent pigments

These pigments are specially prepared sulphides of certain metals, and paints containing them retain an "afterglow" after removal of the exciting radiation. The colour emitted depends on the nature of the metal sulphide used and is of limited duration. Paints made from these pigments contain the minimum amount of binder and have limited durability.

ORGANIC PIGMENTS

A variety of brilliant dyes and pigments can be obtained from plant and animal sources; they were employed in various forms for many years. Their main drawback was lack of permanence, and they have been superseded by the synthetic types.

The modern organic pigments offer a wider range and a greater variety of colours than do the inorganics. Although some are inferior in opacity, they possess good staining power, softness of texture, and brilliance. Some of the simpler types have shown poor lightfastness on reduction, although satisfactory in the pure condition, but developments have led to more complex types with good lightfastness at all levels of reduction. Phthalocyanine blues and greens are examples of pigments with complex molecular structures. The newer types possess very high staining power, and this tends to offset the high cost. Some of the organic pigments contain metals, but a great many are metal-free and these do not present the toxic hazard associated with some of the inorganics, e.g. lead chromes and chrome greens.

Classification

The traditional classification of organic pigments has been into pigment dyestuffs, toners, and lakes. Whilst the classes known as toners and lakes are well defined, there are a number of important modern pigments which are not pigment dyestuffs in the classical sense, such as the phthalocyanine blues and greens. In the present treatment the organic pigments are divided into colour groups, but the number of pigments in each group is such that only a few of the most important can be mentioned. Toners and lakes are defined at this point to avoid repetition later.

Toners are precipitated from solutions of soluble dyes by addition of solutions of metallic salts. Examples are Permanent Red 2B and Lithol Rubine BK. Common metals used are calcium, barium, and manganese.

Lakes If an extender such as barytes, blanc fixe or china clay is mixed with the dye solution before addition of the metal salt solution, the toner is precipitated on the extender. Such precipitates are known as lakes. Lakes have much lower tinting strength than the toners and are not much used in paint manufacture. The term "lake" is sometimes erroneously applied to mixtures of other types of organic pigment with extenders.

Yellows

The best-known and most widely used are the arylamide yellows (which are sold under trade names but are recognizable by the appendage letters G, 3G, 5G and 10G). They are members of the azo class and are characterized by brightness of colour, good lightfastness and tinting strength, with soft texture. On account of their moderate price and non-toxic nature they are used as replacements for the lead chromes. However, difficulties often arise on account of their poor opacity. Other yellow pigments are the benzidines, which have greater tinting strength than the arylamide yellows but only moderate lightfastness, and the more complex flavanthrone yellow in which both tinting strength and lightfastness are of a high order.

Orange

The azo class provides the two most commonly used orange pigments — Permanent Orange G and Permanent Red 2G. They both possess good lightfastness and tinting strength but suffer from poor opacity. Perinone orange is more complex in structure and shows very good lightfastness and resistance to solvents and heat. It is used in motor-car finishes.

Reds and Maroons

This group comprises a wide range of pigments. In the azo class alone, colours range from scarlet to maroon and there is considerable variation in lightfastness. Some of the simpler members are satisfactory in this respect in the pure state, but the tints produced by reduction with white tend to fade rapidly. The purely organic types show poor opacity, but there are a number of toners showing good opacity and moderate to fairly good lightfastness. The following pigments are more complex, and this is reflected in high tinting strength and lightfastness: Quinacridones (red to reddish-violet shades), Perylene maroon, and Thioindigo Bordeaux.

Blues and Violet

The number of paint pigments in this group is relatively small and consists of phthalocyanine blue (page 157), blue toners made from soluble phthalocyanine blue and the Indanthrone blues. They all possess excellent lightfastness and tinting strength. Phthalocyanine blue is the most widely used.

Carbazole violet is the most important pigment of this colour and has very high tinting strength and excellent lightfastness, both in the pure state and in tints.

Greens

The green pigment of most importance is phthalocyanine green (page 158). Other greens are produced by admixture of the appropriate yellow and blue.

NOTE: There are no white organic pigments, and carbon blacks satisfy practically all black requirements.

11

Vehicle Types, Oils and Driers

The vehicle or *medium* is the liquid constituent in paint which serves to carry the pigment in suspension until it is deposited on to the required surface by brush, spray or other means. Once the paint is thus applied in a thin film, the function of the vehicle is then to act as a *binder*, i.e. to hold the particles of the pigment firmly to the surface in a cohesive film. It will be seen that the vehicle must (a) support the suspended pigment without undue settling or detrimental chemical reaction; (b) provide sufficient fluidity for easy application by brush or spray; (c) dry to a tough, elastic skin which effectively binds the pigment in a solid film, and (d) have good adhesion, enabling the film to become firmly attached to the surface on which it is applied. The paint vehicles in use today can be divided into *convertible* and *non-convertible* types, and the important members are set out in the following table. The individual members are described in Chapters 14 and 15 respectively.

Types of Paint Media

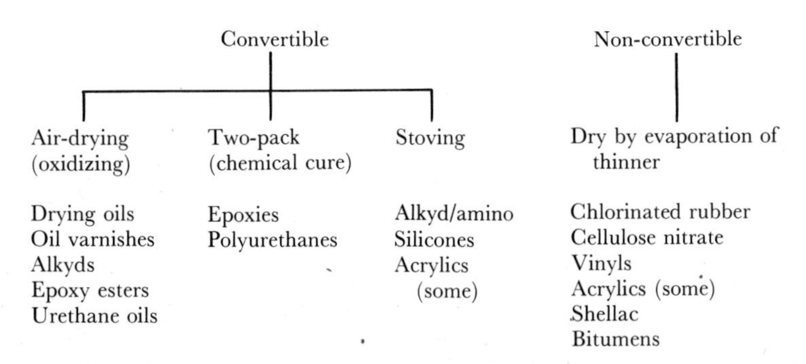

When the convertible type of vehicle or medium is applied to a surface, the thinner first evaporates and the binder which remains then undergoes a chemical reaction and becomes solid (cures). The solid film is a different chemical substance from the original oil or resin and does not dissolve in the original thinner. The chemical reaction can take one of three forms: (a) the absorption of oxygen from the air, which is what takes place with the normal type of decorative paint; (b) with two-pack (cold cure) materials such as epoxies or polyurethanes the two packs or components are mixed before use and a chemical reaction takes place in the film; (c) by heating or stoving the painted article. When the article is stoved, hardening or curing takes place very rapidly, but special types of vehicle have to be used; stoving types are used mainly for industrial finishes.

Non-convertible vehicles dry simply by loss of the volatile thinner. No chemical reaction takes place in the film, and this remains soluble in the thinner in which it was originally dissolved. The film is not converted to any other chemical form. Apart from emulsions (pages 196–7), non-convertible vehicles have limited use in decorative work.

DRYING OILS

Very little paint is manufactured today in which the binder is solely a drying oil. Linseed oil continues to be used in some primers such as red lead, but the amount of this type of primer used is decreasing rapidly. Treated forms of linseed oil, e.g. boiled oil and stand oil, are often used in other types of paint to improve flow properties, whilst both linseed and tung oils are used in the manufacture of varnishes. Linseed and tung oils, as well as the semi-drying oils such as soya bean, safflower, sunflower and tall oil, are sources of the fatty acids used in the manufacture of alkyd resins.

Linseed oil

Linseed oil is expressed or extracted from the seeds of the flax plant, grown chiefly in Russia, India, the Argentine and, to a smaller extent, in Great Britain.

The qualities used in paint are produced mainly by hot pressing and, in a lesser degree, by extraction with solvents. In the first method, the seeds are crushed to a meal, heated to facilitate pressing and to increase the yield, then pressed in hydraulic presses to extract the oil, the residual oil-cake being used as cattle food. Cold pressing gives a lower yield, but produces a better quality oil which is often used in artists' colours. Extraction by petroleum solvents gives a higher yield but the oil is inferior in colour and also has a peculiar odour.

The crude oil is dull and cloudy (due to the presence of mucilage) and requires tanking or refining to clarify it. When filtered and stored in large tanks the mucilage or "foots" settles to the bottom and the oil gradually clears; this is the familiar *raw linseed oil*. If cloudy it should be rejected, unless one is prepared to store it for several months (the longer the better) until all mucilage has settled out and the oil becomes clear; only then is it fit to use in paint. The original method of refining was to expose the oil to sunlight for some months in shallow, glass-covered tanks, the oil being bleached thereby; this produced a pale oil of superior quality, greatly valued by artists but too costly for general industrial purposes. Oil intended for varnish-making is specially refined and in modern practice the oil is heated until it "breaks", when the mucilage coagulates and settles out; it may then be treated with fullers' earth, caustic soda, or sulphuric acid, finally to clarify the oil. Another method of clarifying is by means of stream-line filters. Artificially refined oil is referred to as *refined linseed oil*.

Properly tanked raw linseed oil should be clear, unclouded and free from "foots" and one litre should weigh 0·93 kg (2·05 lb). It is readily identified by its characteristic smell and clear, limpid, yellow-brown colour. Linseed oil dries by *oxidation*, that is to say, it absorbs and combines with oxygen in the atmosphere, so that when brushed out on to a clean surface it gradually changes from a liquid coating to a dry film or skin (this dried substance being technically known as *linoxyn*). Normally, raw linseed oil takes about four days to become surface dry which is too slow for practical purposes, therefore driers are added to hasten the drying.

White paints and pale shades containing linseed oil tend to yellow slightly on exposure, but the yellowing is not noticeable with deeper colours.

Linseed oil is saponified (turned into soap) by the action of alkalis and it is therefore unsafe to use paint containing linseed oil on new Portland cement and lime plaster surfaces which are actively alkaline (p. 223).

Boiled linseed oil

Boiled linseed oil is produced by heating raw or heat-refined oil to a high temperature for several hours and adding a small quantity of driers. The oil becomes much darker on heating and assumes a thicker consistency. By adjusting the conditions of heating and the type of driers used, either "dark boiled oil" or "pale boiled oil" can be made.

The film produced by boiled oil has more body and gloss than that of raw oil and is more resistant to moisture. It also dries more rapidly to a harder surface and therefore requires less added drier. Boiled oil is added in small amounts to several types of paint to improve flow and ease of brushing.

Stand oil

Stand oil is a thickened or "bodied" oil generally made from alkali-refined linseed oil, but mixtures of linseed and other drying oils such as tung can be used. The oil is maintained at a high temperature for a prolonged period, without the addition of driers, until it thickens to the required consistency from "light" to "heavy", the latter approximating to the flow of honey (this thickening is technically known as polymerization). When mixed with driers, stand oils give films of great elasticity but they tend to be soft. They can be blended with resins to give harder films with good gloss and durability. Stand oils are used in varnishes and are added to some types of decorative undercoats to improve levelling properties.

Tung oil

This oil, which is also known as *wood oil*, is obtained from the nuts of two species of tree of the *Aleurites* family which are grown in Malawi, Argentine, the United States and China. Raw tung oil is pale amber in colour and has a rather unpleasant taste and smell. It cannot be used in its raw state and requires very careful heat treatment to render it suitable for paint and varnish media. When heat-treated it is a better drying oil than linseed and is more resistant to alkalis. In association with hard resins it is used in quick-drying enamels, alkali-resisting primers, and boat or spar varnishes. The latter require the high resistance to moisture which is a property of tung oil varnishes.

Semi-drying oils

These oils are slower drying than linseed but give films which have little or no tendency to yellowing on exposure. They are not used as oils but their main use is in the manufacture of non-yellowing alkyd resins which, however, do not show the slow-drying properties of the oils.

Soya bean oil

The oil is extracted from the ground beans of the soya plant by means of solvent which is afterwards removed by distillation. The extracted oil is treated to remove impurities before being used in alkyd manufacture. Finishes based on alkyds derived from soya bean oil are now widely used in the decorative field. They show good brushing and levelling properties, are non-yellowing and have good durability.

Dehydrated castor oil

Untreated castor oil is non-drying and is sometimes used as a plasticizer, but when processed by chemical or heat treatment it becomes a very useful drying oil. Dehydrated castor oil is considered to have properties intermediate between tung and linseed. The dried film is very durable and flexible, having good water resistance and chemical resistance when combined with phenol-formaldehyde resins. Mixed stand oils are produced from castor, linseed and other drying oils.

Tall oil is obtained as a by-product in the treatment of wood pulp for paper manufacture. It is carefully purified and is then used to manufacture non-yellowing alkyd resins.

Safflower seed oil and *Sunflower seed oil* are also used to produce non-yellowing alkyd resins.

DRIERS

The drying of oil paint is of utmost economic importance to the painter, for he usually aims to apply one coat each day and relies on undercoats drying hard enough overnight to stand light sandpapering and recoating next day. When the paint fails to dry, the programme of work is upset, with all the attendant annoyance and inconvenience. The necessity for overnight drying therefore often leads to the excessive use of driers, with consequent detriment to the paint film. Rapid surface drying, for instance, may cause the paint film to skin over, thus denying oxygen to the oil beneath and so preventing its proper hardening; the elasticity of this soft underlayer may ultimately cause the hard surface skin to crack or, on the other hand, it may "sweat" through to form a permanently tacky surface.

Linseed oil dries by *oxidation* (i.e. by absorbing oxygen from the air), but this process is so slow that it takes about four days for a thin film to become surface dry. By adding certain drying agents, however, the process of oxidation is accelerated, and this makes it possible to adjust the rate of drying to a more convenient period, such as 12 to 16 hours (which is considered normal overnight drying). But oxidation does not cease when the film becomes dry and hard; it continues, very slowly, throughout the "life" of the film which, in consequence, gradually loses its elasticity until it eventually becomes brittle and begins to disintegrate, or perish. When, therefore, the initial drying period is hastened unduly, it may be expected that the life of the film will be considerably shortened, and it is true to say, in general, that quick-drying oil paints and varnishes are less durable than those which dry more slowly*.

The drying agents used are compounds of certain metals, mainly lead, cobalt, and manganese, combined with organic acids to render them soluble in paint media. Lead dries from the bottom and hardens uniformly throughout the film, whereas manganese and cobalt both commence drying at the surface; the paint manufacturer therefore uses a mixture of driers to ensure uniform drying of the film. The compounds used are naphthenates (from naphthenic acid, a by-product of petroleum refining) and synthetic octoates. The latter are paler and do not possess the unpleasant smell of the naphthenates; they are preferred for use in low-odour and odourless paints.

With modern types of alkyd finishes the quantity and ratio of the drier metals is critical. The correct amount is put in by the paint manufacturer, and no further additions should be made without reference to the manufacturer.

* This principle, however, does not necessarily apply to media such as the quick-drying cellulose and synthetics. Cellulose, for instance, dries by rapid evaporation of its solvents, whilst the speedy drying of synthetic finishes is largely governed by polymerization of the drying oils and the special characteristics of the synthetic resins, some of which do not "air dry" but are "set" chemically by catalyst additives. Moreover, both cellulose and synthetics usually contain certain plasticizers which remain more or less permanently in the dried film, thus imparting sufficient elasticity to prevent brittleness and early perishing. Quick drying is therefore achieved without having to sacrifice durability.

12

Solvents and Thinners

To render paint and varnish more fluid and workable in consistency, diluents or thinners are added in varying proportions according to the type of material and the degree of gloss required. Although they cannot be regarded as drying agents, thinners do assist the drying of paint in some degree by allowing it to be spread over a large area of surface in a thinner film, thus exposing a given quantity of oil to a greater amount of oxygen. Thinners also assist the penetration of oil into absorbent surfaces, allowing the paint to knit well into the surface to secure a firm grip. This is an important point in the formulation of priming paints for woodwork.

Whilst readily dissolving any oil or varnish with which it is mixed, the thinner should not act as solvent to the dried film, it being very important that undercoats are not softened up by superimposed coatings. This principle, however, does not apply to cellulose and certain synthetic materials designed for spray application.

Thinners of good quality should be colourless and almost completely volatile, that is to say, they should slowly evaporate and form no part of the dried film. A spot dropped on white blotting or filter paper should evaporate in about half an hour or so, leaving no oily stain or noticeable residue.

The number of volatile solvents and diluents available to the paint manufacturer has greatly increased during recent years owing largely to the development of cellulose, rubber and synthetic materials; these are dealt with at length by W. M. Morgans in *Outlines of Paint Technology* and by H. W. Chatfield in *Varnish Constituents*. The main solvent used in decorative paints is white spirit, but in some cases of difficult resins the white spirit is mixed with a proportion of stronger solvent such as xylol, dipentene, or special petroleum solvent.

White spirit

This is a petroleum distillate with a rate of evaporation between that of petrol and paraffin. For a number of years it was regarded as a "turpentine substitute" and this term has persisted to the present day. Genuine white spirit should comply with British Standard 245; any material sold as "turpentine substitute" and not bearing the BS number on the label is probably a cheaper grade of distillate with inferior solvent power.

White spirit has a mild but recognizable odour and should be water-white in colour. The flash-point should be above 34°C (93°F). It has moderate solvent power, and with some media — for example, the medium or short-oil alkyds used in quick-drying paints — it is mixed with a proportion of a stronger solvent such as xylol (*see below*). It is always advisable, therefore, to refer to the manufacturer before thinners are added to modern types of paint. Low odour and odourless paints are thinned with a specially purified grade of white spirit.

High aromatic solvents from petroleum

A number of these solvents are marketed under trade names, e.g. Aromosol (ICI), Shellsol, and Solvesso, and they cover a wide range of flash-points and evaporation rates, some of which are similar to those of white spirit. They have strong solvent properties, similar to the older coal-tar naphthas which they have replaced, but their odour is much more pleasant. They are used as solvents for chlorinated rubber and, on occasions, to strengthen the solvent power of white spirit.

Xylol is an aromatic solvent with a characteristic odour and a flash-point of about 24°C (76°F). It is very flammable and, in common with other aromatics, the burning material produces vast quantities of soot. Xylol is a good solvent for a wide range of resins but it evaporates much more rapidly than white spirit. In consequence of the rapid evaporation rate it is little used in conventional decorative paints to be applied by brushing; it is, however, a very important solvent in spray-applied industrial finishes, both stoving and rapid air-drying. Some non-convertible maintenance paints, such as chlorinated rubber, contain a high proportion of xylol in the thinner.

Dipentene is obtained by a distillation process similar to that used to produce turpentine from the resinous materials present in pine and other coniferous trees. (Turpentine is no longer used in the paint industry, but it continues to be marketed by artists' colourmen.) Dipentene has the smell which is characteristic of terpene compounds; it is an excellent solvent for many synthetic resins and is often used to augment the solvent power of white spirit. It possesses some anti-skinning properties and, in consequence of a low rate of evaporation, it is used to increase the wet-edge time of decorative finishes and undercoats.

Alcohols

The alcohols form a very important group of solvents in the manufacture of cellulose nitrate and other lacquers, but the painter is hardly likely to use any of these except perhaps in the form of special thinners supplied by the manufacturer.

Industrial alcohol is produced either by fermentation of grain followed by distillation, or by a chemical process from ethylene gas. Most of the industrial alcohol is now made by the latter method, as ethylene is available in quantity from the petroleum refineries.

Methylated spirit is ethyl alcohol which has been adulterated or "denatured" by adding a small proportion of mineral or wood naphtha together with a little methyl-violet dye to render it unfit for drinking and to free it from excise duty. It is used as a solvent for shellac in spirit varnishes, French polish and patent knotting.

Methylated finish contains a small proportion of rosin or shellac which enables it to be sold without a licence. It is used for thinning spirit varnishes and for spiriting off in French polishing.

Industrial spirit is available only for manufacturing purposes. It is similar to methylated spirit except that the methyl-violet is omitted and it is less adulterated.

Other solvents

Ethyl and butyl acetates belong to a group of solvents known as esters, which are the result of the chemical reaction between alcohols and acids. They are used in cellulose nitrate lacquers and in two-component systems.

Methyl ethyl ketone (MEK) and *methyl isobutyl ketone* (MIBK) are extremely strong solvents for a number of resins. They are used in two-component and other synthetic types.

13

Varnishes

The craft of varnish-making grew up over many centuries, the "know-how" being guarded very jealously and frequently passing from father to son. Natural drying oils (mainly linseed) and fossil resins were used in varying ratios according to the type of varnish required. These fossil resins varied in quality, and considerable experience was required in selecting the most suitable grade for a particular varnish.

The most widely used fossil resin was *copal* and the name persists in "copal varnish" which continues to be made, though on a very small scale; in fact, very little fossil resin is now used in varnishes. These resins have been replaced by the synthetic types which are more uniform in composition and, in most cases, cheaper; with them, varnish-making has become a scientifically controlled operation and has ceased to be a craft. (For more detailed information on fossil resins and their processing the reader is referred to *Natural Varnish Resins* by T. Hedley Barry or to *Varnish Constituents* by H. W. Chatfield.)

Varnish-making

This is a relatively simple operation as most of the synthetic resins used today are soluble in drying oils at a comparatively low temperature. With linseed oil, it is usual to heat the resin with part of the bodied linseed oil until the mixture is clear. The remainder of the oil is then added in portions and the heating continued until a sample gives a clear solution in white spirit. It is then cooled, thinned with white spirit to the desired solids content, and finally the driers are added.

Some of the softer types of modified phenolic resin are soluble in white spirit at room temperature. These can be made into varnishes by dissolving in the solvent (a process known as "cold cutting") followed by addition of the bodied linseed oil and driers.

The manufacture of varnishes from tung oil is not as simple as with linseed oil. On account of its greater reactivity, tung oil has a tendency to gel unless the temperature and rate of heating are carefully controlled. (For a more detailed treatment of linseed and tung oil varnishes the reader is referred to *OCCA Paint Technology Manuals*, Part 3, or to *Outlines of Paint Technology* by W. M. Morgans, vol. 1 (Charles Griffin & Co. Ltd)).

Oil varnishes

The number and variety of varnishes is very great indeed, suitable types being made for almost every specific purpose. The quality and character is dependent on the type of resins and oil selected and the proportions used, the oil being responsible for elasticity and durability of the film, the function of the resins being to impart gloss and hardness. Varnish dries by evaporation of the volatile thinner and by oxidation of the drying oil, so that the drying time of a varnish depends largely on its oil content; when this is high, the drying tends to be slow and vice versa.

Oil varnishes are referred to as "long-oil" or "short-oil" according to their resin–oil content. Long-oil varnishes have a higher proportion of oil than resin and are therefore slow-drying, durable and elastic, withstanding long exposure and considerable changes in temperature. All *exterior varnishes,* finishing carriage varnishes, etc., fall into this class, the proportion of oil to resin being as much as 3 or 4 parts oil to 1 part resin (by weight). They should become "touch-dry" in eight to ten hours, and dry in 18 to 24 hours. Short-oil varnishes, on the other hand, have a greater preponderance of resin and are consequently harder and quicker drying, having more brilliant gloss but less durable films.

Traditional interior varnishes are of this type, but today many interior varnishes are based on synthetic media such as polyurethanes (page 302).

Boat and spar varnishes (so called because originally used on ship spars) possess maximum water resistance, being made with tung oil and 100 per-cent phenolformaldehyde resin. Such varnishes also possess good resistance to attack by alkali.

Mixing varnishes are usually long or medium oil-length varnishes of low acid value. They are used alone or mixed with alkyds as vehicles for undercoats or primers, and generally contain linseed and tung oils with maleic resin for pale media and phenolformaldehyde for darker colours. They can be used alone as general utility varnishes.

Flat (or matt) and eggshell varnishes are made by incorporating flatting agents (such as extremely fine silicas of very high oil absorption) into medium or short oil varnishes. They may require agitating before use to ensure that any ingredients which may have settled are thoroughly dispersed. Incidentally, this type of varnish makes an excellent medium for decorative painting and murals when used in conjunction with tube colours or flat oil paints.

Goldsize is really a short oil varnish with a high drier content. It owes its name to its use as a "mordant" or adhesive for golf leaf. Goldsize is also used as a binder for fillers and stoppers.

Varnish resins

The resins described here are the hard synthetic types which are used with oils to produce varnishes. In a few instances, rosin (a natural substance) is incorporated so that these resins should be regarded as semi-synthetic. (Other synthetic resins which are used as the sole binders for paints are described in Chapter 14.)

Phenolformaldehyde resins

When phenol (perhaps better known as carbolic acid crystals) is treated under suitable conditions with an aqueous solution of formaldehyde in the right proportions either a solid or a syrup-like resin can be obtained. Baekeland, the inventor of Bakelite, was one of the first to adapt these resins for lacquering purposes, but being essentially of the thermohardening type they needed to be stoved in order to obtain the final finish; they were much in vogue years ago for lacquering brass surfaces, since their defect of yellowing had no distracting influence upon the appearance of the finish. Because of their unique property of setting under heat and pressure, Baekeland later developed from these types of resins the Bakelite moulding powders which led to the establishment of an important branch of the plastics industry, but although so useful in the direction outlined above, these resins in their hundred-percent form are insoluble in drying oils.

In view of the very resistant properties of the stoved lacquers that were obtainable from these resins, it was only natural that further investigations were pursued to ascertain how they could be modified to make them oil-soluble, and it was then discovered that by cooking the syrup-like resin with ordinary rosin at a fairly high temperature in the presence of a sufficiency of glycerine (to neutralize the acidity of the mixture) oil-soluble resins were produced; these are now usually referred to as reduced or modified phenol-formaldehyde resins in order to distinguish them from those resins discussed in the next group below. It should, however, be fully appreciated that this combination of the resin with wood or gum rosin is not a form of adulteration but is an essential processing method which enables oil-solubility to be achieved. Though most of these types of resins can be cooked with linseed oil or stand oils to give tolerably good varnish media, their outstanding properties are best brought to the fore when they are cooked with tung oil, or, at least, where preponderating amounts of the latter are used in the varnish oils employed. Generally speaking the yellowing property of this type of resin does not encourage its use in the better class of varnishes, but nevertheless, they find much scope in the production of goldsizes and undercoating media.

100 percent phenolformaldehyde resins

Although ordinary phenol when combined with formaldehyde is not capable of forming an oil-soluble resin, certain modified forms of phenol are able to combine with formaldehyde and fulfil the conditions of oil solubility, producing the so-called 100 percent phenolic resins which are completely free from rosin. These resins are naturally more expensive than the reduced resins discussed above, but varnishes prepared from them, particularly those cooked with tung oil, have exceptionally good resistance to alkalis and to corrosive conditions such as are experienced in industrial and sea atmospheres. These varnishes are very quick drying since film formation is due more to a process of polymerization than to oxidation, but this speed of set-up should not be interpreted as indicating the lack of durability which some decorators are prone to associate with quick drying. On the contrary, these varnishes are very durable and are much in demand for conditions which are difficult and severe for ordinary surface coatings.

Ester gums

An ester is the result of the chemical reaction between an alcohol and an acid. Ester gums are rosin products which are used extensively in the varnish industry; they are prepared by treating or (as it is technically known) esterifying ordinary rosin with glycerol or pentaerythritol for the purpose of neutralizing the acidity of the rosin, thus making it harder and more compatible with pigments. Ester gums are usually made into varnishes with tung oil. These varnishes dry rapidly to give tough, hard-wearing films and are used in machinery enamels.

Rosin (or colophony)

This is the solid residue from the distillation of dipentene and turpentine. It has a very high acid value which renders it water-sensitive and unsuitable for mixing with many pigments. The acidity can be neutralized by conversion to ester gum (*above*) or by heating with lime to give "lime-hardened rosin". Both these products give useful varnishes with tung oil.

Maleic resins

When maleic anhydride and glycerol are reacted with rosin at a suitable temperature, a very pale ester-type resin of low acid value is formed. It is used with certain drying oils to produce very pale varnishes of good durability. These varnishes are used in metal decorating and are also added to alkyd type gloss finishes, where they are considered to improve the gloss and gloss retention.

Coumarone-indene resins

These resins derived from coal-tar distillates by direct polymerization under the influence of a suitable catalyst. They are soluble in drying oils and provide varnish media of low acid value. They are usually blended with tung oil, and such varnishes are alkali-resisting and do not react with metals; hence their suitability for metallic bronze paints.

14

Synthetic Convertible Coatings

For some time after World War II the word "synthetic" was viewed with a certain amount of mistrust because of its unhappy association with wartime substitutes for the real thing, but nowadays the term is applied to a wide range of resins produced by what is a branch of the chemical industry.

There is no clear line of demarcation between the synthetic types of paint and the older conventional types based on oil varnishes. Some synthetics, such as epoxies and polyurethanes, are made entirely from chemicals, but alkyds, although classed as synthetics, contain fatty acids derived from natural oils. Again, in modern oil varnishes the most widely used resins are the phenolformaldehyde which are synthetic.

In this chapter we are concerned with those synthetic resins which are not made into varnishes with drying oils but which are themselves used as paint media. They are produced under close scientific control and can be "tailor-made" to meet any specific requirements, for example, quick drying, chemical resistance, elasticity, hardness and so on. It should be remembered that synthetic resins can be made quick drying without necessarily a shorter life as with many oil varnishes in which quick drying is achieved by reducing the oil content, with a consequent loss in elasticity and a tendency to brittleness. Most of the decorative paints in use today are of the synthetic type and there is therefore a tendency to drop the term.

During the past half-century enormous strides have been made in the development of synthetic resins, and in addition to the original phenolics and alkyds we now have the large vinyl group of resins, epoxy resins, polyurethane resins, chlorinated rubber, cyclized rubber, silicones, etc. Perhaps the greatest advance has been made in the production of industrial finishes for cars, washing machines, refrigerators and other household goods, but we are concerned here mainly with those products normally applied by the decorator, as discussed in some detail below.

Alkyd resins

These form the largest group of resins used by the paint industry. They are manufactured by reacting a polyhydric alcohol, such as glycerol or pentaerythritol, with a polybasic acid (e.g. phthalic anhydride) and fatty acids of linseed or other oils. Acids from drying or semi-drying oils such as linseed, tung, tall oil, and soya yield alkyds used in air-drying paints, whilst acids from coconut and castor oils give non-drying alkyds which are used in stoving finishes. The products are not the familiar hard resins but very thick syrups which are supplied to the paint manufacturer as solutions in solvent. The long-oil types are thinned with white spirit and are used as the medium for brushing gloss, semi-gloss and flat paints. Short-oil alkyds are thinned with xylol and are used in quick-drying or stoving finishes to be applied by spray.

Most gloss finishes for decorative, maintenance and marine work are based on long-oil alkyd resins owing to their generally acknowledged durability and weather-resistance, gloss, and colour retention. They have good flexibility and flow and, although they are fairly easy brushing, they tend to set up rather quickly in warm weather which reduces their wet edge time; this calls for good organization and speedy application on large areas.

Finishes based on alkyd resins are sensitive to moisture during the drying period (4 to 6 hours depending on temperature) and, to a lesser extent, when dry. If the dried film remains in contact with water for long periods, swelling occurs leading to blistering. Alkyds should, therefore, not be used on surfaces which are to be immersed in water for long periods.

Alkyd resins modified by reaction with styrene or vinyl toluene are used in the manufacture of fast-drying paints. They are also less affected by moisture during drying than are the straight alkyds. Styrenated alkyds allow drying times as short as half an hour to be obtained, but since they are carried in a fast evaporating solvent (xylol), brush application is difficult on anything except very small areas. Vinyl-toluenated alkyds, on the other hand, are soluble in white spirit and therefore allow paints to be made that have brushing properties similar to those of orthodox gloss paints, yet still drying in one to two hours.

By reacting alkyds with certain types of polyamide resin, media are produced with strong "thixotropic" properties. These thixotropic alkyds are described in Chapter 17.

Epoxy resins

These are polymers obtained by reaction between epichlorhydrin and diphenylol propane, both of which are by-products of petroleum refining. The reaction can be so controlled to give a range of products varying from hard resins to syrupy liquids.

Epoxy esters are made by reacting an epoxy resin with the fatty acids from drying oils for air-drying finishes and from non-drying oils for stoving finishes. When based on the fatty acids from castor oil, epoxy esters produce excellent stoving finishes of great flexibility, and these are used on collapsible tubes.

The air-drying types resemble alkyds to some extent but show greater adhesion and chemical resistance. They dry by oxidation, can be applied by spray or brush (although the set-up is quicker than with conventional brushing paints), and can be applied over old paintwork without softening it. Although they do not possess the outstanding resistance of the two-pack epoxy systems, they are suitable for use under conditions of mild chemical attack as, for example, in industrial atmospheres. Epoxy esters are used as binders in gloss finishes, undercoats, and primers for metals.

Two-pack (cold cure) epoxy systems have a completely different form of "drying" or hardening. This is a chemical set-up or polymerization in the film. They are supplied in two containers, one of which contains a varnish or paint and the other a catalyst or hardening agent, usually an amine adduct or polyamide. When the two components are mixed the chemical reaction between them commences and so it is unwise to mix more than is required for one day's working. The pot-life, or time that the mixture remains usable, is of the order of 8 to 12 hours. They can be brushed but are more often applied by airless spray. However, on account of the strong solvents employed, they cannot be used over existing paint films. All equipment and tools must be washed out with special solvents immediately after use.

These cold cure systems should be applied at normal temperature (18–20°C) to ensure curing and may be recoated in 12 to 15 hours, although a curing period of 7 days must elapse before the film attains its maximum chemical and solvent resistance.

Two-pack epoxy systems show outstanding adhesion to most surfaces combined with excellent chemical and solvent resistance. They have a great many uses, among which are coatings for metal and concrete (metal surfaces should be freed from all scale, rust, grease, etc. before painting), adhesives, and flooring compositions. Their use for protecting structural steelwork is steadily increasing and for this purpose they are often supplied in a thixotropic or "high build" form.

Polyurethane resins

Like the epoxies, these are two-pack materials: one component is a polyester resin and the other a solution of an isocyanate. The two components are mixed immediately before use and only sufficient should be mixed for one day's working as the pot-life is limited.

Polyurethanes offer a much wider range of film properties than do the epoxies. By choice of the appropriate polyester and isocyanate, films may be made flexible enough for rubber balls or hard enough for use in floor varnishes and flooring compositions. The films show good adhesion and withstand chemical attack, heat, moisture and many solvents. They are weather resistant and are used to protect ships' superstructures, oil storage tanks and other exposed structures. They are also used extensively where flexibility is required as on rubber articles.

A special form of polyurethane contains both the components in solution in dry solvent. When this is exposed to the air in a thin film, atmospheric moisture is absorbed and then the components react and the film "cures". Such materials are known as *moisture curing* polyurethanes. As a consequence of the reaction with water, these materials are difficult to pigment, as commercial pigments always contain a small amount of moisture. They are therefore used as clear varnishes for application to concrete and other types of floor.

Urethane oils are produced by reacting a di-isocyanate with compounds obtained from drying oils. The products are soluble in white spirit and dry by the normal oxidation process with addition of cobalt driers. The films are hard and tough and they are used as clear varnishes. However, they lack the overall resistance of the two-pack polyurethanes. The hardness is superior to that of conventional alkyds and they are used where this property is an asset, for example, in paints for metal kitchen furniture.

Silicone resins

These differ from the conventional type of resin in containing the element silicon, which makes the resin semi-inorganic. Silicones contain both silicon and carbon, both of which are polyvalent, and combinations of these elements lead to a large number of polymers.

Some silicone resins are dissolved in solvent and are very effective waterproofing agents for brickwork and concrete surfaces. They can be brushed on to the surface and do not alter the appearance in any way.

As far as paints are concerned, silicone resins are valued for their resistance to high temperatures. Their use in heat-resisting paints is discussed in Chapter 17.

15

Non-convertible Coatings

As mentioned earlier, these dry by evaporation of the solvent, and no "cure" or chemical reaction takes place in the film. In consequence they are quick drying and the pigmented and clear versions are frequently referred to as "lacquers". The following are the most common non-convertible binders.

RUBBER DERIVATIVES

Natural rubber is obtained by tapping certain trees in Brazil, Malaya and elsewhere; the latex exudes as a milky-white liquid which when coagulated with acids forms solid "crepe" rubber. This can only be dissolved in highly volatile petroleum solvents, the solutions being so viscous that when thinned out sufficiently for brushing the resulting films are excessively thin and of poor adhesion. Natural rubber is therefore not suitable for paint coatings, but two of its derivatives are important binders.

Chlorinated rubber

When natural rubber is treated with chlorine under heat and pressure it reacts to form a "resin" which is dried and supplied in the form of a white powder. It is soluble in aromatic hydrocarbons such as xylol, esters, and some ketones, but is not soluble in white spirit. The film deposited on evaporation of solvent from a solution is brittle and lacks adhesion. However, addition of a plasticizer results in tough, adherent films with outstanding resistance to moisture and chemicals. The amount of plasticizer used varies from 20 to 50 percent according to the amount of pigment in the paint and the end use. Owing to its rapid set-up, brushing is difficult and it is best applied by spray (especially when re-coating) in order to avoid disturbance of undercoats. As a result of its outstanding resistance, chlorinated rubber has a very wide range of applications. Among these are the protection of steelwork under severe conditions, as in chemical works and ships' hulls, the coating of concrete swimming pools, wall paints in breweries, etc., where high humidity exists. Chlorinated rubber is compatible with some long-oil alkyds and is often blended with them to upgrade their chemical resistance, hardness, and resistance to lubricating oils.

Cyclized or isomerized rubber

If raw rubber is heated with phenol and a catalyst a change in its structure takes place, yielding a product which dissolves in white spirit. When used alone as a binder, cyclized rubber requires a plasticizer, but it is compatible with a number of alkyds and is frequently mixed with these to improve their chemical resistance. Isomerized rubber is also used in printing inks.

VINYL RESINS

The vinyl resins now form one of the most important groups in the paint industry. They are derived from liquid monomers (such as vinyl acetate) which may be polymerized with suitable catalysts to form colourless resins. The resins may be used as solutions in solvent or dispersed in water to form emulsions (page 196).

Polyvinyl acetate (PVA) is derived from vinyl acetate, a chemical obtained by the combination of acetylene gas and acetic acid; the liquid monomer is then polymerized with a suitable catalyst to form polyvinyl acetate. The lower grades of polyvinyl acetate dissolve in alcohols but give films which are brittle and lack adhesion. These faults can be overcome by the addition of plasticizers, but most of the suitable plasticizers gradually leave the film by volatilization or migration into the underlying surface. The film then gradually becomes brittle and fails. These weaknesses were present in the earlier polyvinyl acetate emulsions.

Co-polymer resins are now widely used because of the attractive contribution they make to durability, adhesion, and retention of flexibility. A number of these resins are available, each with its own characteristic built-in properties. They are prepared by reacting (or "co-polymerizing") vinyl acetate with other monomers chosen to give the desired specific properties to the polymer. Many modern emulsion paints are based on the co-polymer of vinyl acetate and ethylene (page 196), whilst the co-polymer of vinyl acetate and vinyl chloride, in solution form, is used in coatings for collapsible tubes.

Polystyrene resins are derived from the monomer styrene, a liquid product of the petroleum industry. It forms resins by polymerization, either alone to give polystyrene (which finds little use in the paint industry) or, with other monomers, to give very useful co-polymers. When co-polymerized with butadiene or acrylic esters it produces resins with high water-resistance and durability which are used as binders in paints for exterior masonry.

Acrylic resins exist in the form of co-polymers made from different acrylic ester monomers. A range of resins is available, varying from rubbery to hard, almost brittle. They are soluble in esters, ketones and aromatic hydrocarbons. Their good adhesion and durability are utilized in a number of products, but in the decorative field they are best known in the form of emulsions.

CELLULOSE FINISHES

The introduction of nitrocellulose finishes brought about some revolutionary changes in painting practice which have become firmly established in the sphere of industrial finishing. They became the standard finish for motor cars, where the rapid dry and spray techniques were admirably suited to mass-production methods. They were also adopted for use on aircraft and on furniture.

In the modern car industry stoving synthetic finishes have replaced cellulose on new cars, but in car refinishing nitrocellulose lacquers continue to be used. Some furniture is still finished in nitrocellulose, but here it is being steadily replaced by polyurethane lacquers.

Both the solid components and solvents used in nitrocellulose finishes are highly flammable and the manufacture, storage, and use of the lacquer are subject to strict statutory regulations. Application is always by spray which is carried out in specially designed spray booths, where the overspray and solvent vapour can be removed by exhaust fans. For these reasons nitrocellulose lacquers have found very little use in the decorative field. It is well, however, that the painter and decorator should know something about them and we therefore give a brief outline of the various constituents and their functions.

Nitrocotton

The basic film-forming material is nitrocotton (also known as *pyroxylin* or *collodion cotton*). It is prepared by treating cellulose, in the form of cotton linters (short cotton fibre) with nitric and sulphuric acids. The nitrated cotton resembles cotton wool and is highly explosive when dry, being akin to gun-cotton. It is supplied to the lacquer maker damped with alcohol to render it safe in transport. When dissolved in solvents, such as acetone, amyl and butyl acetates, etc., it produces a clear varnish-like solution which dries in half an hour or so to a hard, tough, transparent film.

Plasticizers

Plasticizers, or softeners, are generally non-volatile liquids such as tricresyl phosphate, dibutyl phthalate, castor oil, etc., which, added to a cellulose compound, render it more flexible. The particular choice and proportion of plasticizer governs the final properties of the finish, and films of high elasticity are formed when the plasticizer is in high proportion.

Resins

A pure cellulose film has little or no gloss and but poor adhesion; resins are therefore added in varying proportions according to the class of lacquer required, those intended for wood and other absorbent surfaces having a higher gum content generally than the metal lacquers. Here again the choice of resin, like that of the plasticizer, enables the manufacturer to formulate lacquers for any particular field of use. Damar and many modern synthetic resins are used to impart gloss, durability, adhesion, etc.

Solvents and diluents

The purpose of the solvents mentioned above is to dissolve the nitrocotton and keep it in solution, but the cost of using *true solvents* (as they are called) to reduce the lacquer to spraying consistency is very high and unnecessarily extravagant since they evaporate entirely and form no part of the dried film. Pure solvents (chiefly esters, such as amyl and butyl acetates, etc.) are therefore combined with cheaper diluents (notably coal tar and petroleum distillates and alcohols) which help to keep the cost within reasonable limits; these function along with the solvents in controlling the evaporation rate of the mixture from the wet film.

Nitrocellulose lacquer

We see from the foregoing that the constituents of nitrocellulose lacquer are —

(1) *Nitrocotton*, the main substance of the film,
(2) *Solvents and diluents* to dissolve the nitrocotton and keep it in solution,
(3) *Plasticizers* to provide flexibility, and
(4) *Resins* to give gloss and adhesion.

Since drying depends entirely on evaporation of the solvents and diluents, it should be noted that unlike the drying of oil media, which requires the addition of driers, the drying of nitrocellulose is governed by the choice of fast- or slow-evaporating solvent mixtures and plasticizers. In brushing lacquers, for instance, the evaporation rate has to be *retarded* sufficiently to give time for brush manipulation, and therefore only slow-evaporating mixtures are employed.

Coloured nitrocellulose lacquers and enamels By incorporating spirit-soluble dyes, a class of translucent coloured lacquers is produced which make very attractive "lustre" finishes on bright metal. Nitrocellulose pigmented enamels are prepared by grinding suitable pigments in the clear medium, but the amount of "binder" in the dried film (which is extremely thin) is not sufficient to support very high pigmentation; only the best quality pigments are therefore permissible in order to obtain maximum opacity from the minimum pigment content. Extenders, such as barytes, etc., are used only in fillers or other specialities.

Application of nitrocellulose finishes

The ideal method of applying nitrocellulose finishes is undoubtedly by spray. The speed of drying enables several coats to be applied in quick succession, so making possible a rapid film build-up in spite of the extreme thinness of nitrocellulose coatings. The dried nitrocellulose film is soluble in its own solvents and therefore softens up on recoating; thus each superimposed coat becomes "welded" into the undercoat and the result is a completely homogeneous film.

Nitrocellulose finishes can be buffed or polished with abrasives and polishes, and clear lacquers can also be finished with "pullover" solutions in a similar way to French polish. Buffing and polishing is now on the decline, however, due to improvements in synthetic resins, the resultant films drying smoother and with greater depth of gloss. Imperfections in the surface cannot be hidden or filled up by a flowing coat as with oil varnish and enamel; the surface must therefore be well prepared and built up with filler and undercoating and, when sanding down prior to applying the finishing coats, scratches must be scrupulously avoided by using specially fine abrasive papers.

The softening action of the solvents enables touching up or "faulting" of nitrocellulose films to be carried out with ease by spray and when finally polished such "faults" are practically indiscernible. The same solvent acts as a vigorous paint remover on normal oil paints, so that it is unsafe to apply cellulose over existing oil-painted surfaces, or on any undercoating other than that recommended by the manufacturer.

Nitrocellulose finishes are particularly suitable for industrial finishing processes (cars, metal, furniture, etc.) where mass production methods demand the maximum speed both in application and drying time.

BITUMENS AND PITCHES

Bitumens are either natural materials, of which the best known is the hard Gilsonite, or residues from the distillation of crude asphaltic petroleums. They are intensely dark in colour and dissolve in white spirit and other petroleum solvents. The dark colour limits their use in paints to blacks and dark shades of grey, brown and green.

Bitumen films tend to be soft and become still softer on heating. They are, however, very resistant to water penetration and to attack by acid or alkali. Bituminous paints are therefore used for the protection of steel and other surfaces in situations where protection is of greater importance than decoration. Application of a conventional paint over a bitumen film results in the bitumen "bleeding" into the applied paint film.

Pitches are residues obtained from the distillation of tars which may be derived from coal or petroleum. Certain grades of pitch are of interest in being compatible with epoxy and polyurethane resins. Mixtures of pitch and resin will cure on addition of the hardener or second component to give films which show very high resistance to moisture. The epoxy–pitch type is widely used in the form of "high build" coatings for the protection of steelwork, especially in submerged situations. However, these coatings show the characteristic drawback of pitch and bitumen films in that they bleed into any conventional coatings applied over them.

NATURAL SPIRIT-SOLUBLE RESINS

Shellac (or Lac) is obtained as an incrustation on the twigs and branches of certain types of tree occurring mainly in India. The lac insects feed on the sap of the tree and excrete a resinous material in which they finally become embedded and die. The twigs are collected and the resin (crude shellac) is separated and purified. It is marketed in various physical forms known as orange shellac, button lac, garnet lac, etc. and dissolves in industrial methylated spirit to give turbid solutions.

The best-known solutions of shellac in industrial methylated spirit are "knotting" and French polish (which is now little used). Knotting is used to prevent exudation of gummy material from knots in new timber. Shellac solutions are highly acid, and the traditional types of container have been glass or stoneware. Additives are now available which render the solutions safe in tin-plate containers.

Damar is obtained from certain Malaysian trees and is marketed in a range of colours from pure white to black. It is soluble in petroleum solvents and the films produced are very flexible. The best grades of damar are used in paper varnishes and in nitrocellulose lacquers where they improve the gloss and adhesion.

Manila is obtained from several sources in the Far East. The quality varies considerably. It is soluble in industrial methylated spirit and is used as a binder for road marking paints.

16

The Manufacture and Characteristics of Paint

A modern paint factory is, ideally, a two- or three-storey building so designed that the raw materials are stored and charged into machines on the top floor and use is made of gravity in transferring the paints from one machine to another. Filling into containers then takes place on the ground floor.

Basic operations in manufacture

The manufacture of paint consists, basically, of the following series of operations: (a) dispersing the pigment in the medium (this operation was formerly called "grinding", but nowadays the pigments supplied to the paint manufacturer are ground fine enough and no further reduction in size or "grinding" is required); (b) addition of the driers, anti-skinning agents, etc. and, if necessary, tinting to the desired shade; tinting is carried out by the use of concentrated liquid stainers; (c) adjustment to final viscosity; (d) straining through vibrating sieves and filling into containers.

A variety of mills is available for dispersing the pigment in the medium, and the types chosen by a manufacturer depend on a number of factors such as volume of work (size of batches), the type of paint (air-drying or industrial stoving), the degree of dispersion required, and so on. The following mills are among the most commonly used at the present time.

Ball mills are large horizontal cylinders of 230–2300 litres (50–500 gallons) capacity having steel or porcelain linings and containing a large number of steel balls or round pebbles. The mill is charged with the requisite amounts of pigment and media and, when rotated, the pigment is ground by the rubbing or abrasive action of the pebbles against the pigment particles, the fineness of grinding being dependent upon the number of hours occupied in grinding. The advantages of this method are that the mill requires little attention and is suitable for mass-production of standard colours. Being totally enclosed, it is very suitable for cellulose materials, preventing escape of volatile solvents and thus minimizing risk of fire.

Sand mills are now widely used for certain types of production. A sand mill consists of a vertical cylinder fitted with a central steel shaft to which are attached a number of steel discs. The cylinder is partially filled with glass ballotini (fine beads) or carefully graded sand (hence the name). The pigment and medium are premixed in a separate vessel and pumped into the cylinder and, at the same time, the shaft is rotated at high speed. Rapid dispersion of the pigment takes place, in some cases (for example with titanium dioxide), in as little as 10 to 15 minutes. The sand is retained in the cylinder by fine wire gauze which allows the paint to pass through. By using several vessels for premixing the pigment and medium the process can become continuous and is suitable for long runs of one particular type of paint. For small batches, and particularly if frequent changes of colour are required, the ball mill is more suitable.

Single-roll finishing mills are designed for the finishing or refining of gloss paints. Extreme fineness is achieved by removal of all but the finest particles. A special type of single-roll mill incorporates a "double bar" and performs the dual function of dispersing and refining in one operation.

High-speed dispersers are used when the highest degree of dispersion is not required or where pigments are readily wetted, as in emulsion paints and some types of primer. They consist of an open cylindrical vessel fitted with a central shaft to the end of which is attached an impellor of special design. The impellor is, in some machines, enclosed in a cage designed to increase the turbulence caused by the high speed of rotation. The pigment and medium are fed into the machine while the impellor is running, and dispersion takes place rapidly.

After the paint has been in the mill for the prescribed length of time a sample is removed and tested for the degree of dispersion of the pigment. If this is correct, the batch is transferred to a mixer for the next operation.

Mixing, thinning and tinting These operations are carried out in open top containers fitted with mechanical stirrers. The balance (if any) of the medium, the driers and other constituents are added, and after thorough stirring the paint is checked for viscosity and adjusted, if necessary, by adding thinners. Tinting or colour adjustment is done by addition of concentrated liquid stainers.

Testing Every batch of paint is made to a specification. This may be the manufacturer's for his proprietary materials, or it may be issued by the customer.

Before filling into containers each batch is checked to ensure that it satisfies the specification requirements. The tests carried out include viscosity, specific gravity, and drying time. For decorative paints the sample is compared with a liquid standard for ease of brushing, freedom from brushmarks, drying time, and colour. These tests are carried out on prepared metal or wooden panels and the panels dried at 15–20°C. Frequent practical tests are carried out on appropriate surfaces, emulsion paints being brushed on plaster and other building surfaces. It is only by strict compliance with the specified requirements that the manufacturer maintains his reputation for consistent quality.

Straining and filling After final approval has been given by the control laboratory the paint is strained through vibrating sieves of fine wire gauze and filled into containers. For containers of five litres and downwards the filling is usually done by automatic machine.

CHARACTERISTICS OF PAINT

Consistency

Consistency is a rather vague term used to describe the condition or fluidity of paint, "stout" or "round" consistency denoting a creamy material, while "viscous" refers to the sticky condition usually associated with varnishes or enamels. Correct working consistency is best determined by actual trial under practical conditions. Some pigments and media have "slip" and brush easily, even in a round condition, while others work "sticky" or "gluey". Consistency must also vary according to the nature of the surface being treated and the method of application; primers for porous surfaces, for instance, require to be thinner than primers for non-porous surfaces, while brushing paints are much "rounder" than those applied by spraying. When the best consistency of a paint has been determined by practical application tests, it can be given a figure in terms of "poise", the standard unit of viscosity. A paint can be described as having a viscosity of "3·5 poise at 20°C". The temperature is important, as the consistency or viscosity of a paint decreases as the temperature rises and vice versa. The viscosity is measured on a *viscometer* and affords a standard by which batches of paint can be controlled.

Opacity

Technical terms are often loosely applied and lead to confusion. The term "body" for instance, is sometimes used to describe consistency or viscosity (as in "full-bodied" varnish) while at other times it implies covering or hiding power. Only those terms which cannot readily be misinterpreted should be used, and this is especially important in the writing of specifications. The word "opacity" can legitimately be applied only to obliterating or hiding power and therefore seems the safest term to use. The opacity of a paint depends on the nature of the pigment and the amount used in relation to the vehicle. Highly pigmented paints usually have good opacity, even when reduced with extenders, but in gloss paints and enamels (where the pigment content is necessarily low) only pigments having good opacity should be used.

The opacities of two paints can be compared visually by the use of glazed black-and-white chequered cards. Equal weights of the two paints are brushed out to cover equal areas of board and allowed to dry. The relative opacities can then be assessed. A quantitative measure of opacity is possible by the Contrast Ratio method. For details of this method the reader is referred to *Outlines of Paint Technology* by W. M. Morgans (Charles Griffin & Co. Ltd).

Spreading capacity

Here again the term *covering power* is often used when referring to the spreading rate of a paint, but it may apply also to hiding power or opacity. When referring, therefore, to the area covered by a given quantity of material, it is more explicit to use the term *spreading capacity* as this cannot possibly be misconstrued. Theoretically, any two samples of paint of equal volume and equal consistency should be capable of being spread over equal areas. This is not the case, however, since the brushing properties vary widely with the nature of the pigments and media used. Paints which have "slip" (such as those containing graphite, stearates, mica, etc.) may be brushed out very thinly and spread over a very large area, while paints of a "sticky" nature (such as gloss paints and enamels) produce thicker films and therefore do not spread as far. Between these two extremes there are innumerable variations.

Apart from differences in application due to the human element (for no two men will obtain identical results in any spreading test), the spreading rate is largely governed by the surface being painted. The porosities of surfaces vary tremendously, and absorbent softwoods, wallboards, and plasters may easily take twice or three times as much paint as similar areas of non-absorbents such as metal and glass. Texture and surface undulations are also important factors to be reckoned with, the spreading rate on roughcast, plastic textures, relief-decorations (Anaglypta, etc.) and corrugated surfaces being anything from 50 to 75 percent less than on smooth surfaces of similar porosity. It will at once be realized that it is quite impossible to quote reliable spreading rates for all the various surfaces and conditions liable to be met, and the practice followed by leading paint manufacturers of quoting figures obtained on *smooth non-porous surfaces* is probably the only safe and practicable course to adopt. These figures are usually quoted in square metres per 5 litres.

Thixotropic or "jelly" paints need special mention. Many of these can be spread so easily under the brush that there is a temptation, which must be resisted, to spread them excessively and in this way the advantages of the thicker coat are lost.

Durability

The "life" of a paint film will naturally depend on many factors such as the composition of the paint, the nature and condition of the surface to which it is applied, the climatic conditions prevailing at the time of application, and the conditions of exposure to which it is afterwards subjected.

Given ideal conditions such as —

(a) a dry well-seasoned inert surface, correctly primed with a paint suited to the nature of the surface;
(b) well-compounded materials applied in correct sequence, i.e. becoming progressively more flexible towards the final coat;
(c) a glossy flexible finish, having good water-resistance, and
(d) a clean uncontaminated atmosphere;

the paint film may be expected to have a long useful life, gradually weathering to a dull or chalky surface. This natural chalking (when it is not too rapid or excessive) presents the ideal surface for repainting since cleaning and rubbing down is all that is necessary, whereas blistered, cracked or wrinkled paint films generally require complete removal.

The above conditions, needless to say, are very seldom met with, though one does occasionally come across examples of exceptional durability in exterior painting. The average useful life of a gloss paint film in an urban atmosphere, however, may be reckoned as about six to seven years, but on unseasoned timber or in heavily contaminated atmospheres or exposed positions, it may be considerably less. Moreover the life will be considerably shortened by faulty materials, poor workmanship, or adverse conditions at the time of painting.

In trying to assess fairly the durability of his products, the paint manufacturer does his best to fulfil conditions (a) and (b) in preparing his wood and metal panels for exposure tests. But he is not content merely to fulfil condition (d) since he requires to know how his paints will behave under *all* conditions of exposure. He therefore erects exposure stations in various "strategic" parts of the country (and abroad, if he is an exporter); these may be located in the chemically polluted atmospheres of large industrial centres, on bleak highlands, in damp lowlands, in coastal regions and wherever special conditions prevail. Panels are exposed (usually on a roof) at an angle of 45 degrees facing south, thus getting the full severity of sun and weather. The information garnered from these tests enables the paint chemist to formulate paints having a fairly high standard of performance under widely varying conditions.

Conditions affecting durability

A paint film may be adversely affected by (i) the nature of the surface upon which it is applied, (ii) the conditions prevailing at time of painting, and (iii) conditions of exposure. These will be considered when dealing with the application of paints, but they may be usefully summarized here as follows—

(i) *Nature of surface* New building materials may contain considerable amounts of water and in addition may be chemically active. Other instances of chemical activity, often overlooked, are those of exterior painted surfaces in industrial districts where a considerable amount of dust and soot is deposited, which, being of an acid nature, may attack the paint from beneath if not thoroughly washed off. Alkaline deposits left on the surface after washing down with soda water will also play havoc with paint media. Rust on metal surfaces will spread beneath a paint film and ultimately throw it off.

(ii) *Painting conditions* Moisture may be regarded as enemy number one. Most building surfaces are liable to hold condensed moisture in damp or humid weather, although with absorbent surfaces this may not be noticeable. Fog, early morning mists and white frost are also responsible for heavy surface deposits of moisture. Imprisoned moisture impairs adhesion of the film, giving rise to blistering and peeling. A damp atmosphere retards the drying of paint and varnish and impairs the gloss, whilst a warm dry atmosphere promotes good drying. In very hot weather drying may be too rapid, especially in direct sunlight, the surface possibly skinning over and later wrinkling. Cold weather is not harmful if the atmosphere is dry, though painting should not proceed when the temperature falls below 4°C. On interior work stagnant air due to inadequate ventilation will prejudice drying, but so will cold draughts; a gentle circulation of fresh air is the ideal. Daylight, or any light rich in ultra-violet rays, favours the drying of paint whilst, conversely, darkness tends to retard drying; this, combined with lower temperatures, creates poorer conditions at night.

(iii) *Conditions of exposure* In large cities and industrial areas the air becomes polluted with dust and chemical gases (sulphur fumes, etc.) which form corrosive surface deposits on paint films, activated from time to time by rain water (also polluted and generally having a definitely acid reaction). Occasional washing down with clean water would do much to minimize the effects of such conditions, but when repainting is done it is imperative that the surface be *thoroughly washed* and not merely dusted, for such deposits bite into the paint film and cannot be removed by dusting. In addition to corrosive attack on the paint media by chemically polluted atmospheres, certain pigments are also affected, lead paints (including chromes, Brunswick greens, etc.) being discoloured by sulphur fumes whilst ultramarine blue is bleached by acid conditions.

Light plays a most important part in the decay of exterior paints, the ultra-violet or actinic rays promoting the disintegration of the binder. Certain pigments (e.g. pure red oxide) tend to absorb or "degrade" the actinic rays and thus assist in prolonging the life of the film. The industrial haze in large towns and cities also acts as a filter which minimizes the effect of actinic light rays, but this form of protection is, of course, offset by the destructive effect of the polluted atmosphere itself. In coastal regions, where the atmosphere is generally very clear and many more hours of sunshine are enjoyed than inland, the actinic rays have a correspondingly greater destructive effect on paint films. Consequently loss of gloss followed by chalking sets in earlier than in inland and urban atmospheres. Proprietary marine paints are usually based on long-oil alkyd resins pigmented with rutile titanium dioxide. The resistance to moisture is sometimes upgraded by inclusion of a proportion of chlorinated rubber.

Light is also destructive of certain pigments, though it appears that in some cases the fading of colour is not due to this alone. Moisture and atmospheric contamination also play a part. Several pigments which fade under normal exposure are fairly permanent when adequately protected; for example, ultramarine blue fades rapidly under normal exposure but shows good lightfastness when the film is protected by glass or an impervious varnish such as chlorinated rubber.

In addition to the above points, the composition of the paint may be such that its life is prejudiced from the start. When pigments and media are well chosen for their intended purposes and carefully compounded together, good durability may be expected. When extenders or driers are used excessively, however, the paint will have a short life due to early chalking or perishing of the film. Moreover, certain pigments, being hygroscopic, may absorb moisture when stored in powder form, and if these are not properly dried before grinding in oil, trouble is likely to develop when the paint is applied; the drying may be tardy, or blistering may occur when the film is exposed to a hot sun. A reputable manufacturer, however, cannot afford to jeopardize his reputation by selling such materials. Rather he strives to maintain and improve the standard of performance, for research is ceaseless and competition is keen.

To those of his clients who demur at paying for high-grade materials, the decorator might point out three things. First, that a good quality product will put up a good show even under the worst conditions (whereas a cheap material, owing to its composition, cannot last long even under the best outdoor conditions). Second, that the cost of applying a cheap material is just as much as that of applying a good one, and sometimes more. Third, that the cheaper material will require renovation much more frequently than the high-grade product and therefore works out considerably more expensive in the end.

17

Types of Decorative and Maintenance Paints

The paint manufacturer is required to make literally hundreds of different types of paints for every conceivable purpose and in every conceivable finish. Fortunately for the painter and decorator, however, the types of paint with which he is chiefly concerned are comparatively few in number (though the number of proprietary names given by different makers to materials of similar composition is somewhat bewildering). The principal types are outlined here in the order in which they are usually applied to a surface.

Sealers

Sealers are designed for various purposes such as the sealing of materials which are liable to bleed, e.g. tar, creosote, bituminous finishes and certain lake colours, metallic bronze paints, etc., for which "impenetrable" spirit varnishes or stop-tar knottings are used. They are usually transparent, but pigmented sealers are also marketed (referred to as "primer-sealers") which function as suction stoppers, alkali resisters and, according to some makers' claims, even hold back dampness and efflorescence. Whilst the latter may be effective in certain mild cases, it would be unwise to expect too much from them, as it is manifestly impossible for any film to withstand the hydrostatic pressure which may be set up in a waterlogged wall, or to overcome the irresistible force of soluble salts which crystallize on the surface on drying, pushing off anything which lies in their path. This subject is dealt with more fully in the chapter on the priming of plaster surfaces (page 224).

Primers

The composition of priming paint varies according to the nature of the surface for which it is intended — wood, metal, cement and plaster, etc. No single primer is suitable for all surfaces and conditions. For porous surfaces, primers must contain sufficient oil to satisfy the porosity while being thin enough to penetrate into the surface to secure a firm key (though some authorities now consider this unnecessary and prefer to rely on the *adhesive* properties of the medium to form a good attachment to the surface). Cement primers must possess maximum resistance to alkalis.

Primers for wood must adhere well to the surface and must be elastic enough to expand and contract with the timber as it absorbs or loses moisture. The traditional lead-based "pink primer" has now been largely replaced by non-toxic leadless types which possess the properties mentioned above and, in addition, show better flow. Both alkyd and varnish media are used, pigmented with titanium dioxide or metallic aluminium. In recent years acrylic emulsion primers have been introduced and are being used successfully.

Primers for metals Steel corrodes (rusts) rapidly when exposed to air and moisture. Before steel is primed, all rust, scale and grease must be removed and the clean metal should be primed before rusting can recommence. The most effective way of cleaning steel is by shot- or grit-blasting, and if this is done inside, the steel should be primed within four hours. Outside, the priming should be done immediately after cleaning.

Primers for steel should have good adhesion to the surface, good moisture resistance, and must contain pigments which inhibit corrosion. Among the pigments used are metallic zinc, zinc phosphate, zinc chromate and calcium plumbate. Some red lead continues to be used but, on account of its toxicity and slow-drying properties in linseed oil (in which it works best), its use is declining. The media used for modern primers include varnishes, alkyds, epoxies, chlorinated rubber and polyurethanes.

Zinc and galvanized surfaces, if new, tend to show poor adhesion for a number of primers unless the surface is etched. However, calcium plumbate primer adheres satisfactorily to the smooth metal.

The cleaning and priming of metals as well as the priming of plasters and other surfaces is dealt with in detail in Chapter 19.

Fillers and stoppers

Fillers and stoppers consist of silica, slate powder and other materials dispersed in short-oil varnish media of the goldsize type or, for special purposes, in synthetic media such as epoxies or polyurethanes. They are usually supplied in paste form for knife-filling surface indentations, stopping small holes and cracks, etc. They can also be thinned with the appropriate solvent and used as a brush or spray filler. They harden rapidly and rub down to a smooth surface with waterproof abrasive paper. Powder fillers (Polyfilla, etc.) are mixed with water, applied with knife or brush and, when dry, are rubbed down with glasspaper.

Undercoats

Undercoats are highly pigmented paints which possess good opacity and provide a suitable ground for the finishing coat. Modern undercoats are based on alkyd resins with, probably, small additions of linseed stand oil or varnish to improve the "wet edge" time. For whites and pale tints the pigment used is titanium dioxide, which has largely replaced white lead, to give non-toxic properties and improved levelling of brushmarks.

The ideal undercoat would possess the following properties: (a) good adhesion to the primer, (b) good flow and levelling, (c) good opacity, (d) a surface with low gloss to which the finishing coat can adhere well, (e) a colour similar to the finish but possibly a little lighter. Most undercoats are a compromise as it is very difficult to formulate a material possessing all these properties to the maximum degree. For this reason there is appreciable variation in undercoats from different makers, particularly in respect of flow properties and degree of gloss. They are, however, designed to be used with the gloss finishes from the same manufacturer and it is unwise to use primers, undercoats and finishes from different sources in the same paint system.

GLOSS FINISHES

The gloss finishes used in decorative work are based on medium- to long-oil alkyd media in which the alkyds contain fatty acids derived from either linseed or semi-drying oils. Linseed-based alkyds tend to yellow on exposure and are suited only for the deeper colours. Alkyds containing fatty acids from the semi-drying oils are non-yellowing and are used for whites and pale tints in which the pigment is invariably rutile titanium dioxide. Films from these paints set up less rapidly in warm weather (i.e. have a longer wet edge time) than films from linseed-based paints. Some brands of gloss finish also contain a proportion of polyurethane and silicone resins which are claimed to improve hardness and durability.

Alkyd-based gloss finishes have good gloss-retention and durability and are suitable for both interior and exterior work, except where chemical attack or water immersion are involved. They are resistant to very mild acids but are attacked by alkalis and strong acid solutions. Long continuous contact with water causes swelling and blistering, and so these finishes should not be used on surfaces likely to be immersed in water for prolonged periods.

Thixotropic (dripless or jelly) paints

These are based on a medium produced by reacting an alkyd with a polyamide resin. They form firm gels when at rest, and the container can be tilted or even inverted without spillage. The paint flows freely when stirred or brushed, but on removal of the brush it returns quickly to the gel condition. As a result, thick coatings can be applied without risk of runs or sagging, and the required coating thickness can be achieved with fewer coats than with conventional paints.

LOW-GLOSS FINISHES

Semi-gloss finishes

These paints find a limited use in interior decoration. They can be regarded as gloss finishes to which carefully chosen extenders are added to reduce the gloss without detracting from the flow properties.

Eggshell finishes

As the name suggests, these paints possess a lower gloss than semi-gloss but, unfortunately, there is no generally accepted standard of gloss for either. Eggshell finishes are based on highly bodied alkyds or mixtures of these with conventional or thixotropic types. The amount and type of pigment is extremely important as these affect the gloss, flow, and levelling properties. These finishes are used on interior walls, and slight variation in properties becomes evident as "sheariness" or "patchiness".

Flat finishes were formerly popular as interior wall paints and were based on highly bodied alkyd resins. Their use has declined considerably since the introduction of emulsion paints.

EMULSION PAINTS

Resins such as vinyl, acrylic and synthetic rubber are generally difficult to dissolve except in highly volatile solvents giving viscous solutions which, when reduced to brushing consistency, dry rapidly by evaporation, leaving films of very low solids content. Fortunately it is possible to *distribute* these resins in emulsion form in which they are not dissolved but *dispersed* in water. Thus, the cost of expensive solvents is obviated and so is fire hazard. On evaporation of the water, the resin particles flow together or coalesce, a non-reversible process which produces a washable film. Application by brush is easy and drying is rapid, re-coating after two hours or less. There is little or no odour, good alkali-resistance and external durability.

Although the polymer is more solid than liquid, the dispersion has always been known as an "emulsion". A large number of synthetic polymers are produced today, but only a few are suitable for use in emulsion paints; of these the important members are polyvinyl acetate (PVA), co-polymers of vinyl acetate with other monomers, and polyacrylic esters.

Vinyl emulsion paints

The polymer emulsion is prepared by first making an emulsion of the liquid monomer (e.g. vinyl acetate) in water with an emulsifying agent, then polymerizing by heating in presence of a catalyst to give a colloidal dispersion of polymer resin particles.

Polyvinyl acetate (PVA) gives hard, brittle films and it is necessary to add a plasticizer to impart flexibility. The film, however, gradually loses the plasticizer and becomes brittle again with age. If vinyl acetate is co-polymerized with a second monomer such as ethyl acrylate or ethylene, a co-polymer resin is produced. By careful choice of the type and quantity of the second monomer, co-polymers can be produced with any desired degree of flexibility, and no addition of plasticizer is required. These co-polymers are in general use today and show greater adhesion and durability than the earlier plasticized PVA emulsions.

The emulsion paint is manufactured by dispersing rutile titanium dioxide (or coloured pigment) in water, together with various additives such as wetting agents, thickeners, preservatives, etc. The dispersion of pigment is then added to the polymer emulsion and mixed thoroughly to give the finished paint.

Vinyl emulsion paints have little odour and good brushability and they dry quickly, being recoatable in 2 to 3 hours. They dry by evaporation of the water, the pigment being bound by the coalescing particles of the co-polymer. They also have exceptional washability and can be cleaned within a few hours of application.

They show resistance to alkali and can be used on new lime plaster, Portland cement and asbestos cement without a primer. When thinned with a small quantity of water they are used as primer/sealers for hardboard and the various types of absorbent building boards. They can be applied to previously painted surfaces, provided these are efficiently rubbed down, but they should not come into contact with bare iron or steel on which they tend to promote rusting; such surfaces should be primed with a zinc phosphate or other rust-inhibiting primer, whilst non-ferrous metals require a primer based on zinc chrome.

Vinyl emulsion paints are marketed in varying degrees of sheen from matt to semi-gloss, the degree of sheen depending, in general, on the ratio of pigment to polymer. Matt finishes are used on interior walls, but the more glossy types are used on both interior and exterior surfaces.

Because of the quick-drying property of these paints, brushes should always be washed out immediately after use, being first rinsed in cold water to remove the bulk, then washed in warm water and soap, and finally rinsed in clean water.

Vinyl emulsion paint can be applied over hard bituminous or creosoted surfaces without bleeding, but on thick, soft coatings there is, naturally, a risk of cracking.

Acrylic emulsion paints

Acrylic emulsions contain co-polymers of various acrylic esters. By choice of appropriate monomers, co-polymers of any desired degree of flexibility can be produced. The manufacture of the polymer emulsion and of the emulsion paint is carried out by methods somewhat similar to those used for the vinyl types, but there are important differences between the respective paints.

Acrylic emulsion paints show greater adhesion and durability and can be made more flexible. They are not as popular for interior decoration as the vinyl types (probably on account of their characteristic odour) but they have useful applications in other directions. Their flexibility and adhesion render them suitable for use as wood primers, and they also adhere well to new zinc and galvanized surfaces and to some plastics. When used on alkaline and on absorbent surfaces they give, on the whole, better results than the vinyl emulsions. Application properties, drying time, etc. are similar to the vinyl types, and the same precautions should be taken in the cleaning of brushes. In recent years acrylic emulsions have become popular as media for artists' colours where their quick-drying properties are an asset.

PAINTS FOR SPECIAL PURPOSES

Anti-condensation paints are designed to reduce the amount of condensation and the unsightly runs on the walls of rooms where high humidity can develop, as, for instance, in dye rooms, dance halls, and so on. They are made by adding cork flour to a flat finish or emulsion paint, and the mixture is best applied by spray. The cork reduces the heat conductivity of the paint film and so reduces the amount of condensation.

Floor paints are designed for use mainly on concrete floors, both inside and outside buildings. They check the unpleasant dusting from such floors but, in addition, should possess good adhesion, alkali resistance, and resistance to wear. Chlorinated rubber, styrene–butadiene type co-polymers and two-pack epoxies are among the types of media employed. Moisture-cured polyurethanes are also used, but only as clear coatings (page 178).

Masonry finishes "Masonry finish" is a general name for coatings used on exterior wall surfaces. A number of types are in use and the following are among the most popular: (a) styrene-butadiene type co-polymer solutions pigmented to give flat finishes, (b) emulsion paints used alone or mixed with grit to give imitation stone finishes, (c) proprietary cement paints supplied as dry powders and mixed with water for application.

Fungicidal paints These contain compounds of tin or other substances designed to discourage the formation of mould growths.

Anti-fouling paints These are used to protect ships' bottoms from marine growths such as weed, barnacles, etc. They contain toxic substances such as metallic copper, copper compounds, tin compounds, etc. in special media.

Marine paints Decorative paints used in marine environments must also give good protection. They are usually based on long-oil alkyds, often modified with tung oil and pigmented with rutile titanium dioxide. Similar paints are used on ships' superstructures, but for this purpose polyurethanes and chlorinated rubber are also being employed.

Acid-resisting paints The type used depends on the nature and concentration of the acid. Bituminous paint, which can be applied in thick coats, will withstand most acids and is the cheapest type. It is used where colour and appearance are less important than protection, as in battery boxes. For white and coloured paints chlorinated rubber is generally used, but it is important that the pigments should also be acid-resisting.

Anti-corrosion paints Paints intended to come in direct contact with metal should contain nothing liable to react unfavourably with the metallic surface. When this condition is met almost any well compounded paint may be used for secondary and finishing coats, provided always that it combines good water-resistance with flexibility and durability. The all-important coat therefore in any anti-corrosion treatment is the primer or first coat. Primers for iron and steel should contain rust-inhibiting pigments (alone, or combined with neutral or non-stimulating pigments) in suitable media. Primers which prove satisfactory on iron and steel may, however, be totally unsuitable for non-ferrous metals, as for example *bauxite* red oxides which, owing to their alkaline nature, may react unfavourably with aluminium in the presence of moisture. This fact should be borne in mind when specifying paints for various metal surfaces.

Fire-retarding paints These paints have assumed considerable importance as a result of the wide use of flammable lining and partitioning materials in building work. They are designed to reduce the surface flammability of combustible materials by sealing them off from the air for a vital initial period during which emergency measures can be taken. If the source of heat is intense enough the material will ultimately catch fire, but the use of fire-retarding paints delays the onset of general burning. They are important in such places as hotels and ships' interiors where anything likely to check the rapid spread of fire is invaluable. All fire-retarding paints must be accompanied by a certificate of efficiency issued by the Fire Research Station, Boreham Wood, Herts. The two general types in use are the following.

"Flame-retardant" paints, designed to prevent the rapid spread of flame across a surface. One type used extensively contains antimony oxide dispersed in a medium containing chlorinated compounds. Under the influence of heat, volatile antimony chloride is formed and smothers any flame. Patents for these paints are held by Associated Lead Manufacturers Ltd.

"Intumescent" paints contain an incombustible compound such as ammonium phosphate which forms a foam when the film is heated. This gives an aerated coating of low heat-conductivity which protects the surface for some time.

Mordants Chemical solutions, usually of copper salts, used for etching new galvanized iron or other zinc-coated surfaces to obtain a suitable key for painting.

Heat-resistance of paints

The heat-resistance of a paint film is generally referred to when discussing paints to be applied to radiators, steam pipes, boilers, metal chimneys, etc. and relates to the temperature a given paint film is capable of withstanding without discolouring or breaking down.

A paint film may discolour to a marked degree yet still provide a protective coating to the surface on which it has been applied. This discoloration can spoil a colour scheme, however, or affect an identification colour to the extent that a dangerous error could occur through wrong identification during maintenance work. Discoloration usually occurs as a darkening of the medium when subjected to heat, and the degree of darkening will vary according to the type of medium and the temperature to which it is subjected. Certain pigments are also susceptible to heat, but the temperature at which discoloration occurs can vary considerably. It is of the utmost importance, therefore, that the paint be formulated to withstand the temperature to be encountered.

In many cases it has been found that standard paints are suitable for low-temperature surfaces such as hot-water pipes and radiators, etc., and although some discoloration may occur at approximately 66°C (150°F) in white and very light pastel colours the discoloration may not be apparent in darker colours such as browns and black.

Apart from the actual heat-resistance of a paint film it is interesting to note that the radiation and heating efficiency of a hot-water or steam heating system depends upon the type of paint applied to the radiators and heating pipes. Bright metal surfaces, aluminium paint, bronze paints and gloss enamel paints (as will be seen from the following figures) give minimum radiation, whilst matt finishes and metal surfaces in a rusty condition give maximum radiation.

Taking the amount of heat radiated from a new pipe as 100 units, the following relative figures have been observed for the heat radiated under the following conditions from a pipe tested under steam at 14 bars (200 lb) pressure.

Condition	Units
New pipe	100
Fair condition	116
Rusty and black	119
Painted dull white	120
Painted glossy white	100·5
Painted dull black	120
Painted glossy black	101

It would appear from the foregoing figures that the colour of the pipe has little or no effect on the radiation of heat, although radiation varies considerably in relation to the glossiness or dullness of the surface. The figures demonstrate that there is no serious loss in efficiency through making pipes and radiators harmonize with the general colour scheme of the rooms in which they are placed, provided that glossy finishes are avoided.

As the surfaces to be painted are usually heating pipes, radiators and hot water pipes, the method of painting should be such as to keep the elasticity of the film build-up similar, otherwise unequal expansion of the different coats in the paint film might cause cracking and subsequent flaking. This may be achieved by the use of a good adhesive primer such as zinc chromate primer followed by two coats of the required finishing material (i.e. omitting any undercoating in the case of enamel finishes). Before priming, the metal should be thoroughly prepared to remove all rust, dirt and loose material. It should then be subjected, if possible, to mild heat — approximately 50°C (122°F) — to dry off any moisture which has condensed on the surface, and the primer applied while the metal is still warm.

Special heat-resisting paints do not require primers as these may disintegrate at lower temperatures than the finishes themselves; and it is absolutely essential that the metal be perfectly clean and dry before the paints are applied.

Air-drying heat-resisting paints based on silicone alkyds are available in a reasonably wide range of colours. Pale tints are resistant to working temperatures up to 125°C (257°F) whilst dark colours will withstand temperatures up to 175°C (347°F).

For temperatures higher than these, unmodified silicone resins have to be used, but these paints do not air dry. They need to be heated to 200–230°C (392–446°F) for 1 to 2 hours to cure the film, after which the article may be raised to the required temperature. For very high temperatures, such as those existing on steel chimneys, engine exhaust manifolds, etc., aluminium paints based on silicone resins are usually employed. Steel chimney stacks reach temperatures of 600° to 700°C in the lower sections, and these are above the temperatures at which organic binders decompose. In these paint films the silicone resin decomposes at about 350°C (662°F), but the metallic aluminium pigment becomes sintered to the steel surface and affords protection.

Miscellaneous paints

Aluminium paints are used for both decorative and protective work and are expected to retain their bright metallic sheen for a long period. If the paint becomes grey or "leaden" in a short time the cause can be traces of acid in the medium, and this is often accompanied by gassing in the container. These paints are usually based on a tung oil/coumarone or tung oil/phenolic varnish pigmented with leafing aluminium pigment. The particles of pigment are in the form of flat plates, and overlapping or "leafing" takes place in the film. This accounts for the great opacity and high moisture-resistance of aluminium paints. When based on silicone resins, aluminium paints will withstand very high temperatures (see previous section). Aluminium paints should not be used inside buildings where flammable solvents are used, as a spark can be generated if the paint surface is struck with a steel object.

Bronze paints are manufactured in a number of shades ranging from Pale Gold ("gold" paints) to Antique Bronze. The pigments are alloys of copper in flake form, but their leafing properties are not as pronounced as those of aluminium, probably due, in part, to their higher specific gravity. Bronze pigments have poor weathering properties and are very sensitive to traces of acid in the medium. They are usually dispersed in solutions of polymers such as polyvinyl acetate or polystyrene and are used in interior decoration.

Graphite paints contain natural or synthetic graphite in a varnish or alkyd medium. The graphite particles exist in the form of flat plates and have a similar leafing property to aluminium. Graphite paints are not recommended for priming bare steel but are excellent for undercoats and finishing coats as they have high water-resistance. They are rather slow drying but possess exceptional spreading power and opacity.

Non-oxidizing paints are based on petroleum jelly, a material which does not oxidize or harden. It is pigmented and made sufficiently rigid to withstand gentle water movement. These paints are applied in thick coats to iron and steel surfaces which are to be immersed in water for long periods, and they represent one of the cheapest ways of protecting such surfaces.

Multicolour finishes

In the ordinary way spray-spatter effects require two or more spray applications over a prepared ground coat as described on page 131, but there are now several proprietary multicolour paints available which enable these effects to be achieved in a one-coat application. Currently these paints are made in this country under licence to the Coloramic Corporation in Los Angeles, and are based on an ingenious method of manufacture patented in 1952 by John Zola, an American paint technologist.

The formulation of these paints may vary with different makers, but in general they consist of one or more pigmented nitrocellulose lacquers suspended in a protective colloidal sac or envelope (e.g. methyl cellulose) which prevents coalescence. The pigmented nitrocellulose lacquer is poured into the aqueous colloid solution and dispersed with a low-speed paddle mixer at a carefully controlled speed and temperature. High viscosity helps the particles to remain in suspension, but stirring, if necessary, must be gentle to avoid coalescence of the suspended particles. Careful storage is required, and the average "pot life" at time of writing is approximately only six months.

Application is by ordinary spray equipment, using a large nozzle internal-mix sprayhead at low pressure, i.e. 1 to $1\frac{1}{3}$ bars (15–20 lb) on the air line and $\frac{1}{3}$ to $\frac{2}{3}$ bar (5–10 lb) on the pressure cup. Hot-spray and airless spraying techniques are not suitable, nor is brushing or roller-coating. Small amounts of water can be added where necessary to adjust the consistency, but overthinning will produce a poor film and coalescence of lacquer particles. Spray equipment is cleaned with water and finally with nitrocellulose thinners to remove any trace of lacquer.

Multicolour paints produce a thick, tough film which is extremely hardwearing, scratch and abrasion resistant. They are non-static, and resistant to oil, grease, soap and water. They can provide very attractive finishes for walls and are particularly useful for camouflaging rough surfaces, i.e. concrete, rough metal castings, etc. They are used extensively for industrial finishes on metal, plastic and fibre pressings, but are not suitable for foamed polystyrene surfaces. They can be used externally if protected by varnish or clear PVA coating. They are touch-dry in 1–1½ hours, and hard-dry overnight in a warm, dry atmosphere, but humidity and condensation tend to retard drying. Film thickness is 3–4 times that of ordinary paint coatings, and the spreading rate is therefore low, i.e. 2 to 4 sq. metres per litre (12–20 sq. yd per gallon).

Claims are made that these one-coat finishes can be applied direct on new wood and other absorbent surfaces, and whilst this may be true we believe that priming porous surfaces is advisable to improve adhesion and film build-up. Special primers are required for metal surfaces and it is generally necessary to protect previously painted surfaces with a "barrier" coat to prevent lifting of the old film by the action of nitrocellulose solvents.

There is no spray mist owing to the low pressure required for application, but adequate ventilation is necessary to dispel the odour from the solvents, and although the latter have a high flash-point, the concentration of heavy vapours should be avoided to eliminate any fire hazard

18

Miscellaneous Materials

STAINS AND SCUMBLES

The treatment of hardwood and decorative veneers was once the special province of the French polisher, but with the increasing use of flush-panelled veneered doors, and of such applied materials as Realwood panelling, the painter is now being called upon more often to finish these surfaces and will have to exercise due care in the choice of materials. The delicate markings of the natural wood grain can be greatly enhanced by the right use of well-chosen stains, just as readily as they are ruined by unsuitable opaque materials.

There is some ambiguity in the use of the terms "stain", "glaze" and "scumble" and it might be as well to define them before going on to consider the various types —

Stain A true stain is a solution containing soluble dye. The colouring-matter may be water-soluble, spirit-soluble or oil-soluble dyes according to the medium used. Non-soluble semi-transparent pigments such as siennas and umbers are also used for staining woodwork, but these lack the purity and transparency of true stains.

Scumble Semi-transparent pigments in oil or water media, usually containing ingredients which retard the set and flow in order to assist manipulation; generally applied over a painted ground. The painters' familiar graining colour or "graining oil" may be described as a scumble.

Glaze Any transparent or semi-transparent colour applied over another colour to enhance or to modify it. Thus, bright reds are often enriched by "glazing" with lakes. Graining is sometimes "glazed" to give depth and translucency to the work, and light colours may be toned down or otherwise modified by glazes.

The introduction of even small amounts of opaque colour will destroy the translucency of a stain or scumble, and this fact should be constantly borne in mind, especially when matching up to existing furniture. A case brought to the writer's notice, in which the client complained of "muddy" colour in a room where the woodwork had been grained to match some mahogany furniture, illustrates this point. It is difficult enough to match the rich transparency of stained and polished work by means of paint and scumble, but in this case the painter had made things worse by adding vermilion (one of the most opaque of pigments) to his graining colour. Such ignorance is, indeed, lamentable and amply justifies the somewhat tedious statement of elementary principles when these might otherwise be safely taken for granted.

Stains are available in water, spirit, oil and varnish media, though the two latter are usually pigmented and therefore not in the category of true stains. Water and spirit stains tend to strike into absorbent surfaces as a dye and are therefore sometimes rather difficult to manipulate, especially in hot weather when evaporation is more pronounced. Oil stains and scumbles are less penetrative and can be applied more uniformly even on very absorbent surfaces, though scumbles are generally applied over non-absorbent painted grounds which give ample time for surface manipulation.

Water stains

Water stains are coloured aqueous solutions made from water-soluble vegetable dyes such as logwood, indigo, saffron, quercitron bark (Dutch pink), etc. They emphasize the grain, especially that of soft woods, the stain being resisted by the hard resinous parts, but readily absorbed by the soft, porous portion. Semi-transparent pigments such as sienna, umber, Vandyke brown, Prussian blue, etc., are also used for staining, graining, glazing and scumbling in water; they are supplied in pulp form, usually lightly bound with a little water-soluble gum such as gum arabic and having a little glycerine added to keep them moist. When used for graining, they are often bound with stale beer. Water staining on bare wood is liable to raise the grain, roughening the surface and entailing much subsequent glasspapering.

Chemical staining

Chemical staining of wood is sometimes done by the French polisher, using aqueous solutions applied direct, as with permanganate of potash (Condy's fluid), soda and lime, or as a "fume" process, e.g. the fuming of oak with ammonia. There would seem to be little need, however, for the painter to dabble with chemical stains since he is well catered for with materials better suited to his purpose. Moreover, for obvious reasons, chemical stains of an alkaline nature are not to be recommended when oil varnish is to be used for finishing.

Spirit stains

Spirit stains are spirit-soluble dyes in shellac or other solutions, the principal colouring agents being Nigrosine (black), Bismarck brown, Gamboge (yellow), Turmeric (yellow) and various synthetic dyes. They are usually very penetrative and evaporate quickly, thus requiring expert handling to avoid patchiness. Water and spirit stains may be finished with French polish, oil-varnish or cellulose lacquer as required.

Oil stains

Only a limited number of dyes are oil-soluble and so-called "oil stains" therefore usually consist of semi-transparent pigments — sienna, umber, Vandyke brown, Prussian blue and various lakes — ground in linseed oil and thinned with white spirit, turps or solvent naphtha. Some proprietary brands are referred to as "flat oil stains". Oil stains are normally finished with gloss or flat varnish, but they can also be French polished if nothing but volatile solvent is used for thinning.

Varnish stains

Varnish stains are oil stains mixed with a gloss varnish, usually of the short-oil, hard drying type. They are used for the treatment of floor surrounds, matchboarding, etc., but the effect is inferior to properly stained and varnished work.

Wax stains

These are semi-transparent pigments in beeswax thinned with turpentine and provide an economical finish for hardwoods. When dry they can be polished by rubbing briskly with a short, stiff brush (such as a shoe brush) or a piece of coarse jute. They must be thoroughly removed before any subsequent redecoration with oil stain or varnish, otherwise drying may be seriously impaired.

Graining colour

Graining colour (or "graining oil") is composed of semi-transparent pigments such as the earth colours ground stiff in oil and hand-mixed by the painter according to his special requirements. The old-time grainers had their own ideas about the composition of the medium, some using raw linseed oil and turps, others preferring boiled linseed oil and turps. Some sort of "megilp" — prepared from melted beeswax, sugar of lead and boiled oil — was also favoured, with the object of arresting flow and giving the necessary "stay put" properties to the graining colour. Incorporating wax in this way is a risky business unless carefully controlled, and a much safer and more convenient method of arresting flow is to use a small proportion of one of the proprietary transparent glaze media. When boiled oil is used in graining colour (or in any paint) it should be remembered that less drier will be required.

Scumbles and glazes

Proprietary scumbles are designed to take the place of hand-mixed graining colour. They are prepared by mixing semi-transparent pigments in oil with a suitable glaze medium and may be thinned to working consistency with raw or boiled linseed oil and white spirit, or with white spirit only, as required.

Transparent glaze media are prepared from pigments which have little or no opacity in oil, such as certain stearates, china clay, etc. They have a flatting effect on oil media and also prevent flow and retard the setting, thus giving time for surface manipulation such as combing, rubber stippling, etc., whilst retaining the pattern crisp and clean. They may be tinted with oil colours or with the same maker's scumbles. The consistency as supplied is fairly thick, but the glaze brushes out quite easily owing to the "slippy" nature of the pigment. They should be well brushed out, or thinned with white spirit to avoid too thick a coating, otherwise they tend to dry to a soft film with a surface skin. These glaze media have other uses; they may, for instance, be added to flat oil paints to facilitate "blending" operations since they retard the setting without adding any appreciable gloss to the flat finish.

In recent years the supply position has become very difficult and uncertain, but there are still some manufacturers who are producing scumbles and transparent glaze media.

WOOD PRESERVATIVES AND INSECTICIDES

There are many instances where wood requires to be protected from rot and decay, but where periodic maintenance with paint would be out of the question; telegraph poles and wooden railway sleepers are cases in point. Preservatives which are to be effective must possess maximum resistance to moisture and must be proof against attack from various fungi, wood-boring insects, etc.

The materials commonly used for this purpose are (i) coal-tar and creosote preparations; (ii) spirit-soluble metallic compounds, and (iii) water-soluble metallic salts.

Coal-tar pitch

Coal-tar is the main ingredient in cheap tar preparations and is often used in its crude form; wood posts, for instance, which are to be inserted in the ground are often dipped in hot tar to prevent rotting. If the wood is damp, however, tar would seal up the moisture and this would lead to dry rot sooner or later.

Creosote

Creosote and coal-tar derivatives (such as cresylic acid, naphthalene, phenol, etc.) are generally used for outdoor fencing and for forced impregnation of timber which is to be in contact with the ground. Wood preservatives of this class are usually very penetrative but, whilst they make the timber impervious to damp, they do not seal the surface and therefore do not induce dry rot when applied on relatively damp woodwork. In fact, their toxic and antiseptic properties tend to destroy dry-rot spores and also afford some protection against wood-boring insects. They usually have a penetrating odour and because of their acid nature they will irritate the skin and so need care in application.

The characteristic colour ranges from a light brown to a dark antique oak shade, but proprietary preparations of the familiar "Solignum" type provide a limited range of other colours and are specially suitable for unseasoned timber where paint would be unsatisfactory; some of these preparations may be varnished *after they have become perfectly dry*. Ordinary creosote or bituminous preparations cannot be subsequently painted or varnished unless the surface is effectively sealed with tar-proof knotting, otherwise discoloration and defective drying will probably result. Any sealing treatment, however, is unlikely to be satisfactory on *thick* coatings of tar or bitumen, since these generally remain soft and plastic and therefore liable to induce cracking if treated with knotting.

Spirit-soluble metallic compounds

Copper or zinc naphthenates and other metallic compounds soluble in white spirit and/or naphtha give rot-proof solutions which have excellent toxic properties; these solutions are particularly effective against woodworm, furniture beetles, and similar wood pests. Proprietary solutions such as "Cuprinol" should be used in accordance with the maker's instructions.

Water-soluble compounds

Water solutions containing mercuric chloride (which is highly toxic), zinc chloride, copper sulphate or other water-soluble metallic salts, have rot-proofing and fungicidal properties which discourage mould growths. Some of these have a corrosive effect on iron and steel which should, therefore, be protected from them during application.

Both the spirit-soluble and the water-soluble types of preservative are generally harmless to oil paint *when dry*. Surfaces which have been impregnated with these preparations may be painted with safety, provided the surface is quite dry at the time of painting and remains dry afterwards.

GLUE AND SIZE

Glue is an impure form of gelatine obtained by boiling the skins and bones of animals. The best quality cake glue is known as *Scotch glue*; it is generally made from skins and is stronger and more adhesive than bone glue. 0·454 kg (1 lb) of good Scotch cake glue when soaked in cold water for 48 hours or so should swell up and absorb the greater part of 5 litres (1 gallon) of water. *Concentrated size* is granulated or powdered glue which varies greatly in quantity and strength owing to the fact that it is often reduced with adulterants; 0·454 kg of concentrated size, when melted in about 7 litres ($1\frac{1}{2}$ gallons) of hot (not boiling) water should form a trembling jelly of usable strength on cooling, but if a firmer jelly is needed, only 5 litres of water should be used for melting. It is important that the granulated size be first soaked for about three minutes with just sufficient cold water to cover it, otherwise it will tend to coagulate into hard lumps when the hot water is poured over it. Overheating of size will prevent it from jellying, and the practice of melting it over a fire is not recommended.

Alum is sometimes added to glue size to harden it and render it insoluble; when added to jelly size it reduces it to a liquid state. 28 grams of alum added to 10 litres of size or *claircolle* (a mixture of size and whiting) will help to make a following coat of distemper work cool on a "hot" ceiling, but the chief objection to its use is that it renders the subsequent removal of distemper and claircolle very difficult; this does not apply in present-day practice because such distemper is no longer used in the U.K., though it is relevant in certain parts of the Continent where distempers are still in use.

A solution of glue size is rapidly putrefied or decomposed, especially in hot weather, unless some form of preservative is added, such as a small amount of carbolic acid or oil of cloves (about 2 percent of the size content).

In addition to being the principal binding agent in distemper, size is used in the preparation of walls for paperhanging, when it evens up the suction of porous walls and assists the adhesion of the paste.

WAXES

There are various kinds of wax derived from animal, vegetable and mineral sources, but those which chiefly concern the painter are Beeswax and Paraffin Wax.

Beeswax

Beeswax is a secretion of bees from which the honeycomb is formed. The natural colour is yellow, but a white or bleached wax is obtained by refining. It is soluble in turpentine but solution is greatly assisted by gently warming and agitating. Beeswax has been employed since ancient times as a painting medium, the Greeks applying melted wax with heated spatulas in their "encaustic" painting, while in medieval days an emulsion of saponified beeswax, linseed oil and glue in water was used before the advent of oil painting. A similar emulsion is used in some types of washable distemper today, but the principal use of beeswax is in the preparation of wax polishes and stains. A small proportion of beeswax is used in some matt varnishes and flat oil paints (though, in general, other flatting agents are preferred nowadays).

Paraffin wax

Paraffin wax is a product of petroleum distillation. It is a white translucent wax, soluble in white spirit and other solvents, its chief use being to thicken spirit paint removers to make them "stay put" on vertical surfaces and to prevent the too-rapid evaporation of the highly volatile solvents.

PAINT REMOVERS

Paint removers are solvents which soften paint and varnish films, enabling them to be removed by scraping. There are two main types — alkaline and spirit.

Alkaline removers

Alkaline removers are strong caustic solutions, usually thickened to a brushable paste consistency to facilitate application and minimize splashing. They are not recommended for wood, plaster, or other porous surfaces which are to be repainted, since (apart from raising the grain and darkening the colour in the case of wood) the solution becomes absorbed in the surface and is extremely difficult to remove completely. A wash of vinegar or acetic acid is usually recommended to neutralize any remaining alkali, but this does not always guarantee a neutral surface. The only instance where the use of caustic can be justified is possibly on large stretches of ironwork where other means of removal might prove too costly. Bristle is ruined by caustic removers, and only cheap fibre brushes should therefore be used for application; face, hands and clothes should also be protected from its damaging effect. It should never be used on aluminium which is dissolved by strong alkalis.

Spirit removers

Spirit removers are strong volatile solvents thickened with wax or stearates to retard evaporation and to make the liquid "stay put" on the surface. They are much safer and more effective than caustic, but more expensive. Highly flammable types must be used with caution, but nonflammable solvents are also available. Spirit removers may be safely applied with ordinary paint brushes, and can be used with advantage on delicate carved work, window sashes, etc., where burning-off would involve some risk. Several applications may be required to remove thick paint films.

It should be noted, however, that many painters simply do not understand the process and use paint remover in a wasteful manner, as a result of which they get unsatisfactory results and add considerably to the cost of the operation. It is of no use expecting a liquid paint remover to behave like a blow-lamp flame and to soften a paint film directly it is applied. Yet this is precisely what is seen so often on the site — a coating of paint remover is applied to a surface and the stripping knife is brought into play at once, before the film has started to soften; all that happens is that the paint remover is scraped away needlessly, and a further application is required. This is repeated over and over again until the wastage reaches serious proportions, and still the paint film is not properly removed.

The correct and efficient method of using paint remover requires patience. First, a thin coating of paint remover is applied to the surface; this is allowed to remain there for 20–30 minutes until it has begun to etch its way into the film. At this stage a further coating is applied on top of what is already there; this coating is applied as thickly as possible, and it will be found that it clings to the etched surface. The second coating is then allowed to remain undisturbed until the paint film has softened throughout, and this may take several hours; once it is softened, however, it will scrape off cleanly right down to the bare surface. On one occasion the writer saw a quantity of carved woodwork in a Wren church in the City of London being stripped with liquid paint remover. It was estimated that there were some eighty coats of paint on the surface. It took four days for the paint to soften completely, but the reward of patience was that the whole mass of paint came away in one operation, leaving a beautifully clean surface without any damage to the delicate carving — and without wasting vast quantities of paint remover.

When all paint has been removed, the surface should be well washed down with petrol or white spirit to clean off any residual wax which might otherwise retard the drying of subsequent coats of paint. Spirit removers are necessary when the surface is to be subsequently stained since alkalis tend to discolour or darken the wood, and burning off may spoil the surface by scorching.

ABRASIVE MATERIALS

Not the least important part of painters' work is the preparation of a smooth surface, and this entails rubbing down with various abrasives. This can often be a laborious and costly business; but, just as every job is expedited by using the right tools, so can rubbing down be done more efficiently when the abrasive is suited to its purpose. Some surfaces require wet rubbing down, others must be rubbed dry. For instance, a hard enamel surface cannot be rubbed down satisfactorily by a dry process; it requires the presence of wet grit to grind it down to a suitable condition. On the other hand, water-soluble materials such as water fillers must be rubbed down dry since wet rubbing would merely work up the material to a sticky mess. This point should be borne in mind by those who use soda when rubbing down wet in the belief that the softening of the paint will assist the rubbing process. It will not. You cannot rub down soft surfaces properly. If the surface is greasy, however, and tends to clog the abrasive, a mild detergent such as a knob of lime putty or a little ammonia may be used in the water.

Most of the abrasives described in the next few paragraphs are no longer used in ordinary trade practice, but for those occasions when a top-class job is required irrespective of cost, they are still relevant.

Pumice

Pumice stone is a grey volcanic rock imported from the Lipari Islands in the Mediterranean; it is light and spongy in texture and a good sample should float in water. It is purchased in rounded lumps, four or five inches in diameter, and requires rubbing down to a flat face on a wet stone flag to make it ready for use. Large lumps may be sawn in two with an old saw kept for the purpose. Rubbing down with pumice is usually done during the process of washing down. The stone must be kept free of loose grit (which would scratch the surface) and it should be rubbed from time to time with another piece of pumice stone to prevent clogging. One can tell by the sound and feel whether the stone is grinding properly, or merely sliding over the surface.

Powdered pumice is used for "felting down", that is, rubbing down wet with a piece of hard felt about an inch thick. This was generally used for rubbing down new paint or varnish between coats as in coachpainting, but it is now largely superseded by waterproof abrasive paper which is much cleaner and better in every way.

Pumice blocks are made from powdered pumice and usually contain a mild detergent which removes surface grease in the rubbing process. They are suitable for broadsword work such as matchboarding, etc., but are rather clumsy for panelled work.

Cuttle-fish bone

Cuttle bone is sometimes used for the wet rubbing down of undercoating varnishes between coats and used to be popular with coachpainters before the advent of waterproof paper. Cuttle is clean in use and grinds beautifully; the hard outer shell should be removed to avoid scratching the surface.

Rotten stone

Rotten stone is used for the finishing process in rubbing down varnish prior to hand polishing. The polishing varnish is felted down with powdered pumice or flatted with fine waterproof abrasive paper, then followed with powdered rotten stone, using linseed oil as a lubricant and polishing finally with dry flour.

Steel wool

Steel wool is available in various grades ranging from very fine for rubbing down old and new paintwork, to very coarse steel shavings for rough scouring; it may be used either wet or dry. It is often useful in providing a mechanical key to smooth surfaces such as non-ferrous metals, also for removing rust from steel; when used for the latter purpose, paraffin or white spirit should be used as a lubricant instead of water. It is generally advisable to wear gloves to protect the hands when using steel wool.

COATED ABRASIVES

Glasspaper and sandpaper

Glasspaper and sandpaper are used for dry rubbing down and for removing the nibs from paintwork in between coats; they are prepared by coating a stout, rigid, heavyweight cartridge paper with glue and strewing it with powdered glass, sharp sand, flint, etc. The paper should be flexible enough to stand creasing without shedding its grit and should be stored in a dry place, as damp conditions will soften the glue and loosen the grit. There are six or seven grades, ranging from "0" or "00" (very fine) to "No. 3" (coarse), but the grade numbers vary with different makers; some manufacturers have abandoned numbers in favour of initial letters, with F for fine, M for medium, C for coarse and S for strong. The very finest grade is sometimes known as "Flour". All glasspaper and sandpaper tends to clog up easily, as a result of which it is soon ready to be discarded; it also tends to scratch the surface.

Waterproof abrasives and other types of abrasive paper

With the passing of the Lead Paint (Protection against Poisoning) Act which became operative on January 1st, 1927, the dry rubbing-down of lead paint became illegal, and the need for a convenient wet-rubbing process was met by the introduction of waterproof abrasive paper. But decorators are a conservative race, and old habits die hard. The writer literally never saw waterproof sandpaper used during the whole of his apprenticeship, although lead paint was the commonest material employed at the time, and as recently as in 1956, when buying a quantity of waterproof sandpaper at a large decorators' stockist in the Midlands, he was told that no other customer ever asked for this product.

Coachpainters, however, were very quick to realize the value of the newer material. There were many factors leading to their acceptance of it — the fact that both the gritside and the backing are absolutely waterproof, so that the coating bond does not shed its grit nor the backing paper become soft and pulpy; the fact that the cutting action is very positive which makes for speedy working; and the lubricating action of the water which minimizes clogging, so that the paper can be kept clean by occasional rinsing and thus lasts far longer than dry sandpaper. The higher cost of the waterproof paper is therefore offset by its longer life and greater efficiency, and very soon it had completely superseded some of the traditional coachpainting methods of rubbing down, and especially the use of powdered pumice with its attendant risk of grit fouling the fine finish.

Gradually the resistance of the painter and decorator has broken down and the use of waterproof abrasives has increased. For a time there were two distinct types on sale — the painters' type in which flint and garnet were used and which resembled ordinary glasspaper in colour, and the industrial type used by the coachpainter which was dark grey or blue-black in colour and which used an abrasive grit composed of a natural mineral such as emery or corundum or an artificial product such as Carborundum.

In recent years the range of coated abrasives has widened enormously, and there is now a great variety of types available, each with its own particular sphere of use.

Grading

In most cases, abrasive papers are graded by a system of numbering, the system being a very simple one which gives a clear indication of the relative fineness or coarseness of the grit. Instead of the arbitrary 1, 1½, 2, etc., by which sandpaper is graded, the number corresponds with the size of mesh through which the abrasive particles will pass; thus, if the paper is graded 100 it indicates that the particles will pass through a mesh with 100 holes per square inch, if the grade is 400 the particles will go through a sieve with 400 holes per square inch, and so on. Messrs Durex Abrasives of Birmingham make some very fine grades such as Hydro-Durexsil 400 to 600A, which can safely be used on the finest gloss varnishes or enamels without fear of scratching. Some papers are marked on the back with both the grade number and the grit number, and the important thing to remember is that the higher the grit number the coarser the grit is, whereas the higher grade numbers indicate the finer grits. The paper backing is also graded; the finer grades of abrasive particles are bonded to a lightweight flexible paper backing, grade A, to facilitate rubbing down by hand; the coarser grades of grit are on a heavy rigid backing paper, grade D, which is suitable for both hand and mechanical working.

The most important of these abrasives in connection with general painting and decorating is the familiar "wet-or-dry" which, when lubricated with water, is excellent for rubbing down any kind of previously painted work and maintains its cutting properties for a very long time, provided it is rinsed and kept clean whilst in use. It is made from a synthetic silicon carbide attached with waterproof adhesive to a thin waterproofed backing paper (grade A), and it is supplied in five grades ranging from 120 to 600.

Silicon carbide is also used in the manufacture of what are termed "self-lubricating" papers, which have been developed for coachpainting or, as it is nowadays called, vehicle refinishing, and similar work. They are made in the same way as other abrasive papers but are also coated during manufacture with zinc stearate, which serves as a lubricant. They are used for rubbing down paint coatings, filler coats, etc., and although they are used in the dry state they do not clog. Because of their convenience they are being increasingly used in place of the traditional wet rubbing-down. At present they are available in five grades, ranging from 220 to 500.

Garnet papers are still used for dry rubbing-down. The coarser grades ranging from 40 to 100 consist of natural garnet bonded to a grade D paper, and are used for both manual and mechanical preparation of woodwork before painting. The finer grades ranging from 150 to 240 are also of natural garnet, bonded in this instance to grade A paper, and these are used for flatting down between gloss coats and in French polishing. But garnet as an abrasive for these purposes is being superseded by aluminium oxide, and in fact in the finer grades garnet papers have almost entirely given place to the aluminium oxide papers.

A further material for the dry rubbing-down of rough painted surfaces and new timber is composed of tungsten carbide bonded to a thin metal sheet, and this is available in three grades, listed as fine, medium and coarse.

For the etching and degreasing of metals, either in the dry state or in conjunction with a spirit lubricant, *emery cloth* or emery paper is used. As its name implies, it consists of natural emery attached either to a cloth or stout paper backing. It is supplied in fine, medium and coarse grades.

For normal decorating purposes all these papers are supplied in sheets of standard size, but they are also available pre-cut to size in rectangular or circular shape for use with orbital or disc sanders, and in rolls for use with belt sanders.

The curious thing is that all these sophisticated types of abrasive are now available at a time when very little lead paint is used, whereas in the days when the lead dust hazard was an ever-present menace on every site there was only dry sandpaper available.

Mechanical sanding

The use of abrasives need not necessarily be a laborious hand process; modern requirements can be matched by modern techniques. In the woodworking and furniture trades large power-fed drum and belt sanders are commonly employed, and the same principles have been adapted to the production of various portable tools for painting and decorating work. The range of these includes belt sanders, straight-line machines, rotary disc and orbital sanders, powered by electricity or compressed air. For most types of work orbital sanders are the most suitable, but where woodwork is to be stained or treated with a clear finish the orbital sander would mar the work by scratching across the grain, and the straight-line sander would be the better choice. These machines have been described in Chapter 4.

A sanding sealer is sometimes used to fill the grain of the wood without obscuring it before the mechanical sander is used.

19

Priming New Surfaces

The most important part of any building is the foundation, for no matter how magnificent the architecture, if the "footings" are unsound the structure will soon develop faults and may even collapse. And so it is with a paint film. The priming coat is the foundation upon which the subsequent coats depend for their attachment to the surface and, if this foundation coat is unsuitable for its purpose or fails to secure a firm anchorage, the paint film will sooner or later come to grief.

Yet there are many who regard priming as being unimportant enough to be carried out by unskilled labour with any old kind of paint. So long as steelwork is delivered to the site painted with something red and new woodwork is treated with something pink, they are regarded as being adequately primed. The decorator has no means of knowing what materials have been used in such "shop primings" unless the work has been done to a strict specification, and in all cases of doubt he will be well advised to remove as much of the priming as possible by vigorous scouring or rubbing down before applying his own first coating.

New surfaces which the decorator is called upon to deal with each have their own peculiar characteristics; they may, for instance, be absorbent or non-absorbent, wet or dry, chemically active or inert (neutral), corrodible or non-corrodible. It will be seen at once, then, that there can be no single universal primer to meet all these varied conditions satisfactorily and, in choosing or compounding a primer suitable for its purpose, the following points must be borne in mind.

Absorbent surfaces such as wood, plaster, cement, brick, stone, asbestos-cement sheeting, wallboards and paper allow the paint to penetrate and so become keyed into the surface, but the tendency on unduly porous surfaces is for the oil to be absorbed, leaving the pigment on the surface in an underbound condition. The paint should therefore contain sufficient oil to satisfy the suction, while still effectively binding the pigment in a cohesive film which provides a firm foundation for subsequent coats (but see Asbestos-cement sheeting", page 233).

Chemically active surfaces such as new lime plasters, Portland cement and asbestos-cement sheeting require alkali-resisting primers adjusted to suit the porosity of the surface. Acid conditions which cannot be properly neutralized will need to be treated with acid-resisting materials.

Corrodible surfaces such as iron and steel require protection from the corroding elements in water and acids. The primer must therefore be free from those pigments liable to stimulate corrosion, while the media must provide an adhesive impervious film.

Wet surfaces Traditional building methods rely on wet processes which involve long periods of waiting for surfaces to dry out before any permanent decorative treatment may be safely applied. In the old days a house that was not dry enough to decorate was not dry enough to live in, but this has now changed since houses cannot be built fast enough to meet the evergrowing demand. But people expect some form of decoration even on walls not yet dry, and whilst this is readily provided in a first-class waterpaint, many are not content and demand oil paint or enamel finishes, especially in kitchens and bathrooms, etc. Even municipal authorities are not blameless in this respect and one often sees gloss paints specified on new hospitals and schools.

The curious thing is that if you ask any decorator to paint wet woodwork he will tell you that it cannot be done or that it would be asking for trouble to do so, yet he persistently toys with the idea of finding a solution to the problem of painting on wet plaster, encouraged no doubt by widely advertised "sealers" and "damp-resisting solutions". That some of these preparations are *sometimes* successful is beside the point, for the painting of wet plaster (with one important exception) is wrong in principle, and thoroughly unsound practice. The one exception is that of priming Keene's and Parian type cements which are dealt with later (page 229).

It has been estimated that the amount of water contained in the new brickwork and plaster rendering of an average size room is somewhere in the region of one ton, and although in some cases a proportion of this water becomes "fixed", that is to say, combines chemically with the plaster or cement, the bulk of the water must escape by evaporation, a process which ordinarily requires many months even under good drying conditions. It does not require much imagination, therefore, to foresee what is likely to happen if one attempts prematurely to seal the surface with impervious coatings. Much of the moisture will escape by way of exterior cavity walls, but, where internal partition walls are painted or sealed on both sides, the effect is likely to be disastrous both to the paint *and to the plaster*. This latter point is often overlooked, and it needs pointing out with some emphasis that the premature application of oil paints may prevent the plaster from hardening properly, owing to imprisoned moisture. With certain types of plaster this may ultimately give rise to a crumbly or powdery surface which in itself is too weak to support a paint film for long. In addition there is the much more general risk that imprisoned moisture, condensing at the plaster–paint interface, will greatly reduce paint adhesion and result in early blistering or flaking.

With modern building methods an attempt is being made to reduce the amount of wet processes, and even to build entirely by dry methods as in prefabricated structures. Such innovations will bring their own problems, of course, but they will also widen the scope of the decorator in the treatment of new surfaces. The preparation and priming of the various new surfaces likely to be met in present-day practice is dealt with in the following sections.

PLASTER AND CEMENT SURFACES

In no other sphere of activity does the decorator experience more trouble than in the treatment of new plaster and cement surfaces. Very often he is obliged to carry out specifications, against his better judgment, on surfaces obviously unfit to receive paint; but it must also be admitted that many decorators are unaware of the principles underlying the proper treatment of such surfaces.

On many jobs the decorator has great difficulty in obtaining exact information as to the type of plaster and backing used and, even supposing this is given, the number of proprietary plasters on the market in this country is bewildering to a non-specialist. The situation was clarified some years ago by W. R. Pippard, working at what was then the Building Research Station. Bulletin No. 13 (*Calcium Sulphate Plasters*) issued by the BRS included a full classification of proprietary plasters. This has been incorporated in more recent bulletins issued by H.M.S.O., and the British Standards Institution has issued a specification for Gypsum and Anhydrite Plasters (B.S. 1191:1972).

Before we consider the problems of decorating the various plasters let us very briefly examine their origins and chief characteristics under the three main headings, Portland cement, lime plaster and calcium sulphate plasters.

(A) Portland cement

Briefly, Portland cement is obtained by burning a mixture of limestone and clay or shale at a high temperature until a clinkered mass is formed; this is cooled, then ground to a fine powder and packed in bags. When mixed with water (so replacing the water driven off in burning) the cement combines chemically with part of the water and sets to a hard mass. Portland cement is the essential constituent in concrete, asbestos-cement sheeting, etc., and is largely used in backings or floating coats for Keene's and other gypsum plasters as well as for exterior renderings (such as cement stucco). It is highly alkaline and quickly destroys ordinary oil paint when this is applied before the cement is dry. The period of time before a Portland cement surface can be safely decorated varies considerably, and a *minimum* period of twelve months is usually allowed before ordinary oil paint is applied. Surfaces may be strongly alkaline after several years, however, and it is advisable always to test the surface with litmus before painting (see Alkalinity, page 223).

High alumina cements

These appear under various brand names but are perhaps best known by the proprietary name *Ciment Fondu*. They are prepared by burning a mixture of limestone and bauxite, cooled rapidly, and then ground to a fine powder. They are very dark, almost black, in colour and harden very quickly; during setting they develop considerable heat and must therefore be kept damp for at least 24 hours after "placing". A good brand should contain no free lime or soluble alkalis, but it is always prudent to test the surface for alkalinity before painting. Any loose powdery deposit or efflorescence should be removed by using a stiff fibre brush and the surface should be dry at time of painting.

In the early 'seventies there were some instances of major structural collapse in buildings where load-bearing members had been composed of high alumina cement. The widespread alarm caused by these disasters led to an investigation into the safety of other buildings similarly designed, and in some cases it was found necessary to replace beams and carry out other modifications to correct faults which were disclosed. As a result of this, public confidence in high alumina cements was shaken. But although this is important to architects, consulting engineers, and the main contractors engaged on construction projects, it has no bearing on the work of the painter and decorator, which is concerned with the surface treatment of the material wherever he encounters it.

"Roman" cement

Various hydraulic cements are produced by burning a naturally occurring "cement rock" or clayey limestone, the burned mass then being ground to a fine powder. The so-called "Roman" cement is of this type; it was introduced towards the end of the 18th century and was largely used for the exterior rendering of stucco fronts during the Regency period and until late in the 19th century. Natural cements are still available in some areas, marketed either as Roman or Natural Portland cement; they are yellow to reddish brown in colour, quicker setting though slower hardening than Portland. They vary in chemical composition but are generally of an alkaline nature and should therefore be tested with litmus before painting.

(B) Lime plaster

Pure limestone is a soft white rock or chalk (calcium carbonate) which, when burnt in a kiln, forms quicklime (calcium oxide). Quicklime, when water is added to it, becomes slaked lime (calcium hydroxide) or putty lime, which is often used as a thin skimming over a rough lime-sand backing. This skimming slowly hardens and "carbonates" by the action of carbonic acid from the air when it reverts to the original calcium carbonate. In practice the putty lime is usually gauged with small amounts of plaster of Paris to hasten the set (and to prevent "shrinkage cracking"), when it is referred to as "gauged lime plaster". Improperly slaked lime may give rise to defects known as "blowing" or "popping" due to the expansion of particles of unslaked lime when these combine with water. This sometimes continues over a period of years, resulting in a pitted or cratered surface.

Hydrated limes, which comply with the British Standard for lime (B.S. 890:1972, Building Limes), are free from this defect.

Pure chalk lime is said to be relatively harmless to oil paint, but when contaminated with even small amounts of soluble alkalis such as soda or potash it then becomes as destructive as Portland cement. These soluble alkalis are always liable to be present either in the backing materials or in the water used in mixing, which means that in practice *all new lime plaster* must be regarded as being potentially destructive to oil paint.

Lime plaster dries and hardens by evaporation of the water and, as already mentioned, by slow carbonation of the surface. It will be readily seen then that if oil paint is applied prematurely, not only will the paint be attacked chemically, but the plaster also will be prevented from hardening and carbonating properly, resulting in a permanently weak skim. Carbonation to any appreciable depth may take some 9 to 12 months in favourable circumstances, but if drying is delayed, or if further moisture is introduced from any source whatever, chemical activity may continue for a much longer period. When dry and properly carbonated, a lime plaster surface should be chemically neutral when tested with litmus paper and may, under favourable conditions, be safely painted with ordinary oil paint (but see reference to carbonated surfaces on page 230).

Hydraulic limes

Impure limes of the greystone or the Blue Lias type are produced by burning clayey limestone which gives them certain hydraulic properties; that is to say, they combine with some of the water on setting and therefore resemble a weak natural cement. Since they are more or less alkaline, it is advisable to allow them all to dry thoroughly before applying an alkali-resisting primer.

(C) Calcium sulphate plasters

With one exception, commercial plasters are prepared from the mineral *gypsum*, a soft white crystalline rock referred to chemically as *calcium sulphate*. When this is ground to a powder and heated at a moderate temperature (about 170°C or over) part of the water is driven off and common plaster of Paris is produced (known as *calcium sulphate hemihydrate*); when further heated (to 400°C or over) all the water is driven off and we get a class of gypsum cements or plasters known chemically as *anhydrous calcium sulphate*. Thus we have two main groups — hemihydrate and anhydrous — but modifications in manufacture produce gypsum plasters of widely differing characteristics which are classified under five headings as detailed below. The one exception noted above is the mineral *anhydrite,* a natural anhydrous calcium sulphate, which, unlike gypsum, has no water in its chemical composition and therefore requires no heat treatment in manufacture.

There are some fifty or more proprietary brands of plasters on the market, and for a long time there was often no indication as to which group a particular brand might belong. It was laid down, however, in British Standard 1191 published in 1955 that the vendor should "show clearly on each package of plaster the class to which the plaster belongs". For this purpose the plasters were classified as

(a) Plaster of Paris
(b) Retarded hemihydrate gypsum-plaster
(c) Anhydrous gypsum-plaster
(d) Keene's or Parian
(e) Anhydrite.

A recent revision of this same British Standard takes the matter a stage further by distinguishing four classes of plaster according to their relative porosity and hardness; it does so by listing them as Class A, B, C or D, the porosity of the plaster decreasing and its hardness increasing from Class A to Class D. The four classes are as follows:

Class A Plaster of Paris
Class B Retarded hemihydrate plasters (e.g. Thistle)
Class C Anhydrous plasters (e.g. Sirapite)
Class D Keene's plaster.

The hemihydrate group

As previously mentioned, plaster of Paris is the basis of this group. When mixed with water it sets or *hydrates* within a few minutes, which means that the water does not merely evaporate, but partly combines chemically with the plaster and so the mix reverts to its original composition — gypsum. All gypsum plasters set chemically in this way by a process of expanding crystals which unite in a cementing and interlocking action. In this they differ from ordinary lime putty which dries and hardens slowly by evaporation of the moisture and carbonation of the surface. The setting of plaster is usually accompanied by a slight expansion, whereas lime contracts and tends to develop fine hair cracks; hence the practice of gauging lime putty with plaster to minimize shrinkage cracking.

Neat plaster of Paris sets so rapidly that it can be used only for pointing small holes and cracks, etc., but by adding, during manufacture, small quantities of an agglutinant (such as *keratin* dissolved in alkali) the plaster may be retarded sufficiently to enable it to be used for finishing coats which possess the great advantage of being fully set or hydrated within two or three hours of application. Such retarded plasters are known as *retarded hemihydrates*, or, more commonly, "hardwall plasters". The term "hardwall plaster" is somewhat loosely applied and is not confined to the hemihydrate group; for example, Sirapite and Pioneer are sometimes referred to as hardwall plasters although they belong to the anhydrous group. Representative proprietary brands in this class are — Adamant, Aegrit, Faspite, Gothite, Murite, Napco, Pytho, Thistle, etc.

When used throughout for backing and finishing (i.e. without the addition of lime) hardwall plasters of this type are chemically neutral and may be safely painted *when reasonably dry* with ordinary oil paint, or even gloss paint and enamel. Since most of the water combines chemically with the plaster, drying is fairly rapid, especially when they are used as a thin skimming coat on plasterboard or wallboard, and painting may be safely proceeded with in two or three days. From the painting point of view then, *a neat retarded hemihydrate plaster is the most satisfactory to decorate*, since it possesses the fewest potential disadvantages. When applied on a new cement or mortar backing, however, the moisture (and alkalinity) of the backing and brickwork must be taken into consideration and sufficient time allowed for thorough drying before sealing the surface with impervious coatings. If painted too soon, this type of plaster is liable to "sweat out" (fail to harden) owing to entrapped moisture.

Plasterers very often add small amounts of lime putty to hardwall plasters (even when "neat finish" is specified) to render the mix more "fat" and easy working. The decorator should therefore take nothing for granted and should always test for alkalinity before proceeding to paint. If lime has been added (even if only in small amounts) the surface should be treated as lime plaster.

The anhydrous group

The plasters in this group, which includes the well-known Keene's and Parian cements, have small amounts of "accelerators" added during manufacture; otherwise they would set and harden too slowly for practical purposes. The accelerators most commonly used are alum, potassium sulphate, and zinc sulphate. Anhydrous plasters are classified under three headings as follows —

(i) Anhydrous gypsum plaster
(ii) Keene's or Parian
(iii) Anhydrite

The setting action of retarded hemihydrate plasters is fairly rapid when once begun, but with the anhydrous group it is generally slow and continuous since they do not combine with water so quickly. It is important therefore that the surface should remain wet long enough for the plaster to "absorb" as much moisture as it requires to complete its proper setting (or chemical hydration); otherwise defects such as "dry-out" or "delayed expansion" may occur. Such defects may not show themselves till some time after the decoration is completed and the decorator should, in his own interests, study this possibility, since he may be unjustifiably held responsible.

A "dry.out" is evidenced by a highly porous, friable, or powdery condition of the surface and is caused by a too-rapid drying of the plaster skim, due either to evaporation in warm weather or to absorption of water into a too-porous backing, both of which should be avoided by the plasterer. The decorator, however, may unwittingly cause a "dry-out" by using (or advising his clients to use) artificial heat to force the drying in the early stages. The manufacturers of Pioneer (anhydrite) plaster, for instance, state that two or three weeks should be allowed for the various components of the wall to complete the process of hardening and hydration under normal drying conditions (and suitable ventilation) before any artificial heating is applied.

A dry-out may go unnoticed, but if water is later introduced to the surface, either in the decorations or through the backing, the plaster will take up the amount of water required to complete its interrupted setting, the result being that delayed expansion may occur with consequent buckling or cracking of the skim coat.

Anhydrous plasters (or "gypsum cements" as they are commonly called) are regarded as being harmless to oil paint when used neat, since the small amount of soluble salts present as "accelerator" is not considered dangerous; but if lime is added, *even in the smallest amounts*, caustic alkali is formed and the surface becomes almost as destructive to oil paint as Portland cement. Here again, the decorator must satisfy himself that no lime has been added by the plasterer.

The slower rate of setting of the anhydrous group — and more particularly, Keene's and Parian — gives the plasterer ample time to bring them to a very smooth finish and, unless otherwise directed, the good plasterer, with a laudable pride in craftsmanship, will proceed to trowel the surface to a smooth marble-like polish. Such a surface when properly finished in gloss paint or enamel can look extremely well, but unfortunately, closely-trowelled surfaces usually offer very little opportunity for the priming coat to secure a firm attachment. This lack of key (and the presence of moisture in the plaster) often results in peeling or flaking of paint films; so much so that the decorator has come to regard Keene's and Parian type plasters as being particularly difficult to treat (Fig. 19.1).

The priming of these plasters with "sharp" oil paint immediately following the trowel has long been an established trade practice, even though the principles governing the procedure may not always have been properly understood. There is a wider interpretation of the term "following the trowel", some, for instance, taking this to mean "as soon as the plaster will stand brushing", while others consider "any time within 24 hours" as being suitable. There are also wide differences in the composition of the priming

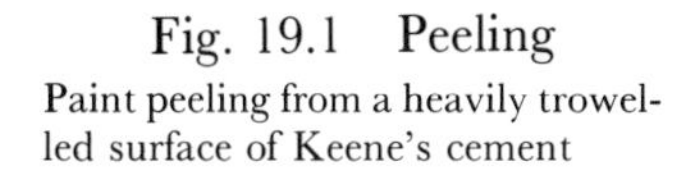

Fig. 19.1 Peeling
Paint peeling from a heavily trowelled surface of Keene's cement

paint. The most serious misconception concerning this practice, however, is the popular belief that the priming of wet Keene's may be quickly followed with undercoating and impervious gloss finishing coats without waiting for the plaster to dry out normally, an error which has naturally led to many serious paint failures.

The subject received little or no scientific investigation until Llewellyn and Eldridge, of the Building Research Station (as then existing), made it the basis of a special study. The results of their work were published by the BRS and were also included in a bulletin issued by what was then called the Paint Research Station (Bulletin No. 29, *The Decoration of New Plaster and Cement*). This valuable document cleared up many obscure points about the precautions needed to make painting on plaster reasonably safe, and the present chapter is largely based upon its findings; the writer gladly makes due acknowledgment both to the authors and to C. L. Haddon of British Plaster Board Ltd, for the great help received in this section. Their work has been incorporated in later H.M.S.O. books and pamphlets on the subject.

The principles involved in priming anhydrous plasters are dealt with under "Priming of Plaster Surfaces" (p. 229). Meanwhile, we will briefly touch on the separate types in this group.

Anhydrous gypsum plasters (trade brands include Sirapite, Statite, Victorite, Glastone, etc.)

Setting is slow and continuous, in common with all anhydrous plasters, which means that too-rapid drying must be avoided. They have moderate porosity and are not so hard as the Keene's type and there would seem to be less need for immediate priming — in fact the makers of Sirapite and Statite stipulate that the surface should be allowed to dry before applying the decoration, whereas the Victorite people recommend priming to follow the trowel. Where immediate priming is applied, however, the same precautions must be observed as with Keene's and Parian. Lime is very often added to anhydrous plasters and the surface should therefore be tested for alkalinity before painting; where lime has been added, treat as for lime plaster.

Keene's and Parian

Keene's and Parian plasters are nowadays substantially the same material and are no doubt vastly superior to the original "cement" patented by Keene in 1836. (The term "cement", though widely applied to Keene's and Parian, should, strictly speaking, refer only to Portland and alumina type cements.) They are the hardest burnt of the anhydrous gypsum plasters, slow setting and capable of being trowelled to a highly polished surface. They set exceedingly hard and generally have low porosity when close-trowelled, but nevertheless when thoroughly dry they often exhibit a fair degree of suction.

Keene's and Parian are used a good deal for running angles, mouldings and sharp arrises, as well as for smooth polished surfaces over strong Portland cement backings. For best results, priming following the trowel is recommended, except, of course, where lime is added as mentioned above.

Anhydrite plaster

This is prepared simply by grinding the naturally occurring mineral and adding the accelerator, no burning being necessary. It is said to give an even harder finish than plasters of the Keene's type, while possessing a fair degree of porosity. The only anhydrite plaster on the market is "Pioneer" the makers of which recommend that, for best results, decoration should be carried out on the dry plaster, although, in certain cases, priming may follow the trowel as with Keene's and Parian. On a highly efflorescent backing, "Pioneer" plaster tends towards quick elimination of the offending soluble salts, and the maker's instructions should therefore be strictly followed, both by the plasterer and the painter.

Lightweight plasters

These can be supplied either as undercoat plasters or finish plasters. They offer various useful features, such as resistance to fire, good thermal insulation, and the capacity to reduce occasional or intermittent condensation by absorbing a certain amount of moisture; it should be noted, however, that if this particular property is desired the plaster should not be painted with an impervious coating.

In the case of one of the lightweight plasters, known as Carlite, both the undercoat and the finishing material are based on retarded hemihydrate gypsum plaster with the addition of a lightweight aggregate consisting of perlite or vermiculite, or both. They both contain a certain amount of free lime and are therefore slightly alkaline, and because they are more porous than other gypsum plasters they tend to hold more water initially and take longer to dry out, especially in winter. If these plasters have not been trowelled evenly they can give rise to problems for the painter; the patchy surface combined with the slow drying-out rate leads to the unequal absorption of emulsion paint coatings, and furthermore, additional coats of emulsion, instead of levelling the surface, may pile up and exhibit the defect known as "sheariness". When this situation occurs, it is suggested that a preliminary coating of a non-aqueous sealer should be used. Another lightweight plaster known as Limelite has an undercoat based on Portland cement and lime, and is in consequence highly alkaline.

Alkalinity

When paint containing linseed oil is applied over an alkaline surface which is not thoroughly dry, the oil is liable to become *saponified* (Fig. 19.2). This chemical attack will usually show itself within a week of application by a

By permission of the Controller, H.M. Stationery Office—Crown Copyright reserved

Fig. 19.2 Saponification

Effect of saponification of oil paint on an alkaline plaster surface. Dark spots consist of yellow soapy liquid formed by the attack of alkalis on the paint medium

softening or stickiness of the paint film and, in acute cases, the film may become liquefied and develop oily, soapy runs or "tears"; the oil in fact has been converted to soap and is now water-soluble. In addition to this the pigment may become bleached or discoloured if it is not "lime-fast"; Prussian blues, lead chromes and Brunswick greens are particularly susceptible (Fig. 19.3).

A surface may be tested for alkalinity by moistening with *distilled water* and applying a piece of pink litmus paper to the wet portion; if alkaline, the litmus will immediately turn blue. Alternatively, a small test patch may be painted with Prussian blue or Brunswick green oil paint and kept under observation. If at the end of, say, a week there is no sign of bleaching or stickiness (saponification) the surface may be judged reasonably safe for painting.

Alkaline surfaces *when thoroughly dry* are not harmful to oil paint. In practice, however, new Portland cement, asbestos-cement sheeting and all plasters containing even small amounts of lime must be regarded as being potentially destructive to oil paint since, even if dry at the time of painting, moisture is always liable to be introduced from outside walls, leaky roofs or

By permission of the Controller, H.M. Stationery Office—Crown Copyright reserved

Fig. 19.3 Bleaching

Bleaching of pigment (Prussian blue) by alkalis, an effect which may accompany saponification of oil paint on alkaline plaster or Portland cement if "lime-fast" pigments are not used

faulty dampcourses, etc. An alkali-resisting primer is therefore always advisable.

Tung oil is more resistant to (though not *proof* against) alkaline attack than linseed oil and for this reason is generally used in combination with natural or synthetic resins in many alkali-resisting primers; other materials in use are heat-treated oils, synthetic media, chlorinated and de-polymerized rubber. Some of these primers produce fairly impervious films and should not, therefore, be applied to damp surfaces for reasons already given. Chemical washes or neutralizers, such as solutions of zinc sulphate, etc., are best avoided since they are liable to cause efflorescence.

Efflorescence

This troublesome defect appears on the surface in the form of a white salty deposit which is readily removed by dry wiping or brushing down, but which may continue to reappear even over a long period of years (Fig. 19.4). Efflorescence may appear on interior or exterior surfaces of brick, cement or plaster, some materials being more prone to the defect than others.

Most building materials contain small quantities of soluble salts and it is often impossible to trace the source of the trouble; bricks and mortar, sand, cement, plaster and even the water used in building operations may all be suspect. Sea-sand is a particularly prolific source and should be avoided at all costs.

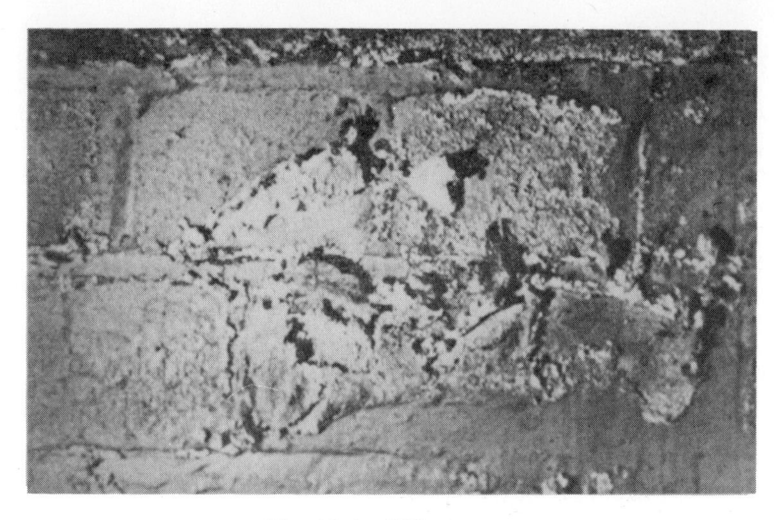

Fig. 19.4 Efflorescence

The defect does not develop until the wall surfaces begin to dry out, when the salts in solution are carried to the surface where they gradually crystallize as a white deposit, sometimes forming quite a thick "growth". If the wall has been decorated with distemper or water paint, the salts will often percolate through and crystallize on the surface where they can be brushed off without doing much harm. When an impervious paint coating is applied, the salts may be held back for a time; but if activated by moisture from the backing, they will concentrate beneath the paint film and eventually disrupt it or push it off in the form of peeling or flaking. It must not be thought therefore that any form of "sealer" will effect a cure, unless the wall is properly dried out and further ingress of moisture prevented.

If soluble salts are present in sufficient amount to cause efflorescence the crystals will usually begin to appear within a few weeks of plastering if drying conditions are good. The surface should therefore be carefully examined before any decoration is applied; if any powdery deposit is present, it should be removed by *dry* stiff brushing or glasspapering, following with a damp sponge to remove the last trace. The surface should then be kept under close observation for a few days, preferably heating the room to encourage the drying out process. If the deposit continues to re-appear after repeated removal, it is advisable to postpone any form of impervious decoration until the trouble ceases. The deposit should not be washed off with copious rinsing since this would merely dissolve the salts which would then be absorbed into the surface, only to re-appear again on drying.

Where the client insists on some form of decoration, however, a pervious coating such as emulsion paint or water paint should be used, thinned with water only. This will give the salts a chance to pass through to the surface where they can be wiped off from time to time as they appear. Not until the walls have become thoroughly dried out and the salts cease to appear should the surface receive any oil paint or impervious gloss finish.

PRIMING OF PLASTER SURFACES

The foregoing sections give some indication of the various characteristics of the different plasters and cements likely to be met in the general run of normal trade practice and the special precautions to be observed in dealing with the respective types. From the decorator's point of view it would seem that a neat, retarded hemihydrate plaster provides a surface with the least potential danger to paint.

Apart from individual requirements (such as in the case of Keene's type cements) the main considerations affecting choice of primer will be the condition of the surface at the time of painting. First and foremost is the question of moisture, for we have seen that alkaline surfaces present little or no difficulty when thoroughly dry. The decorator therefore has to decide —

(a) whether the surface is dry enough to receive oil paint and, if not,

(b) the best alternative treatment.

The first requirement (a) is often very difficult to decide for it is not a simple matter to determine the amount of moisture in a newly plastered wall. The decorator must therefore rely largely upon his judgment and experience, coupled with one or two rough-and-ready tests. Suction, for instance, will give some indication of a dry condition at the surface, though this would not necessarily mean that the plaster is dry throughout. Still, it is sometimes useful to wet the surface and note how quickly or slowly the water is absorbed. If the plaster is dry, it will darken when moistened (in the same way that blotting paper does) but if the plaster is damp there will be little or no change of colour; if the moistened surface remains wet, the plaster is obviously too wet to paint.

Instruments for measuring moisture content

The conductivity meter

A well-known type of moisture indicator consists of an electrical device which automatically indicates whether a newly plastered surface is fit to paint. Wet plaster is indicated when a pointer turns to Red, and partly dry plaster when it turns to Yellow, whilst Green indicates that the surface is dry enough to paint. Similar instruments are available on which the moisture content of the surface is indicated by the position of the pointer on a graduated scale. The drawback to all these instruments, however, is that under certain atmospheric conditions they may be misleading. They work on the principle of an avometer; when the two prongs are inserted they measure the electrical resistance of the surface between the prongs, water being a good conductor of electricity and offering little resistance. They indicate clearly enough when a surface is wet. But at a certain stage in the drying-out process the speed at which moisture can evaporate from the surface is dependent upon the relative humidity of the atmosphere. On a hot day, when the air is relatively dry, the rate of evaporation is very rapid — more rapid, in fact, than the speed at which moisture from the underlying material can travel to the surface; so the actual top layer can be dehydrated whilst the sub-stratum is still moist. On a humid day, when the air is practically saturated with moisture, evaporation from the top surface may be brought to a standstill (in fact, under some circumstances moisture may actually condense on the surface to make it become wetter) and the moisture from below may be travelling towards the surface by capillary attraction more quickly than that on the surface can be dissipated. Thus, according to the relative humidity of the atmosphere on any particular day, the instrument may register the surface as either moist or dry, although its real condition is the same in both cases. It is not wise, therefore, to decide on the basis of only one day's reading whether or not to risk painting. In any case, the surface should also be tested for alkalinity before painting is commenced.

The capacitance meter

This instrument consists of two flat electrodes which are pressed against the wall surface, giving a reading on a scale. It is not very reliable because it only measures the moisture at the very surface of the wall filling; it gives no indication about the state of the plaster beyond about 2 mm depth. The accuracy of the reading is also affected by the presence of soluble salts.

The hygrometer

A more accurate method of determining the moisture content is to assess the equilibrium humidity of an airspace in contact with the wall. This is done by mounting a sealed and insulated box on the wall, with a hygrometer fastened to the box face opposite the wall. A more rough-and-ready method for site work is to fasten a sheet of polyethylene, with a hygrometer inside, to the wall with adhesive tape. It must be realized that in neither case is an instantaneous reading obtained; to be effective the instrument needs to be in position for several hours, so that equilibrium is reached before a reading is taken.

The microwave meter

This is a sophisticated piece of equipment; a beam of high-frequency radio waves is directed at the wall and passes through to a receiver on the other side, giving a reading of the extent to which the presence of moisture reduces the intensity. The reading could be misleading; it is for the whole thickness of the wall, and is modified by the presence of soluble salts. This method would rarely be suitable for use on site conditions.

Measurement by weight

This is the most reliable method of measuring moisture content. It consists of taking samples of the wall filling by drilling, oven-drying the samples, and then weighing them to discover the moisture loss. It is possible to ascertain the moisture content at varying depths, and the result is not affected by the presence of soluble salts. Again, of course, the method is rarely used for routine site tests, but is invaluable when a precise assessment is needed on large-scale work.

Porosity will also determine the oil content of the priming paint, since unduly porous surfaces such as dry lime plaster usually require much more oil than the denser hardwall plasters and gypsum cements. It is as well to mention here, however, that, in some cases, the latter (including Keene's and

Parian) may exhibit fairly high suction when dry (see page 231; but if this is accompanied by powderiness of the surface, a defect in the plastering is indicated, due possibly to a "dry-out" or to killing of the set by over-trowelling. The surface of these plasters when used neat and correctly applied should be smooth, hard and of moderate or low porosity; the decorator should therefore call attention to any defective surface and thus accept no responsibility for faulty plasterwork.

Primer treatment of various surfaces

With the foregoing provisions then we may consider the treatment required for

(a) Partly dry plaster and cement surfaces
(b) Neutral Keene's and Parian (including anhydrite plaster)
(c) Dry alkaline surfaces
(d) Dry neutral surfaces.

(a) *Partly dry plaster and cement surfaces*

Where the surface is wet, nothing whatever should be applied (except in the special case of Keene's and Parian type plasters) because, apart from the points already discussed, the pores of the surface are charged with water and therefore offer no suction or means for any coating to penetrate or gain proper attachment. This is demonstrated when water paint is applied to a wet surface; the water oozing from the pores carries the wet coating down the wall in long streams to form pools on the floor.

Surfaces which are only partly dry may offer sufficient suction for a paint coating to adhere successfully and, in the case of neat retarded hemihydrate plasters, oil paint and even gloss paint is often applied without apparent harm. It has already been pointed out, however, that it is unwise to seal new damp surfaces prematurely with impervious oil coatings or gloss paints because of possible damage to the plaster (such as "sweat-out" defects), as well as probable damage to the paint film (blistering, loss of adhesion, etc.). The danger is greatly increased when both sides of a wall are sealed. Flat oil paints are sometimes recommended on the ground that their high pigment/low oil composition produces a fairly pervious coating which allows moisture to escape. In practice, however, the first coat of such a paint is usually thinned to priming consistency with a fairly high proportion of oil in order to even up surface porosity; moreover, in most cases, the final coat requires to be applied over a uniform semi-gloss ground, and this generally means a three-coat job — the first two coats being fairly oily — all of which tends to seal moisture within the surface.

The safest course is to apply emulsion paint or a good quality oil-bound water paint. Although only lime-fast pigments are used in emulsions and water paints, when selecting a colour for damp alkaline surfaces preference should be given to pale tints rather than deep shades, since the latter show up surface defects and patchiness much more readily.

Emulsion paints and water paints allow the surface to "breathe" and walls may thus dry out normally without damage to the plaster. When the walls are thoroughly dry a more impervious finish may then be safely applied after suitable preparation. Anhydrous gypsum plasters should be either treated as

Keene's or allowed to dry thoroughly before painting, whichever course is recommended by the plaster manufacturer. If decoration is required before the plaster has time to dry, water paint may be applied *but not following the trowel* (see (b) below).

(b) *Neutral Keene's and Parian* (including anhydrite plaster)

It has been established at the former Building Research Station that paint adheres perfectly to smooth, closely trowelled Keene's if painted *when completely dry*, just as paint adheres successfully to non-porous metal or glass. When moisture is introduced from the backing, however, adhesion is seriously weakened and the paint tends to leave the surface in the form of peeling or flaking. In order to secure better adhesion it is customary to take advantage of the relatively slow set of Keene's by applying a "sharp" oil primer *before the set is complete*, i.e. by following the trowel. The reason for this is that during the setting of anhydrous plasters there is a slight reduction in net volume which causes a suction at the surface; when the primer is applied at this critical stage therefore, it is partly drawn into the surface and becomes securely attached.

One advantage of immediate priming in this way (which applies only to the anhydrous group) is that there is less danger of the plaster drying out before it has fully set, and so "dry out" defects are less likely to occur. If an oily primer is used, however, the setting of the plaster is seriously interfered with and a powdery surface may result. It is important, therefore, that the priming should contain a minimum of oil, and the most satisfactory primer would be a reputable brand of alkali-resisting primer, thinned with white spirit to a thin priming consistency. This "sharp" primer should be applied as soon as the plaster is hard enough to bear the brush, or within twelve hours of trowelling. The surface should then be *left until the plaster dries out thoroughly* before applying any subsequent coats of oil paint. If an immediate finish is desired, emulsion paint or oil-bound water paint may be applied after the priming has become quite hard. On no account, however, should any distemper or water paint be used in lieu of the sharp oil primer because, if such preparations are applied following the trowel, the glue or dissolved casein which they contain may completely stop the setting action of the plaster, in which case a badly powdered surface may result.

After Keene's has been thus primed and all moisture has dried out (which may take several weeks or even months, depending on drying conditions) the surface should be rubbed down with fine glasspaper to remove any loose powdery deposit or efflorescence which may have appeared, before going on to apply undercoating and finishing coats.

If a decorator is called in to paint new Keene's or Parian cement too late to apply the sharp primer at the proper time, he should leave the plaster to dry out thoroughly and treat as recommended in (d) below. Where immediate decoration is required, however, emulsion paint or oil-bound water paint may be applied as a temporary finish.

Anhydrite or "Pioneer" plaster can generally be given the same treatment as Keene's and Parian, except that the setting time appears to be reduced; the makers stated that in all cases where "follow the trowel" technique is adopted the primer should be applied as soon as the plaster is hard enough to bear the brush or within four hours of trowelling. They recommended a "sharp" primer of the following composition —

50 kg paste white lead
1 litre goldsize
8 litres white spirit

This could be thinned further with white spirit to the required consistency. In present-day practice, however, it is unlikely that the decorator would have access to the basic materials to make up such a primer, or indeed that he would wish to do so. The usual material for use with the technique of following the trowel would be a reputable brand of plaster primer, well thinned down with white spirit.

The makers insist that the primer must be thin in order that the wall may "breathe" and thus dry out properly before any subsequent coats of paint are applied, and this coincides with our own views. As with Keene's and Parian, any surface deposits or efflorescence which may appear when the plaster dries out must be removed before proceeding with the second coat (see also Efflorescence, page 224).

(c) *Alkaline surfaces (dry)* including lime plaster, Portland cement and asbestos-cement sheeting (see page 233).

As already mentioned, alkaline surfaces are not harmful to oil paint when they are thoroughly dry. There is always the danger of moisture being introduced, however, from percolating rain, leaky roofs, faulty dampcourses, etc., and when this happens the alkalinity is activated and becomes destructive to ordinary oil media. This contingency is best provided for by the application of an alkali-resisting primer, many reliable makes of which are now available.

It is generally advisable to apply two coats, thinned if necessary with white spirit or as directed by the makers; this makes for double protection and ensures that any minute misses in the first coat (e.g. pinholes or other thin portions such as on mouldings and sharp arrises) are adequately covered by the second application.

Before applying the primer, the surface should be well rubbed down with fine or medium grade glasspaper and dusted off to remove all plaster nibs or any loose powdery deposits that might be present. It is important, however, that the priming coat should not be glasspapered since the film may be scratched or otherwise ruptured and so leave loopholes for possible alkaline attack on subsequent coatings.

(d) *Neutral surfaces (dry)*

This category includes neat hemihydrate hardwall plasters, anhydrous plasters (including dry unprimed Keene's, Parian and anhydrite), fibrous plaster, and also alkaline surfaces which have been allowed to become thoroughly dry and carbonated (i.e. chemically neutral). In the case of the latter it must always be borne in mind that, while the surface crust may be quite neutral, the backing material may remain alkaline for years (especially in the case of a dense Portland cement–sand mix), and if water is introduced from any source (such as through porous bricks or leakages) this alkalinity will be reactivated and possibly brought to the surface where it would attack any oil paint present. For this reason it may be considered sound policy to prime *all* new plaster and cement (where alkaline materials have been used) with an alkali-resisting primer, even when the surface is neutral at the time of painting.

Where the building structure is thoroughly sound and not given to damp conditions, the risk of chemical attack from backing materials is small (though always present), and may reasonably be ignored. The chief concern then is the compounding of a primer suited to the porosity of the surface. Porosity depends on many factors (such as the nature of the backing, the amount of trowelling, etc.), but all things being equal it may be said broadly that, taking lime plaster as being the least hard and most porous, the retarded hemihydrate plasters, anhydrous gypsum plasters, Keene's and Parian, and finally anhydrite plaster, increase in hardness and decrease in porosity in that order. Surfaces of low porosity, however, may possess a surprising degree of suction, due to the pores being fine and so exerting a stronger capillary pull than coarser pores; this explains why dry hard Keene's and Parian may exhibit high suction even when the surface is trowelled to a smooth "polish". Portland cement is hard and of little porosity in itself, but Portland cement–sand renderings (stucco) and concrete vary in porosity according to the amount and type of aggregate.

Unduly porous surfaces require a primer capable of overcoming the suction while being thin enough to permit of easy brushing. Proprietary brands of primer–sealers should be thinned according to the manufacturer's instructions. Some manufacturers recommend applying a well-thinned coat of emulsion paint first to satisfy the suction.

In priming exterior stucco renderings which are dry and chemically neutral a reputable brand of plaster primer should be used, although it would generally be safer to use an alkali-resistant primer because of the risk of alkalis being reactivated at a later date. The surface should be well brushed down with a stiff fibre brush, or a wire brush if necessary, to remove any loose, powdery material or surface incrustations.

In the application of priming paints, the primer should be of fairly thin consistency, applied liberally and brushed well; overthinning with white spirit should be avoided, however, since this would tend to carry the oil into the plaster, leaving the pigment on the surface in a more or less underbound condition. Very often surfaces possess uneven porosity, and the difference in suction will soon make itself felt in the brushing. By applying the primer liberally, the surface is kept "wet" longer, and those parts where suction is greatest are able to take up sufficient paint to satisfy suction; the surface should be finally laid off with a fairly dry brush to remove excess paint from the less porous parts and so avoid what might otherwise form a thick, oily coating. In cases of exceptional porosity it is advisable to apply two thin coats of primer rather than to attempt to overcome suction with one heavy coat. The temptation to use glue-size to reduce absorption should be strongly resisted since size is not a sufficiently stable foundation for any paint system, especially where it is liable to be softened by adventitious moisture which may percolate through the backing.

PRIMING OF CONCRETE SURFACES

Concrete is used very extensively in modern building construction. Basically it consists of a mixture of cement, a fine aggregate in the form of sand, and a coarse aggregate of crushed stone or gravel. The three components are accurately proportioned and thoroughly mixed together so that a compact mass is formed, the sand filling the voids between the particles of coarse aggregate and the cement filling the remaining voids between the fine sand particles. The addition of water causes chemical action to take place which sets hard and binds the whole mass together. The predetermined shape of the concrete mass can be formed *in situ* by setting up a shuttering of timber or metal and pouring in the "wet" concrete, the shuttering being removed when the material is set; or it can be precast and delivered to the site in solid state in the form of blocks, slabs, or similar units. Reinforced concrete is a form of construction in which steel rods and mesh are embedded in the concrete, thus greatly increasing its tensile strength. To increase the tensile strength still further and to reduce the tendency of the concrete to shrink and crack on drying out, in modern construction prestressed concrete is often used, in which the steel is stretched, thus putting it into a state of tensile stress until the concrete has set and hardened: the tension is then released and the concrete member is thus put into a state of compression. This technique results in a considerable saving both of steel and cement. The concrete mix can be modified in several ways to increase its strength, its speed of hardening, its water-repellent properties, etc., and colouring-matter may be added.

Further developments have produced lightweight concrete in various forms for structures where a high-strength concrete is not essential. Concrete can be made lighter by introducing air; the fine aggregate can be omitted so that air voids exist between the coarse aggregate particles — this being called "No-fines Concrete"; or a porous material which retains some air in itself can be used as the aggregate. "Aerated Concrete" is produced by introducing air or some other gas into the cement mix so that when the mass is set a uniform cellular material is formed; air can be introduced by mixing-in a stable foam (of the kind used in fire-fighting) or by whipping-in air, or, alternatively, gas can be generated by chemical reaction within the mass whilst in its liquid or plastic stage by adding aluminium powder to the mix, the powder producing bubbles of hydrogen by reaction with lime. In each case the thermal insulation properties of the concrete are increased due to the fact that air is a good heat insulator.

Lightweight aggregate concrete takes various forms. The aggregates commonly used are (a) furnace clinker, i.e., burnt furnace residue fused or sintered into lumps; (b) foamed blast-furnace slag, made by treating molten slag with water to produce a dry cellular product; (c) exfoliated vermiculite, which is a mineral like mica, consisting of flakes formed of many layers, the flakes opening out like an unfolding flower when heated; (d) expanded perlite, made by heating a glassy volcanic rock called perlite so that it expands into a cellular material; (e) pumice, which is a spongy lava used in its natural state; and (f) expanded clay and shale, which are heated till they expand due to the generation of gases inside them. In all cases the thermal insulation properties are higher than those of dense concrete.

The painter and decorator encounters concrete both as a structural material in loadbearing walls and as non-loadbearing slabs or blocks used for the infilling of framed structures and for partition walls. Quite often nowadays building costs are reduced by leaving the bare concrete exposed, even on internal walls, instead of finishing off by plastering. Whilst there is not much difficulty in painting dry and mature concrete, trouble is likely to occur if the structure is wet because of the alkaline nature of the material; this is particularly so in the case of some lightweight aggregate concretes in which the aggregate is highly alkaline. Efflorescence may also be a problem. On new property the concrete should be left as long as is practicable before decoration, in order to ensure as far as possible that it has completely dried out. If decoration is required soon after erection, the most satisfactory materials to use are water paints or emulsion paints. If oil paints are to be used, an alkali-resisting primer should first be applied to decrease the possibility of saponification; gloss paints are more likely to be attacked than flat or eggshell finishes, but in any case it is always desirable to apply an alkali-resisting primer even if the concrete is dry, because of the risk that at some later stage water may be introduced by means of a defect in the building, which would reactivate the alkali. Efflorescence should be dealt with in the manner already described. In some cases there may be difficulty due to the presence of mould-release oil used to help in separating the shutterings and moulds from the concrete after erection; in such an event the oil should be removed with a cleansing agent. This is particularly necessary where the surface is to be hung with wallpaper, and the matter is referred to in the chapter on paperhanging methods.

Lightweight aggregate concrete takes longer to dry out than high-density concrete, and is more porous. Aerated concrete tends to be more friable than normal concrete.

Concrete floors are frequently painted to improve their appearance and to preserve the surface by preventing it from "dusting up". Special floor paints are formulated for this purpose, and it has been found that the most satisfactory application method is by roller.

PRIMING OF ASBESTOS-CEMENT SHEETING

This material varies considerably in porosity, the better qualities being very hard and of low absorption whilst fireproof qualities are often exceedingly porous. They are generally strongly alkaline and carbonate or neutralize more slowly than lime plaster or Portland cement–sand surfaces owing to the sheets being very hard and highly compressed. They should be allowed to dry thoroughly and an alkali-resisting primer should always be used before applying oil paint. The surface should first be brushed down with a stiff fibre brush or glasspapered to remove any efflorescence or other loose material prior to priming.

If there is any likelihood of moisture gaining access from behind, the sheets should be heavily coated on backs and edges before fixing, using either "smudge" or a bituminous composition, to prevent any penetration of moisture; if bitumen is used, however, great care should be taken to keep it off the face side; thin penetrative liquids such as creosote should be avoided since these may seep through to the face side and bleed through the decorative finish.

BRICK AND STONE SURFACES

New brickwork

Brickwork should not be painted with impervious coatings until it has thoroughly dried out because, apart from the undesirability of imprisoning moisture, it is very desirable that any soluble salts present should have a chance to crystallize out, when they can be removed from the surface. Exterior brickwork also acts as a "safety valve" in allowing moisture to escape from interior surfaces. Moreover, lime mortar or cement joints are strongly alkaline and very destructive to oil paint when damp.

When the surface is dry, it should be brushed down with a stiff fibre or wire brush to remove mortar splashes, efflorescence, or other loose material, then primed with alkali-resisting primer, applying liberally and brushing well into the joints; or alternatively, omit primer and apply latex emulsion paint which is resistant to alkalis and performs well on dry rough surfaces.

Where new brickwork is required to be decorated before it is properly dried out, an emulsion or an oil-bound water paint may be applied; this will allow the brickwork to dry out normally without any serious complications and, although the water paint may disintegrate or be thrown off eventually by exuding salts, etc., such defects would be negligible compared with the trouble that an oil paint would cause.

Old brick and stone surfaces (dry)

These surfaces normally do not require painting, but it is often decided to paint them to bring them into line with adjacent painted stucco work or to provide a weatherproof finish on exposed elevations. The first requirement is to clean down the surface and effect any necessary repairs (see "Making good", page 281).

Brickwork may be vigorously wire-brushed to remove all loose surface deposits, any defective pointing being made good with oil mastic. Stonework may be similarly treated unless the surface is badly decayed, in which case an architect or an expert on stone preservation should be consulted. After thorough preparation, the recommendation given for exterior stucco renderings and neutral surfaces (dry) should be followed. Latex emulsion paints are excellent for dry brick, stone and wood-floated cement surfaces since they adhere well to such rough-textured surfaces.

PRIMING OF WALLBOARDS AND SHEETINGS

These have increased in number and variety in recent years and, with the introduction of prefabrication into building practice, they have assumed more importance in the realm of surfacing materials.

In general, wallboards are chemically neutral when dry, and the problems raised in priming are mainly concerned with surface porosity. Boards which have been impregnated with fire-retarding salts, or are bonded with magnesite, should have an oil primer since water paints and distempers may promote chemical activity if applied direct. In the case of boards treated with fire-retarding salts, an alkali-resisting primer is advisable. The raw materials from which building boards are made are usually pulped wood fibre or other vegetable fibre, together with various fillers and agglutinants; some of the more recent types are resin-bonded and closely compacted under high pressure, possessing a smooth polished surface of little or no porosity, e.g. laminated plastic boards and sheetings. At the other end of the scale are the thicker, loosely compacted insulating boards of the type used for acoustic correction, generally with a highly absorbent surface, while between these extremes are the medium-thickness fibre wallboards, some of which are laminated, i.e. built up from several thicknesses of pulped fibre.

Plaster boards are sometimes directly painted upon, although primarily intended as a ground or "lathing" for plaster skimming; they are composed of a layer of gypsum sandwiched between two layers of stout paper or pulped fibre; they are quite safe to receive oil paint if kept dry.

Where wallboards are exposed to damp conditions, such as in lining old damp walls, the back of the sheets should be protected, before fixing, with a heavy coat of smudge or bituminous paint, and some provision should be made to ventilate the cavity formed by the wallboarding and the existing wall structure. Where bitumen is used for protecting the backs, the precautions mentioned under "Asbestos-cement sheeting" should be noted.

For the purposes of painting, then, we can consider building boards broadly under six headings —

(a) Absorbent insulating boards
(b) Medium-pressed boards
(c) Hard-pressed boards
(d) Plastic boards and sheetings
(e) Foamed polystyrene
(f) Fibreglass

(a) *Absorbent insulating fibre boards* (such as Celotex, Ten-test, etc.)

These open-textured surfaces do not lend themselves to a gloss finish and it is more usual to finish them in flat oil or eggshell paint or in emulsion paint. In either case, the extreme porosity of the boards demands special treatment before the finishing coat is applied.

In the case of an emulsion paint finish, the preliminary coating should consist of the same type of emulsion paint, well thinned down with water and applied liberally. When an oil paint is specified, a first coating of thin emulsion paint is equally satisfactory, but an alternative and less expensive treatment is a liberal coating of weak glue size, preferably applied warm. Normally, size is not to be recommended under oil paint, but in this instance it serves a legitimate purpose in protecting the board from the harmful effects of linseed oil on a vegetable fibre (see Fabrics, page 238).

If for any reason an oil-bound water paint is specified as the finish, the first coating should consist of the same brand of water paint thinned down well and liberally applied. When the time comes for redecorating, however, a coat of penetrative primer should be applied, since the porous nature of the surface would probably be unable to support any further coatings without some risk of flaking. Neither emulsion paint nor water paint should ever be applied over glue size.

(b) *Medium-pressed boards* (such as Sundeala, Beaverboard, Upson Board, etc.)

These are generally smoother and of only moderate porosity; sizing may therefore be omitted and a liberal coat of proprietary primer–sealer thinned to the manufacturer's instructions may be applied direct.

(c) *Hard-pressed boards* (such as Insulwood, Masonite, etc.)

Boards of this type generally have a hard, smooth, polished surface and it is rather surprising to find that in spite of their appearance, some possess a fair degree of porosity. Moreover, they sometimes have a tendency to retard the drying of an oil primer or, at least, to prevent it from hardening thoroughly. Specially formulated primers for hardboards are now available.

(d) *Plastic boards and sheetings*

These do not present any great difficulty in painting, provided the surface is free from grease and is given a suitable key. The sheets are liable to become greasy through handling during erection and should therefore be cleaned with white spirit, rubbing down at the same time with fine steel wool and wiping dry with a clean cloth. Some types have a rather friable surface which glasspapers easily. A small proportion of mixing varnish or goldsize may be added to the first coat of paint to assist adhesion, but hard drying, brittle paints should be avoided.

(e) *Foamed polystyrene*

This material is an expanded plastic in which the polystyrene encloses minute air-cells, forming a rigid material mouldable into practically any shape. It is available in very thin sheets (2½ mm) or in slabs of various thicknesses up to 75 mm or more, as required, and is extremely light in weight. It is waterproof and highly buoyant and because of its structure has excellent thermal and sound-insulating properties.

Unfortunately it constitutes a fire risk; when exposed to a flame it immediately liquefies and drops of burning liquid fall on the surface below; on both wall and ceiling surfaces therefore it increases the possibility of spread of flame. Special non-flammable brands are obtainable and these should always be used for the sake of safety.

In sheet form (e.g. Kotina, Poron, etc.) it can be applied to plaster walls with a special adhesive, to prevent condensation or to minimize pattern staining on ceilings; or it can be applied in the form of thick acoustic tiles.

It can be decorated with waterpaint or emulsion paint, but oil paints should not be applied direct since the solvents will have deleterious effect on the material. After priming with emulsion paint it can be finished with oil paints quite safely. It can also be lined and finished with emulsion, or with a final wallpaper, and since the material is rather spongy and easily damaged on lower walls, wallpaper is an added protection.

(f) *Fibreglass*

This material, bonded with polyester, epikote, or polyurethane resin, is being used increasingly for wall sheetings and mouldings of all kinds, including motor bodies, boat hulls, etc. The surface is usually very smooth and often contaminated with mineral oil or other greasy mould-release agent, all trace of which should be thoroughly removed with white spirit before painting, otherwise adhesion will be seriously impaired. A very tenacious primer is required and some makers supply a pigmented synthetic alkyd primer with special thinners for brushing or spraying, and will generally advise on the best procedure. Once the surface is properly primed, it can be followed by normal undercoating and finishing paints, as desired.

PRIMING OF ACOUSTIC MATERIALS

The science of acoustics has developed greatly in recent years and many new sound-deadening materials have been produced for the acoustic correction of auditoria and other buildings. These are generally of two main types: (i) those of a highly absorbent nature, such as thick insulating wallboard (flat or moulded), asbestos blanket, sprayed asbestos, acoustic plaster, etc., and (ii) non-resonating materials faced with wallboard or thin plaster, the surface of which is perforated.

In the first group, the sound-deadening properties are dependent on the absorbent nature of the materials, and if any oil paint is applied, these acoustic properties are greatly impaired, if not destroyed. Such materials are sometimes decorated by spraying with thin waterpaint, but to be on the safe side the decorator should always follow the recommendations laid down by the makers of the particular acoustic material.

The materials in the second group are generally made in the form of rectangular slabs or "tiles" in various dimensions up to a metre square and depend largely on their deep perforations for their acoustic properties. Oil paint may safely be applied without impairing acoustic efficiency, provided the perforations do not become filled with paint. Oil paint or emulsion paint should therefore be sprayed for preference in a fairly thin consistency, but if brush application is used great care will be required to avoid flooding the perforations and so spoiling their acoustic properties.

PAPERED SURFACES

Defective plaster surfaces are sometimes hung with lining paper to provide a good ground for painting; if the plaster is well sized and a good paste used for hanging, the paper will present a surface of fairly uniform suction when dry. The degree of suction will depend largely on the quality of the lining paper, thin calendered paper being much less absorbent than a heavy pulp paper. All moisture should be allowed to dry out before applying the first coat of paint, which should be applied liberally with a 100 mm wall brush, the consistency being thin enough to permit of easy brushing.

A coat of thin size is sometimes applied before painting to prevent the oil from being absorbed into the paper and causing the latter ultimately to become brittle (due to the reaction between linseed oil and vegetable fibre).

More often than not, the coat of size is given purely on grounds of economy. The important point to observe, however, is that where size is used, it should be applied fairly weak (and preferably warm) so that it soaks into the lining paper and does not merely lie on the surface. The first coat of paint will then have a good chance of gaining attachment to the paper instead of lying insecurely on a film of glue-size.

Any good flat oil paint or proprietary undercoating for gloss paint may then be applied according to makers' instructions.

Existing wallpaper which is in good condition and firmly adhering, particularly at the joints, may be safely painted where it is desirable to avoid stripping. In such cases, however, the use of glue-size prior to painting may result in serious trouble; size, or any water preparation, will swell the paper and the old paste may not be able to withstand the strain of the subsequent contraction on drying, with the result that joints may spring open and the whole of the paper may become loose. It is much better, therefore, to rely on a properly formulated priming paint to stop suction and so provide a firm ground for subsequent coats.

Wallpapers are generally printed in distemper colours, some of which are particularly prone to bleeding; reds, pinks, mauves, metallic bronzes and mica are the worst offenders, but any doubtful colours may easily be tested for bleeding by coating a small portion with white oily paint and keeping the patch under observation for a day or two. Colours that are liable to bleed should be insulated with a coat of stop-tar knotting or with patent knotting well thinned out with methylated spirit. The knotting should preferably be applied to the whole surface to provide a ground of uniform suction.

FABRICS

Canvas and other fabrics of the type known as hessian, jute or burlap, are occasionally used for wall decoration and sometimes require painting. Linen canvas (as used for easel pictures) is sometimes applied direct to wall surfaces, as in the case of large mural decorations. When such fabrics are painted direct, the oxidizing effect of linseed oil is to harden or "shorten" the fibre, the fabric eventually becoming brittle and easily torn.

It is therefore necessary to insulate the fabric from direct contact with the oil and this is usually done by applying a coat of gelatine size. Size is generally regarded as being a very unsatisfactory material to apply under paint, but in this case it performs a useful and legitimate function in protecting the fabric from the effects of the oil.

Strong size is not necessary and should be avoided. A fairly weak coat of size should be applied, preferably warm, so that it impregnates the fabric without filling in the texture. This will ensure that the paint has a good key and does not merely lie on a surface-coating of size.

The size should be allowed to dry thoroughly before applying the first coat of paint which should, of course, be designed to retain a degree of flexibility. A proprietary multi-purpose primer or a flat oil paint may then be applied. Where an emulsion paint finish is desired, however, *omit the size* and prime the fabric with a thin coat of emulsion paint, applied liberally, followed by one or more coats of normal consistency.

PRIMING OF NEW WOODWORK

The type of wood most commonly used in joinery work intended for painting is one or other of the resinous pine species such as Swedish fir, Scotch fir, red and yellow "deal", and British Columbian pine (also known as Oregon pine or Douglas fir). These are called softwoods. Hardwoods such as oak, elm, walnut, mahogany, etc. may be French polished, varnished, or coated with clear decorative wood finish to enhance the appearance of their natural grain, but on occasion they too are required to be painted.

Softwoods are the result of fairly rapid growth, whereas hardwoods have grown much more slowly and generally exhibit a fine, close grain in consequence. The grain markings in wood are determined by the light and dark annular rings (which are seen when a tree is sawn through); each pair of rings represents a year's growth, the light portion being formed when the sap surges vigorously up from the roots in the spring, the thinner, darker layer being the result of the gradual fall of the sap in the late summer and autumn. The spring wood is soft and porous by reason of its quick growth, the summer wood being hard, dense and much less absorbent. By virtue of its porous, cellular structure the spring wood holds more water and is therefore liable to much greater shrinkage changes on drying than the summer wood; this is plainly demonstrated when unseasoned open-grain softwoods such as Oregon pine are painted, for, on drying, the soft grain shrinks, leaving the hard grain standing out in comparatively bold relief.

The annular ring structure of the tree (see cross-section, Fig. 27.1) is built around a central pith or "medulla" from which radiate fine ducts (known as the "medullary rays") passing through the heart-wood and the soft outer sapwood to the bark. When certain hardwoods are cut along the medullary rays, the fine decorative grain (such as the figure or silver grain in "quartered" oak) is exhibited.

Seasoning

In the living tree, the wood is more or less saturated with moisture, and, after felling, a period of "seasoning" is required to allow most of this moisture to escape before the timber is fit to use in carpentry work. Shrinkage takes place during the seasoning period, the wood gradually becomes more stable and the sapwood lightens in colour and hardens up as it dries out.

The urgent need of timber (plus the amount of capital locked up during seasoning) results in a great deal of unseasoned wood being used for building purposes.

Sound timber presents few problems to the painter — in fact it provides an ideal surface for painting — whereas unseasoned timber is an unmitigated nuisance both to the carpenter and the painter alike, for shrinkage movements, warping and cracking are liable to occur during the drying process. Moreover, when unseasoned timber is painted on both sides, a certain amount of moisture is imprisoned, which, besides being a fruitful source of blistering and peeling of the paint film, may have the more serious consequences of causing the wood itself to become rotten.

Knots and knotting

Knots are formed in timber where the parent stem (the medulla) pushes out young shoots, some of which emerge from the trunk and grow into substantial branches. When it is remembered that each branch, like the parent stem, has resin ducts running along its length, it is easy to see why knots permit resin and moisture to flow so readily to the surface. The most troublesome knots are usually those from which the main branches have sprung; such knots are generally very large, resinous and often loose, in which case they should be cut out and plugged with sound wood by the joiner.

The resinous matter which exudes from knots is the same gummy substance from which turpentine is distilled, which accounts for the fact that when paint is applied over unprotected knots, the resin is dissolved and seeps into the paint film, causing a brown stain. Such resin stains will bleed through any number of paint coatings; moreover, when exposed to a hot sun, resin will be brought to the surface where it will exude in unsightly tears. To prevent this, all knots should be sealed with two coats of pure shellac knotting, applied thinly and extending about an inch round the knot. All resinous streaks and bluish sapwood should be similarly treated. Thick knotting, applied carelessly, will leave hard ridges which persist in showing through several coats of paint; two thin coats are always better than one thick coat because a single coating is liable to contain minute fractures or pinholes through which resin may bleed.

It is important that only the best knotting be used, i.e. pure shellac in methylated spirit. Cheap knottings are useless since they are generally adulterated with rosin which, being soluble in turpentine, fails to hold back the resinous matter in knots. The smooth, glossy surface of shellac knotting presents a poor key for an oily primer and it is therefore advisable to touch up all knotting with a coat of sharp flat paint. This may seem superfluous, but it is definitely worth while and generally pays for itself by helping to obliterate dark knots which might otherwise necessitate an extra coat of paint. In the old days, troublesome knots were often twice knotted and gilded, either by applying the gold leaf to the second coat of knotting when tacky, or by laying-in with goldsize.

Another way of treating bad knots is to hollow them well below the surface with a joiner's gouge, slightly undercutting; they can then be filled with a stiff mixture of a proprietary filler based on gypsum, leaving it slightly proud of the surface. When quite hard it is glasspapered down to the level of the wood. If carefully done, this can give excellent results since the gypsum filler tends to expand slightly on setting and thus secures a firm hold. Still another method (where price is a major consideration and a minimum treatment is required) is to apply one coat of shellac knotting slightly pigmented with aluminium powder, or alternatively to apply clear knotting and dust with aluminium powder when tacky. The leafing action of the aluminium helps reinforce the sealing property of the knotting.

Priming paints for wood

There is an unfortunate tendency in many quarters to regard the quality of priming paint (and, indeed, its manner of application) as unimportant since it is covered up by the intermediate and finishing coats. Thus we find that joinery work is often "primed" in builders' shops by unskilled labour, or, worse still, "spray-primed" in the larger woodworking factories where doors and windows, etc., are mass produced. The "primer" used is something pink, generally purchased at a low price. Such priming paints will, most likely, contain large amounts of barytes or other extenders of fairly low oil absorption which, being inert, do not combine with the oil. When applied on porous timber, therefore, the tendency is for the oil to soak into the wood, leaving the pigment on the surface in an underbound condition — a very poor foundation for any subsequent coats.

Considered purely from the point of view of providing good protection, there is no doubt that the best all-round primer for exterior woodwork is the traditional *pink primer* made by mixing white and red lead. This is largely due to the ability of these pigments to react with the medium to form a lead soap which imparts a tough, elastic, leathery quality to the paint. Ready-made primers based on white and red lead, produced by the leading paint manufacturers, are modified to conform to BS 2521, and include a proportion of titanium dioxide to improve their opacity and a resin to reduce their drying time; they offer a good measure of water-resistance and are able to withstand a reasonable amount of site exposure before the first coat of paint is applied. Unfortunately this information will soon be only of academic interest as the use of lead-based pigments is declining rapidly, and how long they will remain available to the painter is a matter of speculation. This is because of strong pressures from the EEC countries to get all lead paints banned throughout the Community. A great deal of concern has been expressed about the toxic dangers they present, especially to children who may chew them or lick them — a risk which has been much exaggerated, as the occasions on which children chew structural woodwork are rare indeed. But the result is that although we know what are the best materials for protecting property, particularly in the uncertain climate of Great Britain, we can no longer use them and must perforce make do with materials that are vastly inferior. For regulations concerning the use of lead-based paints see Appendix 2.

On interior work *leadless primers* may be used. A range of non-lead primers with an oleo-resinous or alkyd base is available. They vary considerably in quality. Some of the better ones have a fair degree of water resistance, but even so they are not really suitable for outdoor exposure; the poorer ones are prone to rapid chalking and have poor adhesion.

Aluminium, in a suitable varnish medium, makes an excellent wood primer, especially for "back-priming" those parts of window casings and door frames that come in contact with wet brickwork and cement, etc., where its water resistance is particularly valuable. It is, however, difficult to see why aluminium is put forward as being specially suitable for unseasoned timber, since its waterproofing properties would effectively prevent the escape of contained moisture (particularly if the wood is sealed on both sides) and would thus give rise to those undesirable defects noted elsewhere. In our view, aluminium primer is best applied on seasoned timber. The media used are either quick-drying phenolic resin/tung and linseed oil mixtures or alkyd resins. Sometimes difficulty is experienced because subsequent coats do not adhere very well to an aluminium primer. One great advantage of aluminium primers is their resistance to the exudation of resin from knots. For this reason they are particularly suitable for timbers with a high resinous content such as Columbian pine.

The consistency of the primer must depend on the absorbency of the surface. The main requirement is that the primer should contain sufficient medium to satisfy the suction while still effectively binding the pigment which remains on the surface. For very absorbent woodwork such as soft whitewood the manufacturer's instructions should be followed; for hardwoods it is usual to add thinners in the recommended proportions. The same proportions are also suitable for burnt-off woodwork, which is generally much less absorbent than new timber.

Water-thinned primers for woodwork

To some extent the changing outlook about primers and priming is reflected in the continually increasing use of emulsion primers thinned with water. This of course is part of the general trend in present-day paint production away from the traditional oil media and back towards water-dispersible pigment binders.

Emulsion primers certainly offer a number of advantages. What appeals chiefly to the decorating contractor is their ease of application and the rapidity with which they dry, enabling work to be primed, undercoated, and in some cases finished as well, all in one day, together with the fact that the brushes used in them can easily be washed out after use. Their lack of odour and the absence of any strong-smelling solvent that needs to be evaporated are also useful properties, especially where foodstuffs, etc., might become tainted. In addition, of course, there are the important points that the pigments are non-toxic and the solvent is non-flammable.

When properly formulated, emulsion primers are reasonably durable. The better qualities are based on acrylic polymers which are extremely flexible, being similar in chemical composition to the plastic material Perspex. Some authorities consider them to be equally as durable as lead-based primers. Against this, they are permeable and will not exclude water from joinery which is stacked in the open air; this could lead to trouble at a later stage due to the high water-content of the timber. A great drawback from the decorator's point of view is the impossibility of laying them off to eliminate brush-marks, which mar the appearance of the finished work. They do not adhere well to woodwork which has been treated with a preservative fluid. It is generally stated that when these primers are used they themselves will seal the knots, so that knotting is unnecessary, but experience shows that this claim is rather optimistic; it is safer to apply knotting as well.

A more recent development is the type of primer which consists of a water-repellent preservative fluid which is lightly pigmented, generally pink, and which is used for the factory priming of mass-produced joinery. Some of these primers lack the film-forming characteristics of true primers; others, however, do contain materials of a film-forming type, and although they have very little penetrative power when applied by brush they do, when used as dipping primers, have some preservative action on the end-grain of timber. It is highly desirable, however, that timber primed with these materials should be given an orthodox primer coat by the decorator before any further painting is carried out.

All-purpose primers are now available which are said to be equally suitable for metal or wood, thus obviating the need for cutting-in with two different materials on work such as window frames, where metal and wood are adjacent to one another.

Application of wood primers

Ideally, of course, the painter who applies the intermediate and finishing coats should also apply the priming; the evils of unskilled "shop-priming" would then be avoided, and the painter would know what was going on from the very beginning. When woodwork is sent to a building site in an unprotected condition, it becomes saturated in wet weather, causing endless trouble to all concerned. It should always be properly primed before leaving the joiner's shop, and the "jerry builder" who sends out unprimed doors and windows, etc., should be penalized.

All nail holes are usually punched well below the surface by the joiner, but if any have been inadvertently missed the painter must do this, otherwise the holes cannot be stopped up properly. The work should be rubbed down if necessary, using a fine or medium grade glasspaper in the direction of the grain. Internal woodwork on the site will have sundry splashes or nibs of mortar and plaster; these splashes must be removed by glasspapering and the whole of the work well dusted off, paying particular attention to mouldings and quirks. All knots are treated with knotting (see page 240) and allowed to dry hard before priming.

Architects' specifications sometimes read "knot, stop and prime . . ." which, of course, is all wrong since no stopping or filling should be done until the primer has been applied and allowed to dry hard.

In priming woodwork, the paint should be applied fairly liberally and allowed sufficient time to penetrate before laying off; the wood should take up all the paint it requires and still leave sufficient on the surface to form a good foundation for the second coat. Brushwork should be vigorous and brisk, taking care that all quirks, cracks, crevices and nail holes are well bottomed; if nail holes are not properly painted, the wood will absorb oil from the putty or stopping, causing it to shrink and ultimately fall out. On the other hand, these parts should not be flooded with paint, otherwise unsightly fatty runs will occur; all excess colour should therefore be removed from mouldings and quirks, etc. Avoid fat edges also by lightly laying off all corners and projections with a fairly dry brush.

Pay particular attention to *end grain* wherever it occurs, especially at tops and bottoms of doors where it is usually neglected; end grain, being very absorbent, attracts moisture and conducts it along the grain into the heart of the wood. Where circumstances permit, prime all end grain twice. All "back priming" (i.e. those parts of window and door frames which come in direct contact with wet brickwork, etc.) should also receive two coats, preferably using an impervious primer such as aluminium.

There is much insistence by many authorities that priming should not be left for more than a few days before the next coat of paint is applied, on the ground that if left too long, the priming becomes so hard that the following coat has difficulty in forming a good attachment. There is, of course, some truth in this argument, particularly in the case of a straight red lead primer which, if left too long, may become excessively hard or even commence to disintegrate by chalking owing to its powerful oxidizing effect on the oil. Nevertheless, the fact remains that it is not always possible to follow on with a second coating so soon after priming; indeed it is often quite impracticable to arrange this, for the painter is sometimes called upon to tender for work, such as large new buildings and housing estates, where the joinery work has already been shop-primed many months.

The question of adhesion of the second coat then is surely one of (a) adequate preparation of the primed surface, and (b) proper formulation of the paint to be used for second coating. Thus, if a primed surface has been left some time before the second coat can be applied, it should be thoroughly rubbed down with glasspaper or steel wool to remove all dirt deposits and (where chalking exists) all loose powdery material; it should be thoroughly dusted off and all cracks and nail holes stopped with linseed-oil putty or hard stopping before applying the second coat of paint (if, however, the surface is badly weathered, stopping should be left till after the *second coat* is applied). The adhesion of a straight linseed-oil paint would not generally be so good as one containing heat-treated oils and resins; the undercoatings supplied by manufacturers today take this fact into account and are usually formulated with a varnish medium. They do not normally need any adjustment, though it may be necessary to add a small proportion of "universal medium" to the undercoat if the condition of the surface should warrant it.

The question of brush versus spray application probably concerns those who carry out large-scale shop priming, rather than the painter and decorator; but, nevertheless, the latter may toy with the idea of using the spray if he gets an unusual amount of priming to do. In our opinion, however, priming paints are definitely best applied by brush since they are more likely to secure an intimate bond with the surface than when applied by spray. Intermediate and finishing coats, of course, may be sprayed with perfect results, but we definitely prefer brush application where priming is concerned.

Priming of unseasoned timber

We have already mentioned some of the dangers of painting unseasoned timber and we know that, ideally, we should not be called upon to paint such an unsatisfactory material. We are faced with the fact, however, that unseasoned timber is used to a great extent in present-day building and that it must be treated in some way if only for the sake of appearance.

We have seen that improperly seasoned timber contains water which, under suitable drying conditions, will escape through the pores in the form of vapour, but which, if sealed within the surface by impervious coatings, will endeavour to escape by way of blisters, etc. Moreover, when large amounts of moisture are imprisoned for any length of time, more serious consequences such as wood rot are likely to occur. Yet if we leave the surface unprotected and exposed to the elements, the soft springwood and end-grain may continue to absorb even more moisture. The problem then is to apply a coating that may prevent further ingress of moisture and yet allow the contained moisture to escape. This is not so difficult as it may seem.

All paint and varnish films are considered to be permeable in some small degree; that is to say, they are capable of allowing minute quantities of vapour to pass through them by what might be termed a "breathing" action. Straight linseed oil is known to be much more permeable than heat-treated oils or varnishes. It is reasonable to argue then, that a straight linseed oil paint applied on unseasoned timber would, for a time, provide a finish that would keep out the weather while still allowing the internal moisture to escape and so enable the timber gradually to dry out. Permeability may be increased by first applying to the bare wood a heavy coat of high-grade oil-bound water paint. When dry, the water paint should be well glass-papered down, then followed with two coats of straight linseed oil paint, tinted with stainers as required. In dry weather, it would be an advantage to let the water paint harden for three or four days before applying the oil paint.

It should be borne in mind that the above unorthodox treatment is intended purely as a temporary expedient or "first-aid" measure in dealing with damp, unseasoned timber and should on no account be adopted as a regular practice. The life of such a finish would be comparatively short — say two years at the most on exterior work — but by that time the timber would be reasonably dry and in much better condition for repainting than if a more impervious finish such as gloss paint or varnish had been used. Moreover, the surface would probably be fairly chalky and therefore easy to prepare for repainting (see page 270). It may be argued that such a finish would be poor in appearance and not half so pleasing as a high gloss finish. Granted — but just as in dealing with new plaster and cement surfaces, where it is wiser to avoid the use of impervious gloss finishes until the surface has thoroughly dried out, so the above is put forward as a rational method of dealing with a difficult problem.

Plywoods

Although, in the main, plywoods are used for decorative panelling and are often faced with choice veneers, they are also frequently faced with whitewood, such as birch and alder, which usually requires painting. "Blockboard", a composite board built up from whitewood strips sandwiched between two layers of hard plywood, is also widely used for signs, doors and partitions, etc.

Great strides have been made in the manufacture of plywoods and composite boards; many types are now guaranteed against warping, splitting, or other dimensional changes. This is made possible by controlled kiln drying and seasoning of the timber and by the use of waterproof synthetic resins in cementing or bonding the plywood sheets. Resin-bonded plywood is highly stable and stands up well to stringent practical tests, such as its use in concrete shuttering.

Timber is now also resin-impregnated by first kiln-drying, then forcing liquid resin under pressure into the dehydrated cells and tissues. Such timber is extremely hard and is said to be proof against rot, decay, and insect pests. The advent of materials like this may at first cause the painter some concern as to the future of his calling, but he need not worry in the least. The great drawback to "permanent" materials is that people soon tire of seeing the same finish and demand a change, and the most ready means of effecting a quick change at reasonable cost is by painting. These new materials, therefore, instead of jeopardizing our livelihood, will actually simplify our task by providing sound stable surfaces (free from unpredictable snags and defects) upon which we may work with confidence, knowing that our best efforts will not be sabotaged by any vagaries of the surface. Unseasoned timber, for instance, will one day be a thing of the past and that surely will be an unmixed blessing.

In priming modern plywoods and "processed" timber, then, the painter has a perfectly sound, stable, knot-free surface to work on and his main concern will be to gauge his primer to suit the porosity. Absorbent woods require a high proportion of medium with sufficient white spirit to assist penetration, while non-porous resin-impregnated timber will require a primer with good adhesive properties, i.e. one based on a varnish medium.

Aluminium primer is particularly useful for plywoods faced with Columbian pine and similar woods. A feature of these plywoods is that large areas of soft absorbent whitewood alternate with bands of highly resinous hardwood of a dark-brown colour (this being caused by the way the annual rings of the tree's growth are cut through at an angle as the long continuous layer of the veneer is shaved from the rotating log). Ordinary primer cannot cope successfully with these variations; if it is oily enough to satisfy the softwood areas it will not obliterate the darker bands, nor will it dry on them; and conversely, if it is adapted to drying on the hardwood bands it is unable to penetrate the softer areas. Aluminium primer puts what is virtually a continuous layer of thin metal flakes over the whole surface, which levels up the porosity.

Oily woods

Certain woods such as teak and cedar are of an "oily" nature which tends to retard the drying of ordinary oil paints and also impairs adhesion to some extent. Unless such woods are properly treated in the first place, much trouble can be caused through subsequent flaking and peeling.

A wash of acetone is sometimes advised, scrubbing the solvent into the surface with a stiffish brush to remove the oily residue. If this is done the acetone should be allowed to evaporate thoroughly before applying the primer, since it is incompatible with normal oil paints. A wash of white spirit however, is generally sufficient to prepare the surface for priming.

The most important requirement is that the primer should be *adhesive* and should dry well. A coat of goldsize and white spirit meets this condition and is satisfactory where the finish is to be clear varnish. Where paint is to be used, the primer may be a short-oil varnish (such as a hard-drying church oak), lightly pigmented with aluminium powder and thinned well with white spirit. Such a primer is also suitable for Columbian pine which often gives similar trouble in retarding the drying of priming coats, but it is generally agreed that aluminium primer is the best type of primer for this timber.

PROTECTION OF IRON AND STEELWORK

There is no field of protective painting in which the preparation of the surface, the type of primer used and the conditions under which it is applied, assume more importance than in the priming of iron and steelwork.

Firstly, preparation may be an arduous and costly job and, for this reason, the work is often improperly carried out even by competent workmen. Secondly, the primer must be chosen with especial care, because, while suitable paints help to prevent corrosion, others may actually stimulate it. Thirdly, the advantages of good preparation and a well-chosen primer may be entirely offset by adverse conditions prevailing at the time of painting. Thus, the three conditions for success in protecting iron and steel against corrosion are interdependent.

Corrosion

Iron and steel are the most susceptible of the metals which corrode by the action of air and water. Dry air will not corrode unprotected iron, neither will pure water; but moist air readily induces corrosion. When the atmosphere is smoke-polluted, as in large industrial towns and cities, the corrosive conditions are more acute owi g to the acid nature of dust and soot deposits and the distinctly acid rain water of such districts. Dilute salt solutions, e.g. sea water, are also highly corrosive.

The product of the corrosion of iron and steel is an oxide known chemically as *hydrated ferric oxide* and more commonly as "rust". Its formation is now held to be due largely to electrochemical (galvanic) action, and certain pigments, when in direct contact with iron, either inhibit (i.e. check) or promote this action. Those which inhibit corrosion are said to render the metal "passive" and act by repairing the extremely thin but highly protective natural oxide skin. Alkalis also serve to suppress rusting by precipitating a protective film of iron compounds in close contact with the metal, which is why steel reinforcement is permanently preserved when properly embedded in Portland cement or concrete; expanded metal lathing is similarly preserved by the lime in mortar.

Electrochemical stimulation can occur only when moisture is present; if iron is *thoroughly dry* and free from rust when painted with a good water-excluding paint, no corrosion is likely to occur so long as the film remains intact. In practice, however, the paint film may become permeated with moisture via pinholes, abrasions, carelessly painted joints and bolt-heads, etc., or by disintegration of the film through weathering; more often than not paint is applied over existing rust owing to the great difficulty of removing it by ordinary methods. In these circumstances rust is either promoted or, if already present, continues to spread beneath the paint film, resulting sooner or later in serious flaking.

Inhibitive and stimulative pigments

Pigments which have been found to possess rust-inhibitive properties are

Red lead
Basic lead chromate (orange chrome)
Zinc-rich paints (cold galvanizing)
Zinc chromate
Zinc phosphate
Sublimed blue lead
Calcium plumbate.

Pigments which are known to be rust-stimulative are

Graphite
Lamp black
Carbon black
Ferrocyanide blues (Prussian, Brunswick, etc.)
Venetian red and certain cheap red oxides.

Inhibitive pigments may not always be suitable for intermediate or finishing coats, red lead for instance being unsuitable as a finish on account of its susceptibility to carbonation, quite apart from its garish colour. Conversely, stimulative pigments may be excellent as intermediate or finishing coats when applied over a rust-inhibitive primer, graphite for example being specially good for its water-excluding properties. Good primers may also be produced by mixing rust-inhibitive pigments with neutral or even stimulative pigments, provided that sufficient of the former is used; for instance, 50/50 red lead and graphite; 25/75 zinc or lead chrome and red oxide; 25/75 red lead and red oxide; all these mixtures have been found to give good results.

It is known that straight linseed-oil films are fairly permeable to water, but that by using heat-treated oils and incorporating resins, films of a more impervious nature are produced, although it is recognized that no film is *completely* impervious. Another advantage of heat-treated oils and varnish media is that they have better adhesive qualities, and metal priming paints need to be tenacious.

PRIMERS FOR IRON AND STEEL

The problems associated with the corrosion of steelwork are so great, and the cost of wastage caused by corrosion reaches such astronomic proportions, that a vast amount of research has been devoted to finding ways to combat it, and all the leading paint manufacturers produce a wide range of priming paints, each with its own distinctive properties. The main types of primer for iron and steel are as follows:

Red lead

This is still a primer of major importance, in spite of the introduction of newer materials and in spite of certain obvious disadvantages. It is excellent for site work, and a particularly valuable feature is its tolerance of poor surfaces. Very often the only feasible method of preparing steelwork on site is by manual chipping and wirebrushing, a method which at best is only a compromise; it removes most of the rust and scale but certainly does not produce a clean surface. Even where better methods of preparation are available, few of them can produce a surface that would be classed as completely clean. On such surfaces red lead gives far better results than the more sophisticated types of primer that have recently been developed.

Red lead primer should conform to the standards laid down in BS 2523: 1966, in which three main types are listed; these are —

TYPE A: 80 percent genuine red lead pigment in a linseed oil medium, thinned with white spirit;

TYPE B: 60 percent genuine red lead pigment together with a mineral suspending agent in a linseed oil medium, thinned with white spirit;
TYPE C: 30 percent red lead, 30 percent white lead, and 15 percent asbestine in a linseed oil medium, thinned with white spirit.

It is recognized that some commercial products contain less red lead than is specified in the Standard.

These oil-based primers are slow-drying; there are faster-drying types that do not conform to the Standard, but they are usually less resistant to corrosion, although they may be useful in situations where the drying conditions are poor.

Red lead primer has good properties of build — that is, it produces a good thick film; when overpainted with micaceous iron oxide, which also has excellent build qualities, the resultant film satisfies one of the chief requirements of a paint system for steelwork, namely a film of very considerable thickness. The drawbacks of red lead are its toxic nature, its poor flow which makes it difficult to apply and brush out, and its strong tendency to settle in the can whilst in use — but the latter point is largely overcome by the inclusion of suspending agents. It is not suitable for spray application or dipping.

Calcium plumbate

This is an oil-based all-purpose primer, grey or off-white in colour, fairly slow-drying, and excellent for both ferrous and non-ferrous surfaces. Two main types are specified in BS 3698: Type A, containing 48 percent calcium plumbate, and Type B with 33 percent calcium plumbate, with a linseed oil medium, thinned with white spirit. Other pigments are incorporated such as titanium, asbestine, barytes, etc., but the proportion of calcium plumbate should not be less than 70 percent of the whole mixture.

Calcium plumbate primer possesses good adhesion, good opacity, good flow and good levelling properties; light-coloured undercoats and finishes cover much more easily upon it than they do upon the strong orange colour of red lead. Against this must be set the fact that a number of paints do not adhere well to calcium plumbate; it is therefore necessary to specify an undercoat and finish which are compatible with it.

Calcium plumbate is chiefly used as a primer on hot-dip galvanized iron rather than directly upon steel.

Zinc chromate

This is a primer of greenish-yellow colour, made with a pigment of zinc chromate with an alkyd or oleo-resin medium thinned with white spirit. Some brands contain a proportion of red oxide; the proportion of zinc chromate should not be less than 15 percent, and the better types contain 40 percent or more.

It is a primer with good resistance to marine atmospheres, good flow and good levelling properties, and is quick-drying; it is suitable for brushing, spraying, dipping, and stoving. It is more resistant than red lead to the strong solvents which may be present in subsequent coatings. It is generally regarded as non-toxic, and this fact is very often quoted as a reason for using a zinc chromate primer in preference to red lead. Concern has been expressed, however, about the possibility of harmful dust being ingested by operatives during the scaling and scraping of steelwork in situations where a zinc chromate primer was used, and it is suggested that EEC regulations should be amended to take this possibility into account.

Zinc chromate primer produces a thin film, is less resistant to corrosion than red lead, and is less tolerant of imperfectly cleaned steel.

Zinc phosphate

This is a non-toxic and fairly quick-drying primer, usually white or light grey in colour. The light colour is an advantage in that any early spots of rust developing on the metal surface quickly become apparent and can be dealt with before they constitute a major problem. It is a highly anti-corrosive product and is being increasingly used as an alternative for red lead. High-build zinc phosphate coatings are available which help to produce an adequate thickness for the entire film.

Zinc-rich primer

This is a primer consisting of a high concentration of metallic zinc powder — 90–95 percent of the dry film by weight — in a non-saponifiable medium such as chlorinated rubber or epoxy resin. It gives a high degree of "cathodic protection" to ferrous metals, and is sometimes called a "cold galvanizing paint" although its hardness and abrasion resistance are not equal to that of zinc metal applied by hot-dip galvanizing. Under damp conditions an electrolytic cell is set up with zinc as the anode and the ferrous metal as the cathode, the zinc being attacked instead of the ferrous metal; for this reason zinc-rich paint is called a "sacrificial" film, and this is what is meant by referring to the action of the film as cathodic protection. It should be noted, however, that zinc-rich paints can only function in this way when applied to the bare metal, a direct electrical contact being essential in order to give cathodic protection; the primer should not, therefore, be applied over old paint, and is not suitable for spot priming.

One difficulty with zinc-rich primers is that the zinc content tends to embrittle the medium in subsequently applied coatings, making them liable to break down into a mass of tiny flakes. For this reason it is common practice for several coats of the primer to be applied as a self-finish for structural steelwork, bridges, etc., the zinc-rich paint forming the entire protective system; but where the appearance of the paintwork is an important factor, gloss paints or enamels formulated so as to be compatible with the zinc dust can be used as the finish.

Zinc-rich primer gives a film thickness two or three times as great as a normal paint film, and it dries very quickly indeed. A very important feature is that even if the film is scratched or scored, the cathodic protection is so strong as to prevent the bared metal from corroding for some considerable time, and certainly until such time as the damage can be repaired.

In situations where application difficulties exist and exposure conditions are severe, zinc-rich paints are of very great value. The writer recalls an instance on a contract for the painting of a dock installation where, in addition to the usual problems of preventing corrosion, protection against salt water was required, and where the work had to be carried out in short spells each of strictly limited duration in between tides, each spell becoming a race against time; in this situation the rapid drying and highly protective properties of the material made zinc-rich paint an ideal choice.

It is, of course, an expensive product, especially when used as a complete protective system. On the other hand a "prefabrication primer", which is the type based on epoxy resin, applied in a thin coating over a grit-blasted surface to give short term protection before final painting, is not unduly expensive.

Zinc silicate

This is a two-pack coating in which a zinc pigment reacts with an inorganic silicate solution. It can be used as a primer with other paint coatings superimposed, or can be applied in several coats as a self-finish. It is especially suitable for wet situations such as marine structures, bridges and immersed pipes, and can also be used on hot surfaces. It is not tolerant of imperfectly cleaned metal; a high standard of surface preparation is essential if it is to give satisfactory results.

Etch or wash primer

This is available either as a one-pack or two-pack material, the two-pack type providing greater stability and adhesion. It consists of vinyl resins reduced with solvents to a thin "wash" consistency, lightly pigmented with zinc chromate or other rust-inhibitor, and with an acid additive such as phosphoric acid. In the two-pack type the acid is supplied separately to be added immediately before the material is used.

Etch primers are principally used for the pre-treatment of non-ferrous metals, but they can also be used to give immediate protection to steelwork which has been blast-cleaned; in this case they need to be followed up with one of the other listed priming paints as they themselves give no protection if left unpainted. They are not tolerant of imperfectly cleaned steel and are therefore not suitable for steelwork which has been cleaned manually.

Welding primer

This is a material specially formulated to enable welding to take place without the joint being weakened and without the production of toxic fumes. It is less protective than any of the inhibitive primers, and for this reason the welded area should be reprimed as quickly as possible. Some prefabrication primers are classed under this heading.

Metallic lead

There is considerable pressure from the EEC for the use of this material to be abandoned for environmental reasons, and at one period in 1976 there was every indication that its manufacture would cease, but at the time of writing it is back on the market and is being supplied by Associated Lead Manufacturers. Notes about the properties of metallic lead have been given in Chapter 10, page 161.

Other products

Various materials are available, and have been given prominence in both the popular press and in trade journals, which claim to give protection to steelwork even when rust and scale are present, thus eliminating the difficulty and expense of getting the surface thoroughly clean. These materials are said to react with the rust so as to form a protective coating. However, when the claim has been investigated by experienced researchers it has generally been found to be unduly optimistic. Unfortunately it sometimes happens that a customer has heard of these materials and insists that the decorator should use them. In such a case the decorator would be wise to disclaim (in writing) all responsibility for the outcome of the experiment. He should point out that if there is any difficulty about getting the metal clean, a red lead primer is sufficiently tolerant to give satisfactory results and would be a far more reliable choice.

Preparation and priming of iron and steel surfaces

It is highly important that in the treatment of iron and steel, the preparation and priming should be regarded as a single operation; that is to say, the priming should follow the preparation immediately, since rust forms so quickly on the unprotected metal. Ideally, the surface should be clean, bright and free from moisture at the time of painting, but in cases of severe rusting it is practically impossible for the painter to fulfil the above conditions with the ordinary methods of preparation, namely chipping, scraping, and wire-brushing.

Cast-iron is rather "spongy" in texture and, when thoroughly dry, presents a surface to which paint adheres perfectly. New wrought iron and mild steel, however, are usually covered with a thin layer of millscale. When steel comes off the rolling mill it is covered with a tightly adhering layer of iron oxides (the "millscale"), and if it were possible to prime the metal as it leaves the mill, that is, while still warm and with the millscale intact, excellent protection would be assured, as long as the scale remained intact. This is rarely possible, however, since rolled steel generally has to be cut, drilled and fabricated before painting. Once the scale is broken, galvanic action is set up between the exposed metal and the scale in presence of moisture, and corrosion spreads rapidly beneath the scale, which in time becomes loose enough to scrape off easily. Some engineers specify that steelwork should be left to "weather" outdoors for at least six months to enable the millscale to loosen; it is then removed by chipping and wire-brushing before painting.

Where size permits, de-scaling may be carried out at the works by means of sand-blasting, shot-blasting or phosphate bath. Steel windows, for instance, are usually de-scaled and then given a rust-inhibitive treatment such as "sherardising" or "bonderising" which provides an ideal surface for the reception of paint. For structural members, immersion in hot acid solutions effectively removes rust and scale, and there are several proprietary processes which do this and also form a surface film of metallic phosphates.* Such films, especially if treated afterwards in a chromate bath, provide an

*One of these is well known, under the name of the "Footner process".

excellent adhesive surface for painting and increase the resistance of the surface to rusting. This is of course essentially a factory process, not one for the site.

Steelwork may, however, be delivered to the building site in an unprimed condition or else treated with a primer of doubtful composition before despatch from the works. In the latter case the primer may be a cheap red oxide (possibly with rust-stimulative properties) applied, in all probability, over a surface of discontinuous millscale; in these circumstances rusting is almost certain to proceed under the paint film (no matter how good the superimposed paint may be) and if the job is important the decorator will do well to remove the "shop priming" as far as possible in order to replace it with a good coat of red lead.

Where the steelwork is delivered unprimed, the surface will have to be gone over with chip-hammers and vigorously wire-brushed to remove as much of the loose scale and rust as possible. Rust is most difficult to remove. Its moisture content (both physical and chemical) makes it peculiarly tenacious, and great heat is required to bring about proper dehydration, as in the flame-cleaning process described on page 58. The surface obtained with hand or power-driven chipping hammers and wire brushes is not good enough for the newer anti-corrosive paint systems; where these tools have been used a "tolerant" primer such as red lead should be used.

Mild rusting on machined or polished steel may be removed by rubbing down with engineer's emery cloth or steel wool, using paraffin as lubricant; the trace of paraffin left on the surface after wiping dry should not retard the drying of the primer if red lead is used. There are also proprietary rust-removers and de-greasing agents, such as "Deoxidine", made by I.C.I. (Paints Division), which are very effective in removing surface rust; but it is most important, where such chemical solutions are used, that the manufacturers' instructions be strictly followed, particularly in regard to the thorough rinsing and drying of the surface before painting.

Double priming and stripe priming

Many authorities nowadays advocate that steelwork should be given two coats of primer in all but the mildest conditions of exposure. The reason for this is the absolute impossibility of producing with one coat a continuous film of regular thickness and free from pinholes — the points at which corrosion begins. A double coat of primer is far more likely to prolong the life of the whole paint system than an extra undercoat or an extra finishing coat. It is also more economical, because at a later stage the surface can be kept in good condition by applying a new coat of finishing paint, which is far cheaper and easier than a complete process of cleaning down, priming and repainting the work.

Another point to notice is that paint always tends to recede from the sharp edges and arrises of steelwork and from the crowns of corrugations, and it is at these places that rusting first begins. Even where for reasons of economy a second coat of primer is not considered justifiable, it is a wise precaution to apply a "stripe coat" of primer over these areas before the full priming coat is applied.

BLAST CLEANING OF IRON AND STEEL

Blast cleaning is a very effective method of preparation if carried out correctly. It is chiefly used as a factory process but may be carried out on the site where conditions permit, especially if the vacuum-attachment type of equipment is used to reclaim and recirculate the abrasive and protect the operative. Today, blast cleaning has become a process of considerable importance in relation to industrial painting operations, and the indications are that it will become increasingly important. People who are engaged mainly on the more decorative side of the craft are usually unaware of the factors involved and the highly technical nature of the process. It will help them to understand the process if some of these factors are examined and explained. Typical equipment for blast cleaning has been described in Chapter 4.

Abrasives

It is essential that the right type of abrasive be selected with regard to (a) the condition of the steelwork and the nature of the material that is to be removed, and (b) the type of paint finish required, which in turn has considerable bearing on the quality of the cleaning finish that must be achieved.

The abrasives can be classified firstly according to shape; the term "shot" is applied to small round abrasive particles; these bombard the surface like miniature hammers, producing a *rounded* profile or anchor pattern in the steel; the term "grit" is applied to small angular particles which have a sharp cutting edge to produce an *angular* profile or anchor pattern.

The *types of abrasive* which are employed include chilled iron shot, steel and malleable iron shot, chilled iron grit, steel and malleable iron grit, and non-metallic abrasive substances, together with certain other metallic abrasives such as aluminium, brass and copper which are used on a limited scale for specialized purposes. These are all abrasives which can be used in factory blast-cleaning processes and with vacuum-attached blasting equipment on the site — processes in which the abrasive can be reclaimed and used several times over (in some cases they can be used as many as ten times before being discarded). On site work where vacuum equipment is either not feasible or not available it is practically impossible to re-use the abrasive and the cost therefore becomes prohibitive; in this case a type of abrasive is used which is cheap enough to be regarded as expendable, and for this purpose mineral slag, produced in the course of refining iron and copper, is used. Sandblasting, although at one time fairly widely used for certain kinds of industrial work including external ship painting, is no longer employed on site, and even in factory processes its use is very limited and is strictly controlled because of the health hazard.

The *size of the abrasive particles* is also important, and a grading system is specified in BS 4232:1967, in which the various types are listed according to the size of mesh through which they will pass. Generally speaking, the larger the particles, the harder the surface will be hammered and the quicker the scale will be removed, but the result of using particles which were too large would be to make the surface so rough and coarse that it would be unsuitable for painting, and in trade practice it is customary to use the smallest size of abrasive consistent with bringing the surface to the desired level.

Surface roughness and cleanliness of the finish

It must be realized that the blasting operation does not make the steel surface smooth; on the contrary, its effect is to etch and roughen the surface into a series of tiny indentations, the result of which can be likened to a series of peaks with tiny troughs or valleys between them. This etching of the surface is of great value to the painter because it enables the paint to cling to the surface by anchoring itself into the hollows — in other words, the adhesion of the paint is greatly increased by the blasting process. Incidentally the consumption of paint rises too; the etching effect enlarges the surface area that the paint must cover, and more paint is required for such an etched surface than on a perfectly smooth level surface. Various methods of measurement are now available (these are also specified in BS 4232) to ascertain the vertical distance between the summit of the peaks and the deepest point of the adjacent hollows or troughs; the vertical distance is termed the "amplitude". A "maximum amplitude" of 0·10 mm (0·004 inch) as produced by metallic abrasives is regarded as being acceptable for painting, and in the case of non-metallic abrasives a maximum amplitude of 0·18 mm (0·007 inch) is acceptable. The grades of abrasive particle sizes in the British Standard have been selected to produce a target roughness defined by these maximum amplitudes.

The degree of finish that can be achieved and the surface roughness obtainable depend to a great extent upon the condition of the steel before the blasting is commenced, i.e. whether it is deeply pitted, badly corroded, and so on. The greater the degree of corrosion and rusting, the more difficult it is to gain a satisfactory level.

Degrees of surface cleanliness

Obviously if the operation of blast cleaning were not properly controlled it would be very likely to produce patchy results, with some areas of the steel reasonably cleaned whilst other areas were quite unsatisfactory. The quality of surface finish therefore depends not only on the roughness achieved but also on the general all-over cleanliness. There is no precise method of measuring this cleanliness, but within certain broad limits it can be defined and controlled. BS 4232 specifies three recognized degrees of surface cleanliness, and defines the types of paint system for which each of them is applicable.

The *First Quality* is achieved when the entire surface of the steel shows blast-cleaning pattern, and the bare steel is completely free from contamination and discoloration. This quality is required when an extremely clean surface is called for to prolong the lifespan of chemically resistant finishes in the most exacting conditions.

The *Second Quality* is where the entire surface shows blast-cleaning pattern and at least 95 percent of the surface is completely clean. To satisfy this standard, tightly bonded millscale is permissible on up to 5 percent of the entire surface; up to 10 percent is permissible on any single square of 25 mm side. This standard is defined as the minimum requirement for chemically resistant paint systems such as epoxide and vinyl resin paints, and is also desirable where the best results are required from conventional paints under fairly corrosive conditions.

The *Third Quality* is where the entire surface shows blast-cleaning pattern and at least 80 percent of the surface is clean bare steel; for this purpose tightly bonded millscale is permissible on up to 20 percent of the surface on average, with not more than 40 percent on any single square of 25 mm side. This is the standard recommended for steelwork to be painted with conventional paint systems for exposure to mildly corrosive atmospheres.

A photographic scale illustrating and defining the various stages of rusting and treatment is available at a moderate price on application to the Paintmakers' Association of Great Britain.*

Standards for abrasive cleaning are also specified by the Swedish Standards Organization and by the Steel Structures Painting Council, U.S.A., both being internationally recognized. The Swedish standards are defined by comparing the surface with a set of professionally produced coloured photographs; the American standards, like the British, rely upon a purely verbal definition. On the Swedish scale SA3 corresponds with the British First Quality, SA2·5 to the Second Quality, and SA2 with the Third Quality; the American equivalents are "White metal", "Near white metal" and "Commercial".

It is important that the kind of surface required should be clearly specified; some surface roughness is desirable to afford a physical key for the adhesion of the paint, but excessive roughness can lead to the situation where peaks of metal project through the paint film or at best are only inadequately covered. It is in such places that rust would quickly form.

Procedure

Steelwork to be blast cleaned must be perfectly dry, and the operating conditions must be such that no condensation occurs. Any oil or grease must be removed from the surface before cleaning is begun, and where compressed air is used it must be free from oil and must be dry. It is because of these limitations that flame cleaning is often preferable to blast cleaning for site work on external structures. Where the steel is badly rusted, and especially if scaly rust is present, it will be found economical to clean the work with power-driven tools before the blasting is commenced. Immediately the blasting is finished the cleaned surface should be air-blasted, vacuum cleaned, or otherwise cleared of residue and dust, unless of course the vacuum-attached type of blasting equipment has been used. When all the abrasive material and residue has been removed the surface of the steel should be rubbed over with a nylon scourer or scraper in order to remove the "rogue peaks" (odd peaks of metal projecting above the general level).

Application of the priming coat should follow the blast-cleaning operation as quickly as possible before visible deterioration sets in, and certainly within 4 hours of cleaning. Thin-film prefabrication primers such as etch primers or zinc-rich epoxy paints can be used to give immediate protection after cleaning; these must of course be overpainted with a conventional priming paint as soon as is practicable.

---

*The photographs supplied by the Paintmakers' Association are printed under the title *Échelle Européenne de Degrés d'Enrouillement pour Peintures Antirouilles*.

Protective clothing must be worn whilst blast-cleaning (visor or blasting helmet, rubber or leather apron, abrasive-resistant gauntlets, and industrial safety boots — see Chapter 4).

Chemical site processes

Reference has already been made to phosphating, and this of course is a process that can only be carried out under factory conditions. There are, however, various washes and pastes available which are based on phosphoric acid and which are sometimes recommended for cleaning on the site. They do, unfortunately, present certain difficulties. They often leave a brittle film on the surface, and if used on non-rusty areas they tend to leave a sticky deposit, neither of which makes a good foundation for paint. If the work dries slowly the steel may become rusty again. Washing off is difficult because of the risk of staining the adjoining structure. Some brands are said not to require washing off, but this claim should be treated with reserve.

Flame-cleaning

There are still many situations where the only really satisfactory method of preparing badly corroded iron and steel for painting is by flame-cleaning (see page 58), and where important structural steelwork is involved, such as bridges, piers, crane gantries, etc., it is advisable to enlist the services of a firm specializing in this process. The great advantage of flame-cleaning (or flame-priming as it has been called) is that it dehydrates the rust, that is to say, it drives off the contained moisture in rust, changing it first from the hydrated to the anhydrous ferric oxide and finally to black magnetite; this is a highly stable granulated powder which can be easily removed by wire-brushing, leaving a blue-black appearance on the surface of the metal which is characteristic of the process. All *loose* millscale is also removed by the heat, and any which remains after the intense oxy-acetylene flame has passed over it may be considered sufficiently adherent to resist subsequent flaking. The removal of the oxide and scale is based on the principle of differential expansion, i.e. an intense local heat is applied to the scale which heats up quickly and expands rapidly, while the parent metal to which it is attached remains comparatively cool — thus, there is the differential expansion between the parent metal and the scale which causes the latter to free itself and crack off. Flame-priming enables work to be carried out under adverse weather conditions of damp and humidity which would normally prevent any possible chance of success. The process is also most valuable in situations where the structural steel is encrusted with salt or other chemical residues or is exposed to ammonia fumes, etc. It is recognized in America and in the U.K. that flame-cleaning often represents the only practical method of obtaining perfect conditions for painting iron and steel structures after erection.

The vital part about the flame process is that the primer should be applied immediately, while the surface is still warm and thoroughly dry. This ensures that the paint will flow into any surface irregularities and gain closer contact with the metal. If the surface is allowed to cool, condensed moisture may be deposited from the atmosphere, in which case the advantage gained by heating is completely lost.

Application of primer on iron and steel

In applying the primer particular attention should be paid to rivets, bolt heads, angles and joints, and especially to sharp edges where the film is liable to be extremely thin. According to some authorities the ready-mixed or non-setting types of red lead have less inhibitive value than hand-mixed red lead. This point of view is only of academic interest, however, since hand-mixed red lead is now prohibited by law (except for use in stopping and filling). Nevertheless, genuine ready-mixed red lead has been found to give excellent service, especially when followed up with a good red oxide paint or with good water-excluding pigments such as aluminium and graphite or mixtures of these. Small amounts of aluminium powder may also be mixed with other finishing pigments to provide reinforcement without unduly altering the colour; when added to light colours, the aluminium gives some protection against actinic light rays and also improves water-resistance.

The question of brush versus spray frequently crops up in connection with the application of priming paint and many advocates of the merits of spray painting point to the successful spraying of primers in motor body and other industrial finishing processes. It should be remembered, however, that in such cases the metal is usually chemically pretreated (bonderised, etc.) and that spraying is generally carried out under conditions of controlled temperature. The spraying of structural steelwork outdoors is a very different matter, and under certain atmospheric conditions condensed moisture may be deposited on the surface along with the atomized paint (except of course in the case of airless spraying). Moreover, a sprayed coat lacks the advantage of the valuable frictional effect of brushing which, in the writer's view, results in a more intimate bond between primer and metal. In our opinion then, a well-brushed priming coat is definitely better than one which is sprayed, but when once the priming is dry and hard, intermediate and finishing coats may be sprayed with great advantage in speed of application. The spraying of lead paint is, of course, prohibited on *interior* work.

Priming for steam pipes and other surfaces subject to heating should be carried out with the paint specially formulated for the purpose, colour-coded as laid down in the appropriate British Standard. Aluminium and metallic bronze paints stand up to high temperatures very well, but they reduce the radiation of heat and so cause a loss in heating efficiency.

Other treatments

Certain types of bituminous compositions give better service in some situations than oil paints as, for instance, on buried metal, underwater pylons, ships' bottoms, etc. Above ground, bitumen tends to disintegrate rapidly on exposure to prolonged sunlight, but the addition of aluminium powder to the bitumen gives greater durability by affording some protection against the destructive action of light. Bitumen may be applied over a red lead primer, provided the latter has been applied some two or three weeks and has become thoroughly hard, otherwise the primer may be softened by the solvent action of naphtha which is the usual thinner for such compositions.

Bituminous emulsions may be safely applied over red lead or other inhibitive primer and are often recommended for *direct* use on iron and steel. In this connection, however, it must not be forgotten that some emulsions are very unstable and are easily upset by extremes of temperature; an emulsion exposed to keen frost, for instance, is liable to break down and release free water which, applied to bare iron or steel, would soon promote corrosion. The use of an inhibitive primer under all bituminous emulsions is therefore the safest course to follow. Bituminous materials should be used only where it is fairly certain that oil paint will not be subsequently desired, since it is difficult, especially with thick elastic coatings, to render the surface non-bleeding; application of hard-drying tar-proof knotting over such coatings is liable to result in cracking.

To summarize, the successful painting of iron and steel depends on (a) thorough preparation to obtain a clean, dry surface; (b) a primer chosen for its anti-corrosive properties, applied while the surface is dry and, if possible, warm; and (c) a good moisture-excluding top coat. Conditions (a) and (b) are the most important, and if these are met any top coat of average quality will give good service, but where the greatest durability is desired a paint incorporating water-excluding pigments such as aluminium or graphite should be chosen for finishing.

PROTECTION OF NON-FERROUS METALS

Non-ferrous metals are being increasingly used in modern building practice, especially in lightweight prefabricated construction and in decorative architectural details. They are also widely used as protective coatings on iron and steel and, in addition to the familiar zinc-coated or galvanized iron, we now have coatings of lead, aluminium, and sometimes cadmium. Copper is also displacing lead to a great extent in present-day plumbing.

Normally, non-ferrous metals do not require protection by painting, for although they do oxidize or corrode in ordinary atmospheres, the process is very slow compared with the rapid corrosion of iron and steel. Corrosion, however, may be greatly accelerated in chemically polluted atmospheres, and some form of protective coating, such as varnish, lacquer or paint, is generally required in such circumstances.

In the case of non-ferrous coatings on iron and steel, the protective layer is usually so thin that its power to protect the base metal from corrosion must be of comparatively short duration under adverse conditions of exposure. For instance, unpainted galvanized iron in clean rural atmospheres will generally remain in good condition for several years, whereas in large industrial districts the effective life of the zinc coating may be no more than 12 to 18 months, after which time rusting may begin to appear. Once rusting has begun it may spread fairly rapidly beneath the zinc coating. When such surfaces are properly painted, however, the protective non-ferrous coating is maintained in good condition and its life is considerably prolonged.

Even where structural steelwork has been protected by metal sprayed by the hot-wire process and thus given a coating of extreme durability, the life of that coating is considerably extended if the metallized surface is painted; and if the paint coating is renewed at regular intervals a practically permanent protection is given.

In some cases painting is required merely for the sake of appearance, as when lead and copper waterpipes, etc., are painted to conform to the general colour scheme. The painting of non-ferrous metals has come to be regarded as something of a problem owing to electrochemical and galvanic action causing corrosion of the metal and thus giving rise to insecure attachment of the paint film, with consequent flaking or peeling. In addition to the question of adhesion, certain pigments may react with the surface in the presence of moisture and so stimulate corrosion of the metal as in the case of iron and steel, but pigments that are suitable for priming the latter are not necessarily suitable for non-ferrous metals. It will be seen, then, that the successful painting of non-ferrous metals depends on suitable preparation of the surface to produce a "key" and on careful choice of pigment in the priming coat; the medium should also be of an adhesive nature.

Galvanized iron (and other zinc-coated surfaces)

Galvanized iron is generally produced by "hot-dipping", i.e. by passing the iron sheets through molten zinc in presence of a suitable flux, the finished coating having a bright spangled effect; it is also produced by an electroplating process.

When ordinary oil paints and varnishes are applied to new galvanized surfaces they tend, sooner or later, to peel or scale badly, particularly when exposed to the weather. The cause of flaking is attributed by some authorities to the smooth "greasy" nature of the surface, while others hold that it is due to chemical action between the oil media and the zinc coating which results in the gradual formation of loosely adherent zinc soaps at the interface; probably both conditions are contributory to the lack of key.

The simplest and most effective way to prepare the surface for painting is to allow it to "weather" outdoors for a period of six months or so, but where this is not convenient some form of chemical pretreatment is necessary in order to roughen and degrease the surface. In the past various dilute acids and copper salt solutions have been advocated to etch the surface, but these have not always proved successful, the black deposit of copper which results from the use of copper solution often being so loosely adherent as to be useless in preventing flaking. Again, some acid deposits, if not thoroughly removed before painting, may stimulate corrosion in presence of moisture.

There are, however, proprietary solutions on the market (known as "mordant solutions") which have proved very successful in practice. These solutions react with the metal to form an insoluble zinc compound which roughens the surface and provides a good key for paint, but it is important that the surface should be free from oil and grease or any other matter liable to interfere with the etching process. After removing grease with white spirit, solvent naphtha, or cellulose thinners, etc., the solution is brushed on the surface and allowed to dry (care being taken to avoid contact with the skin, since the solution is of an acid nature); the surface should then be well rinsed with water to remove any loose deposit and again allowed to dry thoroughly before applying the first coat of paint.

Alternatively, after degreasing, the surface can be etched chemically with a special self-etching primer (see page 251).

In priming zinc-coated surfaces, graphite, lead and other metallic pigments (excepting zinc dust) should be avoided because they tend to react with the zinc surface and may stimulate corrosion. Either zinc chromate or calcium plumbate primers are eminently suitable, both having excellent adhesion and providing tenacious, flexible coatings, but in present-day practice calcium plumbate is usually specified (it was for this specific purpose that calcium plumbate primers were originally developed.)

Other zinc coatings

The process known as *sherardising* consists of coating steel by heat treatment with a mixture of zinc dust and zinc oxide in an enclosed chamber; steel windows and other fixtures of suitable size are generally treated in this way. *Zinc spraying* is another method of coating iron and steel with metallic zinc, the advantage of this process being that the molten zinc may be sprayed on structures after erection.

Unlike galvanized iron, sherardised and zinc-sprayed coatings are slightly rough and present a fairly good key for paint. After degreasing, prime with zinc chromate or calcium plumbate primer. Etch primers are also most valuable for use on zinc-sprayed steelwork.

Aluminium surfaces

Many building fixtures are fabricated in aluminium or duralumin; steel structures also can be protected with aluminium coatings applied by hot-dipping or by molten spray.

Aluminium will retain its bright polish indefinitely in a clean indoor atmosphere and under such conditions requires no protection by painting, although it is often varnished or lacquered for decorative effect. However, when exposed to outdoor atmospheres, particularly in coastal regions or smoke-polluted industrial districts, the bright surface quickly dulls and becomes coated with a rough, whitish film. If this is not protected in some way, corrosion will proceed fairly rapidly and, in the case of thinly applied aluminium coatings, the underlying iron or steel will soon be affected.

All aluminium, or aluminium-coated steel, intended for exterior work should preferably be painted or protected by other means. Very often aluminium is *anodized* by various patented electrolytic processes which produce a thin surface coating of aluminium oxide, but although this coating is highly protective (except in severe corrosive conditions) it is usual to reinforce it with paint or lacquer coatings. (The familiar anodized decorative finishes are generally produced by dyeing the oxide coating and finishing with nitrocellulose lacquer or with wax.) Aluminium forms a good surface for the reception of paint, provided it is absolutely free from any greasy matter, for which it appears to have an affinity. Alkaline detergents should not be used for degreasing because aluminium is attacked and dissolved by caustic alkalis. It should be cleaned with petrol or white spirit (applied with a stiffish brush and well "scrubbed"), then rubbed down with emery cloth (never steel wool) to form a key. Alternatively, a proprietary degreasing agent such as Deoxidine, made by I.C.I.

(Paints Division), may be used, in which case the manufacturer's instructions should be carefully followed.

In the priming paint, lead, graphite, bauxite reds (known as Burntisland red) and metallic bronze pigments should be avoided since they react unfavourably with aluminium in presence of moisture. After degreasing and etching, prime with zinc chromate primer, or with an etch or wash primer. As already indicated, etch primers are used far more for this purpose than for steelwork surfaces.

Lead surfaces

Lead normally requires no protection by painting and, in fact, tends to reject paint, by peeling or flaking. Nevertheless it is frequently necessary to paint lead pipes, etc., to conform to a general colour scheme. Moreover, lead and lead alloys are being used for coating steel by hot-dipping and metal-spray processes, and as such coatings are often too thin to give permanent protection to the steel, they may require reinforcement with paint.

Chemical pre-treatment with the proprietary solutions recommended for zinc has given good results; the primer may be zinc chromate or calcium plumbate. Graphite should be avoided in the priming coat, for if the lead coating is ruptured in any way, the graphite would make contact with the steel and (in the presence of moisture) would tend to stimulate corrosion.

Copper, brass and bronze surfaces

Copper and copper-base alloys such as brass and bronze tend to retard the drying of straight linseed oil paints, sometimes for as much as three or four days; moreover, acid drying oils and varnish media may dissolve traces of the metal, sufficient to produce a green stain in the paint or varnish. Degrease and etch simultaneously by rubbing down with emery cloth and white spirit. Never abrade them dry. It is particularly important to moisten the surface with white spirit to prevent dispersal of metallic dust which is liable to cause green or grey spots to develop on any surface where it alights. Prime with zinc chromate or calcium plumbate primer. Ordinary oil paints may be used for finishing coats.

Self-etching primers

When sprayed or brushed on aluminium or zinc-coated surfaces (which have been degreased) these primers chemically etch the surface and leave a firmly adherent "wash" coating of zinc chromate that forms a firm foundation for subsequent finishing coats.

20

Preparing Old Surfaces

The success which attended the efforts of the craftsmen of former days was due in no small degree to the time and care which they lavished on the preparation of the surfaces involved. Nowadays, high labour costs and the client's reluctance to be inconvenienced for more than a few days together, often preclude the necessary amount of time being spent on thorough preparation of the whole of the surfaces, and we are reduced to "making a show" on selected portions of the work, such as on the front door or other place liable to come under close scrutiny.

We must not lose sight of the fact, however, that preparation is a most important part of the job and that within the limits of the time at our disposal we should concentrate on essentials such as securing a clean firm ground upon which to build our sequence of coats. This is the minimum requirement and should apply equally to all surfaces, interior or exterior, irrespective of the material to be applied, be it paint, paper or emulsion.

Yet how often do we see exterior painted surfaces being repainted after a cursory brush down with the jamb duster! This treatment may or may not be adequate in a clean rural district, but it can be entirely inadequate in the smoke-laden atmosphere of some large industrial towns. No matter how well the duster is used it cannot effectively remove the deposit of grime and soot which even now can settle widely, and this surface coating is bound to weaken the adhesion of the new paint. Moreover, this grimy deposit is more than likely to retard the drying of the first coat and, being of an acid nature, may attack the paint film from beneath.

The only satisfactory way to remove a sooty deposit is by *washing down*, but this necessitates leaving the surface to dry thoroughly before we can paint it — a procedure which is apt to be considered too inconvenient and costly compared with merely dusting and painting in one operation. If, however, washing is ruled out on grounds of cost (though it must not be forgotten that washing very often saves a coat of paint), or if it is undesirable to use water (as on iron and steel surfaces where rust is present), an attempt should nevertheless be made to clean the surface by other means, such as by vigorously scouring with steel wool or wire brush. If we do less than this we should not blame the paint manufacturer if the paint fails to dry satisfactorily or to give reasonable durability.

In the preparation of painted surfaces the painter is often under the disadvantage of not knowing the history of the previous treatment, and he must therefore rely largely on his own observation and experience in deciding whether the existing coating is sound enough to provide a firm foundation for the new treatment after proper washing and rubbing down, or whether it requires complete removal.

High-pressure washing and steam cleaning machines have been described in Chapter 4.

PAINT REMOVAL

Paint should be entirely removed when it shows obvious defects such as cracking, crazing, wrinkling, blistering or peeling. It should be removed when it is soft or tacky, a condition which may have been caused by faulty drying when originally applied, or by saponification as on new alkaline plaster. It should also be removed when the film is discoloured by mould, which is usually evidenced by blackish or purple stains, or when there are any other stains which cannot readily be diagnosed. Such stains may be due to some form of local chemical action, such as the green discoloration from copper or brass which often bleeds persistently (bleeding can usually be overcome by coating with tarproof knotting, but it is the best plan, wherever possible, to remove the cause of the trouble).

Paint should also be removed when it is desired to change from a light finish such as white enamel, to a dark finish such as walnut or mahogany graining, or vice versa, i.e. from dark to light. This applies especially to interior woodwork which is liable to get knocked and chipped and where every mark would show up badly; it is not so necessary on exterior work, or on surfaces normally out of reach, since there is not the same danger of the surface becoming so defaced.

What is not so easy to decide, however, is the treatment of paintwork which looks reasonably sound and is free from surface defects. A careful inspection will usually give some indication as to the amount of paint already on the surface; if, for instance, the mouldings and quirks tend to be filled up, one can safely assume that the surface is overloaded with paint and should be removed, for it may not be able to bear the strain of another two or three coats of paint. The painting of overloaded exterior surfaces, particularly on southern elevations, is very liable to result in blistering or cracking and, ultimately, peeling; removal is therefore the safest policy. Chalked surfaces do not normally require complete removal; in fact, a chalky surface is probably the ideal surface to prepare for repainting since it only requires rubbing down well to remove all the loose powdery material.

The removal of old paint is achieved either by burning-off or by dissolving with a paint solvent. Other methods are provided by the Sanderson Lightning Stripper (see page 55) or the Sikkens hot air stripper (page 71).

Burning-off

Burning-off is generally the quickest, cleanest and most convenient method of removing defective paint. The choice between petrol or paraffin blowlamps and LPG or air–acetylene equipment and blowtorches is largely a matter of personal preference (or prejudice), but those who have tried the latter would not willingly go back to the ordinary blowlamp. The LPG blowtorch or the air–acetylene handpiece are light and convenient to handle, and there is no bother in lighting or waiting for the lamp to warm up, nor the nuisance of cooling-off in a wind. Running costs are higher than for the ordinary blowlamp (though with the cost of petrol soaring the gap is closing), but if one sets against this the time wasted in fiddling with the latter, the blowtorch has a definite advantage (see Chapter 4).

Whichever appliance is used, the essential thing is to keep the flame constantly on the move and to avoid playing too long on one portion of the surface. Burning-off should commence at the bottom and proceed upwards so that the rising heat helps to soften the paint above and thus makes removal easier when the flame reaches it. The lamp is held in the left hand and the broad knife or chisel knife in the right; as the flame is played on the surface and moved upwards the knife follows immediately beneath it, sliding under the softened paint film and cleanly removing it. The knife should not be too sharp and should be held at an angle of about 30 degrees or less, in order to avoid stabbing the surface. Mouldings should always be burned-off first as these take longer than the flats and it is very difficult to avoid scorching if they are left to the last. The shave-hook should be used most carefully so that the contour of the moulding is not damaged.

To the beginner, the simultaneous handling of flame and knife is rather awkward at first, but practice soon establishes a sympathetic movement of both hands in unison, and it then becomes almost a mechanical operation. Follow the grain of the wood and do not attempt to burn more than you can immediately remove with the knife; allow the knife to dictate the speed of the lamp, for when once heated and cooled again, the paint becomes harder than before and more difficult to remove. Scorch out all large knots to remove as much resinous matter as possible, but avoid charring the wood.

White and light colours usually burn-off fairly dry, and the droppings may be allowed to fall harmlessly to the floor. If the paint is fairly soft, however, it tends to cling to the knife or stick to the floor if let fall. This may easily damage a polished wood floor or lino, and it is advisable to have an empty paint can or a pail handy in which to deposit the refuse and on which to scrape the knife to clean it. Old paint which is excessively soft or "treacly" is very messy to burn-off, but if a thick coat of limewash is first applied and allowed to dry, the paint is hardened up and becomes more manageable and burns off more cleanly. Beware of the paint catching fire, and always be on the lookout for dry rot in timber; many fires have been caused by a flame creeping through joints in rotten timber, setting it aglow beneath the surface and left unnoticed until a fire becomes well established — often after the painter has finished for the day.

Rubbing down

Where the paint has burned-off dry and cleanly it may be rubbed down satisfactorily with glasspaper before dusting off, knotting and painting. Very often, however, it requires rubbing down wet, using pumice stone or waterproof abrasive paper to obtain a smooth surface fit for painting; needless to say, in such cases, the surface should be allowed to dry thoroughly (particularly in mouldings and quirks), and if the grain has become raised through moisture it will require another rub down with dry glasspaper and dusting off before any paint is applied. If it is desired to paint immediately after burning-off, the rubbing down may be done with pumice stone, using oil and spirit as lubricant (1 part raw linseed oil to 3 parts white spirit), afterwards cleaning off with white spirit and drying off with non-linty rag; this cuts out any waiting for the surface to become dry and, what is more important, avoids the possibility of moisture being absorbed in the woodwork. Alternatively, use a self-lubricating abrasive paper.

Exterior woodwork which has been burned-off (or stripped with paint remover) should always be painted the same day if possible, for if left unprotected overnight it is liable to absorb a considerable amount of moisture from rain, night mist, morning dew, etc. Yet one frequently sees stripped woodwork left unpainted for days, even in bad weather; this is asking for trouble. Granted that burnt-off woodwork is not likely to be quite so absorbent as new woodwork, since the oil absorbed in painting is never entirely removed in burning-off (unless the wood becomes charred), it is nevertheless unwise to leave it unprotected a moment longer than is necessary. If we always regard moisture as a possible source of danger, and make every effort to exclude it from surfaces to be painted, we shall remove one of the most potent causes of blistering and peeling.

Paint removers

It is sometimes desirable to remove paint without having recourse to burning-off as, for instance, where intricate modelled or carved detail has to be cleaned, or on narrow glazing bars where burning-off may involve considerable risk of cracking the glass or damaging the woodwork. Again, surfaces which rapidly conduct heat away from the lamp (such as plaster, slate, iron, etc.) make burning-off a very slow job and if large areas are involved it is more economical to use a paint remover. The action of a paint remover is to soften or dissolve the old paint, which can then be removed with the scraper; this usually makes the job less clean than burning-off, but if the work is done methodically and normal care taken, it need not be unduly messy. It is necessary to protect floors and furniture, etc., against possible damage from accidental drops or splashes. There are a number of proprietary paint removers available and they may be classified into two main types, alkaline and spirit.

Alkaline removers are strong solutions of caustic soda, potash, black ash, etc., usually thickened to a brushable paste with lime or starch, thus enabling it to "stay put" when applied to vertical surfaces and also minimizing splashing. The need to keep this type of remover under control is important for it is very potent and damaging to the skin, clothing, polished floors and furniture, etc. Floor coverings may be protected with Sisalkraft or other tar paper which is impervious to water. Alkaline removers attack the oil in the hardened paint film by saponifying it (i.e. converting it to soap); this enables the softened paint to be removed by scraping or by scrubbing with a stiff brush and water. The surface is then well rinsed with several changes of clean water (with particular attention to quirks and interstices) to remove all trace of the alkali, otherwise the subsequent coats of new paint may also be saponified. A little white spirit in the rinsing water will help to reduce the excessive frothing which usually occurs when rinsing off alkaline solutions. A wash of weak acetic acid or vinegar is usually given to neutralize any remaining alkalinity, but this does not necessarily guarantee a neutral condition and it is advisable to test the wet surface with pink litmus paper (see page 223) to see whether any alkalinity remains. All moisture must be allowed to dry out thoroughly before any paint is applied. Wood surfaces which have been so treated will also require well glasspapering when dry, as alkaline removers always tend to raise the grain.

It will be seen from the above that alkaline removers are liable to lead to trouble if used carelessly, especially if used on woodwork or plaster surfaces where the alkali may be absorbed into the porous surface and thus rendered more difficult to remove. Therefore, except possibly for "pickled pine" effects (which are usually finished with wax), or the stripping of oak which is not to be restained or varnished, it would seem that the only justifiable use for this type of paint remover is in the treatment of iron and steel surfaces or fittings where there is a reasonable chance of getting rid of the alkali by thorough rinsing. It should *not* be used on non-ferrous metal, however, particularly aluminium, which is dissolved by strong alkalis. Only fibre brushes should be used for application since animal hair is destroyed by alkalis.

Spirit removers are more expensive but much safer in use compared with the alkaline type, though it should be remembered that some spirit removers are highly flammable and therefore require the normal precautions against fire. The thickening agent which is incorporated with the volatile solvent (to retard evaporation and make it "stay put" on vertical surfaces) often contains wax which must be thoroughly removed before applying paint; this is best achieved by washing down afterwards with white spirit and at the same time rubbing down with pumice stone or fine steel wool. Wipe off finally with non-linty rag. This done, the surface may be painted immediately; there is no waiting for moisture to dry out and no raising of the grain when used on woodwork.

Notes on the correct and efficient use of spirit paint removers have been given on page 209.

WASHING AND RUBBING DOWN

If the condition of the old paint surface is sound, that is to say, quite hard and free from surface defects such as those enumerated at the beginning of this chapter, it will require washing and rubbing down before proceeding to apply any paint. The object in washing down, apart from considerations of hygiene, of course, is to remove all surface deposits of dirt, grease and grime in order to ensure that nothing liable to prejudice adhesion or retard drying comes between the new paint and the old surface. On smooth glossy surfaces, such as those previously varnished or enamelled, it is desirable to provide a "key" to assist the new paint to secure a firm attachment to the surface, and this is achieved by rubbing down with an abrasive such as waterproof paper or pumice stone.

Washing may be done with soap powder dissolved in hot water (there are several proprietary brands of decorators' soap powders and detergents available), or a moderate solution of decorators' sugar soap applied with an old scrub (a worn-down flat brush); this will soften the grime and will tend to etch the surface slightly after a few minutes, when it should be lightly rinsed off with clean water and the surface then rubbed down with pumice stone, waterproof abrasive paper, or fine steel wool. Finally, the surface should be well rinsed down with plenty of clean water, then sponged off or chamois-leathered, and left to dry thoroughly. The sugar soap should not be too strong, nor left on the surface long enough for the paint to be softened, because a soft film cannot be rubbed down satisfactorily. Many prefer to counteract the effects of grease by adding a little lime putty or liquid ammonia to the rubbing-down water; this is milder in action than sugar soap, though not so efficacious in removing grime.

The merits of the various rubbing-down materials are outlined under "Abrasive Materials" (pages 210–13), and need not be enlarged upon here, except to express the opinion that the waterproof abrasive papers are by far the most convenient and efficient form of abrasive for most purposes and, in the writer's opinion, are certainly worth their higher cost. The choice of grit number will depend on the nature of the surface, an old hard enamel requiring a keener cutting grit than a straight oil paint surface, whilst the rubbing of newly applied undercoats and varnishes requires the finest grades. There are several types of rubbing blocks or pads available for holding these abrasives conveniently; they are made in rubber, thick felt, wood and cork, but a home-made wood block serves almost as well. It should be noted that all surfaces treated with lead paint (new or old) must be rubbed down by a wet process in accordance with Lead Paint Regulations (see Appendix 2). This means, in effect, that *all* old surfaces must be rubbed down wet unless one can *prove* that lead paint has not been used. Hands should be rinsed frequently to remove any white lead particles before they dry on the skin, and it is specially important to scrub the hands well with soap and water before partaking of food.

Rubbing down is at best a slow process (and therefore costly) and the tendency now is for work to be graded in order of importance and treated accordingly. Thus, in a room, the doors and mantelpiece generally have most time spent on them, while other features in the room receive less and less attention as they diminish from view; on exterior work it is the front door that receives most attention. This is a logical and rational procedure to adopt, for if there is to be any cutting down in quality it must be done where it cannot offend the eye. This is not condoning scamped work; it is merely indicating where the craftsman may conserve his energy so that he may concentrate his skill where the result will be seen best and appreciated most.

On those portions of the work, therefore, which do not come under direct scrutiny, all that is required is a clean and slightly etched surface to receive the new paint. A wash with sugar soap, rubbed down with a swab sprinkled with pumice powder and well rinsed off with clean water, will quickly produce this condition. This is perhaps the most convenient treatment for large areas of gloss-painted wall surfaces, but the rubbing with pumice may be omitted on flat oil paint since this already possesses the requisite key or tooth due to its high pigmentation and lack of gloss.

Washing and rubbing down should be carried out in a methodical and workmanlike manner, that is to say, with a minimum of splashing and avoidance of a sloppy mess on the floor. As each section of wall or woodwork is finally rinsed down, the surplus water should be mopped up to prevent it treading about. In washing walls and woodwork, always commence *from the bottom* and work upwards in order to avoid water running down a dry surface, leaving streaks behind; this is specially important in washed and varnished work (or where washing is not to be followed by painting), for if such streaks occur they are difficult if not impossible to eradicate. Sugar soap, soda, or any other alkaline solution *should not be allowed to dry on the surface*, but should be kept wet until the final rinse down, otherwise it may bite into the paint surface and not be removed in the final rinsing. Any trace of alkali left on the surface through insufficient care in rinsing becomes a potential source of danger to the new paint; cases of local softening due to saponification, or discoloration of those pigments prone to alkaline attack, may easily result from slipshod methods in washing down. The final rinsing should therefore be done with two or three changes of clean water.

Importance of drying out

The importance of allowing the surface to dry properly before painting cannot be emphasized too strongly, nor too often, since more troubles are probably attributable to moisture entrapped beneath the paint film than to any other single cause. The peculiar difficulties of painting new plaster surfaces which may contain large amounts of water are well known and usually provided for, whereas the dangers of painting too soon after washing down are not sufficiently realized. The general surfaces may *appear* perfectly dry and may indeed *be* dry, but water may lodge unnoticed in quirks, crevices, and open joints of woodwork, or seep into plaster by way of cracks. Again, where the paint film has been rubbed through so that bare plaster or timber is exposed, there will moisture penetrate. On exterior work, badly perished or chalky surfaces will absorb more moisture than a surface which has not commenced to disintegrate. These points should be constantly borne in mind, and if the surface shows breaks, bare patches or open joints through which water may penetrate, painting should be delayed until the moisture has had reasonable opportunity to dry out.

On interior work, drying may be hastened either by leaving windows and doors open to create a good circulation of air, or by any artificial means at our disposal, such as electric fires, fans, and hot-air driers, etc. On exterior work, however, we are at the mercy of the elements and cannot very well use artificial aids to speed up drying apart from the blowlamp. If, therefore, the condition of the surface is such that it is liable to absorb moisture and if the prevailing weather is not conducive to rapid drying, it may be advisable to use, instead of water, a mixture of three parts white spirit and one part raw linseed oil as a rubbing-down lubricant. In this case, the surface should first be well dusted, then brushed over with the oil and spirit mixture and rubbed down with pumice stone or fine steel wool, finally drying off with non-linty rag. Painting may then follow on immediately as there is no moisture to worry about. The only snag about this procedure is that the grimy surface deposits found in large industrial towns are not so easily removed with oil and spirit as with an alkaline water solution, and the removal of such a deposit will therefore depend largely on vigorous abrasive action in rubbing down. Another point is that white spirit acts as an irritant to a sensitive skin, in which case it should be avoided. This, however, is another instance where a self-lubricating abrasive paper is the perfect answer.

Touching-up

After washing and rubbing down is completed and all moisture dried out, any bare patches should be touched-up with a priming coat, that is, with paint thin enough to penetrate into the surface yet oily enough to stop absorption. Generally, the normal priming paint as supplied by the manufacturer will be suitable, but on non-absorbent surfaces it should be thinned down with white spirit as recommended by the maker (see p. 242); it should as far as possible be of the same tone as the surrounding colour on which it is applied, so that the patches will not tend to grin through the first coat of paint. When touching-up or patch-priming like this, thick edges should be avoided by softening *into* the patch with a dry tool or a hog-hair softener.

PREPARATION OF WOODWORK

The surface should be carefully inspected for defects; where the film is soft and tacky, cracked, wrinkled, badly blistered or peeling, or where it is overloaded with repeated paintings, it should be removed by burning-off, or stripped with spirit remover. The surface may then be treated as new woodwork, except that the primer should be thinned down slightly with white spirit. If blisters occur only over one or two knots while the general paint surface is quite sound, there is no need to remove the whole film and local treatment may be given as described under "Knots", below.

If the work is sound and free from any of the above defects, or if the surface is matt or slightly chalky, it merely requires washing and rubbing down wet to bring it to a fit condition to receive the new paint; such surfaces provide ideal conditions for repainting. Where the paint film has completely disintegrated or perished, however, owing to excessive chalking (as, for example, in cases where the surface is paint-starved through gross neglect over a number of years, or where certain leadless whites have been used outdoors), a good dry rubbing with glasspaper and dusting off will generally be sufficient to remove all loose powdery material, and this will leave the surface ready for immediate painting; wet rubbing would only result in saturating the absorbent and hungry surface with undesirable moisture, some of which may become entrapped by premature painting.

Weathered and perished surfaces such as this should be treated as for new woodwork, i.e. primed with a thin oily priming.

Knots

Where an otherwise sound surface shows stains (or blisters) from knots, scrape away the paint film down to the knot, using a sharp shavehook, and coat with genuine shellac knotting, or better still, cut the knot well below the surface with a joiner's gouge, then fill in level with (or slightly proud of) the surface with proprietary filler and allow to harden; glasspaper smooth and touch up prior to painting.

Persistent blistering is sometimes experienced on exterior woodwork over a period of many years, especially on very resinous or knotty timber; even burning-off and careful choice of materials of low oil content often fail to cure this trouble. In cases like this a good oil-bound water paint has been found effective, but it is important that such treatment should be undertaken in dry weather only, so that the water paint has a chance to become quite dry before oil paint is applied. The old paint should be burnt off, allowing the flame to play on the surface till the resin exudes and is scraped off, but taking care to avoid charring the wood. After rubbing down, a good coat of water paint may be applied evenly and laid off carefully to avoid ropiness. When thoroughly dry and hard, the water paint should be glasspapered smooth and the surface dusted off before applying the first coat of paint. The subsequent undercoats may with advantage be thinned down with a mixture of universal medium (mixing varnish) and white spirit in equal quantities.

Creosoted wood

When surfaces have been previously treated with bituminous preparations or creosote they should not be painted at all unless the creosote is old and weathered, or the bitumen fairly hard and non-elastic, because they require sealing with a shellac preparation to prevent bleeding, and the hard-drying shellac coating is liable to crack if the ground is soft and plastic. Given suitable conditions, however, two coats of tar-proof knotting should be carefully applied to ensure that the surface is effectively sealed. Alternatively, the knotting may be pigmented with aluminium powder in the proportion of about 1 kilogram to 5 litres and applied with care to avoid misses or pin-holes through which the tarry substance might bleed; one coat is usually sufficient in this case because the aluminium acts as a guide coat and its well-known "leafing" properties also make it a good sealer (see also "Bleeding", page 544).

Mould stains

Painted woodwork is sometimes affected by mould growths, especially when exposed to conditions of continual damp, stagnant air and excessive humidity as in greenhouses, etc. In these cases the paint film should be burnt off and the woodwork carefully examined for evidence of dry rot (which should be cut away and made good). Before repainting, the wood should be sterilized (as detailed on page 553) and the priming coat may be thinned with solvent naphtha instead of white spirit, since the former is more penetrative and possesses germicidal properties. Naphtha should not be used in any subsequent coats, however, because its solvent nature would tend to soften up the paint beneath. In very persistent cases of mould growth, one of the proprietary fungicidal paints should be used as an extra precaution.

PREPARATION OF PLASTER SURFACES

In considering the preparation of plaster surfaces we are at once confronted with a much greater variety of conditions than in the case of woodwork; these conditions will depend on (a) the type of plaster surface; (b) the type of decorative material previously applied, and its present condition; and (c) the proposed new finish — whether it is to be a flat, glossy or textured effect, etc.

Without some previous knowledge we cannot tell for certain what type of plaster has been used, but a cursory examination will show whether it is a comparatively soft, porous surface such as we expect in lime plaster, or whether it is likely to be a hard wall plaster or even a smooth closely trowelled Keene's cement having relatively little porosity. The condition of the previous decoration will also give some indication as to the nature of the surface; for instance, the presence of efflorescence (see page 224) shows that the plaster or backing has contained water-soluble salts which have dried out and crystallized on the surface, but which are liable to be reactivated if the surface is wetted. It is therefore important not to wet the surface, but to remove the efflorescence *dry* (by scraping and glasspapering), otherwise the crystallized salts will be dissolved and possibly absorbed in the surface, only to reappear again on drying out.

The surface should then be kept under observation for a few days to see whether the efflorescence reappears. If no further efflorescence appears, the surface may be painted with safety. If crystals do appear, however, it would be very unwise to seal the surface with an impervious paint finish for, sooner or later, flaking would probably occur. The best treatment in this case is to apply a coat of sharp flat paint (well thinned with white spirit) followed with a coat of water-thinned paint; this will give the salts a chance to pass through to crystallize on the surface, from which they can be removed with a dry cloth from time to time as they appear.

In the event of such defects as peeling or flaking, an examination of the back of the flakes will usually give some indication as to the cause (see "Paint Defects", Appendix 4). If the flaking is extensive, it is advisable to remove as much as possible by dry scraping and the remainder by spirit paint remover or with the aid of the Sanderson Stripper (page 55). Oil-bound water paints are particularly difficult to remove and the steam stripper is the only satisfactory way to remove them completely. It may be argued that complete removal is unnecessary and this may be true in cases where the new decoration is to be a flat finish or a texture effect. Where the finish is to be in gloss paint, however, complete removal is the best course if a first-class finish is desired. If removal is ruled out on grounds of cost, the bare plaster patches may be primed with thin oil paint and allowed to dry; the thick edges of the remaining old paint may then be levelled or chamfered off with proprietary filler applied with the broad knife, allowed to dry, glasspapered smooth, and touched up before applying the first coat.

Other surface defects such as wrinkling, cracking or blistering call for complete removal as outlined under "Paint Removal". The cause of blistering should be sought and if, as is likely, it is due to damp or to entrapped moisture, the source of the damp should be eliminated and the surface allowed to dry thoroughly before again sealing with oil paint. If decoration cannot be delayed long enough for the surface to dry, it will be safer to use a water-thinned paint, as this will not lock up moisture within the surface as would oil paint. Another form of surface defect, rather akin to blistering, takes the form of bunion-like protuberances in lime plaster walls. These are sometimes called "blows" and are caused by small particles of unslaked lime which remain active and continue to expand until the surrounding surface swells into a little mound. These should be cut out and the holes made good with Keene's cement or proprietary filler; if very extensive, lining may be advisable.

If the surface has been previously treated with size-distemper, the latter should be thoroughly removed by washing and scraping; cracks and other defective plaster should be made good (see page 281) and all moisture allowed to dry out thoroughly before applying a coat of *thin* oily priming paint. This could be an undercoating for gloss paint, thinned as directed by the makers.

Stains

Stains should be inspected closely in an endeavour to ascertain their cause. Scraping will indicate whether the stain is merely superficial or within the paint film itself, or whether it is more deep-seated and coming from the plaster. Removal of the paint film would eradicate any superficial stain, but stains in plaster may be more difficult; these may be caused by dampness or leakages which have brought water-soluble matter from the backing to the surface, in which case the source of the trouble should be traced and cured and the plaster allowed to dry out thoroughly before painting. When this has been done, such stains will not usually affect oil paint. Sometimes, stains are caused by smoke, either on the surface or seeping through fine hair cracks on a chimney breast; such stains are liable to bleed through oil paint, and after washing to remove superficial deposits, the surface should be sealed with two thin coats of patent knotting.

Stains which are caused by mould growth or mildew should be treated as recommended on page 553.

Pattern staining

Another type of surface disfigurement is that known as "pattern staining" which occurs chiefly on plaster surfaces rendered over wood or metal lathing, but also, in certain cases, on other composite backings (Fig. 20.1). The "stain" forms a complete pattern of the backing construction and is

Fig. 20.1 A typical example of "pattern staining"

sometimes erroneously ascribed to the filtration of dust particles through the porous plaster; with this in mind various attempts have been made to seal the surface with gloss paints and even with lead foil, but without success.

The disfigurement, which is due to the accumulation of smoke and dust particles, occurs on surfaces where the air on one side of the plaster is at a higher temperature than that on the other; for example, on ceilings below roof lofts or flat roofs, etc. In such circumstances there is a natural tendency for heat to flow through the ceiling in order to equalize the temperature on both sides, but since the thermal conductivity of plaster is superior to that of wood, it follows that heat will flow more rapidly through a layer of plaster than through a combined layer of wood lath and plaster. This variation in heat flow creates "warm" and "cold" areas at the plaster face, as illustrated in Fig. 20.2, and although the temperature difference is only slight, it results in more dust particles being deposited on the cold areas (c) than on the warmer areas. Thus we get the pattern of the wood laths "ghosting" through as light bands on a dark

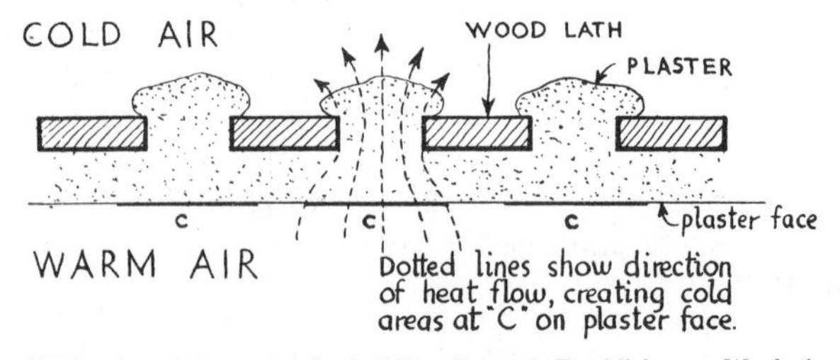

With acknowledgment to the Building Research Establishment, Watford

Fig. 20.2 Pattern staining

Diagram showing how the heat flow through a lath and plaster ceiling (having cold air above and warm air below) results in "pattern staining"

background, or, in the case of metal laths, as dark bands on a light background.

A cure may be effected by equalizing the temperature on both sides of the plaster surface by the provision of artificial heat, but this is costly and not always practicable. If the space above the ceiling is accessible, however, thermal insulation may be provided by lagging the spaces between the joists with slagwool or compressed straw. Where this second method is not possible, the only alternative is to line the plaster face with material of low thermal conductivity such as plasterboard or, better still, a thick wallboard of the insulating type such as foamed polystyrene, available in sheet form from $2\frac{1}{2}$ mm thickness or over, and in "tiles" of various sizes and thicknesses; the insulation properties increasing with thickness.

Pattern staining begins to appear in some cases a few months after decoration is completed, depending, of course, on local conditions; it is so unsightly in itself that it is worth while making every effort to eradicate it.

Another surface disfigurement, which is also a great nuisance, is the local darkening that occurs above radiators and heating pipes. This is caused by dust particles being carried along by air-convection currents and deposited on the wall surface. Wood or metal covers are often fixed above radiators to deflect these air currents, but unless the covers are sunk *into* the plaster and not merely planted on the face, they will fail in their purpose since the warm air often travels up between the cover and the wall surface.

When steps have been taken to prevent the reappearance of these surface stains, the existing stains should be removed by thorough washing before any new treatment is applied.

Surfaces in sound condition

Surfaces previously finished in emulsion paint, oil-bound water paint, or eggshell or flat oil paint will, if in sound condition and free from any of the above-mentioned defects, merely require a thorough washing, but gloss finishes will, in addition, require *wet rubbing down* to provide a key for the new paint (see page 267). Any cracks or other defects of the plaster must be cut out and made good (see page 281); when dry, such repairs should be touched up with priming (tinted to match the ground) before applying the first coat of paint.

Existing decorations such as stencilled ornament, run lines and painted bands, etc., should be well rubbed down to prevent the added thickness showing through the new coating. Previously applied decoration of this sort is often difficult to obliterate, especially when it is in strong contrast to the ground colour (for example, black on white), and the pattern often persists in grinning through faintly even after three or four coats have been applied. If it is likely to prove difficult to hide the pattern with the new colour, it is advisable to coat the whole ground with an intermediate colour such as a neutral grey; this will hide the old design and provide a uniform tone on which to build up the new colour. Colours which are liable to bleed (particularly bright reds and bronze paints) should be coated with patent knotting to which may be added a little aluminium powder (see also "Bleeding", page 544).

PAPERED SURFACES

It is occasionally desired to apply paint over existing wallpaper, either to retain the attractive texture which some wallpapers possess, or to avoid damaging the surface by stripping where the plaster is known to be very old. If the paper is sound and firmly adhering, particularly at the joints and edges, it may be painted successfully if the following points are borne in mind. First, the paper will be fairly absorbent and the decorator may be tempted to stop the porosity with a coat of glue-size for the sake of economy. This may prove false economy in the end, however, for if the paper becomes saturated with size it is almost certain to develop large wrinkles or blisters and, although these usually tighten up and go back to normal again, the expansion and contraction thus set up seriously weakens adhesion and the paper will sooner or later leave the surface. This may also happen when a water-thinned paint is applied direct over old paper (where the paste has lost its original tenacity) and the safest course is to avoid introducing water to the surface in this way but to apply a coat of thin oil paint direct on the paper (see page 238).

The primed surface may then be finished in either eggshell paint or a water-thinned paint. It should be pointed out, however, that if the paper is printed in colours which are liable to bleed (such as reds, metallics or mica), a thin coat of knotting should first be given to seal the surface (see also page 238).

OLD STUCCO AND CEMENT SURFACES

The painting of these surfaces has been fairly common since the late 18th century, when the Italian practice of rendering the exterior façade with stucco became fashionable in this country; before that time, the plastered renderings of half-timbered buildings, and modelled plaster or pargetting, were generally coated with the traditional limewash.

Stucco, an Italian word for plaster, may be composed of "Roman" cement or Portland cement, or lime, gauged with varying proportions of coarse or fine sand according to the degree of finish desired. Houses of the Adam and Regency periods were generally rendered outside with lime or "Roman" cement stucco, and in some cases oil mastic stucco was used; nowadays, however, exterior stucco renderings are almost invariably carried out with Portland cement or lime–sand mixes, and the use of white and coloured Portland cements and cement paints has largely dispensed with the need for oil paint on such surfaces, except where a more impervious finish is desired.

Much of the older painted stucco work is often found to be in a very bad condition, for in some cases paint has been applied regularly every four or five years over a long period, with the result that the paint film may be 3 mm or more in thickness; when it is remembered that coatings of water paint and even limewash were sometimes sandwiched between the coatings of oil paint, it is not surprising that cracking and flaking should sometimes result. The cracking of such a thick film presents a most unsightly appearance, the cracks opening out to form wide fissures with curled and lifting edges and often leading to extensive scaling.

The ideal course to prepare such a surface for repainting would be to strip off all defective paintwork down to the stucco ground before making good and repainting. This is seldom practicable with normal methods of removal, however, burning-off being too slow on account of the heat being conducted away by the cement rendering, while the use of a spirit remover would be much too expensive; caustic removers are not to be recommended for reasons previously mentioned. A possible means of removal would be the steam stripper (page 55), but even this would prove slow and costly where the paint coating is very thick and hard. The only alternative is to remove as much as possible of the old film by scraping, chipping, wire-brushing, etc., and either to rub the thick edges of the remaining areas down to a feather-edge, or to chamfer off such edges with proprietary filler or Keene's cement; if the latter is used, the surface should be wetted before applying the Keene's as in normal plastering practice, but if a filler is used, the bare patches should first be primed with thin oil paint and allowed to dry. A surface treated in this way may present an uneven appearance if finished with normal oil paint, but a texture material, such as one of the imitation stone paints, would tend to minimize any surface irregularities.

If cracking has extended down to the stucco itself, or the rendering sounds hollow when tapped, such parts should be cut out and made good (see page 281). If Portland cement is used for repairs, it must be allowed to dry thoroughly, then treated with alkali-resisting primer before applying ordinary oil paint.

BRICKWORK AND STONE

Previously painted brickwork or stone should be well scraped, or scoured with a stiff wire brush to remove all loose flaky material; very dirty or greasy surfaces should be washed down with lime water or sugar soap solution, well rinsed with clean water and allowed to dry thoroughly. Bare parts should be touched-in with primer–sealer and allowed to dry before applying the first coat of paint.

Where these surfaces have been previously treated with oil-bound water paint, or limewash which appears to be well bound, they should be wire-brushed and dusted down to remove all loose material. Size-bound distemper or loosely bound limewash, however, should be removed as far as possible by thorough washing and rinsing and the surface then allowed to dry. A coat of primer–sealer or masonry paint primer, mixed according to the manufacturer's instructions, should be applied liberally and brushed well into the joints. This will bind any remaining loose particles and will stop surface porosity. If the pointing is bad the joints should be well raked out, pointed up with oil mastic, and allowed to set hard before painting (see also p. 234).

IRON AND STEEL SURFACES

The importance of thorough and conscientious preparatory work cannot be too strongly emphasized when the repainting of iron and steel surfaces is being considered, for upon this depends the efficacy of any subsequent painting. But though it is easy enough to lay down the ideal conditions required for repainting, it is much more difficult to suggest practical measures to ensure them. One often sees it glibly stipulated that all rust should be removed down to the bright metal, yet we know that this is almost a practical impossibility with ordinary methods of scraping and wire-brushing. There are specialized mechanical processes such as blast-cleaning, pneumatic chipping, flame-cleaning, etc., and one or other of these methods (as described in Chapter 4) should be employed on important structural steelwork such as bridges, crane gantries, etc., where corrosion might endanger the safety of the structure; but until some simple form of mechanical abrasive is made generally available, the painter will have to persevere with his present methods of rust removal.

What is far more important than trying to achieve a bright metal surface, however, is procuring a really *dry* surface on which to paint. It is useless, for instance, to prepare the surface one day and defer painting till the next day, even in dry weather; moisture deposited during the night will promote new rusting. Therefore painting should immediately follow the preparation while the surface is dry. This is discussed more fully under "Protection of Iron and Steelwork" (page 247), from which it will be seen that painting should not proceed in damp weather if it can possibly be avoided.

Conditions of damp often exist even on fine days, especially if the atmosphere is moist or humid, and condensed moisture may be deposited on metal surfaces without being visible to the eye. The colour of rust is a useful indication of water content, however, for rust which contains physical* moisture is usually a dark-reddish brown, like burnt sienna, whilst dry rust approaches the colour of yellow ochre (which, indeed, it partly is). If a relatively dry surface is indicated, it may be vigorously wire-brushed, dusted off, and painted immediately. If on the other hand the rust is very dark in colour, it is advisable to employ some form of heat to drive off the contained moisture. The oxy-acetylene flame-cleaning equipment is designed expresssly for this purpose, but if this is not available the blowlamp may be used to heat the surface while scraping and wire-brushing; this cannot possibly be as effective as the oxy-acetylene flame, but it is at least a step in the right direction. All bare metal so treated should be primed *immediately* with red lead while the surface is still warm, if possible, and before any further moisture is deposited.

Painted surfaces should be carefully inspected for signs of rust formation *beneath* the paint film which may be evidenced by brown rust spots or stains. If it is found that considerable rusting has occurred beneath the film, the paint should be completely removed and the surface well scoured or wire-brushed. Ordinary burning-off is not so effective on metal surfaces owing to the heat being too rapidly conducted away, and it requires the hotter flame of the oxy-acetylene burner to do the work expeditiously. Alkaline removers are cheap and very effective on metal work, but they

---

* Rust contains both physically and chemically combined moisture; the first may be driven off by fairly low heat, but the second requires high temperature, such as that of the oxy-acetylene flame, to effect dehydration.

have the disadvantage of introducing water to the surface; if they are used, the surface should be well rinsed off with clean water and allowed to dry thoroughly, then wire-brushed and dusted off before painting. Spirit removers obviate the use of water, but are rather expensive where large areas are to be treated.

Spirit solvents are probably the best means of stripping paint from steel windows, however, and if the surface is afterwards well rubbed down with steel wool soaked in white spirit to remove the rust (and any possible wax residues from the solvent), then wiped dry with a rag, it may be coated immediately with red lead primer. Particular attention should be paid to the glazing compound; if this appears to have perished, or an inferior putty has been used, it should be completely hacked out and the metal well scraped to remove rust. The glazing rebate should be painted with red lead, taking this well into the angles and on to the glass for at least 6 mm and allowing to dry hard before "re-puttying" with mastic cement, or with one of the red lead glazing compounds made specially for steel casements. Cheap putties should never be used as they often contain water and are liable to lead to corrosion of the steelwork.

In all preparation of iron and steelwork, particular attention should be given to nuts, bolt and rivet heads, angles, edges, joints and under-parts, etc., chipping, scraping or wire-brushing thoroughly to remove rust and scale. The same attention should also be given when painting these parts, for it is here that water lodges and finds its way under the film if chinks are left unpainted.

Where the paint film is unbroken and in sound condition, it should be well brushed down with a stiff fibre brush, or rubbed with dry steel wool to remove dust and dirt; grease or machine oil should be thoroughly cleaned off with petrol or white spirit and the surface wiped dry before painting. If rusting or scaling occurs in localized areas while the rest of the work is sound, the defective parts should be warmed with the blowlamp, thoroughly scraped or wire-brushed, then touched up with red lead primer.

Galvanized iron which has been painted sometimes exhibits large areas of peeling or flaking, due to the surface not being properly prepared before painting. Adhesion may be so poor that the old film may often be completely removed by scraping and wire-brushing, but any paint that remains firmly attached after such treatment may be regarded as being safe enough to paint over.

Places where the zinc coating has disappeared and rust has formed on the unprotected iron (generally around bolt heads, etc.), should be thoroughly wire-brushed and touched-up immediately with red lead or zinc chromate. Where the zinc coating is still intact, however, it may be painted with calcium plumbate. The scouring of the old galvanized surface by scraping and wire-brushing should be sufficient to provide a key for the new paint, and a chemical mordant, as required for new zinc-coated surfaces, should not be necessary.

Surfaces previously treated with bituminous preparations may, if reasonably hard, be coated with tar-proof knotting pigmented with aluminium powder to prevent bleeding through ordinary oil paint. If the bitumen coating is soft, however, knotting is liable to crack and should be avoided; in such cases, there is no alternative but to recoat again with one of the proprietary bituminous paints, some of which are available in a limited range of colours. Where rusting has occurred, it should be removed and the bare metal touched up with red lead primer; this should be allowed to dry quite hard before following with any bituminous paint, otherwise it may be softened by the naphtha or other solvents usually contained in paints of this kind.

21

Stopping, Filling, and Making Good

In the normal procedure of preparing a surface to receive paint, wall-paper, etc., the painter must see that all cracks, holes, indentations and unsound parts of the structure are so remedied that such defects are not apparent in the finished job. This involves various operations such as stopping holes and cracks with stiff putties and stoppings; levelling or resurfacing indentations and other irregularities with proprietary or hand-mixed filling compositions; effecting minor repairs, such as cutting out defective plaster and cement work, and making good with sound material. In all these operations it is important that a good key for the stopping material be provided, otherwise failure from poor adhesion will soon mar the finished work. It is equally important that the shrinkage factor in stopping materials is provided against, and that care is taken to avoid the use of highly contractile ingredients.

PLASTER AND CEMENT SURFACES

The repair of plaster and Portland cement surfaces is usually referred to as *making good*. The first step is to remove any defective material such as hollow, crumbly, or decayed plaster, hacking out down to the brickwork if necessary; if the trouble is extensive, the plasterer should be called in. Large cracks should be well raked out with the shavehook and slightly undercut to form a dovetail for the new plaster; if one side of the crack stands higher than the other (as often happens in large settlement cracks) a channel 75 or 100 mm wide should be cut out so that the difference in plane will not be apparent after making good. If the brickwork or lathing is exposed, the patch will first require *roughing-in* with a backing or rendering coat, and the most convenient material to use in this case would be a retarded (hemihydrate) hardwall plaster such as Thistle, Gothite, Pytho, etc. (see page 219). The roughing-in should be gauged with clean coarse sand in the proportion of 1 part plaster to 3 parts sand (by volume); the defective part should be well wetted first and the mix then firmly placed, pressing well home with the trowel and bringing to within 3 mm of the surface. Score with criss-cross lines to form a key for the final skim; then leave to dry for a few hours. The patch may then be skimmed with the same hardwall plaster *used neat* (i.e. without lime or sand); bring this to the general level of the surrounding surface and smooth off by careful trowelling.

By the above method, patches may be roughed-in early in the day, skimmed with "neat stuff" in the later afternoon and primed next day with oil paint. This is possible because hemihydrate hardwall plasters combine chemically with a large part of the water used in mixing, so that setting and drying are fairly rapid and with little or no shrinkage. Moreover they are generally inert (neutral), and therefore harmless to paint. Lime plaster, on the other hand, hardens slowly, is dangerous to oil paint, and tends to shrink on drying. Hardwall plasters, however, should not be used in permanently damp situations, nor should they be painted whilst wet.

For cracks and small holes, several types of stopper and filler are now available. The traditional materials, namely Keene's cement and plaster of Paris, are not much used in present-day practice. *Plaster-based fillers* of the Alabastine type have the distinct advantage of expanding as they set, thus clinging firmly to the surface; they are very absorbent when dry. This type of filler is of course very similar to plaster of Paris, being made from a high-grade form of gypsum — indeed, Alabastine derives its name from "alabaster" which is a pure white form of gypsum. *Plaster/cellulose-based fillers* of the Polyfilla type tend to shrink slightly when drying, but are quite adhesive; they, too, present an absorbent surface when dry. *Synthetic/emulsion-based fillers* of the type sold as "Instant Polyfilla" are for interior use and are supplied in ready-mixed form; they take the form of a smooth paste, and have the same disadvantage of shrinking slightly when dry. *Cement/vinyl-based fillers*, sold as "Exterior Polyfilla", etc., are, as their name implies, for exterior use, being highly resistant to water; they are somewhat difficult to rub down.

The cracks or holes should be thoroughly wetted and filled in, pressing the material well home and smoothing off to the general level. Fine hair-cracks are difficult to fill properly, and on interior work it is generally more economical to size and line the surface with a stout lining paper than to attempt to fill them with a filler composition.

Portland cement and cement stucco renderings are generally repaired with Portland cement and sand, but this cannot be painted immediately. *Ciment Fondu* (high alumina cement — see p. 216) can be safely painted within a few days if applied in accordance with the maker's instructions. Small cracks and patches are more conveniently repaired, however, with Keene's or Parian cement gauged with a little sand and, if necessary, stippled to simulate the surrounding texture of the stucco rendering. Large patches should be primed with sharp oil paint as soon as the Keene's can bear the brush, then left to dry out thoroughly before painting in the normal way. Alternatively, oil mastic cement may be used, first "oiling in" the cracks with boiled oil thinned if necessary with white spirit, then pressing the stiff mastic well in with the trowel. Allow to harden properly before painting. Mastic may also be used for repairs around external door and window frames; it is usually applied with a caulking gun.

STOPPING

In all painted work the stopping of small cracks and nail holes with putty or hard stopping should always be done *after* the priming paint has been applied and has become quite dry, otherwise the unprimed surface would tend to absorb the oil in the stopping material which would shrink and possibly fall out.

Putty is the most commonly used material for this purpose and only the best quality should be employed. Some painters used to make up their own from raw linseed oil and whiting, but whilst this may be convenient in an emergency, it is far more economical to purchase ready-made putty. Hard stopping was usually made by (a) stiffening paste white lead with powdered whiting and tempering with a little Japan goldsize; (b) using a mixture of white lead and red lead similarly tempered, or (c) by a mixture of paste white lead or red lead (or both) with linseed oil putty, stiffened if necessary with whiting.

An excellent hard stopper for ironwork consists of a straight mixture of dry red lead and boiled oil or goldsize; this becomes exceedingly hard. In present-day conditions, though, it would be difficult to obtain the basic materials with which to make these and similar lead-based stoppers. There is a good alternative in the form of a proprietary metal stopper consisting of a mixture of aluminium or iron powder and resin. This is generally supplied as a two-pack material, and it sets extremely hard. As may well be imagined, it is far more expensive than the home-made product.

Various proprietary brands of hard stopper for woodwork are also available.

The putty or stopping should be pressed well into the hole or crack, using the putty knife firmly and with some pressure; level off carefully, leaving no stopping on the surface. If air is trapped in the cavity, the stopping may afterwards bulge out or sink according to the expansion or contraction of the air, and this should be avoided if possible. Do not attempt to fill large holes or cracks in one operation, because shrinkage of the material may cause the stopping to sink; half-fill the cavity and allow to harden before facing up finally. Stopping should always be left to harden properly before glasspapering and second-coating. A small plywood handboard or a tin lid should be provided to hold the stopping.

In certain cases bare wood may be stopped with plaster or patent wood filler preparatory to staining or varnishing, a method generally used by French polishers. Plaster stoppings should be well glasspapered when dry, then touched up with thin clear polish or white knotting before staining, otherwise the plaster may absorb too much stain and appear too dark. A stopping made from whiting or plaster of Paris bound with French polish (stained if desired with spirit stain) and thinned with methylated spirit, is also very suitable for French polish work.

FILLING

After the priming and stopping of nail holes and cracks has been carried out the surface is ready for second-coating, unless it is badly defaced with shallow indentations, scratches, raised grain, etc., or a high-class "coach finish" is desired, in which case the surface will require *filling*. Various proprietary materials are used in present-day trade practice. *Synthetic emulsion-based grain fillers*, of the type sold under the name of "Fine Grain Polyfilla", are suitable for both interior and exterior painting. They are supplied as a smooth paste in ready-mixed form and have the advantage of setting very quickly. *Oil-based "Spatchel"* type fillers are also intended for both interior and exterior work; they are supplied in ready-mixed form, and are formulated to serve as primers and undercoats on wooden surfaces. Both these types of filler can be applied either with the filling knife or with a flexible filling blade, which is a fine steel blade with a broad finger grip, and which is cheap enough to be discarded if and when the blade is damaged.

For high-quality work the choice lies between Japan filler, oil-based paste filler, or water filler. The first two involve wet rubbing-down, while the third may be rubbed down dry, thus being quicker and less expensive; but the better finish is obtained with the first two methods. The filling of wood grain prior to varnishing or polishing is dealt with in Chapter 26 under "Wood Staining".

Japan filler and oil-based paste filler

The filler used to be made up from paste white lead, powdered whiting and slate powder or fine sifted pumice, tempered to a stiff consistency with equal parts Japan goldsize, rubbing varnish and white spirit. There are, however, several proprietary filling compositions available, which are formulated to give smooth spreading, hard drying and quick cutting down; these are generally to be preferred to a hand-mixed filler since, with the latter, it is extremely difficult to exclude coarse particles which are liable to cause scratches when rubbing down.

Filling is a craftsman's job, requiring skill and patience. The work should be done methodically, remembering that the composition is setting all the time; without due care, partly set or hardened material may get mixed with the freshly applied stuff and cause trouble in smoothing off. The broad filling knife should be clean, pliable and have a perfect edge; knives which have been used for stripping wallpaper or for burning off are useless for filling. On large surfaces, a 300-mm (12-in) strip of thin metal, or a bevelled plastic set-square, is useful for levelling and smoothing off; either of these may be moistened with water (to prevent the filler sticking) and lightly dragged over the surface as the composition is setting.

If necessary a coat of paste-filler may be applied first with the knife to fill up the surface indentations, followed when dry with 3 or 4 coats of liquid filler applied by brush or spray to the whole surface, including mouldings. The brush coats may be applied at about 4-hour intervals, but sprayed coats can generally be followed after an hour. Brushed coats are laid off alternately in horizontal and vertical directions to avoid "ropiness" and to distribute the filler more evenly. When sufficient filler has been applied, it should be left to harden thoroughly for at least 24 hours (longer if possible) before rubbing commences.

Rubbing down

For best results a "guide-coat" should be applied as in high-class coachpainting practice. This is an extremely thin coat of ready-bound drop black thinned with white spirit — perhaps it would be better described as white spirit tinted with a little drop black. A thin wash of this is given and allowed to dry (which takes about half an hour). Then rubbing down may commence, using cold or tepid water and pumice stone, or a medium-grade waterproof abrasive paper over a cork or felt block to obtain a true level surface. The paper is more expensive but more convenient than pumice stone since it can be used equally well on flat or moulded surfaces; it does not scratch (if kept clean) and cuts down more quickly. If the abrasive is dabbed occasionally on a piece of white soap (especially when rubbing first begins) the slight lubrication facilitates rubbing, but only the *merest trace* of soap is required. The abrasive paper should be rinsed frequently to prevent clogging. Rubbing should continue, using even pressure in a circular movement, until all the guide-coat is removed; this will ensure a smooth level finish. Great care must be observed to avoid picking up grit; if the rubbing block is put down carelessly on floor or bench it will certainly pick up dirt or grit which, if not rinsed off, will cause scratching and thus largely defeat the object of filling. When rubbing is completed, rinse off with clean water to remove all debris, paying particular attention to mouldings and quirks. Then — and this is important — leave till next day to allow all moisture to dry out before painting. Work of this quality cannot be hurried.

After rubbing down it may be observed that a certain amount of loose powdery material remains on the dry surface in spite of careful rinsing; a light dry rubbing with fine glasspaper, followed by careful dusting-off or wiping down with a damp (not wet) chamois leather, will remove this. The surface will then be ready for "oiling-in", for it must be remembered that the thick coating of filler is fairly porous and requires properly binding down; this is best done with a coat of universal medium or hard-drying mixing varnish thinned well with white spirit, taking care to use a clean brush and a clean can. When dry and hard the surfaces may be treated with undercoating and finishing materials as required, but from this point onwards, having taken so much care in preparing the surface, every effort must be made to keep the surface, the paint, the paint cans and brushes — everything — scrupulously clean and free from specks of grit, for in no other way can a superlative finish be obtained (see also "High-Class Finishing", page 297).

Water filler

A high-grade oil-bound water paint mixed with fine dental plaster (equal parts by weight) makes a very good filling composition which hardens quickly and rubs down well. Paste water paint alone is not suitable since it tends to crack when applied in a thick coating; moreover, it is desirable that the filling composition should be fairly porous so that when "oiled-in" it becomes thoroughly impregnated and virtually turned into an oil filler. There are some excellent proprietary fillers which rub down beautifully; they should be used in accordance with the manufacturer's instructions. Water fillers are applied in pretty much the same way as oil fillers, the composition generally being less sticky in use and therefore more easily managed. Knife filling should never be applied too thickly in one coat as this may result in cracking; where a thick filling is necessary two or more applications are recommended. Smoothing off will be facilitated by moistening the broad knife (or a plastic set-square) with white spirit or paraffin and lightly dragging over the surface as the material is just setting; the white spirit or paraffin allows the smoothing edge to slide over the surface without plucking up the composition. When thoroughly hard, that is, in two or three hours, the filler may be rubbed down, first with medium-grade glasspaper; then with a finer grade for finishing. The only drawback with dry rubbing down is the amount of dust created, especially when it comes to dusting off the surface; steps should be taken beforehand to confine the dust as much as possible, otherwise it will float or tread into other rooms. If a vacuum cleaner is available, the business is greatly simplified since mouldings, etc., may be cleaned out with the small hand tools. When all dust has been removed from the surface, the filler should be "oiled-in" as described above and allowed to dry hard, after which it may be treated with undercoat and finish as required.

22

Undercoating and Finishing

In the foregoing sections we have dealt with the principles and practices involved in the priming of various new surfaces, the preparation and touching-up of previously painted surfaces, and the stopping, filling and making good of surface defects prior to application of the undercoating. In the present chapter it will be assumed (a) that all new surfaces have been properly primed or sealed with materials suited to the porosity and condition of the particular surface; (b) that all previously painted surfaces have been properly prepared to obtain a clean surface with a suitable key for painting, and that where this has necessitated the complete removal of the old paint by burning-off or other means, all such bare parts have been touched-up or suitably primed; and (c) that all stopping, filling and making good of surface defects has been completed and, where necessary, primed or touched-up to give even suction.

The number of undercoats required will depend not only on the type of finish desired, but also on the condition of the surface. For instance, new woodwork or new dry plaster will require a minimum of four coats (priming, two undercoats and one finishing coat) to obtain a gloss paint finish of average quality; anything less than four coats would lack body and appear "paint starved". On the other hand, wood or plaster previously finished in gloss paint and in reasonable condition may require only one undercoat (of the same or nearly the same colour) and one coat of finish to provide a really excellent job. The reason, of course, is that old work which has been repainted several times becomes well bodied-up to a fairly thick film, whereas a mere four coats on a porous surface like wood or plaster would produce an extremely thin film in comparison (the thickness of an average paint film, comprising four coats, being in the region of 0·25 mm).

An undercoating material which has become extremely popular in recent years for the painting of structural steelwork is *micaceous iron oxide*. It is based on a linseed oil/phenolic medium and micaceous iron oxide pigments which are flaky in texture and are chemically inert. The particular value of these products lies in the fact that they produce very thick coatings, one coat being the equivalent in thickness of two coats of conventional paint, and this is an important factor in the painting of steel where the "build" or film thickness contributes substantially to the durability of the paint system. Another significant advantage is that they give extra protection to the angles and sharp edges of the metal, because unlike most paints they exhibit little tendency to recede or withdraw from the edges and arrises. They can be stored satisfactorily for long periods because the pigment does not settle to any marked extent. In many cases they are also used as finishing coats, when they produce a pleasing metallic sparkle which tends to increase as weathering proceeds. Micaceous iron oxide paints were used as finishing coats on the Forth Road Bridge.

Sequence of coats

The old rule about alternating paint coatings so that two flat coats or two oily coats were never placed together applied only to hand-mixed paint. In present-day practice proprietary paints are invariably used and have completely superseded linseed oil paints. The undercoats produced by the manufacturers are supplied in ready-to-use consistency, and if any adjustment is necessary it should be done by adding white spirit strictly in accordance with the manufacturer's directions. But the fundamental rule that a hard-drying coat should never be applied over a ground which is softer or more elastic than itself is more important today than ever, in view of the wider use of hard gloss finishes. This principle holds good whatever the medium — straight oil paint, hard gloss paint, synthetic or cellulose — and if departed from, cracking or other surface defect is liable to develop sooner or later (*controlled* cracking, of course, is sometimes specially desired, as in industrial "crackle" finishes). It may be thought, in view of this, that the high proportion of oil necessary in priming coats on porous surfaces constitutes a danger, and indeed this may be so if the primer is not properly adjusted to the porosity of the surface, or is so unskilfully applied that it forms a soft elastic coating. When properly formulated and properly applied, however (that is to say, when (a) sufficient hardening agent such as red lead is present to harden up the high proportion of oil and (b) the primer is applied in such a manner that only sufficient is left on the surface to form a thin well-bound film after suction is satisfied), the priming coat will form a hard firm ground, suitable for any subsequent coats that may be applied.

In preparing grounds for graining the same principle might well be employed or, at least, a *part* of the oil normally used could be replaced by a hard-drying mixing varnish; we should then see fewer examples of badly cracked grained work. The aim should be always to work from a hard priming or first coating, through hard yet flexible undercoats to a relatively elastic finish. When this principle is reversed, even in small areas, as when knotting is applied over a painted ground, or where glue-size or flour paste is inadvertently allowed to dry on new paintwork, cracking may be expected to occur.

BRUSH APPLICATION

The application of paint must needs vary according to the condition of the surface and the type of paint being used; and the well-meant advice, often given to the young apprentice, "less paint — more painting", certainly requires some qualification. Take, for instance, the painting of a porous surface such as new wood or plaster, from priming to gloss finish. The thin primer should first be applied *liberally* to enable the surface to absorb as much paint as it requires to satisfy suction, but it should also be well brushed out to ensure that no surplus paint remains on the surface. Intermediate undercoatings will be applied more sparingly, well brushed out to avoid a thick coating which may dry on the surface yet remain soft beneath. The finish — varnish or enamel — will be applied fairly liberally, but in view of its flowing properties, careful judgment and handling are required to avoid "curtaining" and runs. Or again, take the painting of non-porous surfaces, such as metalwork; here the priming and undercoatings will be applied *sparingly* and well brushed, the only liberal coat being the final gloss finish. In the case of flat oil paints on walls and ceilings, all the coats — priming, intermediate and finishing — require fairly liberal application to secure best results, but the consistency of the material and the brushwork should be such that a thick soft film is always avoided. Paint should always be allowed to dry hard before following with another coat, otherwise the first coat may be prevented from hardening altogether. The expedient of applying a quick-drying coat in an effort to speed up the drying of a tardy undercoat is thoroughly bad in principle and can only lead to eventual cracking or other surface defect.

The amount of paint in the brush should never be excessive; the tip should be dipped about 25 mm into the paint, the bristles then being gently patted against one side of the can to remove surplus colour and to distribute it among the bristles. A well-trained painter does this automatically. Notice how he keeps his paint-can clean by methodically brushing the sides down from time to time to prevent the paint drying and skinning thereon (nothing is more unsatisfactory than working from a paint-can with sides loaded with a thick paint skin). Notice too that he keeps the handle and stock of his brush clean — that whenever he puts down can and brush, he lays his brush across the top of the can so that the bristles always rest on the same side, leaving the other side clean for the handle. Observance of little points like these go to the making of a clean, methodical workman. Good brushwork depends on good wrist-work, and this can only be acquired with practice. The paint brush is generally held with the stock between thumb and fingers, whilst the larger flat wall and distemper brushes are held by the handle, but close to the stock to secure good balance and avoidance of undue fatigue.

Paint should be applied methodically. Large areas require, first, a sufficient quantity of paint distributed at short intervals, covering a panel or a section of wall about one metre by half a metre; then, by careful brushwork, the paint is spread evenly, crossed and recrossed to eliminate brushmarks, and finally laid off lightly. The next section is similarly treated and brushed into the first, the aim being always to keep edges alive (i.e. wet) so that each section can be laid off into the preceding one before it sets. Ceilings should be painted in narrow widths 300 to 450 mm (12 to 18 inches) wide and laid off lengthwise, wall surfaces being laid off vertically, and narrow members laid off along their lengths. Woodwork is laid off in the direction of the grain. All edges and corners should be lightly laid off with an almost dry brush to eliminate "fat edges".

The apprentice should be taught the logic of doing things in proper sequence — the why and wherefore of doing some things first while leaving others to the last. We remember seeing an apprentice loudly upbraided by a foreman for "getting more paint on himself than on the job". The unfortunate boy had been told to prime a rather intricate wood-frame structure with red lead; the unthinking youth, taking the line of least resistance, painted the easy part first, i.e. the outside, but when it came to doing the inside of the frame the trouble began, for in struggling to reach the more remote parts the boy got a liberal coating of red lead on his hands, face, ears, hair and overalls — a sorry object, indeed. In this case, of course, the foreman was more at fault than the apprentice for not first pointing out the best way of tackling the job.

In all well-planned brush-work, the difficult-to-get-at-parts are done first, the easy parts last. Operations will generally commence at the top and proceed downwards; this avoids inadvertent splashes (or, worse still, dust and dirt) falling on wet painted work beneath. The sequence to be followed in painting interior work is as follows.

Ceilings

Commence near window and work away from it; cornices and friezes are worked from right to left, laying off into the last section and not away from it.

Walls

Start each flank at the top right-hand corner, working each section well into the next and finally laying off downwards.

Woodwork

Commence with windows, then picture rail, doors, mantelpiece and skirting; in this order the brushes will be clean and the paint just right when the door is reached, since final adjustments to the consistency should have been made by then. The skirting should always be the last item because of the risk of picking up dirt and grit from the floor and thus contaminating the paint in the can; for this reason, only a small amount of paint should be in the can when the skirting is done and any material left over should not be "boxed" with the clean stuff, especially in the case of varnish or enamel.

Windows Paint sash windows in the following order: raise the lower sash and pull down the top sash so that the meeting bar can be painted; push the top sash up and wipe surplus paint off the dividing ribs. Next paint the top soffit and runners (pulling the cords out so that they are kept free of paint) and finish the top sash. Paint the top part of the architrave before descending the steps, then complete the lower sash, architrave and finally the sill. Runners and meeting bars, etc., should only be sparingly painted to minimize the possibility of sticking. A small sash tool or a nicely broken-in 25-mm tool is used for "cutting in", the brush being placed on the glazing bar and gently manoeuvred until the bristle just reaches the glass before drawing it along the bar. An easy brushing consistency will facilitate clean, slick cutting-in.

Casement windows are more straightforward. The back edges of the opened casement are first painted and the casement is then closed while cutting-in the glazing bars; it is opened again to finish off the sides and should remain slightly open to allow rebates, etc., to dry and so prevent sticking. The fixed casements are then painted, followed by the architrave (if any) and finally the sill.

Doors Here again the first items to attend to, before painting the door face, are the edges and rebates. If the door opens *towards* you, paint the

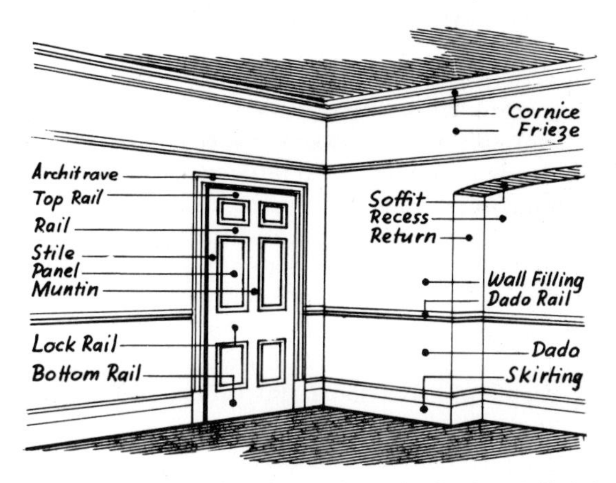

Fig. 22.1 Domestic interior features: principal subdivisions of walls and door

Fig. 22.2 Domestic staircase: names of main features

rebates and opening edge of the door, leaving the back rebate and hanging edge, i.e. the hinged edge. If, however, the door opens *away* from you, paint only the back rebate and hanging edge of the door, taking care always to wipe clean any paint which finds its way through the hinges, etc., to the other side. Edges and rebates done, the top panels come next, painting first the mouldings, then the panels; the bottom panels are treated similarly, then top and bottom *muntins* followed by top, middle and bottom *rails* and finally the outer *stiles* (Fig. 22.1). Mouldings and quirks should be well "bottomed" but not flooded with paint, and a sharp look-out should be kept for any gathering of paint in the mouldings likely to develop into runs.

The trend in modern properties, however, is to do away with panelled doors and replace them with flush doors which do not present such a dust-trap. The painting of flush doors, especially with modern quick-setting paints, demands a different technique, somewhat akin to the painting of a wall area, the aim being to keep the edges alive; generally the top part of the door is painted and laid off and then the lower half is rapidly painted in and joined to it so that there is no gap between the two.

Wherever possible, all door furniture such as finger-plates, locks, keys, escutcheons, etc., should be removed to give a clear run for painting; cutting round such items takes time and is rarely satisfactory. Wipe all edges clean where necessary and leave doors ajar till dry.

HARD GLOSS AND ENAMEL FINISHES

These materials are more viscous than undercoatings, and consequently the brushing is usually more tough; moreover the fairly quick set and the properties of flow require considerable skill and judgment on the part of the painter if first-class results are to be secured. The necessity for extreme cleanliness and care at every stage of the work is dealt with under "Varnishing", the same principles being equally applicable to all high gloss finishes. The fact that pigmented gloss finishes are of stouter consistency and tougher working than varnishes means that even more care must be taken in securing even distribution and avoidance of flooded mouldings and quirks.

Some painters seem rather apprehensive of the flowing properties of these materials and tend to apply a sparing coat, brushing out excessively in their anxiety to avoid runs and "curtain" effects; in consequence their work looks bare or "starved" and lacks the fine flow of the best work. Curiously enough this over-cautious application often produces the very defects it is desired to avoid, for when gloss paints are applied too sparingly it is difficult to brush one section smoothly into another owing to the material setting; this results in more material being applied in one part than in another and it is this uneven distribution that gives rise to creeping and curtaining.

The correct way to apply gloss paints and enamels is as follows. First apply a liberal coat to a door panel or a small section of wall surface and distribute this evenly by firm crossing and re-crossing. Then scrape out the brush on the edge of the can and with it remove any excess material from the surface; one should be able to draw the brush firmly and smoothly along the surface without any feeling of "slip". If the brush slips in one place and drags in another the distribution is uneven and this must be corrected by firm brushwork. Lay off finally with firm, even strokes. On large surfaces, work in small sections, joining up as quickly as possible so that laying off can be done before the material begins to set. The secret of successful glossing lies in being able to judge when the right amount of material has been applied, and this can be cultivated only by getting the "feel" of the material under the brush and becoming sensitive to the slightest drag or slip. The aim should be to apply the material quickly but unhurriedly, and to secure even distribution without over-brushing. Excessively heavy coats should be avoided since these may result in surface drying and subsequent wrinkling.

Some of the newer materials of the synthetic type are inclined to brush out more easily than the natural resin or stand oil types and tend to produce thinner films in consequence; for this reason it is often an advantage to apply two coats of the gloss finish.

All gloss finishes are more sensitive to temperature changes and humidity than undercoatings, and the conditions that give rise to blooming, etc., should be avoided as far as possible (see Chapter 23). If for any reason it is desired to ease the brushing of gloss paints or enamels, a very small proportion of white spirit may be added unless the maker's directions state otherwise; on no account should raw linseed oil be added.

Flat enamel finish

So-called flat enamels generally dry with varying degrees of "sheen" ranging from eggshell to a pronounced semi-gloss finish. They are not so full-bodied as gloss enamels and work much easier under the brush. A flowing coat must be given, the same precautions against "flashing" being

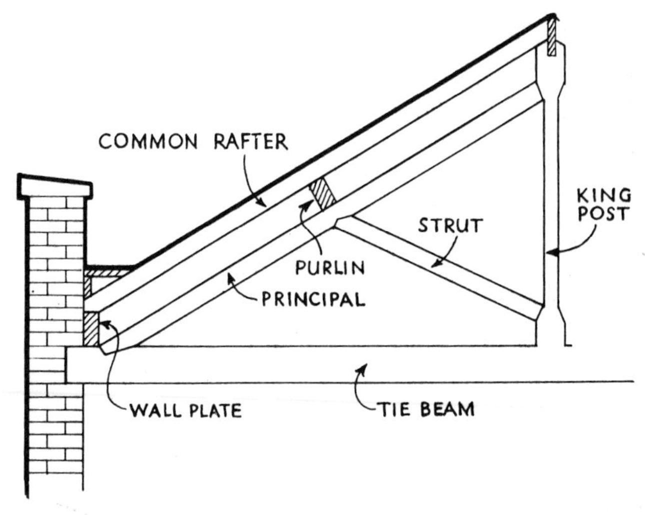

Fig. 22.3 Building terms

observed as in the case of flat varnishes (see page 302). Unlike flat varnish, however, flat enamel must be applied over a flat or eggshell undercoat.

Flat wall finish

As already indicated, modern wall finishes have superseded the painters' hand-mixed flatting as a flat wall finish. This is not surprising when the technical difficulties associated with the application of the old-fashioned flatting on large flank walls and ceilings are called to mind; in comparison, the application of present-day eggshell finish is child's play!

Modern wall finishes are often supplied in a fairly round consistency and are intended to be applied in this condition. The essential point about their application, whether by brush, roller or spray, is the necessity for keeping the edges alive, otherwise an undesirable "flashing" and gathering of paint will ensue. If the application is by brush, good flat wall brushes should be used; the 100-mm or 125-mm sizes are the most suitable.

Some notes on spray painting techniques have been given in Chapter 6.

23

Varnishing

The greater part of interior and exterior varnishing is carried out under everyday working conditions — practical conditions over which the painter has only a limited control. It is easy enough in a textbook to outline ideal conditions that it may be impossible to achieve on a normal job. Nevertheless, it is the purpose of the textbook, no less than the duty of the teacher, to state the ideal conditions under which materials should be applied to secure the best results, sc that the apprentice and immature journeyman painter are provided with a standard — a high standard — at which to aim.

The great danger of working under constantly adverse conditions is that we may come to regard them as being the *right* conditions, with the result that when the extra-special job comes along — such as when a fastidious client wants his doors finished like a mirror — we are incapable of "delivering the goods". It is of no use to say that it "can't be done"; the old coachpainters did superb work, often under trying conditions.

We therefore propose briefly to outline the procedure which, as far as our own experience goes, appears necessary to produce (a) a good-class finish under normal working conditions, and (b) a high-class "coach" type finish. The painter will then be able to judge how far the existing conditions on any particular job will allow him to conform to such procedure. But first, a word or two about general conditions likely to affect the behaviour of varnish and enamel.

Conditions

Varnishes and paints containing varnish media, such as hard gloss paints and enamels, are most sensitive to temperature changes, moisture, gas fumes, greasy surfaces, etc., and show the slightest contamination from dust, either in the material or on the surface. They should be stored in an equable temperature of about 18°C. and, wherever possible, they should be applied at this temperature. When used outdoors or in unheated rooms in cold weather, they should not be taken direct from the store and used immediately, but should be allowed to become adjusted to the lower temperature for a day or two before use. A lowered temperature will generally cause the material to work "tough" and to be difficult to spread, with the result that a heavy film is applied; this may wrinkle badly if exposed later to a hot sun. Heavy application should therefore be avoided in cold weather.

The greatest danger, however, comes from moisture and humid conditions (as in warm, damp weather or close, badly ventilated interiors) which should be particularly avoided (see "Blooming", Appendix 4). Apart from the danger of condensed moisture through excessive humidity, the surface itself may carry *invisible* moisture, especially where it has been felted down wet prior to varnishing, or where grained work has been glazed in water colour; ample time should therefore be allowed for the moisture to dry out thoroughly, particularly from mouldings and quirks, before varnish is applied.

The surface to be varnished should be thoroughly dry and hard, but it should not stand too long before the varnish is applied, otherwise it tends to develop "greasiness" in which case the defect known as *cissing* may be expected (see Appendix 4). Where a long interval is unavoidable — say after three days or more — the surface should be moistened with water and lightly papered with No. 400 fine waterproof abrasive paper, leathered off and allowed to dry, or alternatively, wiped down with a wash-leather damped in water to which a little fine whiting or fullers' earth has been added; either of these treatments will prevent cissing.

General procedure

Before commencing varnishing operations, all dust should be removed as far as possible, both from the surface to be varnished and also from floors and other surfaces in the vicinity. On interior work the floor may be sprinkled with wet sawdust and gently swept to gather the dust without raising it; or, better still, have the floor washed. If this is done, the humidity of the room will tend to rise whilst the floor is drying and varnish applied at this stage may be affected; the window should therefore be left open to encourage a good circulation of air, and varnishing should preferably be delayed until the floor is reasonably dry. Dust being the enemy of all gloss finishes, every care should be taken to avoid raising it; each room should therefore be isolated from all other operations whilst varnishing is in progress and the painter should move about the room quietly, locking the door from the inside to prevent any violent disturbance of air that might be caused by anyone hastily opening the door. Such isolation, of course, cannot always be secured in corridors and public places, but the craftsman who takes pride in a clean gloss finish will try to arrange it wherever possible.

It is assumed that the work has been properly prepared and brought up to the finishing stage by following the principles laid down in the preceding chapters, and that the ground is hard-dry and flat (having been lightly sponged and papered down with fine waterproof abrasive paper, leathered-off, and sufficient time allowed for all moisture to dry off thoroughly).

Varnish should never be shaken or handled violently as this tends to cause frothing; white spirit or turps in the brush will have the same effect and the brush should therefore be vigorously twirled between the hands to expel any spirit after rinsing. Use a scrupulously clean can (some painters apply a coat of knotting to the inside to fasten any loose particles which remain after cleaning out) and pour out only a small amount of varnish at a time so that, if any dirt is accidentally introduced, a large quantity of varnish will not be affected. Work the brushes thoroughly into a little varnish, then dump this and pour out a fresh lot before commencing operations.

Varnishing a room will proceed in the same order as that outlined for painting woodwork (page 290), so that by the time the door is reached the brushes will be well worked-in and clean. It may be argued that one should start right away on the door with fresh stuff and clean brushes, but experience goes to prove that this does not always produce the best results. Brushes are not necessarily clean after rinsing out in white spirit or turps and it is therefore advisable to try out the varnish and brushes first on less important parts of the room; when they are found to be satisfactory and giving clean results, then — and only then — should the door be tackled. Skirtings and any other parts where dust is liable to be picked up should always be left till the last.

As with enamels (page 292), varnish should be applied liberally to facilitate even distribution, any excess being taken off with a fairly dry brush before the final laying off. Application is generally much easier and more rapid since varnishes are not so stout in consistency as enamels and therefore do not work so tough under the brush. The materials should be applied confidently with firm easy strokes, judging the right amount by the "feel" of the material under the brush — too little being indicated if the brush drags and too much if it slips. When the passage of the brush along the surface is smooth and even, there need be little fear of crawling or runs. Mouldings and quirks should be carefully watched for signs of filling up, any excess being taken off with a small fitch. Do not attempt to touch up any portion of the work when once the varnish is set; if any serious defects arise, the varnish should be either removed immediately with white spirit, or allowed to dry and harden, when it may be rubbed down and revarnished. Where two coats of varnish are to be applied, the first should be hard-drying undercoating varnish, allowed to dry hard, then flatted down before applying the finishing coat. This also applies where the finish is to be a flat or eggshell varnish. Finishing gloss varnishes are generally elastic and are not designed for rubbing down.

On exterior work, the varnisher is at the mercy of the elements and cannot arrange the conditions exactly suitable for his purpose; he must therefore choose the most promising periods of the day in which to apply his varnish, avoiding where possible damp and cold on the one hand and dusty winds on the other. Like the farmer, he must try to anticipate possible weather changes and plan his work accordingly.

HIGH-CLASS FINISHING

In the production of a high-class "coach" finish, every stage of the work must be done with the final object clearly in mind — to produce a smooth mirror-like surface, free from nibs or scratches. The processes connected with the building-up of the ground will therefore need to be carried out with more than normal care; undercoatings will be carefully strained, brushes well cleaned and all traces of grit or powdery debris removed

Courtesy: Briton Chadwick Ltd

Fig. 23.1 Flat metal-bound "coach" varnish brush

from the surface after rubbing down operations. In addition to priming and stopping, the surface is usually *filled* to secure a true, level ground free from waviness, indentations, or other surface defects. After the filler has been rubbed down and oiled-in as described on page 285, two or more hard-drying paint undercoats are applied, followed by two or more coats of undercoating varnish (sometimes called "rubbing varnish"). Each coat when dry is lightly flatted down with a fine grade of waterproof abrasive paper, lubricated with water, an operation that requires the utmost care. Abrasive papers with the grade numbers of 400 to 600 will flat down the finest varnished surface beautifully without scratching or clogging, provided the varnish is thoroughly hard.

When all the above operations are completed, the work is ready to receive its final coat of finishing varnish and we must now consider what precautions may be taken to ensure that all the patient labour that has brought the work to this stage is not spoilt by flaws in the final coat. It must be remembered that the smoother and more perfect the ground, the more those odd nibs are likely to show up. Elaborate precautions must therefore be taken against contamination by dust and grit. If the following procedure seems unnecessarily fastidious or too much trouble, we would assure the reader that they are everyday practical measures that are normally taken by craftsmen engaged on specialized finishing of the type being considered.

Dustproof room

A dustproof area is essential if flawless work is to be produced and shops specializing in high-class finishing usually have a special varnishing room for this purpose; every painting class in our technical colleges should be so provided. Where work of this type must be done on the site, however, a suitable screened area can be provided by a *dust barrier*, which is a light portable frame structure covered with a translucent open-weave fabric impregnated with non-drying tacky adhesive which arrests the most minute particles of dust. Thus a dustproof corner, a tunnel, or an inner "room" can be formed, complete with tacky floor and door mat if desired. For those who wish to provide their own dustproof room the following may be of interest.

An inner room should be made as dust-tight as possible, ventilation being provided by having one or two panels at top and bottom cut out and covered with fine wire gauze on both sides, leaving an air-space between to trap the dust. Walls and ceiling should be plastered smooth or covered with wallboard and finished in hard gloss paint or enamel; when thoroughly hard the ceiling and walls may be rendered tacky by applying a coating of paraffin wax and castor oil to hold any floating particles of dust. In cold weather the room temperature should be kept at 18°C or thereabouts by means of hot water pipes or electric tubular heaters fixed to the skirting board. The warm air rising will carry the dust upwards and cause it to adhere to the tacky ceiling and walls. The wax coating can be removed with white spirit and renewed periodically as necessary. The floor should preferably be covered with lino, heavily waxed. This may be rather slippery to walk on, unless pumps or goloshes are worn, but it is better than using water to lay the dust; water will tend to make the warm air humid and that is to be specially avoided when varnishing or enamelling. Before commencing work in the dustproof room, the floor should be rubbed over with a tacky rag to pick up dust. If double doors can be arranged to form an airlock, so much the better. Doors should be opened slowly on entering or leaving to minimize air disturbance; they should be locked on the inside during operations to ensure that no one enters carelessly.

A small work bench is desirable, also a pair of rigid trestles to support panels or doors in a horizontal position whilst being varnished. A rack or shelves may be provided to hold varnished panels during drying.

Overalls and loose clothing such as jackets, dust coats and aprons, should be discarded before entering the varnishing room since these tend to waft dust about; it is usual to work with shirt sleeves rolled during operations, moving about the room as quietly as possible to minimize air disturbance, and it is preferable that no more than one person should occupy the room at any one time.

Brushes and cans

The provision of scrupulously clean cans and brushes is essential if the work is not to appear "bitty". Two receptacles are necessary for "super finishing" of this kind, the first to hold the clean varnish and the second to scrape the brush into. The brush should never be scraped into the first pot because if the brush contains any bits the clean varnish will be contaminated. A tapered enamelled beaker or "pot" is the best receptacle for the clean varnish since these contain no corners where particles can lodge; before pouring in the varnish, wipe the inside of the pot with a non-linty "tack" cloth to pick up any fluff or bits. The varnish pot may be suspended by small ears in the larger can so that when taking up varnish from the inner pot, the brush may, if necessary, be conveniently scraped on the edge of the outer can to remove excess; in this way the varnish in the inner pot is kept clean and any bits from the brush go into the outer can. The varnish in the larger can may be used for less important work.

Only varnish brushes which have been nicely "broken in" on ordinary work are used for high-class finishing; after a thorough washing in petrol or white spirit, they should be twirled vigorously between the hands to remove the spirit (all this preparatory work and "making ready", of course, being done outside the varnishing room).

When ready to begin operations all necessary materials are taken into the varnishing room and the door is locked. A few "brush-out" sheets (stout lining paper previously prepared by knotting or sizing) will be required for "working-in" the brush. Take up a little varnish on the brush (omitting to pat the side of the pot as in ordinary practice) and make a trial brush-out on one of the sheets, working the brush well into the varnish. Lay off, allow the brushmarks to flow out, then observe whether the rubbing is clean; if there are any specks or "bits", scrape the brush out well on the outer can, take up more varnish from the inner pot and make another rubbing. Several rubbings may be necessary before the brush is clean enough to proceed with the work in hand, but if the rubbings are "bitty" or "sandy" after three or four trials, the brush must be washed out and the same procedure gone over again. Not until a clean, flawless rubbing is made should varnishing be proceeded with. It will be realized from this that when once a brush is *really clean* it must be jealously guarded to keep it so. After use, it must be placed in a covered dustproof brush-keeper (see page 83), suspended in varnish or boiled oil, rinsing out afresh in white spirit each time it is used. If the brush is not to be used for some time, it should be rinsed out in white spirit, thoroughly washed in warm water and soap, dried and wrapped in greaseproof paper secured by a rubber band before putting carefully away.

Application of the varnish

After the final rubbing down prior to varnishing, there is always liable to be a fine, loose "down" left on the surface, no matter how carefully rinsed and leathered off. Some craftsmen remove this by running the palm of the hand over the surface, but this is not to be recommended because hands tend to be greasy. It is better to brush the surface with a tacky badger, or to wipe down with a tack rag. A clean dry 25-mm tool or fitch with the tip rendered tacky is useful for going round projections and quirks to pick up any loose particles lodging there; if these parts are not thoroughly cleaned out, the varnish brush will pick up such particles and the whole finish will be spoiled. The tack rag ensures that the dust is picked up and not merely removed to another part of the surface.

It may be said, when the above points have been attended to conscientiously, that all reasonable precautions have been taken to secure a clean job. The varnish is clean, the varnish pot is clean, the brushes are clean and the surface is clean — or should be. It is when it comes to the actual varnishing that skill and craftsmanship are tested to the full. Finishing carriage varnishes are of the long-oil type, slow drying and possessing fine flow. The art of varnishing consists of using these flowing properties to the limit without allowing the varnish to "curtain" or run.

The varnish is carried to the surface straight from the pot (i.e. without dabbing the brush on the side of the pot); three or four liberal dips are rapidly applied and quickly distributed over a section of the surface, laying the varnish lightly to avoid frothing or bubbles. The varnish is allowed to flow out or settle while the next section is being laid-in, then any stray bits or specks which may appear are carefully removed with a small fitch or a pocket knife. The brush is scraped out, the varnish is crossed and re-crossed, then laid off lightly, scraping the brush out after each stroke. Laying off should never be attempted with a loaded brush. On vertical surfaces, laying off should be done with an upward stroke, taking care not to draw the brush across any bead or moulding at the top; the top 25 mm or so should be finished off with the lightest possible horizontal stroke, to prevent any curtaining beneath the bead or moulding.

Our previous remarks on page 297 about sensitive brushwork — judging the right amount of varnish by the "feel" under the brush, coupled with even distribution and deft laying off — probably sum up the essentials of successful varnishing. There may be many failures before that fine flawless "super" finish is obtained, but it is worth while spoiling a few practice panels in order to gain confidence and experience in handling varnish. The man who is so afraid of his varnish running that he applies it sparingly will never produce that super finish.

Polishing varnished work

In some industrial wood-finishing processes the polishing of the final varnish coat is an established practice, but many decorators also pride themselves on their "handing up" or polishing methods. For those wishing to produce a really first-class finish the requisite materials are still available. Success depends on a well-prepared ground and correct choice of materials, for it is of no use trying to polish an ordinary elastic, long-oil finishing varnish. The ground may be primed, filled and painted (or if hardwood, stained and filled); in either case the surface is bodied-up with three or four coats of hard rubbing or polishing varnish, then felted down with pumice and water or fine waterproof abrasive paper, leathered off and allowed to dry. The surface is then rubbed with oil and rotten stone, using a rubbing wad of cotton wool wrapped in soft cambric; this "rubber" is moistened with linseed oil, lightly dabbed on powdered rotten stone, and applied to the surface in a circular motion. Prolonged friction tends to heat up and soften the varnish and this must be avoided; the work must therefore be done in easy stages, allowing short rest intervals, rather than continuous or concentrated rubbing. The final polishing is done with dry flour, some painters using the palm of the hand (hence the term "handing up"), others using a loose wad of cotton wool. A final brisk light rubbing produces a brilliant polish. Great care must be taken in the initial cutting down to avoid scratches since these will show up badly in the finished work.

Flat or eggshell varnishing

Flat or so-called "eggshell" varnishes vary greatly in consistency between one make and another owing to differences in formulation. Some work almost like a normal gloss varnish under the brush, whilst others are of extremely thin and easy brushing consistency. Both types should be applied rapidly in a fairly liberal flowing coat, but taking great care to avoid runs. Each panel or section must be cut-in cleanly and finished as the work proceeds; any going back over a finished portion will result in glossing-up or "flashing". In varnishing doors and other panelled work, panels and mouldings must be done first without encroaching on the stiles and rails; the latter must then be varnished in such a way that edges are always kept alive so that flashing is avoided.

For best results flat varnishes should be applied over an undercoating varnish or a hard "church oak", after lightly cutting down with fine waterproof abrasive paper.

CLEAR WOOD FINISHES

Under present-day conditions, clear polyurethane lacquers are very often used as wood finishes for interior surfaces. They offer many useful features, and one of their greatest attractions is their exceptional hardness and abrasion resistance; in this respect they are quite outstanding. They also have the property of being able to withstand continual washing with detergents, and are unaffected by heat, water, alcohol, petrol and most chemicals, which makes them suitable for the finishing of bar tops, floors, etc., and for the decoration of panelling, etc., in industrial premises.

Many varieties are available; the types which are most popular at present are the two-pack variety, supplied as either a gloss or satin finish, in which a curing agent is added to the base material; but one-pack systems are also in use, and there are pigmented coatings available in both gloss and eggshell finish. Application may be by brush, roller or spray, though spray application is attended by a certain degree of risk because of the toxic nature of the free isocyanates when inhaled. Attention has been drawn to this risk by the greatly increased use of sprayed polyurethane coatings recently in the vehicle painting and furniture finishing trades. Decorating craftsmen who are considering spray application methods with these materials should notice that a respirator with a separate air supply should be worn; the normal type of face-mask does not give protection against this hazard.

When two-pack materials are being used it is essential that the base material and the curing agent should be mixed in precisely the correct proportions, as detailed by the manufacturers. The contents of the smaller tin, the curing agent, are added to the base, and the mixture is thoroughly stirred. When the mixture has become clear and free from bubbles it is ready for use. On new woodwork, the wood is cleaned and carefully rubbed down with a fine grade of abrasive; it is then dusted off or cleared with a tack rag, and sealer coats applied according to the manufacturer's directions, usually by brush (some manufacturers recommend one sealer coat, others two). The sealer is made by thinning the mixture with the recommended solvent. In warm conditions the sealer coat is generally dry and hard enough to recoat after four hours. Further coats of clear unthinned lacquer are then applied until the requisite degree of finish is obtained. It is important that no more than 24 hours should elapse between the separate coatings, otherwise the adhesion of the coatings may be impaired; if for any reason there is a delay which prevents this, the surface must of necessity be flatted down with 400 grade abrasive lubricated with water before the next coat is applied. The pot life of the mixed material is quite short, and it is therefore advisable to mix only the quantity of lacquer that can be applied within a three- or four-hour period. In extremely warm weather conditions it may be necessary to thin the lacquer slightly, but this must be done strictly in accordance with the manufacturer's instructions.

Previously varnished woodwork should be washed down with detergent or white spirit, and it is particularly important to remove residual dirt, wax and grease. If the previous coating is crazed or shows signs of lifting, it should be completely removed. There is always the danger that the polyurethane lacquer may act as a paint stripper due to the strength of the solvent, and it is advisable to ascertain first of all on a small trial patch whether or not the previous coating is liable to be disturbed.

One-pack lacquers are cured by the action of moisture in the atmosphere, which promotes polymerization. They are touch-dry in one to two hours and can be recoated after five or six hours.

When the lacquer is fully cured, there is considerable difficulty in getting subsequent coats to adhere. It is chiefly for this reason that the use of polyurethane lacquers on exterior woodwork is not advisable. Experience has shown that it is practically impossible to maintain an exterior coating in good condition because of this extreme hardness, and lack of adhesion of the subsequent coats.

Acid-catalysed clear finishes

Another group of clear wood finishes for interior work is based upon acid-catalysed resins. Two-pack materials of this kind are subject to the same conditions as polyurethane lacquers, in that the proportions of the base material and the activator or curing agent are critical; unless they are mixed in the correct ratio, the properties of the dried film will be adversely affected. Similarly, the pot life is short; any thinning should be done strictly in accordance with the manufacturer's instructions; the material should not be applied at a temperature lower than 8°C, otherwise the curing process is very much retarded; and there is a distinct possibility of the material disturbing and lifting any previous existing coating.

Clear lacquers based on epoxy resins are resistant to acids, alkalis, detergent fumes and liquids and to mechanical abrasion, and they have the advantage of providing good film build — in other words they produce a film of substantial thickness. Those based on formaldehyde resins offer the advantages of hardness, rapid drying (thereby reducing the risk of being marred by dust), and resistance to heat, chemicals and solvents. There is a choice of either full-gloss or eggshell finish, but if eggshell is selected it should be applied on top of gloss coatings; two consecutive coatings of eggshell would pose problems of adhesion unless the earlier coat were well flatted down. Epoxy-based coatings lend themselves to brush or spray application, with the proviso that if the spray is used, a respirator must be worn. Coatings based upon formaldehyde can also be applied by brush or spray, and for some purposes roller application is recommended, but in this case it is important not to over-roll the material, especially if it is beginning to set up, otherwise bubbles may form that will spoil the finish. Although these clear finishes are not toxic, they give off very powerful fumes as the solvents evaporate, and it is therefore important that the work area should be adequately ventilated, particularly in confined spaces.

These coatings are especially suitable for situations exposed to chemical attack, such as the working surfaces in chemical laboratories where solvents are liable to be spilled, and they are also available as floor varnishes for areas exposed to difficult conditions. When used on floors they are often given a protective coating of wax polish or emulsion polish.

One-pack epoxy ester varnishes are available in both eggshell and full-gloss finishes, and they offer resistance to mild acids, alkalis, detergent fumes and liquids.

Clear wood finishes for exterior work

Some varieties of timber are widely used in modern construction for the external cladding of buildings, especially in the case of domestic properties, and this has led to an increased demand for some form of treatment that will help to preserve the timber whilst at the same time maintaining or even enhancing its natural colour and character. When polyurethane lacquers first came on the market they were hailed as the ideal materials for this purpose, but, as has already been indicated, the extreme hardness which polyurethane achieves causes problems in that the surface becomes covered with unsightly flakes; and the position is complicated by the dimensional movements of the timber when exposed to atmospheric conditions. It is now generally agreed that the traditional clear gloss varnish is much the better treatment, especially if a high-grade alkyd or tung oil/phenolic varnish is used. But even the best varnish has only a limited life-span, and unless it is renewed before the film has begun to break down the maintenance of the treatment is an expensive business. It should be noted that the number of coatings in the film has a direct bearing on the durability of the film, and research has proved conclusively that however long a three-coat varnish system lasts, a four-coat system will last at least twice as long. Varnish being an easy material to apply, the labour costs involved in its application are low; for this reason a four or even five-coat system is generally recommended. It is important, too, that the back and end grain of the boards should be sealed with either varnish or aluminium paint to prevent the ingress of water.

Madison sealers

These are penetrating compositions, marketed by all the leading manufacturers under a variety of trade names; they are used as an alternative to clear varnish, lacquer, or clear wood finishes. They are based on what is termed the "Madison formula" which originated in North America, and they consist of a mixture of boiled linseed oil, waxes, resins, fungicidal agents and white spirit, together with a small quantity of transparent pigment to match up to the colour of natural woods. They are easy to apply, and dry with a pleasing eggshell gloss; being relatively cheap, they work out low in material and labour cost.

Madison sealers provide a finish which, unlike most transparent treatments, prevents timber from assuming a whitish or greyish discoloration due to bleaching of the natural colouring of the wood, while at the same time preventing it from darkening due to the accumulation of mould growths. On planed weatherboarding, a single application of the sealer is often sufficient, although sometimes a second application may be desirable. The spreading capacity is generally between 9 and 11 square metres per litre. Under normal conditions of exposure the coating can be expected to last for about three years, at the end of which time the recoating is a simple matter of making a single further application, because there is no surface film to have developed faults such as cracking or flaking; the sealer allows the timber to "breathe". Normally there is nothing to cause discoloration, but the choice of brand is important; unfortunately some brands tend to remain sticky, so that the surface retains dirt.

The choice of a clear treatment is largely dictated by the type and nature of the timber in question. For many woods clear varnish is the most suitable, but there are certain woods which for various reasons reject varnish. Tests have shown that clear varnish behaves extremely well on such woods as Douglas fir, afromosia, makore, sapele and utile. Western red cedar on the other hand contains oil which interferes with the drying of varnish, and of course this particular wood is very widely used for external claddings in modern construction. Theoretically any coating is superfluous, as red cedar is said to be the only exterior softwood cladding that needs no preservative treatment, but in practice it is desirable to provide some form of treatment if only for the sake of appearance. Hardwoods such as keruing give off a resinous exudation which interferes with the adhesion of varnish. For these and similar woods the Madison formula is highly suitable. Madison sealer is applied by brush, and the surplus material wiped off with a cloth after approximately two hours. The cloths after use should be burned immediately as they present a serious fire hazard. Opened tins of the sealer should be kept out of reach of children and household pets, because of the fungicide content.

24

Emulsion Paints, Water Paints, and Distempers

The terms "emulsion paint", "water paint", and "distemper" are often used rather carelessly, and it is necessary to define them in order to point out important differences in the composition of these materials. The main difference between water paint and distemper is that whereas distempers are usually based on common whiting bound with glue size or casein, water paints generally contain a high-grade pigment and are oil-bound with an oil-in-water emulsion. Emulsion paints, introduced shortly after the end of World War II, are not in the strict sense of the word emulsions at all. They consist of synthetic resin polymers and high-grade pigments dispersed in water. Further confusion is caused by the fact that some manufacturers now refer to the cheaper grades of emulsion paint, which are only suitable for interior use, as "vinyl distempers".

DISTEMPERS

Distempers, as formerly in common use, are of two main types — non-washable and washable, but though these terms are useful in distinguishing between the two types, they should not be taken too literally, for they mean no more in practice than that one is fast to dry rubbing and is non-washable while the other will stand a certain amount of wet rubbing before it is removed from the surface. The degree of washability of the latter varies with different makes, and although some of the colouring-matter is removed fairly easily by light washing, it is often very difficult or even impossible to remove the entire coating from an absorbent plaster surface.

For all practical purposes, distempers can be regarded as obsolete in the U.K. as far as the painter and decorator is concerned, but they are still widely used in certain parts of the Continent. It is for this reason that they have been included in this book.

Non-washable distemper

This is the old-time size-bound distemper (sometimes called soft distemper or colourwash). Proprietary brands are available, but it can also be prepared by hand in the following manner. Whiting is broken up into fairly small lumps in a clean bucket, covered with cold water and allowed to soak for half an hour; the unabsorbed water is then poured off and the whiting beaten to a smooth thick batter. Stainer is then added; colours such as lime blue, lime green, zinc chrome, red oxides and earth colours such as ochre are used because of their resistance to alkalis. The stainer is soaked in cold water until thoroughly wetted; it is never added in its dry state, because this would make it impossible to secure proper dispersion, and small particles of unmixed pigment would float about in the distemper, causing unsightly streaks when applied to the wall. Distemper dries several tones lighter than the wet state, and this needs to be taken into account when mixing. A sample of the distemper with stainer added is brushed on to a piece of lining paper which is then dried in the sun or close to a fire to ascertain whether the dried tint is satisfactory; this should be done before any size is added because size tends to darken when force-dried.

Distemper in unbound condition would dry as a loose powder which would rub off at the lightest touch; glue size is commonly used as the binder, and some skill is needed to judge the right amount to use — if too little is used the distemper will still rub off easily, but if too much size is added the distemper will tend to flake off. A nugget of alum may be added to the mix to help harden it, although this makes the distemper insoluble and difficult to remove when redecoration is required. Carbolic acid can be added as a preservative.

Home-made distempers of this type are, with the exception of limewash, the cheapest form of interior wall finish, and before the advent of modern water paints they were often used with great skill by the old-time craftsmen. When applied on a good well-prepared dry wall, they provide a perfect matt finish of clear luminous tints, often likened to a "sheet of blotting paper". Like blotting paper, however, they absorb and are disfigured by moisture and are ruined by damp conditions, the size being readily attacked by mildew and other mould growths.

Soft distemper cannot be successfully applied over itself, the result being "scuffy" and patchy owing to the ground colour working up and "piling". It is therefore necessary to wash off previously distempered surfaces and to prepare the wall with a coat of size or claircolle.

Claircolle (or clearcole) is a straight mixture of size and whiting which satisfies the porosity of the surface and makes a good ground for the distemper coat. The old-time painter used to pride himself on having his own recipe for a surface preparation, using strange concoctions such as boiled linseed oil emulsified with soft soap or lime.

Application of distemper should proceed expeditiously but unhurriedly, starting at the window side and working away from it. It should be applied liberally and left as soon as possible, laying off generally towards the light so that brushmarks do not show. Windows and doors should be closed while distempering (to retard evaporation) and the work should proceed in fairly narrow bands with the object of keeping the edges "alive". In warm weather a little glycerine will retard the drying to facilitate application over very large surfaces. Every care should be taken to avoid misses, for distemper cannot be easily touched up without showing. As soon as the work is finished, doors and windows should be thrown open to assist the drying, the best results being obtained when this is fairly rapid with free ventilation.

Scene-painting and large-scale temporary decorations are usually carried out in distemper, the complete absence of sheen and excellent light reflection being ideal for the brilliant lighting effects of the theatre. When executed on canvas back-cloths which have to be rolled up, the distemper is often plasticized with a little glycerine or treacle to prevent cracking and flaking off.

Washable distemper

There are various so-called washable distempers made, some in paste form and others in dry powder state, and these are used to some extent on the Continent. The binding agent may be casein or glue or a mixture of the two, and the pigment base is usually Paris white (whiting), barytes, barium carbonate or some other mineral white.

Casein, which is usually the binding agent in dry distempers, is a constituent of milk; it is produced by adding hydrochloric acid to skim milk, the casein being precipitated and dried as a granular powder. Casein is insoluble in water but soluble in alkalis, and this property forms the basis for its use in dry distempers. Powdered casein, lime and borax are incorporated with dry whiting, barytes, stainers, etc.; when mixed with hot water, the casein is dissolved owing to the presence of lime, but when applied and exposed to the air, the lime gradually carbonates and releases the casein in its original form so that it again becomes insoluble or "washable".

WATER PAINTS (OIL-BOUND)

Home-made size-bound distemper is now a thing of the past, except where a removable coating is specially desired, and in painting and decorating practice it was superseded many years ago by oil-bound water paints. These materials, which offered many attractive features, are still used to a limited extent, and it is important that the decorator, if he is to get the best out of them, should understand their composition. The first step is to consider the vehicle or medium, which is an oil-in-water emulsion.

The nature of emulsions

Emulsions are stable combinations of liquids which normally will not mix (e.g. oil and water); the most familiar example is milk — a natural dilute emulsion of fat in water. To produce a permanent artificial emulsion of one liquid in another, a small amount of *stabilizer* or emulsifier is required (such as soap, casein, glue, starch, etc.) on which the life of the emulsion depends. If linseed oil and water are shaken vigorously together in a bottle, the oil is split up into large drops, but when the mixture is allowed to rest, the drops of oil quickly coalesce (or unite) and float as a layer on top of the water. If this is repeated with a trace of soap solution or caustic added, the oil is split up into a multitude of microscopic droplets or globules and the liquid becomes quite milky and no longer separates into two layers. Each of the tiny globules is now covered with a thin film of soap which has the effect of preventing coalescence and the oil is said to have become emulsified in the water. When two liquids are emulsified in this way, light is no longer able to pass right through but is reflected and scattered by the numerous droplets so that the liquid appears milky white.

By using suitable stabilizers, oil may be emulsified in water, or water in oil, the essential difference being that the first may be diluted with water and the latter only with oil or oil solvents. In water paints the drying-oil medium is emulsified in water by using stabilizers such as ammonia, borax, casein, glue, Irish moss, etc., so that the paint is then thinnable with water. Thus the coating consists of minute oil globules cemented together with the stabilizer in which the particles of pigment are dispersed; moreover, it dries with a pleasing matt finish which is recoatable without softening up and, after a short period, becomes washable.

Formulation and manufacture

In some of the cheaper types of water paints the vehicle may contain boiled oil, limed rosin and glue, or even boiled oil and varnish foots, with soft soap as the emulsifier, whilst the pigments are usually common whiting and barytes. In the highest class oil-bound water paints, however, the emulsion usually contains an oil varnish that is specially prepared for the purpose and is rendered miscible with water by heating together with the alkaline emulsifying solution in steam-jacketed pans, afterwards passing through an elaborate emulsifying plant. The emulsion is then incorporated with a high-grade pigment, namely lithopone or titanium dioxide, together with a proportion of extender (such as barytes or other mineral white) and lime-resisting stainers, by grinding in edge-runner mills.

Properties of water paints

Water paints dry by the more or less rapid evaporation of the water and the slower oxidation of the drying oil and this fact should be borne in mind when considering washability, for whilst the emulsion dries hard enough to bear recoating in 24 hours or under, the oil content usually requires several weeks in which to age and harden thoroughly before it attains maximum washability.

The properties of a good water paint may be summarized as follows:

(i) It is smooth and easy to apply by brush, roller or spray.
(ii) It has good spreading and hiding powers.
(iii) It has good adhesion.
(iv) It has sufficient tolerance of damp to allow the surface to breathe, thus permitting its use on new plaster without risk of affecting the drying or carbonation of the plaster.
(v) It is reasonably resistant to alkaline attack, which again makes it suitable for use on new surfaces.
(vi) It gives a firm coating which can subsequently be redecorated with any type of finish without the necessity of stripping the water paint.

A good emulsion remains stable for a long period under favourable conditions but may break down if exposed to extremes of heat or cold.

Water paints should therefore be protected from frost during storage, and when in use should not be melted over a fire or mixed with hot water, as this may liberate the oil in the form of sticky dark liquid.

Thinning

Water paints are thinned to working consistency with water or with a special "petrifying" liquid supplied by the manufacturer. The latter is usually a thin emulsion similar to the binder in the water paint, and the obvious advantage of using such thinners is that they increase the total oil content of the water paint, thus improving its brushing and spreading properties as well as its ultimate washability. Their use in exterior water paints is essential. Before adding any thinner, the soft paste or gel should be thoroughly whipped or beaten up to a smooth creamy batter, the water or petrifying liquid then being added very gradually while stirring until the desired consistency is obtained. Overthinning with water should be avoided as it impairs the opacity and reduces the binding properties of the emulsion; if a very thin glaze is required (as in the production of decorative "Chamotex" effects), only petrifying liquid should be used for thinning.

Application of water paints

Water paints may be applied by brush, roller or spray; in the case of brush application the coating should be freely and liberally applied with a minimum of brushing, the strokes being laid off towards the light. There is not quite the same urgent need for keeping edges alive as with ordinary distemper; nevertheless, the best results are obtained when work proceeds methodically with as little delay as possible in joining up wet edges; when once a ceiling or a flank wall is begun it should be carried through until a convenient break is reached. In warm weather, windows and doors may be shut to retard evaporation whilst working, but should be thrown open immediately on completion to provide a free circulation of air which facilitates drying and hardening. Splashes should be wiped up as soon as possible, for they are extremely difficult to remove when once they are allowed to harden. The practice of wetting floors before commencing operations certainly keeps splashes soft until they can be wiped up, but the danger of using scaffolding on wet floors should not be overlooked. Brushes should not be left in the material during meal-times, but should be rinsed out in cold water immediately to remove the bulk of the material and prevent it hardening in the stock or heel of the brush. When washing out finally, use only warm water and soap. Hot water may cause stickiness in the brush (due to oil being liberated from the emulsion); if this should happen, rinse the brush in white spirit and wash out again in soap and water, rinsing finally in clean cold water and hanging up to dry.

New surfaces Water paints are specially valuable when the immediate treatment of new lime plaster and cement surfaces is called for, as they can be safely applied on relatively damp (not wet) surfaces without fear of imprisoning moisture; their permeable structure allows the surface to "breathe" so that the drying out of moisture from the wall is not impeded and the plaster is thus able to proceed with its normal carbonation and hardening. Moreover, although oil-bound water paints are liable to be affected by alkalis, the extent of saponification is restricted by the comparatively low oil content of the emulsion. Nevertheless, in such cases it is unwise to increase the oil content by the addition of petrifying liquids, and water paints applied on alkaline surfaces should therefore always be thinned with water only.

New surfaces which are quite dry and chemically neutral (such as fibrous plaster, or lime plaster and Portland cement which has been allowed to age for a minimum period of nine months) may be primed with oil primer if necessary to overcome any undue porosity. Very absorbent wall-boards should also be primed before applying water paint.

Old surfaces should be thoroughly prepared as detailed in "Preparing old surfaces" (Chapter 20), especially if previously treated with size distemper. After washing and scraping to remove as much of the old coating as possible and allowing all moisture to dry out, a coat of thin oil primer (as supplied by the maker of the water paint) should be applied. The object of this primer is to penetrate through any old loosely bound material which still remains either on the surface or in the pores of the plaster, and thus to bind the material thoroughly and provide a secure foundation for reception of the water paint. The primer should be applied *very liberally* in order to satisfy and even-up the porosity of the surface, but at the same time the brushing should be vigorous enough to prevent an excess of primer from lying *on* the surface, especially where absorption of the plaster may not be very pronounced. When quite hard, the primer is normally followed with two coats of water paint.

Water paint, being also applied liberally with a large brush, naturally tends to produce thicker coatings than oil paint; moreover, it is by nature more or less contractile according to the amount of glue in its composition. Repeated coatings over a number of years (often with different makes of water paint) tend to build up a film of considerable thickness which imposes a great strain on the adhesion of the first coat; a point is often reached when the absorption of moisture and the contraction on drying of an additional coat proves too much for the initial bond, which gives way with consequent peeling or flaking. A coat of thin sharp oil paint (i.e. oil paint liberally thinned with white spirit), or a coat of the manufacturer's special primer interposed between every four or five coats of water paint, will impart sufficient flexibility to the whole film to prevent flaking, provided the original bond is sound. When in doubt as to the number of coats of water paint already applied, it is safer to prime the surface with a coat of thin paint in this way, followed by one coat of water paint, rather than risk a possible complaint from a new client. It is essential, however, that this oil coat be thin enough to penetrate well into the surface, and that it should dry with a minimum of gloss; an oily coat which lies on the surface in an elastic film is very undesirable for, besides causing cissing when the water paint is being applied, it will also lead to cracking when the water paint hardens. Under no circumstances should water paint be applied over a coating of glue size.

Water paint stainers A range of pure stainers is supplied by some makers in addition to their standard colours. They are ground in the water paint emulsion medium and although intended primarily for tinting water paints and emulsion paints, they are ideal for decorative painting and murals, stencilling, lining, spray decorations, etc. They are also widely used for posters, exhibition and display work, including scene-painting. In some cases they are supplied in tubes. They are invaluable for use in technical colleges; every painting and decorating class should have a full range for teaching practical colour matching and colour theory. These stainers usually have great tinting strength and should therefore be used very sparingly.

EMULSION PAINTS

Emulsion paints have very largely replaced all the other forms of water-thinned material in present-day practice. The name emulsion paint is another example of a term that could be misleading and yet has passed into common usage. Strictly speaking, these materials are not emulsions in the sense described in the preceding pages; it would be more correct to define them as dispersions, since they consist essentially of resin particles together with pigment particles dispersed in water. The resin particles are virtually solid and when, after application, the water evaporates from the paint, these resin particles coalesce, that is they flow together to form a film; in so doing, they also bind the pigment particles. Of the many synthetic polymers available today there are only a few that lend themselves to use in emulsion paints. The most important of these are PVA (polyvinyl acetate), co-polymers of vinyl acetate with other monomers, and polyacrylic esters.

Properties of emulsion paints

There are many reasons why emulsion paints are so popular and so widely used today. From the contractor's point of view one of their most important features is the ease and speed of their application, which makes for reduced labour costs. They may be applied with equal facility by brush, roller or spray, and it is significant that the fastest type of spraying — airless spray — is particularly well adapted to the application of emulsion paints. The saving of time is enhanced by the fact that they do not need laboriously beating up into a paste before they can be thinned and used, but are ready for immediate application when stirred. Another attractive feature is the rapidity with which they dry; in a well ventilated atmosphere they dry within one hour and are ready to be recoated after a short interval, i.e. from two to four hours. This means that further coats can be applied without undue delay, and the premises brought back into use in a short space of time, which is an important point both in domestic work and in the decoration of hospital wards, restaurants, hotel work, etc. They are resistant to alkaline action, making them eminently suitable for use on the decoration of new property. Another point in connection with their use on new construction is the fact that they have a certain amount of tolerance for damp, i.e. they produce a permeable film which allows the passage of moisture. It is not advisable, however, to press this particular point too closely; there is a tendency among both contractors and property owners to use them on wall and ceiling surfaces that are still quite wet, which is unsound trade practice and is bound to lead to trouble through lack of adhesion. It must be stressed that surfaces to be decorated with them must be reasonably dry — the drier the better.

Because emulsion paints are thinned with water there is no objectionable residual paint smell whilst they are drying. They are usually pigmented with titanium dioxide which is extremely opaque, and for this reason they have markedly good covering power. At one time it was necessary in the manufacture of emulsion paints to add a plasticizer to PVA to improve its flexibility, and this had the drawback of becoming brittle as the coating aged, but the difficulty has now been overcome by co-polymerizing the vinyl acetate with another monomer, and the resultant products do not become embrittled and possess improved adhesion (although their properties of adhesion are still less than those of oil-bound water paints or oil paints). Present-day emulsion paints are equally suitable for indoor or outdoor use, and can safely be used in the steamy atmospheres of bathrooms and kitchens. They have the ability to withstand washing and scrubbing within a short time of application.

Emulsion paints are resistant to bleeding, in that they can be applied over roofing felt, bitumen, creosote, etc., without becoming discoloured. It should be noted, however, that they do not in themselves seal the surface, and if oil paint were to be subsequently applied bleeding would take place through the emulsion paint film, causing the oil paint to discolour.

There are some purposes for which they are definitely not suitable. They have no power to bind down loose or powdery material, and they are quite unsuitable for use on bare metals; they contain free acids which will attack non-ferrous metals, and their water content together with their ability to permit the passage of water promotes the rusting of iron and steel.

Preparation of surfaces

On new constructions, no particular preparation is required beyond ensuring that the surface is reasonably dry and that any dust, grit, or splashes of mortar or plaster are cleaned off; any defective work should of course be made good. In the case of new concrete it may be necessary to leach out or remove with detergents any deposits of oily release agents from the surface. The emulsion paint may need thinning if the surface is porous; thinning should be in accordance with the manufacturer's instructions. For use on building boards, hardboards, etc. it should be thinned to an easily workable consistency. A clear water-based additive is supplied by some manufacturers for use with emulsions when treating highly absorbent surfaces. On no account should glue-size be applied before emulsion paint is used.

On previously decorated surfaces the condition of the existing surface will be the guide to the treatment. Distempered surfaces should be thoroughly washed off and bound down with a penetrative sealer (nowadays often termed a "stabilizing solution"). Any loose or powdery surface of any description needs an appropriate primer–sealer to bind it down. On a surface previously coated with oil-bound water paint which is still in good condition, it is usually sufficient to apply the emulsion paint direct with the paint slightly thinned; if, however, there is any doubt about the stability of the coating, or the previous history of the surface is not known, it is safer to apply a penetrative sealer coat first; the water paint should not be washed down, as this would tend to loosen it and weaken its adhesion. In the case of water paint in bad condition, the film should be scraped back to where it is sound, and bound down with a penetrative primer; in very bad cases it is desirable to remove the whole of the previous coating, possibly with a steam stripper. In all cases, cracks and holes should be made good. If emulsion paint is to be applied over old wallpaper it will need to be thinned, but such a treatment is technically unsound because of the possibility of loosening the paper. For use on painted surfaces, the manufacturers very often recommend applying the emulsion paint directly upon the existing coating, but it is most desirable that the painted surface should be well cut down with abrasive first. Emulsion paints are suitable for priming any of the varieties of building boards including fibre insulation boards, provided they have not previously been treated with water-soluble fire-retardant chemical preparations. As mentioned above emulsion paint should never be applied over coatings of glue-size.

Application of emulsion paints

When emulsion paint is to be applied unthinned by brush, the brush should be well loaded and the paint applied liberally with a full, even coating; the paint should not be brushed out bare. If application is to be by roller, the paint should be well stirred to a smooth, even consistency before it is poured into the paint tray. All tools and appliances should be washed out directly the work is completed.

Other grades of emulsion paint

The cheaper grades of emulsion paint, sometimes referred to rather loosely as vinyl distempers, are produced for interior work; they are not suitable for exterior use and if so used would be subject to very early breakdown; nor are they fitted for use in conditions of severe condensation or high humidity, which rules them out as far as bathrooms and kitchens are concerned.

Acrylic emulsion paints, which tend to be more expensive than PVA emulsions, dry more rapidly than PVA and can be recoated sooner.

Emulsion paints are supplied in a wide colour range, but they can also be tinted as required with the water-based stainers which are obtainable for the purpose.

Emulsion varnish or emulsion glaze

The degree of sheen of an emulsion paint depends on the proportion of polymer resin binder to pigment, and this varies with different brands from matt to a semi-gloss or satin sheen. Most manufacturers supply the polymer medium separately, and this can be added to emulsion paint to increase the amount of sheen and incidentally to improve the washability still further. The clear emulsion glaze can also be used by itself, either as a clear coating over emulsion paint or water paint, or over wallpaper to produce a washable surface (although with the increasing use of washable wallpapers this is no longer a very significant part of trade practice). The degree of gloss or sheen can be modified by thinning with water, but even when the glaze is used full strength the finish is never a high gloss.

Protection from cold

It is most important that any type of emulsion paint formulation should be protected from frost, and that the paint should not be applied on cold surfaces or at temperatures of below 4°C.

Fire-retardant emulsions

Some manufacturers produce fire-retardant paints based on a PVA co-polymer and containing fire-resisting pigments such as titanium dioxide. They are supplied for the painting of combustible materials such as paper-faced and unfaced insulation boards, in order to improve their resistance to flame spread. It has been found that two coats applied in accordance with the maker's directions can lift a material of the lowest standard up to the level of a Class I material of very low flame spread. Application is by brush, roller or spray (conventional or airless). Normally no pre-treatment is required, and the first coat is applied unthinned while the second coat may be slightly thinned if so desired. In the case of insulation boards which have been supplied and fixed ready finished with a white facing, it is sometimes found that the white facing tends to be loose and powdery, and that before the emulsion paint is used it is necessary to bind down the surface with a primer–sealer.

It is pertinent at this point to mention that all building materials are assessed according to the ease (or otherwise) with which a flame will travel across the surface, as measured by a stringent BSI test. There are four classifications, ranging from materials of rapid flame spread (Class IV) to materials of very low flame spread (Class I). The application of a surface coating will not render a flammable material flameproof, but it *can* make such a material less easy to ignite, so as to delay and minimize the spread of flame or to prevent the material from continuing to burn when the source of heat is removed.

OTHER WATER PAINTS

Water-thinned enamels

These are a special class of emulsion paints which, although thinnable with water, dry with a high gloss having all the properties of wear and washability of orthodox interior enamel paints. There is one important difference, however, for whereas ordinary enamel paints have considerable pull under the brush and are slow in application, these water-thinned enamels work remarkably easy and can be applied as quickly and easily as distemper.

Since these paints are produced by only one or two firms their formulation is closely guarded, but one patent refers to a polymerized drying-oil ester gum varnish pigmented with titanium dioxide plus driers and emulsified in water with a special emulsifying agent containing ammonia.

These paints brush so easily that there is a danger of brushing them out too much and so producing a "starved" film. A liberal coat should be applied with a minimum of brushing, crossed once, and laid off. Overworking must be avoided and once a section is laid it should not be touched again or it is likely to "roll up" under the brush. Drying is by evaporation of the water and oxidation of the drying oils, and damp or humid weather tends to retard drying somewhat.

Cement water paints

These are made by adding certain materials to Portland cement to accelerate the setting and impart suitable working properties. They are supplied in powder form and are made ready for use by mixing with a stipulated amount of water, becoming more waterproof and washable as they age and harden. They are suitable for virgin concrete, stone, plaster and brick, but may not be applied over existing oil paints, water paint or distemper and, being alkaline, may not be followed with oil paint unless first sealed with an alkali-resisting primer. In fact, if cement paints have been used it is usually risky to apply any other material on top of them, as there is a strong tendency for them to become powdery, causing the newly applied material to flake off. Generally, once a cement paint has been used the choice lies between using the same material for future coatings and undertaking the expensive operation of removing it before redecoration.

25

Texture Paints

The interest in textured surface treatments has been self-evident for many years among the higher.priced wallpapers and there is no doubt that both plain and broken-colour effects are enhanced by judicious low relief modelling of some sort. Relief textures in paint, however, always prompt inquiries as to their dirt-catching properties, and many hygienically minded people will not tolerate them on that account. It must be admitted that when one remembers some of the unfortunate efforts perpetrated by enthusiastic but inexpert members of the craft, there is some justification for this antipathy to surface textures.

The above reference to textured papers perhaps points a moral, for whereas many people will not consider plastic paints for one moment, they often accept surface textured wallpapers without question. The reason for this is probably in the quality of the texture itself, for besides being low-relief, it is usually soft and rounded, having no hard sharp edges or points on which dusters or clothing are liable to catch. If only our plastic enthusiasts would emulate wallpaper designers in this respect we venture to think there would be less objection to texture treatments in paint.

Another important point to bear in mind, of course, is the question of *scale*, for obviously what may be suitable for a cinema auditorium would be totally out of place in a domestic lounge. Yet how often have we seen most inappropriate textures in small rooms, where the material has been literally "laid on with a trowel". When we consider plastic paint effects, therefore, we must always think in terms of the position and purpose for which the effect is intended.

The facility with which plastic paints may be applied perhaps accounts for the weird and wonderful effects sometimes produced, and the ingenuity of some operators in utilizing all sorts of home-made gadgets to secure their effects knows no bounds. Whilst we would be the last to discourage such inventiveness we would plead for a modicum of restraint in applying it. A decorator has a responsibility to his craft besides that to his client, and when he plasters meaningless squiggles and squirls over a client's walls in a permanent material like plastic paint he is letting his side down badly. An insistent pattern or motif, oft repeated, is bound to tire after a time, no matter how attractive it may have seemed at first, but whereas the offending pattern may easily be removed in the case of a wallpaper, it is more or less permanent when carried out in plastic paint. That is

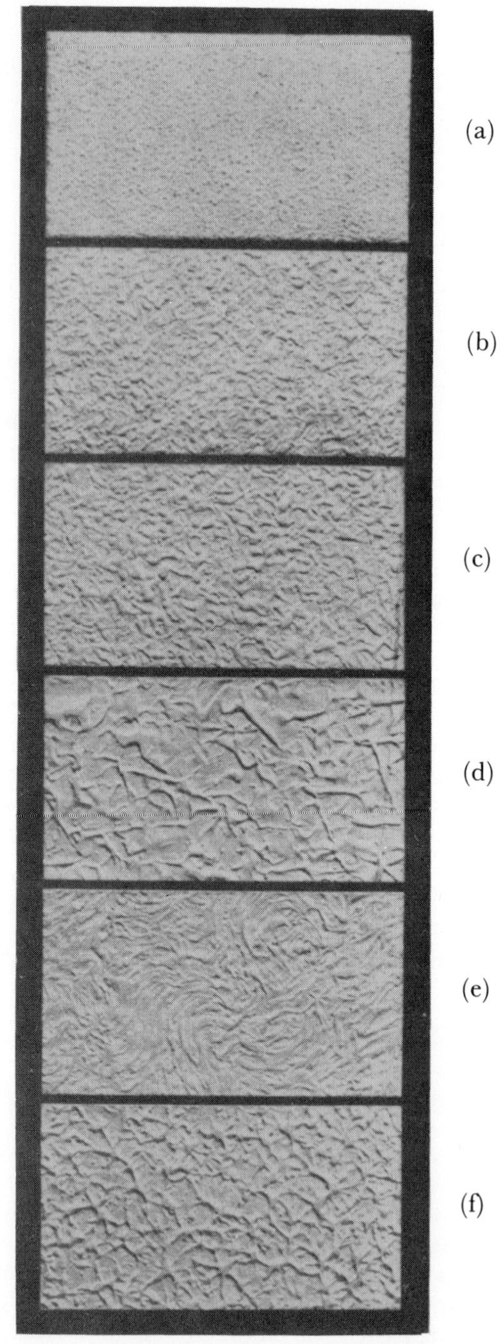

Courtesy: Crown Decorative Products Ltd

Fig. 25.1 Various textures in plastic paint

(a) to (d) produced with fine, medium, coarse, and extra coarse rubber stipplers respectively, (e) by brush flogging, and (f) by suction lift with a handboard

another good reason why people hesitate to introduce plastic decoration in the home.

Where relief decoration is to be carried out then, whether it be in the home, café, hotel or cinema, let the texture form a pleasing, unobtrusive all-over background treatment, and let the decorative pattern or motif be superimposed by painting, stencilling or spraying, rather than permanently worked in the relief. The texture may vary from an "orange-peel" stipple or a low flogged-brush texture appropriate in the home, to a heavy roughcast or pebble-dash effect in the cinema. Such surfaces can be decorated again and again in different colours, new interest being given if necessary with newly applied motifs. The objection to permanent decoration which "dates" is then largely overcome.

Types of texture paints

Some decorators pride themselves on making up their own compositions, usually by adding glue-size, starch, or other retarders to plaster of Paris, but these mixtures rarely have the smooth working properties of the majority of proprietary materials and there is always a risk of cracking where size is used in thick coatings. The compounding of a texture paint is not a simple matter when one considers the requirements — good adhesion, hard drying overnight without brittleness or undue shrinkage, easy working minus flow. The last point demands ingredients with a certain amount of "slip", yet with sufficient "stay put" properties to retain the crispness and character of whatever texture may be applied.

The recognized types of texture paints used by the decorator comprise *plastic paints* in paste or powder form, *imitation stone finishes*, and *anti-condensation paints*. Crackle and crystalline finishes (depending on controlled cracking and wrinkling) are widely used in industrial finishing, but they are usually stove-dried and therefore outside the scope of the ordinary decorator.

PLASTIC PAINTS

Hand applied

When applied in paste form the consistency is very similar to that of a good water paint — a firm gel which should beat up easily into a stiff smooth batter, suitable for medium textures in light stipples and other low-relief effects. The paste may be thinned slightly with water, if necessary, for low textures, but for heavy textures it may require stiffening with fine plaster or a special powder which some makers supply for the purpose. This course is usually necessary when light relief modelling is required, the composition being applied through a stencil and sometimes finished by hand-carving after the material has become set but not hardened (see Fig 25.2). A little glycerine will retard the hardening and help to keep the composition "cheesy" for a day or two in cases where intricate modelling is involved. Certain oil-bound water paints tempered with fine dental plaster (in the proportion of equal parts by weight) may also be used as texturing compositions.

The varieties supplied in powder form differ widely in composition and

Fig. 25.2 Carved relief in plastic paint by Reg Ball

The plastic composition was applied by trowel through a stencil, allowed to set, then glasspapered smooth, carved and modelled

performance, some drying hard in a few hours, while others may take as long as two days if the rate of evaporation is retarded by damp or humid weather. The proportion of water to powder, as stipulated by the makers, should not be exceeded, and mixing is done by adding the powder to the water very gradually while stirring briskly until a smooth paste is formed. In some cases the composition works better after standing overnight, but in others (such as those of a cement nature), the material must be used within an hour of mixing owing to the quick set. This type does not give much time for texture manipulation on very large surfaces.

Any texture treatment depends on the play of light and shade in giving effect to the relief. High relief casts longer shadows than low relief and therefore gives an effect of deeper tone. For this reason it is important that plastic paints be applied as *evenly* as possible, that is, in a uniform thickness over the whole surface. This is not easy to achieve, especially when two or more men are applying the material on the same wall, but nevertheless, it *can* be done (when men are used to working together as a team) and should always be aimed at. When the material is unevenly applied, the relief will be higher in some parts than in others and consequently will appear darker because of deeper cast shadows, the result being that the wall looks patchy and shady instead of presenting a uniform tone throughout.

Application is generally by brush, a half-worn 100 or 125 mm flat wall brush being the most convenient; very rarely will it be necessary to apply the composition by trowel. The material should be applied liberally, laid off fairly evenly, then textured as desired. Fig. 25.3–25.6 show some effects obtained by different tools, and these may be supplemented in numerous

Fig. 25.3 Swirl effect produced by twisting the brush

Fig. 25.4 Texture produced with a rubber stippler

Fig. 25.5 Relief pattern produced by twisting sponge

Fig. 25.6 Producing texture with the handboard (suction lift)

Courtesy: Crown Decorative Products Ltd

ways by an inventive operator. A short flat wall brush will be required, together with steel and wood combs, an open textured sponge, fine and coarse rubber stipplers, and a small handboard with rounded edges for suction lift effects. A rubber squeegee is also useful for removing glaze colour from the relief texture. When the composition is sufficiently set (say in about an hour, or less) it may be dragged with a plastic set square, or lightly rolled with a wide wooden roller (as used by paperhangers) to smooth down the sharp points of the relief texture. Any tendency for the composition to pluck off the surface by sticking to the roller may be overcome by smearing the roller with white spirit. Some decorators allow the composition to harden before rubbing down the sharp points with glasspaper, but this method is very slow and needlessly dirty owing to the amount of plaster dust floating about.

Absorbent surfaces which are to be textured with plastic paints should be primed with thin oily paint, otherwise the plastic composition will tend to work "hot" and set too quickly, giving no time for proper manipulation. When the plastic composition is thoroughly dry and hard it will require to be bodied up with at least two coats of oil paint, the first coat being a thin, oily priming coat and the second a hard semi-gloss finish, which may then be glazed with one of the proprietary glaze media, tinted as desired. It is better to thin the glaze with white spirit and to apply it fairly liberally, rather than to attempt to brush out the thick glaze medium as supplied. The tinted glaze should be stippled with a hair stippler (Fig. 28.1) to secure an even distribution of colour before wiping off the highlights with soft, non-linty rag or (if sharper contrast is desired) with a rubber squeegee.

A very economical decorative effect can be obtained by tinting the plastic composition with water-paint stainers, applying and texturing in the usual way. When quite hard, a thin coat of water paint of different colour (thinned with petrifying liquid) is applied and allowed to set slightly, then removed from the highlights with a damp washleather. This inexpensive two-tone effect can be quite attractive when the colours are carefully chosen and has the added advantage that, since it does not seal the surface like an oil finish, it may safely be applied to new alkaline plaster, provided the latter is reasonably dry.

New property and contract work

On new building construction, plastic paint is extensively used as a means of providing a decorative finish whilst at the same time economizing on labour costs. Price has always been a major consideration in any kind of building work, and it is recognized that in new housing, whether it be in the smart villas on the speculative builder's "executive" plot or the rows of identical little boxes on a local authority's estate, every possible measure will be taken to reduce the cost. One of the obvious ways of doing this would be to decorate directly on to the surface of plasterboards and wall sheetings without going to the expense of first skimming them with plaster. This of course is not a new idea; it occurred as soon as plasterboard was invented. The only drawback is that people tend to be conservative in their outlook; if they are buying a new house they expect to see the ceilings and walls plastered and they would not accept them otherwise. Even though they knew that money was being saved by not plastering, they would feel that something essential was missing. Eventually, however, the possibilities of plastic paint were exploited, and because they were presented in a positive manner they succeeded. Had the approach been on the lines of "we are cutting out the plastering to save money and are covering the deficiency with a layer of paint" there would doubtless have been a tremendous barrier of resistance. But very shrewdly the suggestion was put forward that here was an attractive decorative material which would enhance the appearance of the room, and the advantages of the treatment were stressed instead of the fact that it was replacing plastering. As a result the idea was adopted enthusiastically, and it has now come to be regarded as the norm, especially since the advent of the Industrial Training Boards, when the C.I.T.B. made this one of the basic techniques for new property.

There are various firms providing plastic paint specifically for this purpose, and one in particular has become a household word in this field. The plastic paint itself is tinted, so that a fairly large colour range is available, and no further painting or glazing is required, the play of light and shade on the textured surfaces being sufficient to give variety and interest to the finish. The firm concerned produces a great deal of attractively presented literature, showing examples of the sort of textures that can be achieved and giving precise instructions for dealing with the various kinds of substrate — whether they be plasterboard, concrete blocks, or any other structural material. It can be used of course for both walls and ceilings, although in practice it is more frequently confined to ceiling work. The most popular textures seem to be those produced by combing in its various forms.

Sprayed plastic paints

The development of plant suitable for the spraying of plastic paints effected considerable speeding up in the application of surface textures, and this has proved most advantageous in the treatment of large areas of wall surface; so much so, that modern cinemas and leisure centres are usually decorated in this way. Nor is speed the only advantage, for it is possible to exercise greater control in application, and uniform results may be obtained on large surfaces even when two or three operators are simultaneously employed.

Spraying plastic paints are obtainable in paste or powder form, and in some cases the same material may be used for both hand and spray application; the composition is similar to hand-applied plastics and they may be thinned with water to the required consistency. The height and character of the relief is determined by the consistency of the material, the size of the spray-gun nozzle, and the air pressure used. When ordinary paints are sprayed with an air pressure of between 2 and 3 bars (30 and 45 psi) the paint is atomized in the form of a fine mist spray, but when the air pressure is reduced to between $\frac{1}{3}$ and $\frac{2}{3}$ bar (5 and 10 psi) the paint leaves the gun in the form of an open spatter. This principle generally applies whatever the material being used, but in the case of plastic paint

Fig. 25.7 Sprayed plastic paint finished with spray spatter

the consistency is such that air pressure is required to force it to the gun nozzle before it can be sprayed. The material is therefore placed in a pressure-feed container from which it is forced by air pressure (between $\frac{2}{3}$ and $1\frac{1}{3}$ bars (10 and 20 psi)), depending on length of hose and consistency of material) through a pipeline to the spray gun. Thus two pipelines are required, one carrying the material to the gun (the fluid line) the other carrying the air which sprays the material through the gun nozzle (the air line). The pressure container usually holds from 10 to 25 litres, and some types have agitating gear to prevent settling of the material.

Sprayed plastics may be applied in a close "pebble" texture over the entire surface and manipulated, if desired, by combing, stippling, brush flogging, etc. The spreading capacity generally varies between 0·5 kg and 1 kg per sq. metre according to the type of texture. The usual (and the most economical) way, however, is to apply it in the form of an open spatter which leaves part of the ground colour exposed and forms an attractive two-tone effect. The colour of the plastic paint should preferably be *lighter* than the ground, for then the maximum effect of relief is obtained; if the plastic is darker than the ground colour the effect of relief tends to be lost and one might just as well spatter the surface with ordinary paint or distemper. Tinted plastics applied over a ground of different colour (in water paint or flat oil paint), may be further enhanced by finally spattering with gold bronze or aluminium, and various "shot" effects can be produced by mist-spraying the texture from opposite sides in different colours or metals.

It is important that the spray equipment should be cleaned *immediately* after use, for if the material sets hard in the pipeline a new length of hose will be required. The pressure containers should be emptied of all plastic paint, then filled up with warm (not hot) water, which is sprayed through the pipeline until it is quite clean. The spray gun, of course, should always be thoroughly cleansed whatever material is used. It should be mentioned that special nozzles are required for spraying plastic paints, but they are generally available from all the leading makers of spray equipment (see Chapter 6).

Texture paints in industrial premises

Before using texture paints in industrial premises it is wise to consult H.M. District Inspector of Factories to discover if there is any objection to the material. The writer was recently involved in a case where the corridors and staircases in a large office block occupied by a firm of international standing were treated with a proprietary brand of texture paint. The Inspectorate served an Improvement Notice on the firm under the terms of the Health and Safety at Work Act; it was alleged (a) that the textured finish might be a hazard with the possibility of causing injuries to employees' hands, and (b) that the texture would prevent the surface from being effectively washed down. It is laid down in the Factories (Cleanliness of Walls and Ceilings) Order, 1960 (SI 1794) that the painted wall and ceiling surfaces of industrial premises shall be washed or otherwise cleaned at stipulated intervals, and in this case the local inspector considered that the terms of the order could not be met by this material (although on appeal the improvement notice was rescinded).

Spray equipment in constructional work

The techniques of spray painting can be readily adapted to certain aspects of constructional work, and equipment is available which caters for various operations carried out in the constructional crafts. There are, for instance, heavy-duty plaster-spraying units which are used to apply pre-mixed plaster finishes both for external renderings and for internal plastering; the final setting coat is sprayed on to the wall or ceiling surface, and a small team of plasterers follows it up using smoothing floats to plane it to a smooth finish. The same technique can be used for the preparatory coats, the roughing mix being sprayed on to the surface and then worked over by the plasterer using a scratch float to produce the necessary texture for adhesion of the final rendering. For this type of work the plant is usually fitted with a rolling press to expel the plaster from the plastic or paper sacks directly into the hopper.

An extension of this process is the use of plaster-spraying units to spray cement screeds and various kinds of thin finishes in flooring operations, thus speeding up the pace of construction.

Heavy-duty spray equipment is also used to spray thermal-insulating materials or lightweight plaster finishes to reduce the heat loss in buildings. In this case a thick layer can be sprayed on to a level aerated concrete base without the necessity of following up with a smoothing float.

Many kinds of textured finish can be sprayed with this equipment; Fig. 25.8 shows the Wagner SP 40 machine in use on external rendering work. The usual procedure is to spray a sand mix or a levelling coat on to the surface, following this up with a steel float to provide a key for the finishing material. The finish can be any kind of textured rendering, including the popular "Tyrolean Finish", or a pebbledash finish with either natural or synthetic aggregate, all of which can be applied rapidly by this method without the risk of producing those ridges which so often mar the appearance of manually applied work. Fig. 25.9 illustrates some of the textured finishes that can be applied by the Wagner range of heavy-duty spray plant.

Courtesy: Gray-Campling Ltd

Fig. 25.8 Spraying a textured rendering on external work
The roughing coat has also been sprayed and levelled with a scratch float to provide key

IMITATION STONE FINISHES

There are various proprietary texture paints designed to imitate the texture of newly dressed stone. They usually contain a granular aggregate such as stone dust or silver-sand, together with lithopone and stainers suspended in a flat-oil medium or an emulsion vehicle. They dry with a pleasing matt stonelike texture for interior work, but for exterior use the finish generally has a slight sheen owing to the higher oil content necessary for withstanding outdoor conditions.

These stone paints are generally supplied almost ready for use, but the consistency is so stout that the first impulse is to over-thin them to ease the brushing. To do so, however, is fatal, for the stone or sand particles seem to need the support of a thick medium to ensure even distribution

Fig. 25.9 Examples of textured finishes and wall renderings produced by heavy-duty plaster-spraying units (*slightly enlarged*)

Courtesy: Gray-Campling Ltd

over the surface; when thinned beyond the maker's recommendations the aggregate tends to gather in patches whilst brushing bare in other places, and distribution becomes very uneven. The material should be applied liberally (as with plastic paints) in as even a coating as possible and should be stippled with a fairly stiff bristle stippler. Two coats are necessary for best results, each coat being stippled, and where the material is applied over porous surfaces a coat of oil primer should be given first.

The spreading capacity of these paints is naturally very low owing to the thick consistency of the material and the necessity of applying a liberal coating. On normal smooth non-porous surfaces, the spreading rate may be approximately 250–270 sq. metres for every 50 kg for the first coat and 150–165 for the second coat. The rough surface produced by the first coat is of course responsible for the reduced spreading rate of the second, and this point should be borne in mind when estimating for this type of finish.

These stone paints may be applied successfully by spray, using the same type of equipment as for sprayed plastic paints. The material should be stirred or agitated frequently to avoid settling of the aggregate.

Ashlar treatment The surface may be bonded to form an "ashlar" effect by scoring out mortar lines with a narrow blunt chisel or a small screw-

Courtesy: Crown Decorative Products Ltd

Fig. 25.10 "Ashlar" stone effect in stippled texture paint

driver, before the material sets hard (Fig. 25.10). A few panel pins, or nails with heads filed off, driven into the underside of a wooden straightedge will hold the straightedge about 5 mm off the surface and so prevent damage to the texture while the mortar lines are being scored. When the material has dried hard the mortar joints may be run in with flat white or other appropriate colour, using a narrow lining fitch and straightedge.

Sand finish A fine representation of natural stone can be produced by throwing sand on to a tacky surface and scoring out the mortar joints as described above. Porous surfaces should first be primed, but painted surfaces may be treated straightaway with a coat of heavy-bodied oil paint made from white lead, stand oil and goldsize (or white lead and oil-goldsize); a liberal coat is given and allowed to develop a strong tack, then dry river-sand or silver sand is thrown by hand on to the tacky surface. The ashlar joints are scored out before the vehicle hardens and are afterwards run in with white or colour as required. A pleasing variation in the colour of individual "stones" is obtained by using different coloured sands, when a most convincing effect of natural stone is produced, very appropriate to the interior decoration of churches and other public buildings. This treatment may be applied to exteriors, but in this case it is advisable to finish the sand texture with two or three coats of oil paint which produces a good weather-resisting finish.

Modern masonry paints

The materials so far described in this chapter have been almost entirely for interior work, but in modern trade practice there is a big and growing demand for masonry paints intended for exterior use. Until fairly recently the only materials available for the treatment of external rendering such as pebbledash, cement rendering and so on, were the conventional oil paints and water-thinned paints, cement paints, or limewash. But over the past few years one of the biggest developments in paint production has been in the field of masonry paints, and every firm of paint manufacturers now offers its own distinctive line — or in some cases a whole range of lines — each with its own individual properties. Such a material will obviously have a number of useful properties; what the manufacturers have to decide is the relative importance of these various properties. Some have concentrated chiefly on a very high level of durability, others on toughness, or adhesion, or resistance to water penetration, or the ability to mask cracks and defects in the structural surface.

The properties of any given material depend partly on the liquid medium and partly on the solid matter suspended in it, and from these two factors a great many permutations are derived. The liquid medium may be a drying oil or a synthetic resin or a combination of these, similar to the medium of an oil paint; or it may be a synthetic resin of the kind used in water-thinned emulsion paints. Some of the finest quality materials with extreme qualities of toughness and durability are of the first type, but it is clear that these are only suitable for certain types of surface, and that surfaces of a highly alkaline nature or surfaces which still contain moisture may be better coated with the water-thinned emulsion type. One of the resins produced by a well-known firm of tyre manufacturers working in conjunction with paint manufacturers is said to be unique in that the resin particles are considerably smaller than those of any similar material, a factor which imparts very great powers of adhesion and penetration to the paint. The solid particles of masonry paints are also varied and include sand and granulated stone, both of which are claimed to give great cohesion to the paint; fine aggregates of extremely hard material such as granite, which is claimed to give great durability; combinations of sand and mica, mineral quartz and mica, granite and mica, etc., in each of which the mica is added to give improved properties of adhesion; and nylon fibre which, being much smoother than stone particles, is claimed to give a paint which is far easier to apply and which remains clean much longer than other paints because there is no roughness to trap the dirt. For some masonry paints it is claimed that there is a gradual and controlled wearing away of the components by means of which the surfaces remain perpetually clean in appearance.

Sealing coat A difficulty with any treatment for external renderings is the tendency of some renderings to develop a mass of fine cracks. Many masonry paints are thick enough in their composition, and applied in such thick coatings, that they go a long way towards filling up the cracks. But when the cracking is severe even the most robust paint cannot be expected to mask the defects, let alone provide a permanent cure. To meet such conditions some of the manufacturers of masonry paint produce a preparatory sealer composition along with which a thin membrane of glass fibre can be used. The sealer is brushed on to the surface and the membrane is forced in while it is wet; and then, when the material has dried out, a further coat of sealer is applied. In cases where the cracking is widespread the glass fibre can be applied over the entire wall surface.

Durability Since differing brands of masonry paint vary considerably in composition they are also quite dissimilar in price. In these circumstances a comparison of prices only becomes meaningful when it takes into consideration other factors such as the covering capacity of the paint, the ease or otherwise of its application, and most importantly, the expected life of the coating. In this respect, some brands of masonry paint claim a very high measure of durability; it is asserted that certain brands will last ten years, twelve years, even fifteen years and upwards in some cases. There is one point, however, which is often overlooked. In the case of a house the outside of which is being completely redecorated, given a brand of masonry paint which is confidently predicted to have a minimum life of ten years, it can be argued that the high initial cost of the material is outweighed by the fact that it is averaged out over a ten-year period. But if at the end of five years the remainder of the paintwork, on the doors, windows and gutterings, etc., needs renewing, it is most unlikely that the masonry paint will not be renewed at the same time; even if the coating were still in excellent condition and miraculously unsoiled, the physical act of repainting the remainder will cause it to be so much marked and soiled that renewal will be necessary. So the initial cost is averaged across five years, not ten. And therefore the great durability of the paint is very rarely a matter of major importance on the actual site, since in most cases the coating is renewed before its life expectancy is finished.

Masonry paints applied by manufacturers' agents

There are certain firms which specialize in the production of masonry paints and which, instead of supplying the products in the ordinary way to be bought and applied by the general public, supply them solely as a full application service through approved distributors; application is by high-powered spray and is carried out by operators trained in this particular process. The materials are polyester-resin based, with filling agents which include glass fibre, asbestos, perlite, mica, zinc oxide, titanium, and fungicidal matter; their composition is such that they could not be applied by manual methods. They form a very thick single-coat film with a heavily textured finish, the film being extremely tough, elastic and tenacious.

ANTI-CONDENSATION PAINTS

Cork and other textures

Granulated cork is sometimes applied in the manner described above as an anti-condensation treatment for interior metal and plaster surfaces. This "corking" is carried out extensively in ships to minimize the effect of condensation from steel bulkheads and partitions, iron cold-water pipes, etc. A red lead primer is first applied and followed when dry with the adhesive coat, which may be either a slow-drying enamel or the white lead–stand oil–goldsize mixture mentioned above. The granulated cork is thrown on to the tacky surface by hand, or blown on with a special spray gun, and the adhesive is then allowed to harden before the corked surface is finished with two or three coats of oil paint, the final coat often being a gloss enamel.

Cork, or other non-conductive material, also forms the basis of proprietary anti-condensation paints, the application of which should be carried out in a manner similar to imitation stone paints described above; that is, they should be applied *very liberally* in the stout consistency in which they are supplied, avoiding any attempt to brush the material out thinly. The efficacy of these anti-condensation preparations largely depends on the provision of a thick non-conductive coating which acts as a buffer between the warm moisture-laden air and the cold surface of metal or plaster; two coats are therefore usually recommended and each coat should be stippled to ensure uniformity of texture. As with stone paints, the spreading capacity is low and no more than 5 to 6 sq. metres per litre may be expected from the first coat if correctly applied, while the spreading rate of the second coat drops to about 4 sq. metres per litre. Although the purpose of these paints is purely functional, the granular texture obtained is rather attractive and forms an ideal ground for broad decorative painting in flat oil paint, which may afterwards be worked over with wax crayons.

Paint Harling is another type of functional texture treatment which has been used fairly extensively in Scotland for the exterior protection and decoration of steel-clad houses. The process is patented and consists of throwing painted granite chippings on to a tacky surface as in the case of sand and cork treatments. The steel surface is primed with red lead and allowed to dry hard; a thick coat of tinted white lead–stand oil–goldsize paint is then applied and allowed to stand for two or three hours, when it becomes sticky enough to hold the chippings which are thrown on by hand (as in pebbledashing). The granite chippings are previously coated with the same composition in a concrete mixer. Great durability is claimed for this process and steel surfaces so treated were found to be in excellent condition after nine years' exposure to severe weather conditions.

Decorative uses

The application of dry materials to tacky surfaces has many decorative possibilities. For instance, an unusual decorative treatment for applied motifs is obtained by laying-in portions of a design with oil–goldsize and dusting with pre-coloured sawdust when the right stage of tackiness is reached (the sawdust may be pre-coloured by tinting with thin water-paint stainers and allowing to dry before use). Other parts of the design may be left smooth and treated with gold or aluminium leaf as a contrast to the rough coloured texture.

A suede or felt-like surface is obtained by dusting wool flock over a tacky ground. These flock surface treatments are often useful in finishing small-scale interior models where it is desired to simulate upholstery and soft furnishings.

26

Wood Staining

The structure of wood is such that the grain forms a pattern of varying porosity, the hard portions being much less absorbent than the softer parts of the grain. This is very much in evidence in quick-growing trees such as Columbian pine (Douglas Fir) in which the grain is a broad open pattern of hard and soft wood, caused by the seasonal rise and fall of the sap. The soft wood represents the vigorous spring growth, while the hard, dense portion is the more slowly formed autumn wood. When a coat of stain is applied, the soft wood absorbs more stain than the hard wood and therefore appears darker. This is very pronounced in the case of water stains which strike deeply into the soft porous spring wood while being resisted by the hard grain, which barely holds any colour at all; hence the vigorous contrast of light and dark grain markings. The timber of slower growing trees such as oak, elm, beech, etc., is more close-grained and the difference between the spring and autumn wood is much less pronounced, which results in the stain being absorbed more slowly and evenly.

Timber which is not properly seasoned may take stain very unevenly, especially if there are any bluish streaks of sap-wood present. Such parts are likely to absorb too much stain and consequently appear very dark, particularly when a penetrative water or spirit stain is used. The general appearance of stained work will therefore depend not only on the variety of wood but also on its condition, and the type of stain used.

When wood is to be stained it should be left clean from the joiner's plane, unsoiled and free from greasy finger marks. If it requires glasspapering, it should be rubbed in the direction of the grain, using a fine-grade paper. When rubbing lock-rails, etc., care should be taken to avoid shooting across the grain of the stiles as this may result in dark scratches when stained. If the wood has been rubbed *across* the grain by the joiner (as is often done when preparing wood for painting) the surface should be remade with a cabinet maker's scraper.

Application of stains

The stain — whether it be oil, water or spirit — is best applied direct without previous preparation so that it penetrates the surface and allows the greater absorption of the pores to bring out the full character of the natural grain. In the case of well-prepared hardwoods there is seldom any reason to depart from this practice, but the more common softwoods often present certain difficulties such as unequal absorption, knots, sappy wood, etc. A coat of thin size is often recommended to even-up porosity, but we prefer a thin coat of lacquer made from shellac knotting well thinned with methylated spirit. This may be applied to the whole surface in the case of excessively knotty yellow pine and common red deal, or it may be applied locally only to absorbent portions of sapwood. Alternatively, undue penetration may be minimized by using oil stain mixed with a proportion of one of the proprietary transparent glaze media, the amount added depending on the absorption of the wood being treated. The strongly contrasting grain of Columbian pine, for instance, can be satisfactorily toned down in this way, the glaze holding the stain largely on the surface and so reducing penetration or "striking" into the grain. For this reason, however, glaze medium should not be introduced in floor stains since good penetration is essential for withstanding the hard wear and tear of foot traffic. Incidentally, a good quality black Japan, well thinned with white spirit or solvent naphtha, makes an excellent rich brown stain for untreated floors, penetrating well and drying hard; it can be finished either with wax polish or varnish.

On softwoods stain should be very thin and applied liberally, using oil stain for preference as this satisfies the porosity and, being less penetrative than water or spirit stain, can be applied more evenly. On hardwoods such as oak, elm, ash, etc., the stain can be brushed on fairly liberally and left for a few minutes to soak into the pores, then wiped off with a soft rag across the grain; this leaves the stain in the pores and brings out the full beauty of the wood grain.

Owing to their penetrative nature, water and spirit stains require to be laid quickly and methodically, avoiding overlapping or double-coating which would result in a very patchy appearance. End grain should be knotted before staining, otherwise these parts will appear much darker than the rest of the work.

Nail holes, etc., may be stopped with plastic wood, fine plaster, or proprietary filler, before staining; any surplus should be removed (when hard) by rubbing with fine glasspaper and the stopping should be neatly touched with thin knotting to prevent it from absorbing the stain and appearing darker than the surrounding colour. When oil stain is used, stopping is best done after staining, using tinted putty or wood filler.

Filling

Various proprietary wood fillers are available both tinted and transparent. These are usually in oil medium and may be applied with the knife or thinned slightly with white spirit and applied by brush. In the latter case, a liberal coat is applied and allowed to set for ten minutes or so, then rubbed off *across* the grain with a wad of scrim or jute. Proprietary brands of water-thinned tinted stopper are also available.

Hardwoods with open pores may be filled with fine plaster or other filler, mixed with water to a thick paste, and carefully applied with the filling knife; when quite hard it should be well rubbed down with fine glasspaper (in the direction of the grain) until all surplus filler is removed from the surface and remains only in the pores. When stain is applied it strikes into the plaster and darkens the pores, but if a stronger contrast is required the plaster filler itself may be tinted with a little dry colour. This plaster filler is useful for open-grained mahogany or baywood when a high-gloss finish is desired.

Limed oak effects

Plaster or other filling may also be used in producing limed oak effects. The surface is first stained with spirit stain and given one or two thin coats of clear spirit polish (shellac varnish). If a light effect is required omit the stain. Apply the plaster filler with a broad knife and while still wet remove the surplus with a rubber squeegee leaving the plaster in the pores. If the squeegee is moistened with paraffin or white spirit it will leave the surface cleaner. The work may then be finished with wax polish.

ORNAMENTAL STAINING*

Many decorative effects are obtained by the skilful use of staining, and the following processes may be briefly recapitulated here.

Patterns may be stencilled in knotting or spirit varnish upon a panel; the portion thus stencilled will resist water stain which may be applied to the remainder of the panel, resulting in a pattern of clear wood upon a stained ground. Colour may be mixed with knotting or varnish and a polychromatic effect produced in the same way. A pattern may be put upon the bare wood in deep rich varnish stains blended by stencilling, and the whole panel afterwards stained with water stain to the required depth. A pattern may also be stencilled upon the bare wood in solid body colour, or in gold or silver, the panel being first clear sized, or not, as the worker prefers. When this is dry the panel may be stained all over with various stains in water, and the superfluous stains wiped off with a leather, leaving the gold or other surface clear and clean. A panel may be sized twice with clear size, then decorated in any desired manner upon the sized ground, the wood being allowed to show through the painted ornament. When the painting is dry, the panel may be washed with warm water to remove the size and then stained to any desired depth, the stain being allowed to penetrate; the superfluous stain is then wiped off clean, leaving the painting clear and effective.

Another method of decorating in stain is to pounce the design on the bare wood, outline the pattern with a fine brown or black line, then stain between the lines with oil, spirit, or water stains, and finally varnish. The brown outline will keep the stains from impinging on each other. If a light outline is desired, the panel will require first sizing, then outlining in Brunswick black; when dry, wash with warm water to remove the size, then stain with water stain and finally remove the Brunswick black by a free use of white spirit.

Several different depths of stain may be obtained upon one panel by commencing with the bare wood and using water stains. Stain the whole panel first with the lightest stain required. Coat the portions which are intended to remain in that depth with white hard spirit varnish, and again stain the panel with the next deeper stain; when this is dry cover the parts that are to remain this depth with the spirit varnish and allow it to dry; then stain all over again with the next depth of stain, and again varnish the parts that you desire to be

---

*This section is taken from the original *Painting and Decorating* by W. J. Pearce. These techniques are no longer seen or employed in normal trade practice, but are included here because they are sometimes required in the maintenance and restoration of historic buildings; they may also suggest interesting possibilities to the decorator searching for unusual effects.

finished in that depth; and so on, till the whole of the stains are in. Finally, remove the varnish by applying methylated spirit with a sponge (which will not affect the stain) and polish or varnish in the ordinary way.

A design may be gilded upon wood and afterwards the woodwork may be stained (of course it is assumed that the wood is sized before gilding, first the ground with glue-size, and afterwards the design with Japan or oil goldsize). The glue-size must be washed off with warm water after gilding is done, and then the panel can be stained all over.

These ideas can be extended, elaborated, and used in conjunction, their scope being limited only by the invention and resource of the decorator.

Decorative stain effects should be somewhat conventional in design, as the grain of the wood showing through the stain, while adding beauty and imparting luminosity to the work, makes any attempt at naturalistic painting unsuitable to the material and to the method.

Patterns which have the effect of inlay are admissible, but good judgment is not consistent with slavish attempts to imitate inlay, marquetry, or intarsia work, especially as, when freed from the trammels and limitations that surround the practice of these arts, much greater scope is afforded the designer. Stained ornament upon natural woods is so beautiful in itself that it is quite superfluous to attempt to make it appear other than what it is by taking advantage of its superficial resemblance to inlay. The aim of the decorator should rather be directed to taking full advantage of the freedom possessed by the brush as contrasted with the saw, and of the ease and cheapness of the manipulation when compared with that of inlaying.

FINISHING

Stained work may be finished in various ways to bring out the full beauty of the wood grain. On general stained and varnished work, such as the treatment of pitch pine in churches, the usual method is to apply two coats of glue-size and one coat of varnish. When this method is adopted, it is of utmost importance that the size should be thoroughly dry before varnishing, otherwise varnish defects (such as blooming) may develop. Size has the merit of being inexpensive, but that is all that can be said in its favour; it is not a good foundation for paint or varnish, being softened and swelled by moisture which could find its way into woodwork through top and bottom edges of doors, proximity to wet brickwork, etc., when unprotected. It is a much sounder practice to omit size and apply one or two coats of hard undercoating varnish and finish with an elastic exterior or hard interior varnish as required. Church seats, of course, should always be treated throughout with hard church oak varnish which is specially formulated to dry hard without resoftening on warming. When flat or eggshell varnish is used, it is best applied over a gloss undercoating varnish slightly felted down. Varnish undercoats should not be flowed on as in finishing coats, but should be brushed out well so that the under films dry hard and firm.

Water and spirit stains (and certain proprietary flat oil stains) may be French polished which, of course, is a specialized job. A satisfactory interior finish may be obtained, however, with two or three coats of thin "brush" polish (shellac varnish) and this is a method which the painter can employ with advantage in finishing open-grained hardwoods such as oak, elm, etc., which often do not require a very high degree of gloss. Spirit varnish should always be laid quickly and evenly without crossing or overlapping, using a soft-haired brush.

A most economical one-coat interior finish for oak and similar hardwoods is obtained with certain proprietary wax stains; one coat of stain is applied and allowed to dry hard when it may be "polished" by vigorously brushing with a short-haired shoe brush, resulting in a fine eggshell finish.

Woods such as pitch pine and teak, being of a resinous or "oily" nature, seem to have an antipathy for oil preparations of any kind and are best treated with spirit or water stain, finished with a short-oil varnish of the church oak type; alternatively, a coat of goldsize and turps or white spirit may be given, followed when dry with one or two coats of gloss varnish as required.

27

Graining and Marbling

Authors' Note—W.J. Pearce held strong but usually very sound views on all aspects of the craft, and his discussion on the ethics of graining and marbling is as appropriate today as when first written. We therefore retain his "Imitative Painting" together with his original chapters on graining and marbling in substantially his own words, except for some minor editing and new illustrations.

IMITATIVE PAINTING

Where doctors differ who shall decide? The whole question of the artistic legitimacy of purely imitative graining and marbling has been periodically discussed in the past and though it has been denounced time and again as a sham, its utility cannot be denied. In order to arrive at a fairly correct judgment on the point, let us try to examine the question impartially, irrespective of the personality of those who take sides on the matter. First, what is graining? Is it an attempt to deceive the observer? Second, what is the result? Third, why is it done? Fourth, do these reasons commend themselves to our common sense?

The replies to these questions appear to be these — Graining is an attempt to represent the superficial appearance of something other than the material painted. It cannot deceive the observer who has a knowledge of woods any more than a painted leaf can be mistaken for a real one. The utmost result in the direction of imitative suggestion is that it conveys to the mind the abstract idea of wood. It is used artistically because it conveys this idea of material, in the same way that bronzing and gilding convey the idea of panels, or that certain patterns convey the idea of strength. It is used commercially because, owing to its broken colour surface, it is extremely serviceable and little liable to show slight injury. Therefore, if the proper limitations are observed there appears to be no solid argument against its use. The limitations that should be observed may be set down as follows.

Graining should be used only where it is usual to employ and desirable to suggest the employment of wood constructionally; marbling must be governed by similar laws. Graining should be used only in cases where it is not expedient to employ the real wood, but where the employment of the real wood would be quite possible and rational. No more should be done than is necessary to suggest the wood intended. It is evidently not only vulgar, but also inartistic, to crowd into the work more features than would be likely to occur in the natural wood. The practice of filling the graining with markings is akin to that of third-rate actors, who, because of the cheapness of sham gems, crowd themselves with more jewellery than would be worn by the characters represented and thus loudly proclaim the falsity of their representations. In this connection the words of Pope may be cited as being particularly apt —

> First follow nature, and your judgment frame
> From her just standard.

This view of the subject suggests many doubts as to the actual importance of graining and marbling and leads to the conclusion that if these limitations are studied there are many other equally good methods of obtaining the end aimed at. There is a great deal more graining done than there is the slightest necessity for, and much work is grained that would be better otherwise treated. Before proceeding to that part of the subject, however, it is well to see who are the persons who have led the attack against imitative painting and what they have suggested in its place.

First, there are certain art critics and designers, men whose opinions on all art subjects are worthy of serious consideration, but who, not being thoroughly conversant with the commercial or technical advantages of graining for everyday work, hastily assume graining to be merely a sham. From a purely aesthetic view, if it be granted that graining is an attempt to deceive, they are quite correct in their denunciations. But they apparently start with the wrong premises and they are unaware of the merely utilitarian value of an irregularly broken colour surface.

In the next place, a few personal remarks as to the real intent of the grainer and the effect of graining upon those outside the trade will support the statement that the practice is not an attempt to deceive, that it is intended as a conventional symbol rather than a portrayal.

When a man paints a flower, however well it is done, no one takes it for a real flower, or looks upon the painter as a base deceiver. The very same objections which are raised to graining appear to be equally applicable to veneering, inlaying, gilding and enamelling, oxidizing, galvanizing or plating, in fact to any method of altering the appearance of a surface, if it is assumed that it is done for the purpose of deception.

The aesthetic morality appears to depend entirely upon the artistic intention. There is a great deal of inconsistency among men who condemn graining as an imitation. One prominent denunciator of the practice may tolerate "marble" papers for bathrooms, which appear equally open to condemnation; others defend the use of articles gilt and lacquered, or veneers, or the fashion of staining woods in any and every colour that is opposed to nature and out of harmony with the material, as blue ash, metallic green mahogany and many other equally inconsistent practices.

Following the logical result of these conclusions, there appears to be no reason why so slavish an attempt should always be made to imitate the actual markings of the wood. If the suggestion of woodiness and the broken surface of colour are retained, the actual markings leave scope for the artistic faculties. It is the colour and texture, the light and shade in the wood that charm, and if these qualities are well represented, the actual grain markings might well be treated decoratively or "conventionalized". There are some positions, of course, in which such departure may involve too great a loss of dignity and repose.

Positions suitable for graining and marbling

There are occasions when, failing actual marble or wood, a very near attempt at imitation is required to give the necessary architectural force and character.

Take the case of Corinthian columns in a large hall of classic architecture. No other treatment will give the requisite fitness, stability, and dignity to those columns than could be obtained by the suggestive use of marble. Of course, the use of actual marble is preferable if it can be used solidly. It is, however, very questionable if the use of slabs of marble placed edge to edge round an iron column to form an apparently square and solid pier is not more objectionable from a really artistic point of view than suggestive marbling.

Some of the proper places upon which to use graining and marbling will at once occur to the reader; situations in which a suggestion, more or less conventional, of wood or stone is called for by the architectural setting. They may be safely left to the selection of the decorator, but it is well to point out that not only should the article itself be suitable, but the design and detail of it also. Take the case of an iron mantel. It may be designed to appear either as carved and moulded woodwork, as cast-iron, or as stone or marble, according to the character of the detail and ornament. In this connection a little architectural knowledge and an acquaintance with builders' work will assist the judgment. The mouldings are generally a good guide in the matter.

Another point of importance as a matter of taste is to know how far to give imitative quality to the work. This must be governed by the circumstances of each case, and the student is recommended to incline to *conventional* rather than to purely *imitative* work; to give some definite amount of originality and design to the details, and to lean towards simplicity and regularity. If close adherence to nature appears desirable and is attempted, it must be justified by really good work. It is better to execute a careful and simple stipple suggestive of wood and good in colour than to perpetrate a poor imitation of the finest specimen of natural wood procurable.

The painting of a flower naturally must be superlatively well done to pass muster, which is much less than to give pleasure; it is the same with grained work or marbling. The more frequently a flower is repeated, the less natural should it be in design, and inasmuch as every grainer by mere force of character repeats himself in his work, the same dictum may be well applied to graining.

The treatment of graining as a sketch or suggestion of wood, rather than as an attempt to represent actual wood, gives a wide field for inventiveness and resource and enables the less talented to be contented to do well what is within their power and capability. Every one of the various processes, as combing, stippling, flogging, mottling, overgraining and veining, can be utilized either separately or in combination, to obtain simple and interesting wood-like effects without claiming to represent any particular wood.

As illustrations of the thoughtless manner in which these techniques have been used, one may instance the fact that cast-iron rainwater pipes are sometimes grained; skirtings and bases are grained when occurring beneath a marbled wall; and baths have been seen marbled inside and grained on the outside, or marbled one colour inside and another colour outside. And even doors have been marbled before today! This is painting about as remotely removed from art as is possible.

Varied methods of graining

The methods and processes adopted by grainers for the production of the grain, curl, mottle, and other effects that go to make up the appearance of a wood, vary much in different parts of the country and in different schools of graining. These differences are the result of various men working out their own ideas by means of their own devising. Some of them are highly ingenious. It is not within the scope of this work to explain minutely the *modus operandi* of graining each particular wood or marble, but the following sections will deal briefly with the various woods and stones usually imitated and the colours and tools which will be found to represent them in the simplest manner.

To the student who aspires to master the art of graining and marbling we would say, first, *learn to draw well*, for good drawing is the basic requirement. The delineation of the various grains in wood and veinings in marble demand, in their way, as much skill as the drawing of ornament, or even the figure; moreover, one must be able to draw equally well with sponge, mottler, crayon, pencil, overgrainer, fitch, graining horn or thumb-nail, a range of "tools" that would bewilder and confound the average art student. Yet to see an expert grainer manipulate these tools, to watch his confident and spontaneous "drawing" of delicate wood grain or marble vein on a wet ground is to realize the importance of good draughtmanship. No room here for cautious 'setting out" or timid niggling!

But there is more to it than mere facility in handling or drawing with the tools; it is knowing *what* to draw that matters. And here we would strongly condemn the methods of "teaching" graining in some schools. A panel is rubbed-in and the boy is told to copy an example after a brief demonstration by the teachers. This is not good enough. We would never dream of asking a student to render some painted ornament without some previous study or practice in drawing. The student should therefore be encouraged to make careful pencil studies of wood grain in exactly the same way that he studies plant form. The keen student will jot down in his sketch book any interesting natural grain he might come across in his daily work; on open-grained hardwoods such as oak, elm, etc., one can often take a quick "rubbing" by placing a sheet of thin lining paper on the surface and rubbing across the grain with a soft lead pencil. The grain will appear in reverse on the paper (like a photographic negative) with all the characteristic markings. Such rubbings form very useful references.

When the student has become familiar with wood grains in the above way, he can then practise drawing them from memory with chalk on

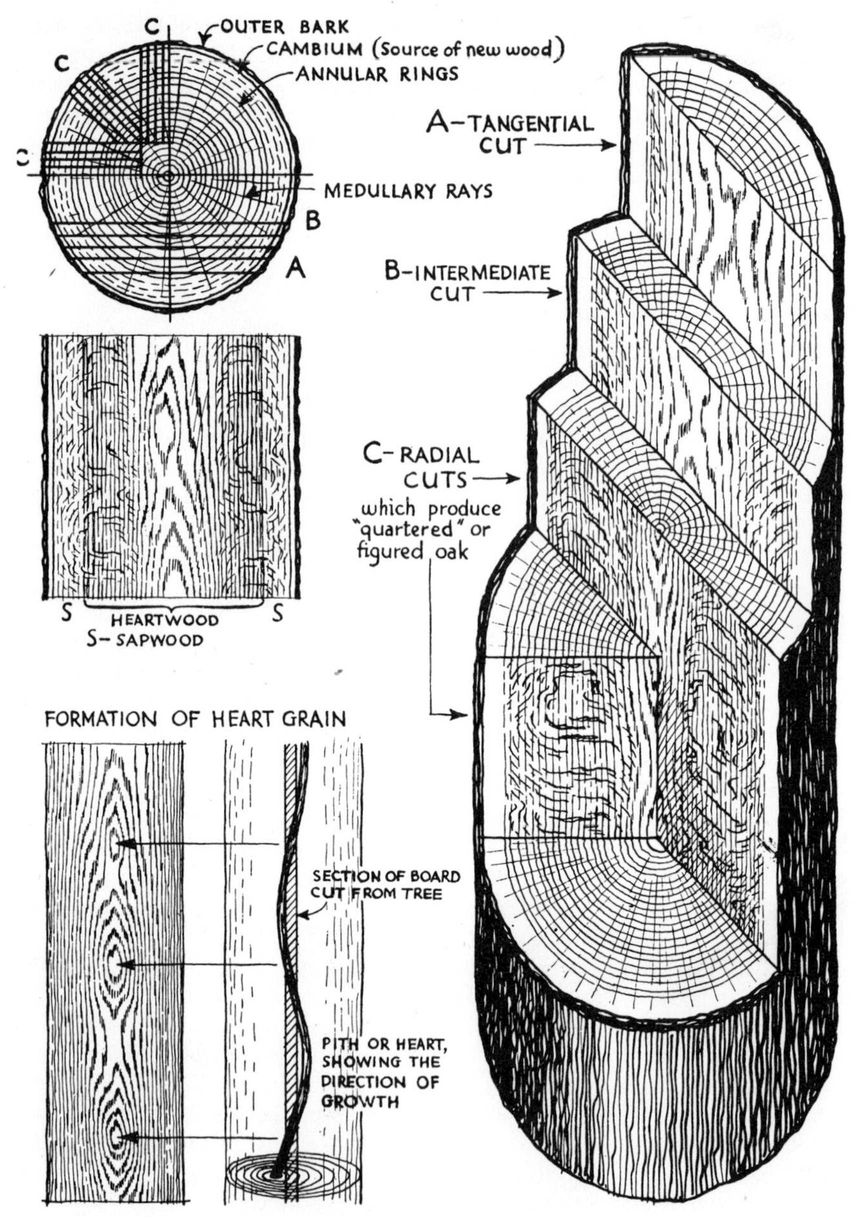

Fig. 27.1 Oak grains

How the various grains occur in oak; and methods of cutting up the log

blackboard; marble veining can be done with thin whiting and water, using writing pencil or fitch. In this way, his knowledge of the subject will be tested and he will gain confidence both in drawing and in handling pencil and fitch; when he shows some proficiency in the above methods he can then start with colour on a painted ground.

An experienced grainer rarely, if ever, works from a "copy". That is because he has made a deep study of the various woods he is interested in, by close observation. Thus he gets to know his subject by heart, so that when he comes to grain a panel, for instance, he has a clear picture in mind of what he wants to do. Therefore, we say, let the student be encouraged from the very start to make an independent study of the woods he sees around him from day to day. When once his interest is really aroused he will take steps himself to procure good examples of the rarer woods and marbles.

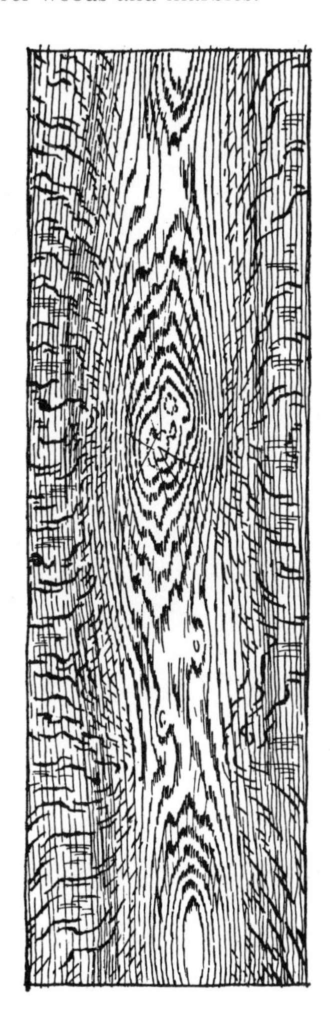

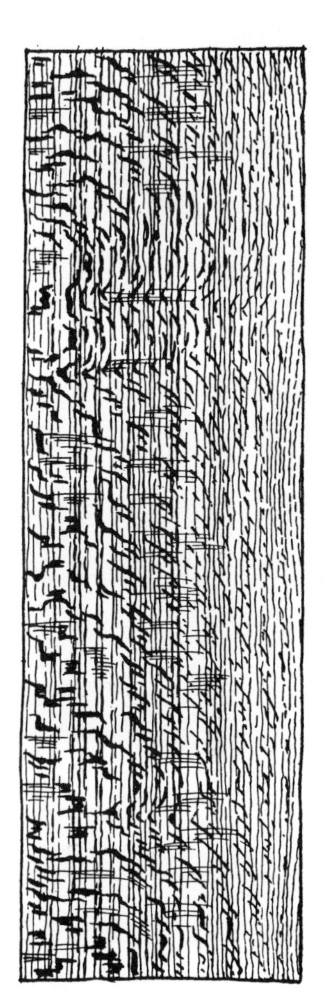

After drawings by Zeph Carr, F.B.I.D.

Fig. 27.2 Oak grains

Heart grain "Silver" grain or quartering

GRAINING TECHNIQUES

Graining, or painting in imitation of woods, must in all cases be done upon a good non-absorbent surface. The preparation and painting of the requisite grounds has already been described in previous chapters and it is only necessary to state here that for graining, the grounds must be perfectly smooth, hard and preferably of an eggshell gloss. The woods most usually attempted by the decorator are oak, walnut and mahogany, each of which has several varieties; but there are numerous other decorative woods and it will be possible to mention only a few of them here.

Oak

A ground is prepared to match in tint the lightest part of the wood it is intended to imitate. For medium oak a mixture of yellow ochre and white lead (or other white pigment) with a touch of raw umber, will produce a suitable colour. It may be remarked, in passing, that the prevalent fault of modern grainers is to use too bright and glaring a ground colour for all woods. For dark oak grounds more umber and a little Venetian red may be added. For rich mellow oak, burnt umber and burnt sienna with ochre and a little white can be used. For green-heart timber or new oak, white, raw umber and a little black will make a good ground. The ground should be kept on the deep and sombre side, cool rather than hot in tone.

When the ground colour is hard-dry, a graining colour is made from burnt umber, to which a little raw sienna or black may be added (for light and deep woods respectively); a little liquid drier is added to the colour, and the proportion of thinners should be about two-fifths white spirit to three-fifths

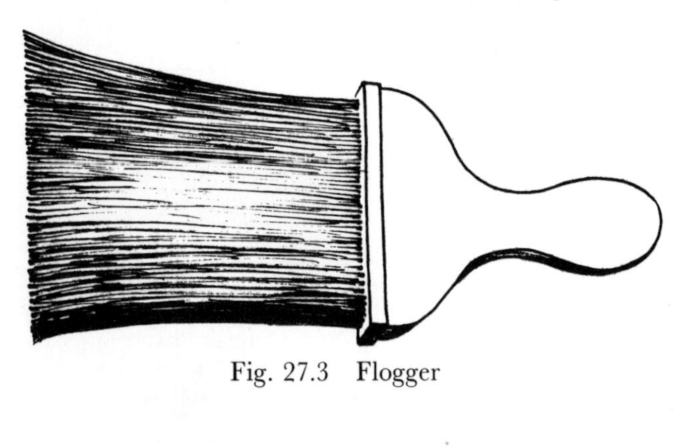

Fig. 27.3 Flogger

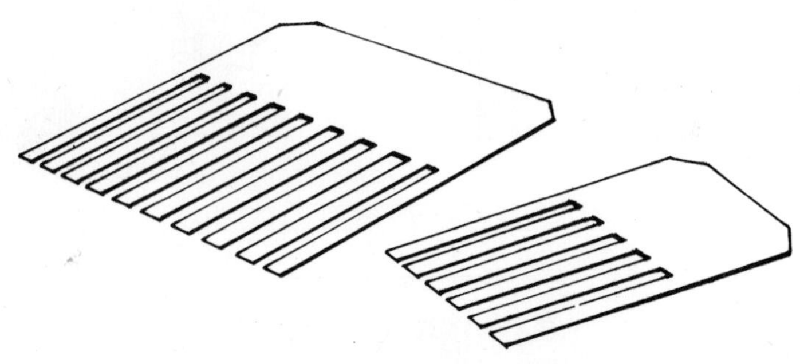

Fig. 27.4 Steel graining combs

boiled linseed oil, or rather more white spirit for internal work; this is scumbled over the work and laid off regularly to a depth of colour representing the average appearance of the wood that is being matched. The graining colour is now "raked" out in streaks with a thin straight-haired brush known as a "flogger" (Fig. 27.3) and combed with steel graining combs (Fig. 27.4) varied sometimes by the use of combs made from leather or rubber. The degree of combing will be regulated by the appearance aimed at by the grainer; sometimes the flogger is used without the comb, and vice versa.

The prominent light markings (known as the figure) are then put in, either with a graining horn or with the thumb nail; the horn or nail is covered with one or two thicknesses of soft rag to allow it to wipe out the marks cleanly without leaving hard edges. The spaces between these marks are then mottled to show the undulations of the grain and the shadows that lie side by side with the light markings; any little touches or softening necessary to complete the likeness to the wood are also put in. The whole is then allowed to dry, when it may be overgrained with colour ground in water.

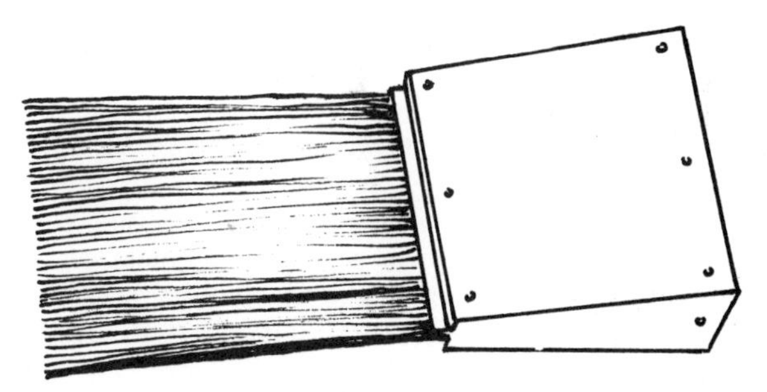

Fig. 27.5 Hog-hair overgrainer

A little blue-black, with or without burnt umber, is recommended for overgraining, slightly tempered with stale beer or milk (to bind the colour) and thinned with water to a mere wash. This is laid on with a mottler (Fig. 27.6) and then softened with a badger-hair softener (Fig. 27.7) so as to represent the general light and shade of the wood. Gum, glycerine, sugar, and fullers' earth are sometimes used as substitutes for the beer. Fullers' earth is favoured by the writer as the least likely to be detrimental to the appearance of the varnish.

Sometimes the work is varnished prior to overgraining and again afterwards, and often the work is overgrained twice to enhance the depth and translucency. In any case, the work must be varnished after the final overgraining. Deeper markings are sometimes added during the graining process by the use of a writer's sable and a fitch.

In the past, such expedients as rainwater, melted beeswax, whiting, yellow soap jelly (made by dissolving soap in boiling water), lime and other materials were added to the oil graining colour to cause it to "stay put". These practices are to be strongly condemned as they destroy hardness and durability of the work and affect the varnish adversely. Nowadays,

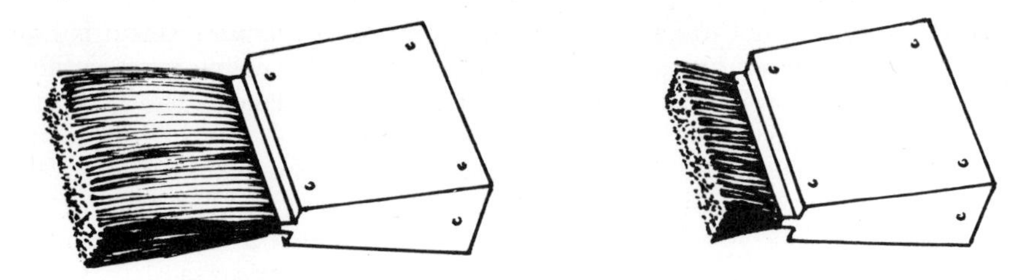

Fig. 27.6 Hog-hair mottler and hog-hair cutter
Brushes of similar design but with a different length of hair

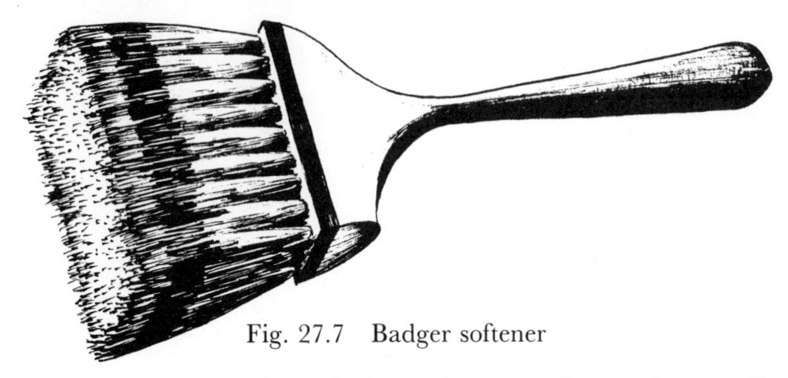

Fig. 27.7 Badger softener

the grainer has at his disposal the various proprietary glaze media to impart "stay put" properties to his graining colour.

There are a number of more or less mechanical appliances used for the purpose of imitating the grain of oak. The effect of combing is very well obtained by the use of the patent combing rollers (Fig. 27.8). The rollers contain a number of notched zinc discs which revolve irregularly and disperse a number of streaks upon the work, which are more like the actual open grain of the wood than the marks produced by steel graining combs. They are used with distemper colour supplied from a mottler, as illustrated in Fig. 27.8.

Fig. 27.8 Oak combing roller

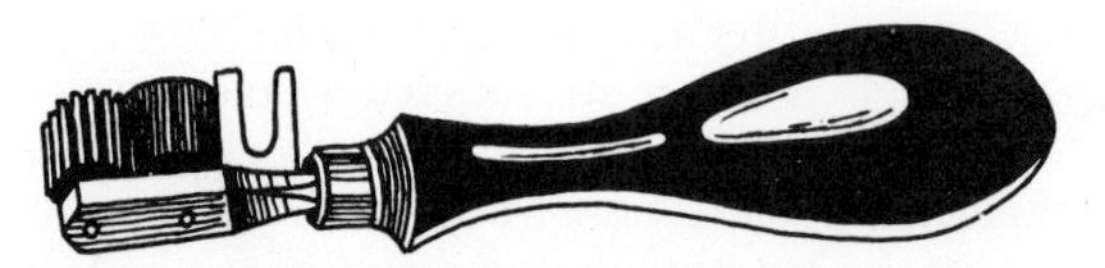

Fig. 27.9 Oak grain finisher

Oak graining used frequently to be done in water medium, but it is not often the case now, except for japanners' work. The work done in distemper has a very clean and sharp appearance. Proceed as follows. Ground in the ordinary way. The graining colour is made from burnt umber ground in water and bound with stale beer. Lay on the colour with a full mottler or a sponge and flog evenly. Comb in the ordinary way or drag with the flogger. Take out the light figuring with a wet leather drawn over the finger or thumb nail and pencil in deep veins with umber. If the colour sets too quickly add a few drops of glycerine. Give a coat of thin varnish or goldsize and white spirit before proceeding to overgrain, or overgrain in oil colour.

Pollard oak

Pollard oak is usually grained in oil, using burnt umber on a fairly dark ground; the colour is applied and hair-stippled and the knotty grain is then taken out with a tiny sponge wrung out in white spirit. When dry, it is glazed and mottled with Vandyke in water to bring out the curliness and richness of the timber, then finally varnished. The student is strongly advised in all graining to copy nature as closely as possible and to work directly therefrom until he is able fairly to imitate the real wood, after which he may launch out on his own lines with a chance of success.

Mahogany

Mahogany is usually grained in water colour on a ground composed of Venetian red and burnt sienna with a little ochre. The graining colour is a mixture of mahogany lake and Vandyke brown, or mahogany lake and blue-black. The tools used are a thick hog-hair mottler, a sponge, a short camel-hair mottler and a badger softener.

The colour is laid on with a distemper tool and manipulated into form with a sponge and mottler, then softened and lightly flogged with the side of the badger to produce the fine grain or texture of the wood. It is then allowed to dry and afterwards overgrained with a thin overgrainer and Vandyke brown in water.

Walnut

Walnut is another popular wood, grained in both water and oil, also in a combination of both. For the ground use yellow ochre and burnt umber; for graining, burnt umber and Vandyke brown, or burnt sienna and blue-black. First lay in the ground with the graining colour used sparingly, and with a wet leather wipe out the light and mottle in a rough representation of the disposition of the light and dark parts of the wood. Allow to dry and then put in with a fitch and an overgrainer the main markings, knots, etc., and work them up with the badger softener and a piece of soft rag; put in finer veinings with the sable pencil and blend together frequently. Allow to dry and finally overgrain with a pencil or separated overgrainer (Fig. 27.11) and a camel-hair mottler. Oil colour may be used for the middle process and water colour for the other two.

Walnut and other rare and valuable woods lose in effect if used in a

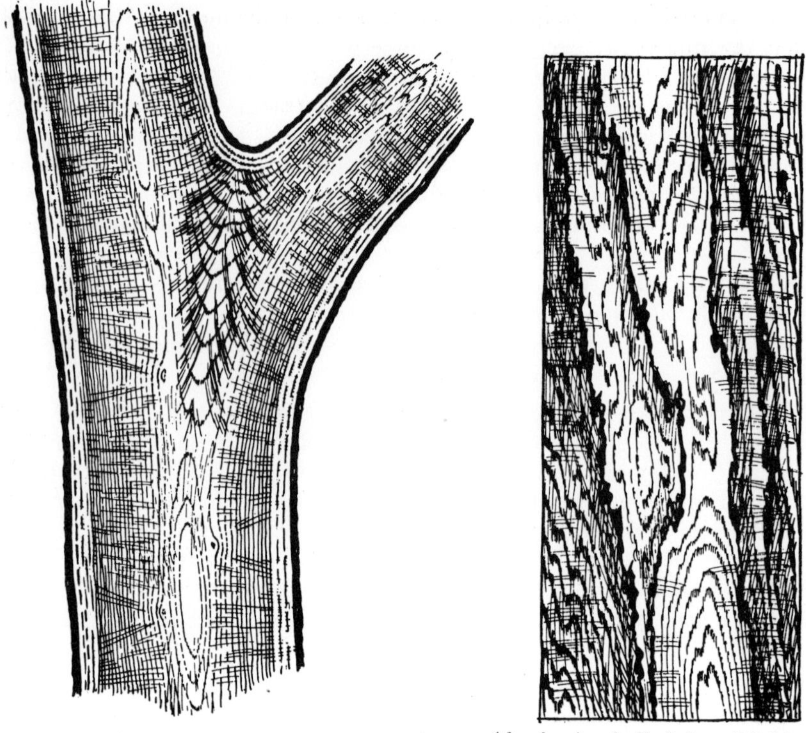

Fig. 27.10 Wood grain

(a) The occurrence of "feather" in mahogany, formed by the separation of grain at a "crutch" or juncture of the tree, and (b) an impression of English walnut

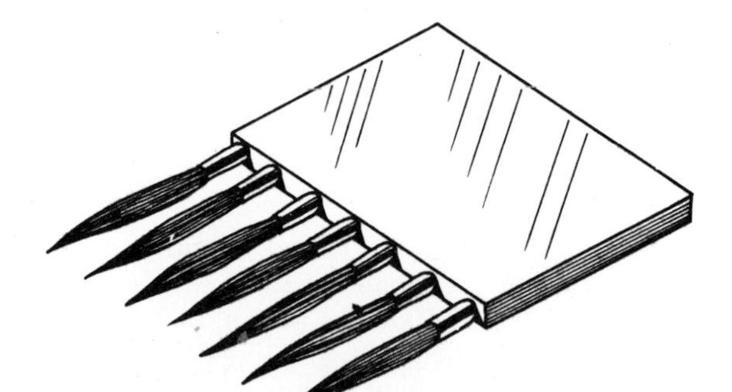

Fig. 27.11 Pencil overgrainer

wholesale and reckless manner which is inconsistent with the probabilities of the use of the same woods in construction.

Pitch pine

Pitch pine is probably one of the easiest woods to imitate and may be grained in oil or water. The ground should be made from ochre and a little Venetian red; the graining colour of burnt sienna, raw sienna, and a little raw umber.

The large heart markings are put in with a fitch or veining tool, the outer and smaller ones with a pencil overgrainer, softened with a badger softener. By keeping the pencilling open so that the softener will not run the lines into each other too quickly they may be driven into a remarkable similarity to the natural grain of the wood. Glazing and mottling with Vandyke in water brings out the characteristic light and shade of the wood.

Rosewood

Rosewood is a very richly coloured ornamental wood; the ground is made from Venetian red and a little ochre, the graining colour being Vandyke brown and crimson lake, or madder brown, with overgraining of ivory black or blue-black. The first process is to lay in with a sponge a roughly modelled effect of the general disposition of the grain. This is then pencilled up by the use of sables and overgrainers — a thin flat overgrainer, separated by a pocket comb, is best for the purpose, as it gives irregular lines — but little softening is required if the overgraining is carefully done. The grain of the wood must be seen to be understood, as it is very intricate, varied, and without much apparent principle in its grouping and direction.

Maple

Bird's eye maple is the most beautiful and popular form of this wood. It is grained upon a white ground, usually in water colour. The graining colour is made from raw sienna, burnt sienna, and blue-black. A mixture of these, with the blue-black predominating, is first mottled over the ground, softened crosswise and allowed to dry. This mottling is done by first scumbling the panel with a distemper tool, then taking a wet chamois leather in both hands, twisting it slightly, ropewise, and using it *à la* skipping rope, striking the panel with it. This produces irregular light markings having a common relation to each other and following a curved disposition. These are then softened from the centre outwards in a horizontal direction. A few highlights are then taken out with a pointed hog-hair tool and a mixture of warmer colour used to put in the "eyes" or dots. The position of these in relation to the mottle will be seen in the natural wood. When all is dry the fine markings of grain are added with a pencil or crayon, using a redder tint and working from the centre or heart and round the eyes unequally. The work is then varnished and afterwards glazed with blue-black in water and mottled to give the silky light and shade which gives the peculiar character to this wood. The student should notice that the shadows are curvilinear, not angular, as in some other woods.

Satinwood

Satinwood and birch may be imitated by a process similar to that given for mahogany, using the real wood as a guide to form and colour. Ground for both is a yellowish white; graining colour, raw sienna, burnt sienna and raw umber, overgrained with blue-black or ivory black.

Ash

Ash is grained upon a yellowish drab ground. The process used is similar to that for pitch pine, but the work is finer and more curly in character. Graining colours as for satinwood. Overgrain in water with Vandyke brown and blue-black. Hungarian ash is the same in colour but much more free, curly and short in the grain. American ash is straight and simple in grain.

Pollard oak, burl ash or pollard ash, root of walnut, and root of birch are richer and deeper, and full of intricate workings, knots and whorls. They should be worked from actual examples if possible.

Other woods

Olive, tulip, and other decorative woods can all be represented truly by the use of the tools and processes already referred to, the chief point being to match the colour carefully.

The commoner woods, as pine, spruce, cypress, etc., are not much imitated, but the grain can be reproduced by the same methods as the more beautifully grained woods. There has been a vogue for "pickled pine" in recent years (representing common pine from which paint has been stripped); this is grained in water using raw umber on a white ground, glazing and mottling with blue-black to achieve a cool effect and finishing with flat varnish. The close and even-grained woods, as cherry, pear tree, box, and a host of others, are little used. They are sometimes, however, very closely imitated in plain colour by the decorator, and there is no reason why more use should not be made of them, as a mere flogging and finishing with the badger would give a fair representation with very little labour. Cherry is a particularly effective wood. Some of our garden trees, as the acacia, laburnum, chestnut, and yew, also give fine colour and grain, of which due advantage is not taken by modern grainers.

Patent graining methods

Various types of transfer graining paper have been marketed at one time or another, but these have generally found application in industrial finishing processes rather than in house painting practice. Probably the most satisfactory method of "mechanical" graining is that in which Gransorbian photograining paper is used since the painter and decorator can apply it in his everyday work. This is not strictly speaking a transfer process, but a method of reproducing wood grain by taking off wet colour from the surface.

Gransorbian paper is a rather thick absorbent paper which has a slightly embossed pattern representing a wood grain (reproduced by photo-engraving). It is made in rolls 600 mm wide up to 100 metres long in a variety of wood grains, the patterns repeating about every 2 metres. The work, having been grounded in the usual way, is rubbed-in with oil or distemper graining colour, hair-stippled and, if desired, combed or raked; the graining paper is cut to size, then laid upon the surface and lightly rolled or brushed into close contact. The embossed pattern at once absorbs the colour (as would blotting paper) and when the paper is removed the pattern of wood grain is left upon the surface; a light badgering softens and completes the work, but if desired the graining may be glazed or overgrained as in normal practice. The paper can be used several times over. The process is simple and can be used by a painter having no experience in graining, but in the hands of a good grainer it is capable of some first-class results.

Other mechanical graining methods are concerned with cutting or incising the grain, either in the paint film as in the Daven process, or in the actual surface of the wood as in the Berries process. Both these processes are limited to the imitation of oak grain.

MARBLING

The imitation of marbles differs materially from that of woods inasmuch as, in the case of woods, it is usual to do the greater part of the work in glazes applied in water colour, whereas the nature of marbles often demands a more solid and opaque treatment. Consequently, marbling is generally executed in body colour, but glazes are used to add depth and translucency where required and water colour is sometimes used for the sake of its rapid drying.

White marble

One of the simplest marbles to execute is white or Sicilian marble. The ground required for this marble is a dead white. When the ground is dry and hard, a thin coat of zinc white in oil is rubbed over it and the veins are put in with a crayon; a warm grey crayon is used for the inner veinings and a soft blacklead pencil or black conté crayon for the more prominent ones. The spaces between the veins are then tinted slightly with grey and green and a few touches of yellowish grey, all very sparingly used, and the whole softened with the hog-hair softener (Fig. 27.12).

Fig. 27.12 Hog-hair softener

Sienna marble

The same ground is used as for the white marble, and while this is still wet it is irregularly painted with two or three tints of yellowish cast, made from white and raw sienna. The veins are then put in either with a black crayon or charcoal, or a soft lead pencil, and softened into the ground. When this is dry, additional shadows, etc., are glazed in with raw sienna and burnt sienna and the veins are emphasized with a little blue or lake. Over all, a few white veins or spots are run and a few lights put on in the interstices between the dark veins.

Italian pink marble

Italian pink marble is used in place of sienna, being about the same depth of tone, but pink, as its name implies. The ground required is the same as above. The ground is scumbled over with pink made from ochre and Venetian red, then shaded-in with greyer tones. The veins are put in with purplish red and the whole blended and softened with the hog-hair softener. Finally, a few white veins crossing the deep ones and a few blotches of white, with here and there rose pink glazings, are added.

Black and gold marble

"Black and gold" used to be a popular marble for plinths and string courses, chimney pieces, etc. The ground is black. The larger veins are a gold colour made from ochre and red and may be varied in colour indefinitely; they are put upon a dry ground with a pencil and oil colour. Very fine distinct white and yellow veins run from the main ones, splitting up the black ground into fragments. The black spaces are then shaded and lightened by the use of grey tints. A few particles of gold leaf or metal put into or upon the gold colour veins improve the effect. Another method is to work in sienna upon a white ground, badgering and blending various golden red and yellow hues together; allow this to dry, then paint in the intervening spaces with black and grey. The peculiarity of this marble is in the intricate ramifications of the veining.

Grey marbles

Grey marble, dove and slate are all worked from a white ground. A feather is used to put in the veins; by this method the colour is irregularly spread over the whole ground. All the veins must run in one general direction and specks and dots must be added in brighter tints with shells and fossils in lighter greys and white.

Red marbles

Red Derbyshire, porphyry and Irish red are all marbled off a bright red ground. Venetian red and vermilion with a little chrome are used in varying degrees of depth. The marbling is done by first glazing over the ground a coat of crimson lake, then breaking it up by the use of a feather and white spirit with a little black. White or grey dots are added and veins in very thin white.

Green marbles

Egyptian green and *verd antique* are green marbles which are worked upon a black ground. Chrome, Prussian blue and white make the marbling colours, varying degrees of colour being used. Fossil spots and rings are added in white, cream, etc., while the innermost ground shows spaces of black.

Lapis lazuli

Lapis lazuli is used for special little medallions, etc. It is obtained from a pale blue ground; ultramarine and gold leaf are used for the marbling and veining respectively. The veins are very fine and broken.

Graniting

Red and grey granite may be imitated by spotting a ground of either colour with white, red, grey and black. The dotting may be done with a "graniting" brush, or a sponge.

Devonshire marble

Devonshire marble is a conglomerate mass of ochres, reds and browns, with white markings. It is represented upon a terra cotta ground by the use of feathers, sponge and rags, the veins being put in with a veining fitch or pencil.

Alabaster

Alabaster is a favourite marble for church decoration. It may be wrought upon a creamy white ground in light red, white and lake. It is a soft stone with undulating veins and is readily imitated.

St Anne's marble

St Anne's and other black and white marbles are worked upon black grounds with white markings. Grey is also used for the middle tints.

In the imitation of all marbles great attention must be paid to the shape of the masses and the direction of the veins. The character and distinctiveness of all marbles rest principally on the form that these take and not on their scale or size. Colour is also important, although every class of marble will present samples widely different in colour as well as in scale.

Many of the most beautiful effects seen in marbles may be imitated by the use of turpentine, which when sprinkled on the wet colour, opens it out in fantastically shaped forms of great beauty and renders that translucent appearance common to the richer marbles.

Amber and other very translucent substances may be imitated successfully by the methods common to marbling. Repeated varnishing and re-glazing is the means adopted to produce great depth and translucency.

Many exquisite suggestions and revelations of colour may be obtained by the examination of fragments of rough marble and mineralogical specimens under the microscope.

The attention of decorators is specially called to the New Zealand and

American marbles, some of which are more beautiful than those from Italy and are much used in building. Though it is usual to restrict painters' marblings to about half a dozen Italian marbles, it should be remembered that at least 250 kinds of marble are known and in use. Almost any colouring may be obtained. A decorative suggestion of marble may be produced without slavish copying of one particular kind; but to imitate the depth of colour and the formation of real marbles, work should be done from samples of the actual material.

28

Decorative Effects

In the matter of interior finishes there are two distinct camps. In the first camp are those who prefer an absolutely plain finish and in the second those who desire some kind of broken colour or other surface interest. The preference for plain off-white, cream, or pale pastel tints is often shown by many people of taste and discrimination, mainly because they have seen so many ghastly attempts at broken colour effects carried out by — yes, let us admit it — members of the craft. Some of these efforts have earned the title of "corned beef" effects and other uncomplimentary names, not entirely without justification. Yet broken colour can be charming and delightful.

One of the many virtues of broken colour work is that it often camouflages surface defects that would be shown up to disadvantage in a plain finish and, similarly, finger marks and other superficial defacements are not nearly so noticeable on a broken ground as on a perfectly plain surface. Success in the achievement of good broken colour effects depends not so much on technical ability as on artistic sensibility, and a man who is sensitive to colour will find inspiration in many unexpected places, such as on moss- or lichen-covered walls, rusted iron surfaces, autumn leaves, foam-flecked water and in the markings of flowers and insects. The study of marble and the practice of marbling usually stimulates a keen interest in broken colour. Even in their everyday work, observant students of colour will not fail to note the often charming accidental effects produced on the paint-chest lid where the paint brushes are usually rubbed out daily.

So much for sources of inspiration. Now for technique. Broken colour can of course, be produced in almost any material and by almost any means which provides for the juxtaposing of one or more colours over a pre-coloured ground, but there are certain well-established methods which the student would do well to master before trying out more unorthodox ways. Oil or water media may be employed and, though many prefer to use their own mixtures, the proprietary scumbles and glaze media, flat paints and water paints will generally be found excellent for the purpose.

SCUMBLING IN OIL

Scumbling consists of applying one or more semi-transparent colours over a dry ground, and the first requirement is that the ground should be quite hard and should not soften up when the scumble is applied; a hard eggshell finish or a flat enamel is therefore desirable. The second requirement is that the scumble or glaze colour should remain "open" long enough to enable a fairly large area to be coated and manipulated; moreover, it should "stay put" so that whatever pattern or markings are produced do not flow out and become lost.

Proprietary scumbles and transparent glaze media are still available from small specialist firms such as J. H. Ratcliffe & Co. (Paints) Ltd, of Linacre Street, Southport; these materials are specially designed to fulfil the above requirements, and they generally function very satisfactorily when one has acquired the right technique in using them. They are supplied in a fairly thick consistency, but owing to a certain amount of "slip" they brush very easily and are capable of being spread fairly thinly, drying overnight with a slight eggshell sheen. They are available in a range of tints, but the writer prefers the clear transparent glaze as this can be tinted to one's own requirements with oil stainers, or even with flat oil paints where a certain amount of opacity is desired.

Application

Assuming that the ground colour is already applied and hard enough to receive the scumble, tint the clear glaze medium with the required stainer and make a small trial to test for colour; if stiff oil colour is used for tinting, it is advisable to strain the glaze before use to avoid any unmixed pigment appearing on the work, but this is unnecessary if liquid oil stainer is used. The small amount of stainer used will generally require no additional drier except in cold damp weather, or where a notoriously bad-drying pigment is used (such as Vandyke brown or lamp-black) when a little liquid drier should be added.

If the surface to be treated is not very large (as, for instance, when a wall is broken up into panels) the glaze may be applied in the round consistency as supplied, taking care to brush out well, for if too thick a coat is given it is liable to surface dry or skin over and thus remain permanently soft underneath. If a large flank wall is being treated, however, it is advisable to thin the glaze with white spirit and to apply the glaze more liberally, using a 100-mm wall brush in order to cover the surface as rapidly as possible. If necessary, the setting of the glaze may be further retarded (without raising the gloss to any appreciable extent) by thinning with one part raw linseed oil and two or three parts white spirit.

Before applying oil scumbles over metallic paints such as gold and copper bronze, aluminium, etc., a thin coat of clear lacquer (white knotting thinned 50/50 with methylated spirit) should first be applied and allowed to dry. The reason for this is that in most metallic paints the metal tends to float to the surface of the medium (hence the brilliant lustre) and when the oil glaze is applied directly over it, the metal may pick up and float in the glaze; the clear lacquer prevents this by acting as a "buffer" coat, thereby isolating the metal from any solvent action of the scumble.

Manipulation

The glaze may now be manipulated as required by rubber stippling, rag-rolling, combing, etc., but whatever treatment is to be given it is preferable to hair-stipple the glaze first in order to ensure an even distribution of colour. Hair-stippling forms an attractive finish in itself, especially for woodwork, but it must be done very carefully to obtain a uniform finish, the stippler being first used vigorously to distribute the colour, then more lightly to finish off. Coarser effects are obtained with rubber stipplers which are made in fine, medium, coarse and extra coarse grades. Brooke's Radial rubber stippler and Eyre's rubber stippling tools also provide interesting and characteristic effects, but these, which at one time enjoyed a ready sale, are now no longer obtainable. There are of course many people in all parts of the country who still possess these tools, and there are probably many paintshops where such tools are stowed away in a cupboard waiting for some enthusiastic young decorator to rediscover them.

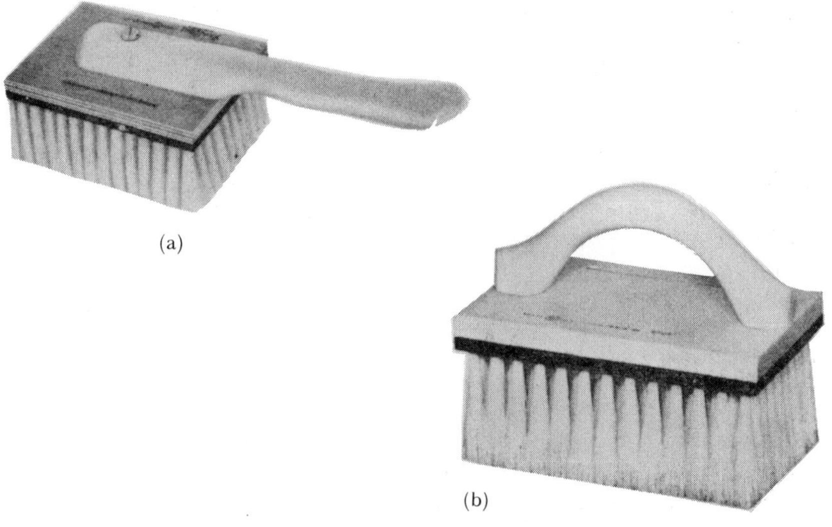

(a)

(b)

Fig. 28.1 Stipplers

(a) With reversible handle (b) Bridge handled type

Combing may be done with steel or rubber combs, the latter removing the colour very cleanly; as in graining, the combs are wiped with a soft rag after each stroke so that the colour is not returned to the surface again. The comb should be used with a firm even pressure and where fabric or plaid effects are being attempted, the combing may be first done vertically, then horizontally; alternatively the scumble may be combed vertically and allowed to dry, then re-glazed in a different colour and combed horizontally. To facilitate the combing of straight lines, a wooden straightedge may be used if panel pins are hammered in the underside so that the straightedge is kept about 5 mm away from the wet surface (see also "Colour Combing", page 367).

Rag-rolling may be done with waste absorbent twill sheeting or with soft scrim or mutton-cloth, used dry or wrung out in white spirit; a large wad of cloth is crumpled up and rolled over the wet scumble, to produce

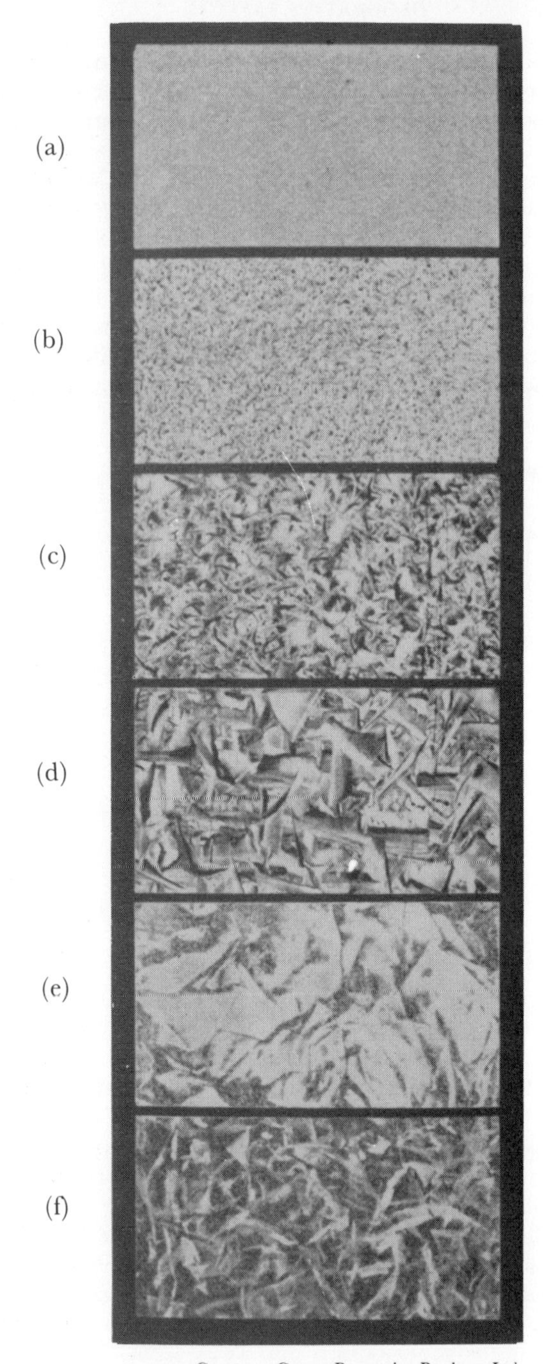

Courtesy: Crown Decorative Products Ltd

Fig. 28.2 Stippled glaze effects

(a) to (d) were produced with fine, medium, coarse, and extra coarse rubber stipplers, (e) and (f) with crumpled paper an irregular pattern, taking care to obtain a fairly uniform effect over the whole surface (incidentally, rags used for this purpose should never be left lying about carelessly at the day's end, as they are liable to catch fire through spontaneous combustion; they should either be spread out to enable them to dry harmlessly, or be dumped in an enclosed metal bin). Crumpled paper may also be used to produce pattern, the effect being generally harder and more angular than that produced with a soft rag.

Unlimited variations are possible both in the application of the glaze and in its manipulation afterwards. For instance, various polychrome effects are produced by using two or more different glaze tints, applying them rather haphazardly in patches and roughly blending them together with a hair stippler, then rag-rolling or combing, etc. A better way of doing this is to apply first a coat of clear glaze medium, then to put in the various tints separately with fitch or sash-tool and blend into the wet ground. On a white ground, combinations of primrose, pink, pale blue, etc., can be applied in this way, blended into each other, then rag-rolled and so on. Alternatively the ground itself may be a fairly strong polychrome or multi-stipple effect and the glaze tinted either with opaque white or transparent Vandyke; in either case the rather bizarre effect of the strong polychrome ground is softened and toned down.

Another method is to "open out" the tinted glaze by splashing or sponging with solvent such as turps, white spirit, methylated spirit, etc., different effects being obtained according to the particular solvent used and also by varying the composition of the scumble itself with varying proportions of boiled oil. The sponged solvents may be clear or tinted, and some attractive effects are obtained when a trace of bronze or aluminium powder is also incorporated.

Scumbling textured grounds

Surfaces which have been textured with plastic paint are often finished by painting and glazing. Two or even three coats of paint (depending on the porosity of the plastic composition) may be required to body up the work and as with plain surfaces, the ground colour should dry to a hard semi-gloss or egg-shell finish. The scumble glaze is applied as described above and hair-stippled to distribute the colour evenly; the glaze may then be removed by wiping the surface with soft rag or, if a more contrasty effect is desired, a rubber squeegee may be employed. Thus the glaze colour is wiped off the relief portions but remains in the hollows, thereby enhancing the texture; by using the squeegee in this way even low relief may be made to look quite bold and imposing if the colour is suitable. Rag should be used for softer and more intimate effects.

Glaze, blending and wiping

Ombré (blended) and "wiped" effects are readily obtained by the use of clear and tinted glaze medium. An old-ivory effect can be produced on panelled ceilings or walls by the following method. Bring the work up to a hard eggshell gloss finish in off-white or ivory colour. Apply a liberal coat of clear glaze medium (thinned with white spirit) to the centre of the panel, then run a broad band of *tinted* glaze (glaze medium tinted with a little raw umber) round the edge of the panel and work the tinted glaze gradually into the clear glaze towards the centre until an even blend is obtained. Finish by hair stippling, working from the centre to the edges. Cornices may be coated with the tinted glaze and hair-stippled, then wiped with a soft cloth, leaving the glaze in hollows and enrichments. Very fine antique effects can be achieved in this way. The same means can be used to produce more colourful effects by tinting the glaze with viridian, crimson lake, or burnt sienna, etc., but blending always into clear glaze; the effect is soft and restful when the colour is nicely judged and makes an appropriate treatment for bedrooms.

Doors and wood panelling may be coated all over with tinted glaze and hair stippled; the panels may then be wiped from the centre with a soft cloth and softened off towards the edges by gently patting with the cloth or stippling with a badger. Mouldings and enrichments may be wiped, leaving the colour undisturbed in the recesses.

SCUMBLING IN WATER

Pulp colours in water may be tempered or lightly bound with fullers' earth or stale beer and used as a scumble glaze over a hard, plain-painted or textured ground. In hot weather (or when working on large areas) a little glycerine may be added to retard the drying. The colour is rubbed briskly over the surface with a large well-worn brush, hair-stippled to produce an even tone, then rag-rolled, or better still rolled with a wash-leather which produces a cleaner, crisper pattern. Thin glazes of Vandyke brown or raw umber may be used over colours which need toning down to produce antique finishes such as parchment and old-ivory effects. Bronze and aluminium paints may be effectively glazed in this way with Vandyke or blue-black to obtain "oxidized" copper and antique silver effects, the colour being finely stippled with a badger and wiped off appropriate parts with a soft cloth. Incidentally, water colour is the most convenient medium for glazing metallic paints since it is not necessary to lacquer the surface before glazing, as with oil scumble. If a washable finish is desired, water scumble may be flat varnished, gloss varnished or finished with a clear cellulose lacquer if the ground colour is suitable.

Oil-bound water paints thinned with petrifying liquid may be used as a thin glaze and rolled with a chamois leather, over a ground of water paint or a hard flat oil paint. The washleather should be rinsed out frequently in clean water so that the colour is picked off cleanly and a muddy effect is avoided. The colour may be allowed to harden for at least 24 hours when another thin glaze of water paint may be applied (preferably in a paler colour) and leather-rolled as before; this produces a very soft finish and is capable of some rather subtle effects. A bright sunshine yellow, for instance, can be glazed and rolled, first with pink or green or blue, then finished off with white. This process, of course, is possible only with a high-grade oil-bound water paint having good re-coatability.

The glazing of surfaces which have been textured with plastic paints is best done over a hard flat or semi-gloss oil paint ground, the water scumble being applied very thinly and hair-stippled, then removed with a damp chamois leather; this leaves the colour lying in the hollows and enhances the plastic texture. In some cases where it is undesirable to seal the surfaces with oil paint (as on new plaster), a thin glaze of water paint may be applied direct over an unpainted texture surface and then removed at once with a damp washleather; or alternatively, the glaze colour may be allowed to dry, then removed from the relief parts by rubbing with glasspaper over a wood block.

SPONGE STIPPLING

The application of colour by means of a sponge is a method well known to the decorator, but for those who have not tried it, here is a brief outline of procedure. A suitable sponge is selected — a fairly large, open-textured one for wall surfaces, or a small, closer-knit one for intricate decorative work. The sponge is sliced in two with a very sharp knife (such as a large carving knife) in order to obtain a flat stippling surface.

The colour, either water paint or flat oil paint, is mixed to a rather creamy consistency and brushed on to a glass or slate palette, from which it is picked up on the flat face of the sponge and stippled on to the wall. It is usual to work on a middle-toned dry ground, first sponging with a darker colour, then over-sponging with a lighter tint; thus a buff ground may be first sponged with a deep old-gold colour, then over-sponged with cream.

The aim in using the sponge — especially in background work — is to produce a fairly uniform stipple, and in order to achieve this the sponge should be held lightly, the wrist being continually turned to left and right in order to avoid any set repetition of the sponge pattern. If the sponge is held rigidly in one position, its texture will be reproduced in a rather insistent way which can look very unsightly. A flexible wrist should therefore be cultivated so that the sponge is half "twirled" in the hand between each dab.

Countless colour combinations are possible, and experiments with two, three or more colours can become quite fascinating. Fairly strong colour contrasts (of equal *tone* value) may be used, for as with sprayed plastics and spatter finishes, the effect of visual intermixture tends to produce a "neutral" colour when viewed from two or three metres away (see "Colour", pages 363 and 517).

A rather intriguing polychrome effect may be produced by sponging over a varicoloured ground. Suppose you make three tints, say primrose, rose pink and eau-de-nil; apply these separately in large patches in a haphazard manner (using 100-mm wall brushes) roughly blending together whilst wet by dragging one colour into another. Allow this ground to dry, then sponge-stipple with white or pale grey; the finished effect is one of very subtle colour gradation. The varicoloured ground before sponging may look quite unbearable, but the effect of the sponge stipple is to soften and harmonize the contrasting colours. Similarly, a strongly defined horizontal banded scheme can be softened and toned down by over-sponging with suitable colour.

By sponging on a *wet* ground a much softer effect is obtained. If flat oil paint is being used, a little clear glaze medium will keep the paint "open" so that the sponged colour softens or blends slightly into the ground colour. In the case of oil-bound water paint it should be applied fairly liberally and preferably over an oil primer to enable it to remain open long enough to receive the stipple in the same way. The colour should be applied by one man while another follows on immediately with the sponging. If the water-paint ground colour is applied in a fairly round consistency, a light relief texture is obtained by sponging with firm pressure, and if at the same time a slight twist is given, a sponge swirl effect is produced similar in appearance to some of the low-textured wallpapers; thus colour and texture may be obtained in one operaton. When sponging on a wet ground in this way it is necessary to pick up fresh colour from the palette more frequently than is the case when stippling on a dry ground, because the sponged colour is constantly being modified by the wet ground colour.

SPATTER OR SPECKLE EFFECTS

The production of spatter or speckle effects may be achieved in several ways and may be applied either in the form of a continuous and uniform all-over treatment, or as "accidental" spatter on local areas to enhance and give further interest to other broken colour treatments. In the first case the spatter is usually applied by spray as it is extremely difficult to achieve a uniform spatter by hand-applied methods. In the second case colour may be spattered on to the surface by dragging a palette knife against the bristles of a sash tool or nail brush, or even a flat distemper brush for certain broad effects. Only the tips of the bristles need to be charged with colour; the brush is held so that the bristles hang downwards and the knife or scraper is drawn along the tips of the bristles towards the operator, that is, away from the surface being spattered. As the knife passes over the bristles, they are pressed back, then released with a spring which throws the colour on to the surface in the form of irregular small splashes, so producing a speckled effect. Broad effects are obtained when the brush is well charged with colour, but as the brush empties the speckle becomes extremely fine. A paint can or a pail should be held under the brush so that any surplus colour which runs off the scraper falls into the pail; after a little practice one can easily hold the brush and can or pail in the left hand whilst manipulating the scraper with the right hand. One stands about 450 to 600 mm (18 to 24 in.) away from the wall according to the springiness of the brush being used. It will be obvious that this method is only practicable for uneven spatter or accidental effects, but nevertheless it is capable of producing some very attractive results on scumbles which have been "opened out" with solvents (see Scumbling) or on sponged or multi-stipple grounds.

Sprayed spatter work

Sprayed spatter work may be done with most types of spray gun, but some firms supply a special Decorators' Spray Head which has different nozzles for spattering, misting and other effects (see Chapter 6). Whatever make of equipment is used, the essential requirement is that air pressure on the gun be cut down to between $\frac{1}{3}$ and $\frac{2}{3}$ bar, so that the colour, instead of being atomized, comes out in the form of small droplets or splashes (see page 132).

Spray spatter is often done over a sprayed plastic ground to provide further colour interest, and a certain amount of sparkle is given to the texture when metallic paints are over-spattered in this way. But spray spatter work over a plain painted or distempered ground is also capable of producing some very attractive decorative wall finishes and forms a useful alternative to sprayed plastic work, which is often objected to on the grounds of its alleged dirt-catching texture (see also "Multicolour finishes", page 201).

Colour

The effect of innumerable spots or specks of colour deposited on a different coloured ground as in sponge-stipple, tinted sprayed plastic and spatter work, is that when viewed at a short distance the separate colours tend to merge together and assume the appearance of single composite colour. This optical effect is known as "visual intermixture" and was the basis of the French Pointillist method of painting pictures (see Chapter 36 on "Colour"). There is great scope for experiment in this direction, and our knowledge of colour theory may be put to the practical test in this way. If, for instance, we take two contrasting colours, such as blue and orange, red and blue-green, or yellow and violet and reduce them with white until they form *tints* of approximately equal tone value, then sponge or spatter one over the other in about equal proportions, we find that the two colours tend to neutralize each other when viewed at some little distance, appearing almost as a neutral grey tint. But this neutral tint, produced by visual intermixture, will have a vibrant quality and surface interest that could never be obtained by actually mixing the two colours together.

There is no end to the interesting ways in which these colourful yet "neutral" backgrounds may be produced, and there are many possibilities to be explored in the subtle use of *discords* for the "unusual" scheme. The study of birds' eggs in the local museum can be a fruitful source of inspiration, and many examples of really delightful discords will be found, such as brown speckles on pale duck-egg blue; deep brownish ochre on pale violet and so on. The eggs of the blackbird, rook, golden plover, raven, thrush, sparrow hawk, skylark and red grouse are all worthy of note.

An effect of relief similar to a sprayed plastic texture is obtained by spattering a middle-tone ground first with a darker colour and next with a lighter tint; in this manner a charming "oatmeal" effect may be produced in flat oil or water paint by spattering a ground of cool stone or putty colour, first with cinnamon brown and then with cream.

Another practical use of the spatter is the possibility of introducing the dominant note of a scheme into the wall colour while still keeping the background more or less neutral (as a background should be). For instance, in a scheme where the ceiling and walls are ivory or off-white and the dominant note of carpet and other furnishings is blue, a pale blue tint may be spray-spattered over the ivory followed with a final spatter of gold bronze or silver. This produces a completely harmonious scheme and is equally applicable in other schemes where the furnishings may be green, old gold, rust or any other colour, provided a *tint* of the dominant colour is used, with or without the metallic spatter.

MULTI-STIPPLE EFFECTS

Attractive broken colour can be produced by so-called multi-stipple methods by means of which three or four colours are stippled on to a wet ground of flat oil or water paint. Three or four tints are mixed in separate cans and best results are obtained when these are about the same *tone* value as the ground colour. For instance, suppose the ground is to be a pale stone or biscuit colour and the stippling colours are to be contrasting tints such as pale pink, primrose, pale blue or eau-de-nil. Mix the ground colours to the same consistency, then test for tone value by brushing out a little of each side by side (touching); when viewed with half-closed eyes, the desired effect should be a uniform tone. If one colour stands out more than another it should be reduced, otherwise when stippled on the wall along with the other tints it is liable to stand out too prominently and to form a rather insistent pattern.

When the colours are ready, the ground colour is applied with a flat wall brush in a liberal coat by one man while another man follows on immediately with the stippling colours, applying each in separate patches with a well-charged sash tool. The colours are not painted in the ordinary way, but stippled on to the wet ground colour with some force, using a separate sash tool for each colour. The colours blend into the wet ground and into each other, producing the effect of a soft polychrome mottle. The Leyland Paint and Varnish Co. once produced an ingenious multi-stippler for this purpose, which now is no longer obtainable, consisting of four small stipplers mounted on a single base, and with a special can divided into four compartments into which the separate colours were placed, allowing the four-knot stippler to pick up four colours with each dip, thus expediting the whole process. The point to watch was that the multi-stippler had to be dipped into the can the correct way each time it was recharged, otherwise the colours got hopelessly mixed and produced a muddy result.

As with sponge-stipple and other broken colour work, multi-stipple effects may be further enhanced by finally spattering with colour or with metallic bronze.

BLENDING IN FLAT OIL PAINT

The blending of one colour into another is often carried on in flat oil paint and looks very well when expertly done. More often than not, however, blending is badly carried out, which suggests that insufficient care is taken in organizing the work; for that is the whole secret of success in blending operations — *organization.*

The operation would be a simple matter if the paint would remain open long enough to play about with until a perfect blend is obtained, but if oil is added to slow down the setting the paint dries with an undesirable gloss. As the final coat of flat oil paint, if thinned at all, is only thinned with white spirit, the setting is naturally fairly rapid, but if a proportion of clear glaze medium is added (say one part of glaze to eight parts of paint) the setting is considerably retarded without materially affecting the matt finish. When very large surfaces are to be blended, however, the setting may be further retarded by adding a proportion of raw oil in addition to the glaze medium (up to 1 part of oil and 1 part of glaze to each eight parts of paint, the remaining thinner being white spirit). This will produce a slight sheen but hardly sufficient to be objectionable.

The work should be bodied-up to a solid finish in the lightest of the blending colours, the last undercoat being thinned with two parts linseed oil or mixing varnish to one part of white spirit, so that it dries with a certain amount of gloss; it should be allowed to dry hard before applying the blending coat. If there is to be much contrast in *tone* of the blending colours (for instance, pale cream to deep tangerine) the last undercoat should also be roughly blended so that the final coat will appear solid, for if a strong colour like tangerine is applied over a cream undercoat the light ground would grin through and appear unsolid when stippled.

Blending procedure will depend partly on the area of the surface to be treated, but the principle involved is the same in all cases. The lightest and darkest colours are first made and a proportion of each is "boxed" together to form a middle tint. These three tints will be sufficient for normal domestic walls (say 4 to 5 metres long by 2½ or 3 metres high) or where the wall surface is broken up into panels of manageable size. For larger wall surfaces, however, and more especially when these are fairly lofty, five separate tints should be made by boxing some of the middle tint with the lightest colour and some with the darkest colour, taking care to adjust the five tints so that they graduate evenly from light to dark. The working consistency, of course, should be the same for each tint.

And now we are ready for blending. Apply a broad band of the lightest colour along the top and bring it down to a third or a fifth of the wall surface (depending on whether three or five tints are being used). Next apply a band of the second colour and work this up and into the light colour, *blending the two together by careful brushwork.* Apply the third band of colour and blend into the second in the same way and so on. Then, with a large hair-stippler, commence stippling at the top, working down gradually and evenly to the middle colour, and with another stippler complete the lower portion of the wall. In the case of fairly high walls, a third stippler may be necessary for the middle colour. The colour should be applied very liberally to ensure that it remains wet long enough to be easily blended and stippled. 100-mm or 125-mm flat wall brushes are required for laying in the colour quickly.

Large-scale work should be planned carefully so that application, blending and stippling proceed smoothly without any hold-up or stoppage through having to move scaffolding, etc. One man should commence at the top with the light colour while a second follows on with the next colour and blends into the first. The first man then commences stippling from the top, whilst the second man carries on with the third colour, blending this into the second and so on. Alternatively, two or three men may be employed in application and stippling whilst one man does nothing but blending; it is desirable that one man should do all the blending so that uniform results may be obtained. If the wall surface is first divided into three or five equal parts by snapping chalk lines the work of applying the respective colours in uniform bands is facilitated and guesswork is reduced to the minimum.

Many decorators are under the erroneous impression that perfect blending can best be achieved with the spray-gun, but on large-scale work, at least, this is definitely not the case. Spray-blending on a large wall is extremely difficult and it requires the utmost skill of an expert to avoid patchiness. Blending by brush is under greater control, however, and perfect graduated blends may be obtained by the average intelligent craftsman after a little practice. The whole secret (if secret there be) lies in the skilful and dextrous brushwork in blending the colours together before stippling commences. Most failures are caused by leaving the blending entirely to the stippler, for it is practically impossible to obtain perfectly graduated blends on large surfaces in this way, the result generally being that the colours merge too quickly into each other instead of gradually and imperceptibly.

Blending need not consist merely of light to dark effects, for two or more colours of different hue often make attractive blends. Choice of colour is of great importance, however, since colours of contrasting hue will neutralize each other and produce either tertiaries or greys when blended together. Thus, rose pink blended to pale blue-green would form a neutral grey in the middle, but the grey produced will be darker than both the pink and the green and will therefore tend to appear rather dirty and unsatisfactory. Even harmonious colours of equal tone-value tend to produce slightly darker intermixtures. Reference to the chapter on Colour will provide information on this point. Some three-colour combinations that have proved satisfactory are —

(1) White blended through primrose to pale apple green.
(2) Ivory through rose pink to pale blue.
(3) Cream through pink to lilac or mauve.
(4) Primrose through pale green to turquoise blue.

BLENDING IN WATER PAINT

This is a much more difficult proposition since water paints dry largely by evaporation and it is therefore not possible to control the material to the same extent as with flat oil paints. Nevertheless, satisfactory blends are sometimes carried out in water paint and the resourceful decorator will not hesitate to employ unorthodox methods to secure good results.

On one occasion the writer, with the aid of a young apprentice, blended a café wall measuring 8 metres by 3 metres high to produce a ground for a landscape decoration. The wall had been previously hung with a lining paper and it was a warm day. First, all windows and doors were shut to minimize evaporation and the lined wall was thoroughly wetted with clean water; the floor was also well sprinkled to create a damp atmosphere. The colour was to be cream, blended to a deep peach. The cream and peach were mixed in separate pails and after a broad band of cream had been applied along the top of the wall a small amount of peach was added to the cream, well stirred and the second band of colour applied; this colour was only slightly deeper than the cream and therefore blended into it perfectly without requiring stippling. This was repeated, each band being slightly deepened and blended into the one above until the deepest colour was reached at the bottom. By working quickly in narrow bands in this way and applying the material very liberally with 175-mm flat brushes, the edges were kept alive long enough to blend one band into the other. The lining paper came up in blisters, of course, owing to this soaking, but the walls had previously been well sized and a good flour paste used, with the result that the paper tightened up again and the blended wall dried out beautifully.

In this case the job turned out satisfactorily, but blending in water paint or distemper must needs be a chancy business and good results cannot be guaranteed. Where walls are divided into panels of manageable proportions, however, there should not be much difficulty in blending, particularly if the porosity of the wall surface is first stopped with a coat of oil primer before applying the water paint.

COLOUR COMBING

In addition to the combed effects in scumble glaze (mentioned on page 357) there are great decorative possibilities with other materials such as flat oil paints, water paints, tinted plastic paints, and metallic paints. In all cases it is essential that the ground should be quite hard and preferably a semi-gloss finish, in order to withstand the firm pressure of steel or hard rubber combs.

Rubber combs such as those made by Ridgely may be used, or the decorator may easily make his own, either from soft squeegee rubber or from rubber flooring of suitable thickness. All that is necessary is to cut V-shaped notches at equal, graduated, or irregular intervals as required. Similarly, regular or graduated steel graining combs may be altered to produce different patterns, either by breaking off unwanted teeth or by inserting thin metal strip or cardboard in such a way as to throw the unwanted teeth out of contact with the painted surface.

The action of combing through wet colour, even when thinly applied, is to leave a slightly "ridged" surface, but when thick colour is combed, a noticeable relief effect is produced with its delicate interplay of light and shade. This may be put to good account, especially if the combing colour dries with a semi-gloss, for then, by combing the surface all over in different directions, the light is caught in innumerable ways which change with one's viewpoint, and the effect is not unlike that of "shot" silk. A more regular change of light is produced by combing vertically and horizontally in alternating squares; if reticent colour is used and the squares are kept fairly large — say from 300 to 600 mm, according to the scale of the chamber — an imposing yet quiet effect is produced, appropriate to an entrance hall or board room.

Some very intricate and ingenious designs have been produced in the past by colour-combing enthusiasts, but whilst paying tribute to the wonderful skill and craftsmanship displayed, the writer prefers to see colour combing used as an all-over surface treatment of a not-too-insistent or ambitious character, believing that all interior surface decoration (apart from obvious focal points such as motifs or mural decorations) should be in the nature of an unobtrusive background and entirely subordinate to the architectural setting. Combed work, consisting as it does of multiple lines which are essentially directional in character, must be introduced carefully if architectural stability is to be preserved. For instance, diagonal combing would have a most disturbing and unpleasant effect unless stabilized by cross-combing in the opposite direction to form a diamond or diaper pattern. Vertical combed bands may be used to give an effect of increased height to low rooms, just as horizontal bands will tend to reduce the height of tall rooms. Again, a broad check pattern might well form a background panel to a piece of furniture or may be used as an intimate treatment to a dining recess or divan corner.

The remarks on colour in sponge-stipple and broken colour work generally apply equally to colour combing, and here again there is a wide field for experiment. The essential technical requirements are that the paint should remain open long enough to enable fairly large areas to be combed cleanly, and that when once combed the pattern should not flow out but "stay put", crisp and clean. Both these conditions are met by using a proprietary clear glaze medium, tinted as required with oil stainers or flat oil paint.

29

Wallpaper and Other Hangings

Hangings of one sort or another have been used for the covering and embellishment of wall surfaces from the earliest times. Probably the first kind were the skins of beasts, but with the introduction of weaving came coarse fibrous cloths of hemp and flax which, by a process of gradual development, blossomed into beautiful handwoven tapestries, stamped and embroidered velvets, silks and rich stuffs of other material. Embossed and stamped leather was a further development, possibly suggested by its greater durability and the ease with which it could be cleaned.

The use of hangings was doubtless originally suggested by the necessity of keeping out draughts and cold in the caves and crudely fashioned buildings of early date. Dwellers in tents still use them for this purpose, as is exemplified to the present day by the nomadic tribes of the Sudan. With the advent of papermaking on a commercial scale and the introduction of block printing, it was not long before the decorative possibilities of printed paper came to be realized and used as wall decoration.

The first patent for the manufacture of wallpaper in the United Kingdom was granted in 1692 during the reign of William of Orange, but there are records of decorated papers being used in the time of Henry VIII, and indeed the earliest example of European wallpaper ever found is known to date from 1509, the first year of Henry's reign; this historic paper was discovered in Cambridge during restoration work at Christ's College.

Wallpapers can be broadly divided into two classes, hand-printed and machine-printed. A few are still coloured and painted by hand, principally floral and landscape papers of French or Chinese origin. Others are block-printed, hand-stencilled, or silk-screened, although within the last few years block-printing has been supplanted almost entirely by screen-printing. It may seem strange that any hand-printed papers at all should be produced in today's highly industrialized society, but there are important reasons why this should be so — reasons that are best appreciated by considering the methods of production.

In *machine printing*, wood-pulp paper is fed into the machine from a continuous reel of very great length, and all the colours are printed simultaneously; it is possible to print a design with as many as twenty different colours all at the same time. The paper passes round a large drum. The pattern is printed upon it by cylindrical rollers, a separate roller being required for each colour. The rollers are in contact with the face of the paper as the drum revolves, and they in turn are fed with colour from a belting of blanket felt which passes through a trough of colour. The printed paper is conveyed through a drying chamber, and as it emerges it is cut off into single rolls of the required length. The essential feature of machine printing is that when the machine is set up, a reel of paper which may be as much as three miles in length is printed in one operation.

In *hand-printing*, too, a separate frame is needed for each colour in the design, but the difference lies in the fact that the paper is cut before printing into separate pieces, each piece the length of a roll of wallpaper. The paper is laid on to a flat surface and each roll of paper is printed individually. Not only that, but each colour is printed separately and each must be dry before the next is printed. The point which is immediately obvious is that there are physical limits to the quantity of paper that can be printed in this way, dictated by the available amount of space and manpower. Hand-printing is therefore more costly, though perhaps not as expensive as might be expected because the operation is carried out with remarkable speed. It must be remembered, too, that machine-printed wallpapers, like any other mass-produced item, are only cheap when very large quantities are involved, because of the considerable capital outlay that is required to get the material into production, and the greater the quantity produced of any particular pattern the cheaper it becomes.

It must be emphasized that hand-printing represents only a minute fraction of the total production of wallpaper, but it satisfies the constant demand for a supply of individualistic papers produced in strictly limited quantities. Such papers can be used when a comparatively small quantity of some particular design is required — and indeed, in situations where a unique effect is wanted, or where a pattern printed in a special colour scheme is needed. A further point is that it is possible to produce a pattern with a bigger repeat by means of hand-printing, because the size and diameter of the rollers used in machine-printing put a definite limit on the length of the repeat.

The difference between machine-printed and hand-printed papers can be discerned by examining the margin, which in the hand-prints shows the register of the repeat of each block, but of course most papers are now supplied without a selvedge. The finish of the pattern can also be observed at the ends of the piece, because in a hand-print there is a portion of plain ground left clear of pattern at each end, whereas in a machine-printed pattern the continuity is unbroken. An expert will also detect the difference in colour surface as between screen-printed and roller-printed colour.

There is now an enormous range of wallhangings available; at no time in the history of wallpaper has there been such a variety of colour, design and texture, or so many different types of both natural and manufactured material used in their production. At any given time there are thousands of different patterns on sale, each with its own distinctive qualities. The skill of the paperhanger consists of knowing how to prepare a given surface in the most effective manner for the sample in hand, what adhesive to use and at what consistency, and which technique of hanging will display that particular paper to the best advantage. There are bewildering numbers of variable factors. Before considering methods of application it is necessary to identify the main categories, defining the peculiar qualities of each.

PRINCIPAL VARIETIES

Basically, wallpaper consists of wood-pulp paper upon which colour is applied either in the form of a regularly repeated pattern of specific size or as a random effect without any regularly repeated feature. The cheapest kind of patterned wallpapers are called *pulps* ; the pattern is printed directly on to the raw wood-pulp paper, the natural colour of which forms part of the design either as a background or ornament. As a general rule, however, the whole surface is machine-coated with a ground colour first and the pattern subsequently printed on to this; such papers are called *grounds*. *Moiré* papers have a delicately engraved surface which gives them a watered silk effect. *Soirettes* are self-colour satin papers with a slight relief and are traditionally used chiefly as bedroom papers. *Jaspés* are printed in varicoloured "veins" which merge and intermingle. *Micas* are papers in which either the ground or the pattern is printed with mica to produce a silvery satiny sheen. *Ingrain* papers are pulps of substantial quality with a textured surface, obtained not by embossing but by mixing wood chips, chopped straw, or fibrous matter of various kinds into the wood pulp during manufacture; they are sometimes known as *oatmeals*. Some have a pattern printed on them, others are finished with a random broken-colour effect. At the present time, ingrains are being very widely used as a base paper to give a textured surface to a wall, the paper being afterwards coated with emulsion paint.

The chief methods of printing and manufacture fall into the following categories–

- (1) Surface printing
  - (a) Cheap pulps, ungrounded
  - (b) Grounded papers
- (2) Relief printing and embossing
  - (a) Simplex
  - (b) Duplex
- (3) Rotogravure

Surface printing

In this type, the printing surface is in relief, and prints show the "edge squash" and mottled appearance characteristic of relief printing. Ground colour and design can be printed in one operation, and a light emboss may be applied as an extra process. Most low-cost papers are produced in this manner.

Relief embossing and printing (Simplex and Duplex papers)

The grounded paper in a single layer (Simplex), or a double layer (Duplex), is passed between inked steel rollers with the pattern in relief mating with a backing roller of compressed wool and paper (referred to as the "bowl") under considerable pressure. Thus the pattern is printed and embossed in a single operation, and the colour printed in the recessed parts gives greater depth to the relief.

Simplex emboss is not so pronounced as Duplex emboss and tends to lose some of its relief on pasting and hanging, whereas with Duplex the grounded paper is laminated to a backing paper with a starch adhesive before inking and embossing, and accommodates a deeper emboss than is possible in a single layer of paper.

Duplex papers generally have more weight and thickness than Simplex and are considered to be of higher quality. They are used extensively for fabric and texture embossed effects.

Rotogravure

In rotogravure (or rotary photogravure) printing, the design is etched in the surface of a chrome-plated, copper-faced cylinder by a photoengraving process; ink is applied to the roller and any excess removed by a steel doctor blade, leaving ink in the etched design which is transferred under high pressure to the grounded and smoothly calendered paper. Up to six colours can be printed in one pass, producing subtle tonal and additive colour effects and an impression of high quality.

Vinyl wallpapers

Vinyl papers were introduced soon after the development of PVC-coated fabrics and are now the most important single group of wallpapers in present-day production, constituting the biggest area of growth in the industry. In 1969 they represented 6 percent of the total quantity of paper made and 6 percent of the retail trade; four years later they accounted for 20 percent of the total production and 40 percent of the retail trade, and this rapid expansion has continued unabated. Vinyl wallpapers consist of paper coated or laminated with a tough waterproof film of PVC (polyvinyl chloride) and they are printed with specially produced inks and dyes that can be fused into the surface. The coating is highly resistant to chemical attack and difficult exposure conditions; so, obviously, they are capable of withstanding washing and scrubbing with detergents, and they have completely superseded the old Washables, Sanitaries and Varnished papers. In addition, they withstand abrasion and mechanical damage far better than ordinary wallpapers. When the time comes for redecoration they are readily stripped by peeling off the vinyl face, leaving the paper backing on the wall as a lining for the new wallcovering, in which case the joints on the backing paper should be rubbed down with medium grade glasspaper. Vinyl is now applied to many wallhangings that appear to be quite delicate but which are in fact extremely tough, an outstanding example being *vinyl flocks*.

There are certain limitations to the use of vinyls because they are prone to discoloration in sulphurous atmospheres, due to a chemical reaction with the PVC coating. White and delicate tints are affected by gas fires and tobacco smoke, and the discoloration has a permanent effect on the coating which cannot be washed off. There may also be problems under certain conditions of central heating and lack of ventilation.

Washable papers

These papers have supplanted the old-fashioned sanitaries for use in steamy areas such as bathrooms and kitchens; they are normally printed wallpapers with a tough waterproof surface coating of transparent plastic. They can be washed or scrubbed when they become soiled, and are resistant to knocks and abrasion. When redecoration is needed they are easily stripped by abrading the top lamination of plastic followed by the normal stripping process, though many of them are now supplied with easy-strip coatings.

Salubra is an oil-printed washable paper supplied in plain tints or with a varnished finish. It has been very largely superseded by the present-day washable papers and vinyls.

Pre-pasted wallpapers

These have an adhesive coating on the back which is activated by contact with water. They are supplied with a plastic trough in which the water is placed; the paper after immersion in this is drawn upwards till the required length has been withdrawn and is then smoothed down with a sponge. They are made to be stripped off easily by lifting the two bottom corners of the paper, the whole length being pulled steadily off the wall from bottom to top.

Pre-pasted papers are immensely popular in America where they are used more than any other kind of wallhanging, but when first introduced into this country in 1961 they made little or no impact on the trade, being regarded as solely for the amateur market. They were re-introduced ten years later with a sales promotion which successfully launched them as a major factor in this country's wallpaper trade. The skilled paperhanger prefers his traditional methods, but the use of pre-pasted wallpapers is growing rapidly, and certainly when properly handled they offer many attractive features.

Non-traditional paper

A still more recent development is the type of wallpaper which is neither pre-pasted nor pasted on the board by the traditional paperhanger's method. With this product, it is the *wall* that is pasted, with the special adhesive recommended by the manufacturer, and the dry wallpaper is hung straight from the roll and pressed into contact with it. The whole operation is designed to be extremely simple, as the wallhanging is made of a foam film which is extremely light in weight. It is suitable for kitchens and bathrooms, being resistant to steam and water vapour, and it is easily stripped off for redecorating. It is easy to apply behind pipework and around fitted units. Being so light in weight, it is easily applied to ceilings. The pioneer in this field, and by far the most important product of this type, is the well-known ICI material called Novamura.

Hand printing

Hand prints consist of three main types — Block prints, Flock prints and Screen prints.

Block prints are produced by using flat boxwood blocks upon which the design is hand-carved or machine-routed to give a pattern in relief. A separate block is used for each colour; the colour is picked from an inked blanket, and the pattern is transferred by pressing the block against the grounded paper. Panoramic or scenic designs stretching over several normal wallpaper widths may be produced in this way, and designs involving more than 200 separate printing operations have been made.

Inks are similar to those used for machine printing, and festoon drying is used between each printing.

Flock prints are a special kind of block print in which the design is printed by wood block, using an adhesive medium which is then "flocked" by dusting or blowing the surface with fine fibres of wool, silk, rayon or nylon. Heavy flocks have a pattern with varying degrees of relief, built up by repeated flockings; they are sometimes supplied in white so that they can be painted after hanging. Flock papers have always been popular at every stage in the history of wallpaper and although very expensive they are still much in demand today. They give a room a most opulent appearance; the raised pile pattern simulates the rich effect of a Genoese velvet hanging.

Strangely enough, when flock paper is hung on a wall it does not attract dust; on the contrary it repels it, very much as the close-packed pile of a Wilton carpet throws off dirt, and for this reason it retains its rich effect for a very long time without beginning to appear dingy.

Screen prints are produced by forcing printing ink through a silk screen "stencil" with a rubber squeegee on to a heavy quality grounded paper, and are characterized by a heavy ink layer of great solidity. The design is transferred photographically to the screen which is coated with dichromated gelatin; then, after exposure, the soluble gelatin is washed away, leaving the insoluble design on the screen in the form of a "stencil". Although this is mainly a hand process, machines have been developed in which the squeegee is operated mechanically. The great virtue of screen printing is that extra large-scale designs can be produced, and short runs of special colourings can be made with comparative ease, but the papers are necessarily expensive because of the handwork involved.

Relief materials

These are decorative patterns and texture effects in various degrees of relief supplied under registered names, as follows.

Lincrusta is a solid-backed low relief material made from a putty-like composition coated on to a paper backing, the relief pattern being impressed in the composition while it is still soft. The composition is made from blown linseed oil, oxidized to a "gel" by a special process similar to linoleum manufacture (hence the name). This product is supplied in a putty colour in a variety of wood effects, decorative patterns, and textures which can be painted and scumbled as required.

Anaglypta (Fig. 29.1) is a hollow-backed relief material which is produced by laminating a white kraft-type paper with a pulp-paper backing by a duplex process, the two papers being passed while still damp between a steel embossing roller and a paper "bowl" to form a shallow relief. Because the emboss is made in the damp material, as opposed to the normal process of pressing the emboss into a dry paper, it follows that the shaping of the emboss when dry is the natural form of the material, which therefore has much less tendency to stretch and consequently retains its relief pattern even when soaked with paste. Until recently, Anaglypta was supplied in high-relief panels and ornaments, incorporating heraldic devices, decorative swags, and replicas of

Courtesy: Crown Decorative Products Ltd

Fig. 29.1 Anaglypta Tanglewood

various architectural features, but these are no longer being made and the material is now available only in low-relief effects.

Anaglypta is supplied in white only, and is usually decorated with emulsion or oil paints after hanging.

Supaglypta (Fig. 29.2) is a hollow-backed high-relief material produced by embossing a pulp "web" made from cotton linters, China clay, rosin size and alum. While the web is still moist it is passed between a steel embossing roller and a gutta-percha "bowl" to produce a high-relief pattern which has greater depth than Anaglypta and possesses excellent emboss retention. As with Anaglypta it is usually decorated after hanging, although sometimes it is supplied in ready-decorated finish. The very high-relief panels which were originally made from the same type of pulp in hydraulic presses are now discontinued, having been largely superseded by fibreglass and plastics.

Vynaglypta, a solid-backed low-relief, is similar in appearance to Lincrusta and is produced by embossing a thick vinyl coating on a pulp paper backing

Courtesy: Crown Decorative Products Ltd

Fig. 29.2 Supaglypta Olympus

(Fig. 29.3). It is supplied in a white washable finish, but can also be painted with emulsion paints. This is a fairly recent production, but it would seem that with further development it might well supersede Lincrusta.

Metallized plastic panels are square panels which can be used singly or, for large areas, in groups. There are plain panels with an all-over pattern giving the appearance of a hammered metal finish, and there are also ornamental panels with relief motifs which simulate such things as Celtic metalwork designs. They provide a permanent finish that is easily kept clean by sponging down. Moulded thermoplastic panels of small size provide a range of ornamental relief patterns in various colourways.

Metallic papers, as their name implies, are papers in which the pattern on the ground is printed in an imitation gold or a tinted bronze powder; real gold leaf is used in some of the high-class productions. *Tekko* is the proprietary name given to a particular kind of continental wallpaper, although similar materials from other sources are available under different names. It is a metallic silk-faced paper which gives the effect of rich damask and which reflects the light in

Courtesy: Crown Decorative Products Ltd

Fig. 29.3 Vynaglypta Derwent

a very subtle manner, providing a continual interplay of light and shade. It is a very lovely material; it is very expensive, being priced by the metre rather than by the roll, but is surprisingly hardwearing and does not deteriorate with age.

Woodgrain papers are photographic reproductions of fine wood grains, printed on good heavy-quality paper. *Marble papers* are similarly reproductions of fine marble markings now produced by the rotogravure process.

Flock papers. As already indicated, the finest quality flock papers are hand printed, but there is also a good range of machine-printed flocks available, and vinyl papers with a flock effect are also obtainable, so that the rich appearance of this delightful material is not restricted to the very wealthy, but is within the reach of practically every pocket.

Japanese grasscloth is one of the most expensive wallpapers on the market and has been popular for a number of years. It originated as the traditional material for the decoration of shrines and temples and was produced as a cottage industry, being made from the inner core of the long sinuous stems of wild honeysuckle, dyed with vegetable colouring-matter. The stems were woven with metallic thread and mounted on a rice-paper backing. Each piece was an individual production, and no two pieces were alike. When this exotic material was seen in other countries it was widely acclaimed, and its manufacture was developed along commercial lines with a more robust backing paper and less fugitive dyes. No attempt can be made to match up the weave along the joints, and in fact the random effect produced is highly prized as an attractive feature, although sometimes the joints are covered with beading. When first hung it looks very pleasing, but in a relatively short space of time the edges begin to curl and fray, and the general appearance becomes tatty and unkempt.

Oriental silk cloths originated in Japan and Thailand, and consist of very thin silk which is woven and laminated on to a coloured base-paper. They are very delicate, unsuitable for areas where they would be subjected to hard wear, and are easily marked and damaged during application.

Cork papers and similar cork wallhangings have been popular for some years, possibly because they are associated in people's minds with the well-known southern European holiday resorts. They are made from natural cork from Spain and Portugal, obtained from the bark of the native cork-oak tree. The cork is boiled and then sliced into extremely thin veneers which are then mounted on to a backing paper coated with a coloured adhesive material. The attraction of these papers lies in the variations of colour and texture from one piece to another, so that within the bounds of one room there can be considerable variety; this is an endless source of wonder and pleasure to smart interior decorators and designers. But the beautiful deep smoky-brown colouring that looks so delightful when the paper is first hung soon bleaches out and fades to a wishy-washy beige of peculiar ugliness.

Lining papers

The purpose of a lining paper is to provide a good foundation of completely uniform porosity for the reception of good-quality wallpaper or relief materials. It can also be useful for improving uneven or badly cracked walls or ceilings before paperhanging, or to form a base of regular absorbency for the reception of paint or emulsion paint.

Lining papers are supplied in white and in colours and in various weights and qualities, such as common pulp or smoothly calendered. Coloured linings are used for delicate fabrics, Japanese grasscloths, and silkcloths. Strong brown linings are supplied for rough walls, and for heavy relief materials such as Lincrusta-Walton. Tar and pitch papers are used for damp surfaces, with the pitch side to the wall. Calico or muslin-faced linings in brown or white are made for lining badly cracked walls and ceilings or for jointed partitions, but the supply position is rather uncertain. If the lining is unobtainable, the wall surfaces will have to be scrimmed where necessary (see page 393).

Discontinued lines

There are several kinds of wallpaper that were on sale until recently but which have now been discontinued by the big manufacturers. There is the possibility that for quite a long time to come a few odd rolls from old stock may still be seen, and for this reason some mention of the techniques of hanging them is included in the next chapter. They include satins (with a polished distemper ground), polychromes (lightly embossed with many colours blended but no definite pattern), plastic paper (embossed texture effect), plastic prints (smooth-backed, with a relief pattern formed with a friable gesso material), and leatherettes (embossed, coloured and bronzed to look like tooled leather).

Expanded polystyrene

This is a material with very low thermal conductivity which is used in situations liable to condensation. It can be in the form of thin sheeting, supplied either in rolls or in large rectangular sheets, and in this form can be hung on a ceiling or wall surface prior to paperhanging; it can also be supplied in square panels of greater thickness, or in large rectangular panels of 50 mm or 75 mm thickness. The square panels are often used to form an all-over covering for ceilings to prevent condensation, and after application the panels can be left undecorated or else coated with emulsion paint or water paint. On no account should oil paint be applied directly to them because of its solvent action. Normal polystyrene panels present a fire hazard; care should be taken to choose the type specially made to eliminate the fire risk (cf. p. 236).

Wood veneers, etc.

Realwood and *Flexwood* are selected wood veneers, derived from a wide variety of hardwood grains, mounted on thin flexible card backings in large sheets, and in rails of various sizes. After application with a special adhesive, they are stained and varnished. *Japanese wood papers* are extremely thin silky wood veneers mounted on a paper backing, supplied in rolls and hung like wallpaper. They can be stained and varnished, or polished, after application. *Rollywood* is a Swedish material consisting of long and very narrow strips of natural wood, woven together with strong cotton-thread. Various kinds of wood are used, including elm, mahogany, birch, walnut, pine, obeche, etc., and many of the hangings consist of a combination of two or more woods. A brown lining is hung, and when it is dry a latex glue is pasted on the surface and the wood material unrolled and pressed into contact with it. The cut ends of thread are sealed with a touch of quick-drying glue to stop them from fraying (see Chapter 30).

FABRICS

The fabrics used for decorative purposes range from the most expensive silk hangings at one end of the scale to cheap coarse hessian and sacking at the other end, with a considerable number of materials such as linen and canvas, in backed or unbacked form, dyed or in natural colour, in between.

Silk, the costliest material, is too delicate to be subjected to a wet pasting process; the method adopted is to mount the fabric on light wooden frames which are then firmly attached to the wall, the silk being carefully stretched over the frames and secured with tacks; care is needed to see that the stretching does not distort or wrinkle the silk.

The *very cheap hessians* and *sackcloths* are not much used today. Their main purpose is to provide a highly durable covering in situations where very rough wear-and-tear is expected and where appearance is only a minor considera-tion. They can be left in their natural colour or painted after application. But today, fabric wallhangings are an important decorative medium in their own right and there are many types on the market. They are popular because they provide an excellent background for pictures or prints or for the display of glass and china, whilst at the same time they are hard-wearing, resistant to mechanical damage, and relatively inexpensive.

Plain hessian is very attractive; it has subtle gradations of colour, and the texture is interesting because of its irregularity, the occasional knots and twists giving variety. In some cases two separate tones are woven together to give extra interest. Normally there is no difficulty to the householder in cleaning it, but it would not be a suitable choice for the kitchen or bathroom because contact with grease or oil causes staining. *Canvas* is also available in natural form; so, too, is *slubbed woven fabric* made of *linen*, and while these also are susceptible to oil and grease they are more resistant than hessian.

All these fabrics are available in a wide range of very attractive colours as well as in natural form. The fabrics are supplied in a variety of ways. Sometimes they are sold like ordinary rolls of cloth, without any backing, in which case the wall surface is pasted and the fabric unrolled and pressed into contact with it. Alternatively they can be mounted on a backing of latex or stout brown paper, which makes them easier to handle. The latex-backed kind can be pasted and hung like wallpaper or pressed into contact with a pasted wall; the paper-backed kind is pasted and hung like ordinary wallpaper. Another form in which the fabrics are supplied is in rolls of self-adhesive material, which makes for a very clean operation with the edges easily butted so that the joints are barely visible. There are some Swedish fabrics mounted on a rigid backing sheet which is so stiff that it cannot be folded; the advantage of this rigid material is that it makes perfectly butted joints and the weave has no tendency to sag, no matter how much force is used in the application.

Coated fabrics

The original coated fabric was *oilcloth*, a heavy linen coated with white lead and intended for covering tables and shelves, but which was occasionally used to provide a tough durable wallhanging in situations where extreme durability was required. From this the more attractive materials like *Lancaster cloth* and *American cloth* were developed, and these occasionally are still to be seen. Then came a material called *Rexine*, a fabric coated with nitrocellulose, embossed and coloured to resemble leather and intended for the purposes of upholstery and car trimming but later adapted also for use as a wallhanging; it had the drawback, however, of being a flammable material which presented a fire hazard, and efforts to find a non-flammable material to replace it led to the development of a fabric coated with PVC (polyvinyl chloride). This was found to be extremely tough and resistant to damage, even when exposed to the most rigorous conditions, coupled with the advantage of being able to withstand both moisture and chemical attack. The new material was an obvious choice for the decoration of rooms and corridors in hotels, ships, schools and other places subjected to rough usage and harsh cleaning methods, though in such areas an attractive appearance and a feeling for modern design trends are also desirable.

One development led to another, until today there is a first-class range of *PVC-coated fabrics* of great beauty and charm and with all the qualities of resistance to abrasion and cleaning that make them ideal in both the domestic and the industrial fields. The first such fabrics were untrimmed and very wide, and were somewhat difficult to handle. Present-day patterns are available in ready-trimmed rolls of easily manageable size as well as in lengths suited to large-scale industrial use. They are soft and flexible and they handle very easily; even if severely crushed or screwed up whilst in use they do not retain the wrinkles, but stretch out flat and free from marks when applied, a quality which is of great value when awkwardly shaped areas are being decorated. They are available in the British Standard range of colours and in a wide range of very attractive patterns and designs; there are also some striking imitations of natural materials, including some particularly fine wood-grain effects.

Padded and quilted PVC fabrics are primarily intended for shopfitting and for use on bar fronts, pelmets and upholstery, but are sometimes applied as wallhangings; in this form they are, of course, extremely robust, and because of the padding they possess high thermal insulation and acoustic properties; they may be supplied in rolls or backed with hardboard for panelling purposes.

Other decorative surface treatments

All the materials so far discussed come well within the scope of normal paperhanging practice, but there are many materials used in present-day decoration which call for quite different techniques of fixing and hanging. Any ambitious young decorator who wants to keep abreast of modern trends, and who aims to be able to supply whatever his customers may demand, ought to be aware of these materials and of the application techniques they require. There are, for instance, many kinds of panelling which are available in large sheets and which are fixed by securing them on wooden battens; they include *hardwood plank panellings,* consisting of a solid core of timber with a veneer of fine hardwood, often grooved to simulate tongued and grooved boarding, and finished with wax polish, polyurethane lacquer, or stain and varnish; *sculptured timber panellings* giving the appearance of unplaned planks and similar rustic effects; moulded *thermoplastic panellings* in which the panel and stiling shapes are pre-formed and the colour incorporated in the plastic; *perforated claddings* of aluminium, anodized in gold, silver or black; *marble finishes, glass-fibre panels,* and several others. The materials supplied in smaller units include *ceramic wall tiles* which may be plain, patterned or hand painted, *concrete tiles, vinyl wall tiles, cork tiles, concrete panels* moulded into decorative mural motifs, and *glass-fibre* panels and murals. The various technical journals carry feature articles about modern developments of this kind from time to time.

WALLPAPER DIMENSIONS

The standard size of a roll of machine-printed English wallpaper is 10·05 metres long by 533 mm wide when trimmed, in line with Continental practice. Hand-printed papers do not necessarily conform to these dimensions; they are usually 11 metres long but may be supplied in other lengths, in which case a clear indication of the size is given by the manufacturer

American manufacturers work to a different standard; their papers are usually supplied in 7-yard rolls (6·4 metres) or as double rolls of 14 yards length (12·8 metres), the width being 533 mm. Some decorative wallhangings are made considerably wider; in the case of some fabric hangings, grasscloth, etc., a greater width is desirable in order to reduce the number of vertical joints on a wall and to give greater continuity to the weave. Borders, although made up in rolls of standard size, are sold by the metre, according to the length required.

MEASURING UP

Calculating the quantity of paper required for a ceiling

The first thing is to determine in which direction the paper is to be hung; normally this is parallel with the wall in which the window is situated. In a room where the window or windows are on one wall only, this is straightforward. In rooms with windows on two (or even three) walls, it is necessary to find which of the windows is the major source of light, and the paper will be hung parallel with the wall in which this occurs.

Having decided on the direction of the paper, measure the length of the ceiling in this direction, and determine how many such lengths can be cut from one roll of paper, allowing for waste.

A roll of paper is now taken along the wall which is at right angles to the window wall, to measure how many times the width of the paper fits into this dimension, counting any fraction of a width as a full width. This indicates the number of lengths required, from which the number of rolls can be calculated.

Calculating the paper required for the walls

Measure round the perimeter of the room with a roll of paper, ignoring the short lengths over the doors and under the windows, and calling the rest full lengths; this will indicate the number of full lengths required. Then measure the height, to determine the number of lengths that can be cut from each roll, allowing for waste. Divide the number of full lengths by the number each roll will cut, and this will give the number of rolls required. Any fraction of a roll must be reckoned as a full roll. Usually the surplus will be sufficient to provide for the short ends over doors and under windows. In the case of a room with some significantly large feature needing paper of different lengths (e.g. a fireplace wall), calculate the number of such lengths separately and add this to the total.

Large patterns cut to waste much more than small patterns, and this must be borne in mind when measuring up. Similarly, angular shapes such as attic and staircase walls will require more allowance for wastage than straightforward work. In measuring up for panelling or stiling borders, always allow for wastage in mitreing.

What has been described is the practical craftsman's method of measurement, which is invariably used on the actual job. An alternative method is to calculate by the area method. For this purpose the perimeter of the room is measured and multiplied by the height of the wall, giving the superficial area in square metres. This is divided by 5·356 which is the number of square metres in one roll of paper. The result is the number of rolls required to cover the whole surface, from which deductions are made for windows, fireplace and doors, and 10 percent added for wastage and matching. This method is suitable for working from floor plans and elevations when the actual site cannot be inspected, but in the ordinary course of events it is not a method that can be recommended, as the possibility of error is too great.

SELECTION OF WALLPAPERS

The selection of wallpapers cannot, of course, be reduced to a formula of hard-and-fast rules, but there are certain basic principles which are applicable to all forms of interior decoration and which should be observed if the result is to reflect good taste. Many decorators are content to send round a pattern book and leave the choice of wallpaper entirely to the client. This is plainly shirking one's responsibility, for the decorator's job is not merely to carry out the work, but to advise and assist his client in every way to make a wise choice. In this he must be something of a diplomat, for whilst he should give first consideration to his client's wishes and avoid any appearance of imposing his own views, he must nevertheless try to prevent his client from making any choice which he knows from practical experience would be unsatisfactory.

The public is now fairly widely informed (often misinformed) in matters decorative, from countless articles in magazines and the daily press by so-called decorative "experts", and the decorator must always be ready to discuss the merits or demerits of any particular scheme about which his client may have read. It is not enough to speak disparagingly of some novel suggestion; the client will respect his opinion more if the decorator is able to give an intelligent explanation as to why a suggestion may be good or bad in practice.

Let us examine some of the guiding principles involved in deciding upon a scheme of wallpaper decoration. The room itself and the purpose for which it is used will impose certain conditions. It may be a large room or a small one, bright and sunny or dark and depressing. It may have good architectural features of a definite period, or it may be a stark, box-like apartment entirely devoid of any wall subdivisions. If there are existing furnishings, such as curtains, carpets, upholstery, etc., the colour of these will be a deciding factor in choosing the wallpaper.

Colour

A room with a northern aspect receives no direct sunlight and tends to appear cold unless warm colour is introduced in the walls. Rooms facing south get most sunlight and may therefore have cooler colour, such as blue or grey if these are desired. Colour preference is such a personal matter, however, that it generally over-rides any consideration of aspect. It will be found, for instance, that in this country, warm colours are more generally preferred than cool colours (due no doubt to our not-overabundant supply of sunshine) and this natural preference will weigh more than the question of aspect in deciding the choice of colour.

The suggestive use of colour is dealt with more fully in a separate chapter (page 519), but the following points may be noted here. Pale blues, greens and greys suggest coolness and distance. The warmer colours — pinks, buffs, fawns, browns, etc., are more intimate and cosy. Light colour tends to make rooms look bigger, while strong or dark colour has the opposite effect. Colour plays a big part in creating "atmosphere" which may be bright, cheerful and stimulating as in entrance hall and dining-room, or quiet and restful as in lounge, study, and bedroom.

Pattern

When choosing wallpaper from the pattern book it should be remembered that a pattern looks larger in a small sample than it does in the piece or when hung on the wall. Scale should therefore be well considered since a small pattern would tend to look monotonous in a large room, whilst a large bold pattern would be too assertive in a small room.

Pattern looks much more interesting when used in conjunction with a plain or semi-plain paper instead of covering the four walls uniformly. The reason for panelled effects coming back into fashion at regular intervals is probably connected with this fact, i.e. that it satisfies our sense of architectural "rightness" to see a free-flowing or meandering all-over pattern stabilized in some sort of frame such as a stiling border. Any border used with a patterned paper should be severely simple, sufficient only to form a pleasing division between the pattern and plain stiling. Panelling, however, tends to introduce a note of formality which is quite out of place in some rooms, such as, for instance, a child's bedroom or a nursery. A sense of the appropriate should always be preserved; busy restless pattern should not be used in lounge or bedroom, nor should there be any assertive or distracting pattern in office or study.

Where walls are free from restricting subdivisions and projections such as picture and dado rails, greater freedom in the use of plain and patterned papers is possible. For instance, the pattern may be reserved for a single large wall or alcove to form a rich background to a bedhead or a divan, the remaining walls and ceiling being treated in a light plain or low-texture effect. This gives a light airy appearance to the room and also enhances the effect of the patterned paper, giving it a more "precious" quality. Similarly, a dual-purpose room such as a bed-sitting room, or a dining-cum-living room, might have just one corner in a pattern or a rich texture in an otherwise plain scheme, thus creating more interest and defining its dual function.

The "selective" treatment of one complete wall in a pattern, woodgrain, or rich texture effect, in contrast with other walls in light plain pastel tints, is now an established departure from the orthodox, capable of some very fine results when handled with tasteful discrimination.

Texture

Lightly embossed papers such as low plastic textures and fabric effects give more "surface interest" to plain and semi-plain papers, but they often give rise to objections from the housewife on account of their alleged dirt-catching propensities. Such objections are often greatly exaggerated, for the emboss invariably flattens out somewhat on hanging, while still retaining the effect of the original texture. Where such objection has to be met, however, it is the *character* of the texture that should be considered. For example, an ingrain, though extremely low in texture, is more likely to hold dust (due to the "clinging" nap surface) than would a normal low plastic emboss which usually has a "rounded" texture, free from sharp edges or points.

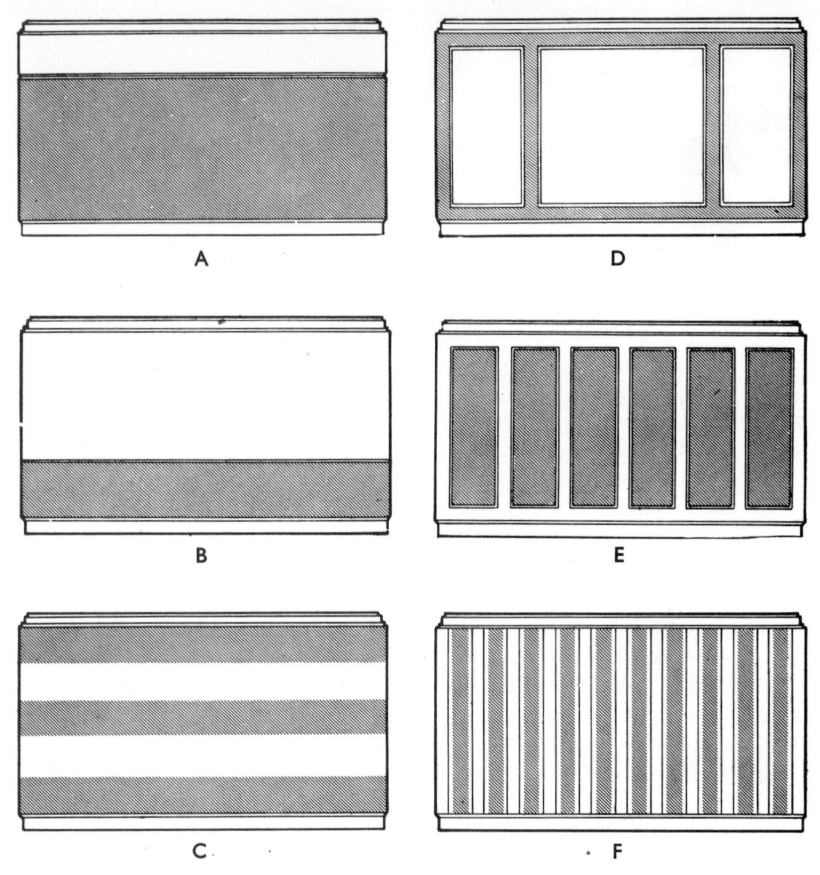

Fig. 29.4 Subdivision of wall surfaces

Horizontals give an effect of increased width and reduced height, whilst verticals tend to suggest increased height

Proportions

Vertical stripe patterns will tend to give height to a low room while, conversely, horizontal bands, friezes and dados all tend to reduce height. Thus ill-proportioned rooms may be improved by the judicious use of vertical or horizontal subdivisions. Where a non-pattern paper is required, height may be suggested by vertical panels; for example, a fairly narrow panel at each end of a wall with one large panel in the middle (Fig. 29.4(D)). Similarly, a tall room may be apparently reduced in height by using two papers of similar texture but slightly different tone and hanging them horizontally to give the effect of alternating light and dark bands (Fig. 29.4(C)). This is more in keeping with modern tastes than subdividing the wall into frieze, filling and dado. A coloured ceiling will also have the effect of reducing the height of a room, especially when the wall colour is lighter than the ceiling. It is important to remember that any subdivision of the walls — whether as frieze, dado, or panelling — tends to make rooms appear smaller.

30

Paperhanging

TOOLS REQUIRED

The paperhanger will require the following tools: a good pair of paperhanger's scissors with blades of about 250 mm length; a felt-covered roller of between 175 and 225 mm width; a 25 mm boxwood seam roller and an angle roller (Fig. 30.2); a papering brush, metric rule, plumb bob and chalk line; a small trowel and a plasterer's small pointing tool, a sponge and a supply of clean cloth. He will also need a small hammer, a screwdriver, and a pair of pincers or pliers to cope with the removal of small fittings and the inevitable nails which require either drawing or hammering in. These are the necessities. In addition, there are various useful accessories which the conscientious man will wish to provide. Many paperhangers like to have two papering brushes, a sturdy one for dealing with heavy relief papers and the general run of wallpapers, together with a slender one with soft bristles for hanging the really delicate papers. Virtually all papers are supplied ready for immediate use, but for those that may still require trimming a pocket trimmer is necessary. For the very high-quality papers which may still be supplied with a selvedge the paperhanger will need a bevelled steel straightedge 2 metres long, a zinc strip 75 mm wide, and a trimming knife, or alternatively a Ridgely patent Trimmer and Straightedge (see Fig. 30.6, page 397). A casing knife, sometimes called a corner knife, and a casing wheel with a circular cutter, will also be useful in some circumstances (Fig. 30.5). For cutting around light-switches and similar obstructions, some paperhangers like to equip themselves with a pair of manicure scissors.

It is important that papering tools should be washed frequently and kept scrupulously clean. Paste which has dried and hardened on the bristles of a papering brush will cause scratches on the surface of the next paper to be hung, and if paste is habitually allowed to dry on the scissors it will, because of its acid nature, eventually cause corrosion and pitting of the blades.

PREPARATION OF THE SURFACE

This is a matter of the utmost importance. Before applying wallpaper it is essential that the surface should be properly and thoroughly prepared.

The ideal surface for papering is one with a slight but uniform degree of suction, free from cracks, nibs and other blemishes. It should also be completely dry, otherwise the colours used in the printing may be affected;

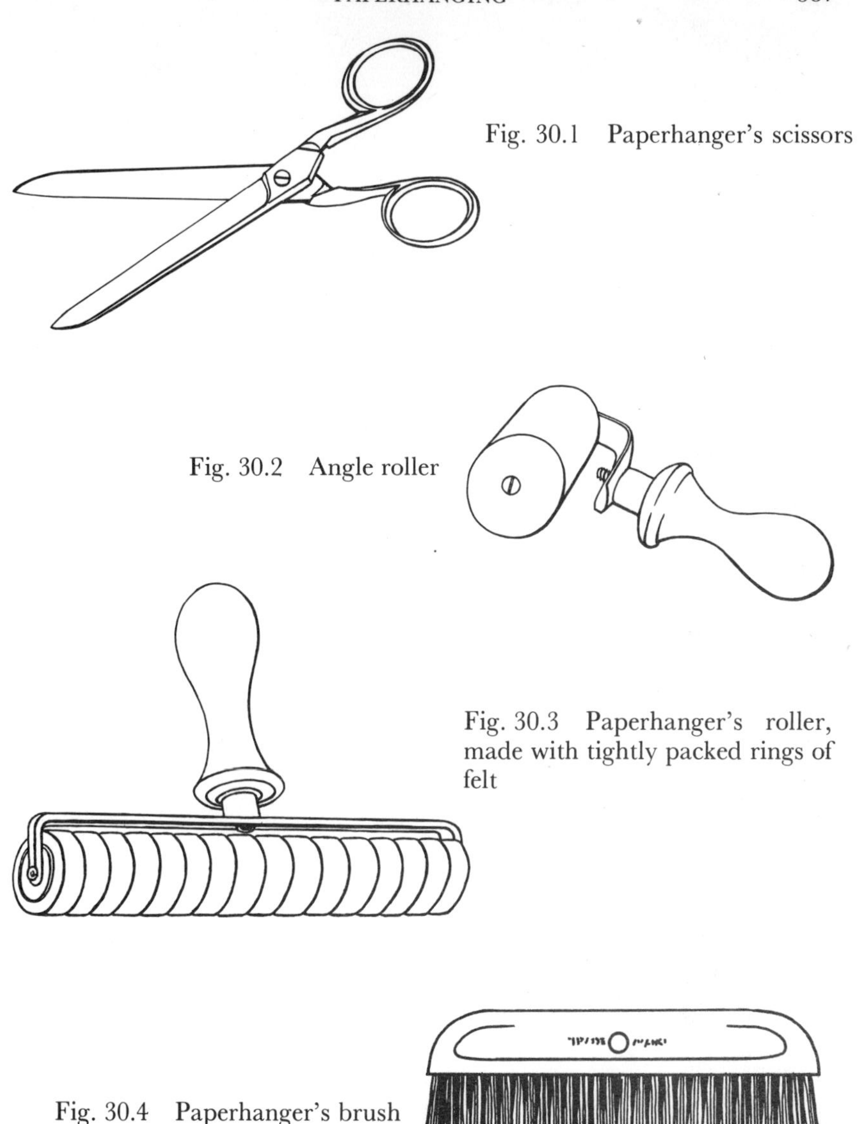

Fig. 30.1 Paperhanger's scissors

Fig. 30.2 Angle roller

Fig. 30.3 Paperhanger's roller, made with tightly packed rings of felt

Fig. 30.4 Paperhanger's brush

moreover damp conditions cause size and paste to putrefy, encouraging the growth of mould or mildew.

If the porosity of the surface is not uniform, the paper dries out in a patchy manner which may cause the joints to open because of uneven shrinkage. To help in achieving uniform porosity, part of the preparation includes sizing the surface and, in certain cases, lining.

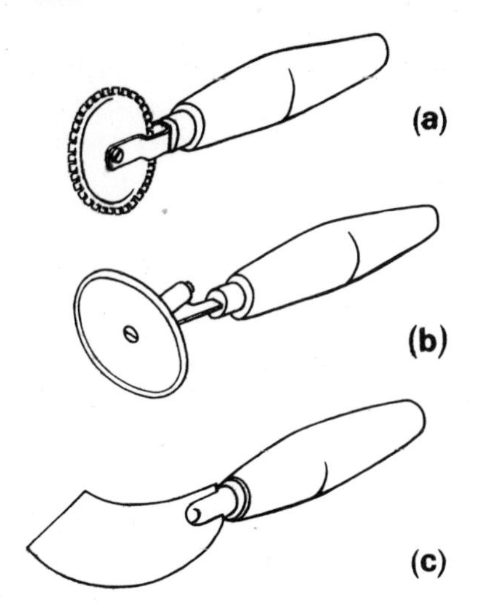

Fig. 30.5 (a) Serrated corner wheel; (b) normal corner wheel or casing wheel; (c) corner knife or casing knife

Sizing

The application of a coat of weak jelly size helps to equalize the porosity and also gives some degree of "slip" to the paper, enabling the paperhanger to slide it into the required position; without it, the paper clings to the surface as soon as it touches, and cannot be properly manipulated. The size should be mixed with hot, but not boiling, water, otherwise it will not gell properly when cool. Unduly absorbent or "hot" surfaces may require two thin coats of size. But the common error of making the size too strong must be avoided, as this seals the surface and causes the paper to remain wet for too long a time, leading to excessive shrinkage.

For all normal purposes glue size is perfectly satisfactory, but not when materials such as vinyl or washable wallpapers are being hung. Because of their impervious nature these papers take much longer to dry out than ordinary papers, and this may cause glue size to putrefy; it is customary to use a coating of thinned cellulose paste for sizing, and the manufacturer's instructions should be followed.

Lining

Lining paper is used wherever a high quality of paperhanging is required and whenever an expensive wallpaper is being hung. It provides the ideal surface for the reception of wallpaper, a surface of exactly the right degree of porosity and one which of course is entirely uniform. Lining paper is also used to cover uneven walls, to mask hair-cracks and other blemishes, and to counteract any failures of adhesion on excessively smooth hard surfaces.

PREPARATION OF NEW SURFACES

Lime plaster

The plaster must be completely dry before papering is attempted. The surface should be tested with litmus paper for alkalinity. If it is chemically neutral it should be cleared of any mortar or plaster splashes, any cracks or irregularities should be filled, and it should then be sandpapered down to remove nibs, and given a coat of weak glue size. If the surface is chemically active it should, after the sandpapering, be given a coat of alkali-resisting primer, followed when dry by weak glue size.

Retarded hemihydrate plaster

The plaster must be dry. Mortar and plaster splashes are removed, cracks and irregularities filled, the surface sandpapered and then given a coat of weak glue size. If any lime was added to the plaster it should be treated as a lime plaster.

Anhydrous plaster

Plasters in this group are often highly trowelled and present a very hard polished surface; they are also liable to a fine powdery film of efflorescence. The plaster must be dry, irregularities and cracks filled, the surface sandpapered to remove splashes, nibs, efflorescence and other loose material, and a coat of weak glue size applied. Where a coat of sharp priming paint has been applied, as in the process of "following the trowel" (see pp. 220 and 229), it is usually desirable to line the surface after sizing. If a good-quality wallpaper or in fact any form of wallpaper is to be hung, it is wiser to hang a lining paper first because of the smoothness of the finish.

Lightweight aggregate plasters

The plaster must be dry, cracks and irregularities filled, splashes and nibs removed with sandpaper, and weak size applied. Since these plasters are usually highly absorbent, it may be necessary to use *two* thin coats of size.

Wallboards and sheetings

The surface must be clean and dry, and any nail or screw heads touched in with zinc-chromate primer. Wallboard and plasterboard surfaces need special treatment if the joints are not covered with wood moulding or beading. The joints are usually left slightly open to allow for expansion, and they require scrimming to prevent the paper from splitting at the joint. Strong calico scrim or linen cover-strip is available for the purpose, or alternatively a good quality of 75-mm wide cotton bandaging may be used. A coat of size or paste is first run down each joint in a band about 100 mm wide and allowed to dry. A second coat of stiff flour or starch paste is then applied and the bandage or scrim pressed into contact with a paperhanger's roller. Alternatively the surface may be sized and then hung horizontally (i.e. across the joints) with reinforced calico-faced lining, if available.

A further point to note is that these boards are often insufficiently battened at the back (that is, the wood supports are spaced too far apart) and there is a danger of the wallboard surface being warped or buckled when hung with wallpaper, due to absorption of moisture from the paste and contraction of the paper when drying. There is also the difficulty in the case of paper-faced boards that when at some future time the wallpaper has to be stripped off, the water used for stripping may soften the paper facing of the boards. To overcome these two dangers it is advisable to seal the surface of the wallboards with priming paint before the surface is sized and papered; the priming paint will help to prevent moisture from penetrating the face. Even so, it must be recognized that stripping will need to be done carefully so as to avoid damaging the face of the boards.

Asbestos cement sheeting

The surface must be clean and dry, nail heads or screw heads touched in with zinc-chromate primer, and the joints and nail or screw heads filled with a proprietary filler which when dry is sandpapered. Because of the danger of the alkaline asbestos-cement affecting the colours in the wallpaper, a coat of alkali-resisting primer should be applied, and this will facilitate the removal of the paper when the time comes for stripping. This is followed with a coat of weak size.

Concrete

Difficulty may arise because of the mould-release oil used to help in separating the shutterings and moulds from the concrete, the oil film left on the surface being liable to stain the wallpaper. This oil must be washed off with detergent solution. The surface must then be allowed to dry completely. Any blow-holes in the surface must be filled, and the surface then sandpapered down, dusted off and coated with weak size.

Lightweight aggregate concrete

The same conditions apply here, with the additional factors (a) that this type of concrete requires a longer period of drying out than normal high-density concrete, and (b) that it is more porous than normal concrete and may require two thin coats of size.

Aerated lightweight concrete building slabs

These slabs, because of the type of aggregate used in their manufacture, are far more friable than normal concrete and are susceptible to damage when wallpaper is being stripped from them. They must be perfectly dry; they are brushed down vigorously with a stiff fibre brush and coated liberally with wallboard primer, thinned down to assist penetration, and when this is dry they are coated with weak size.

PREPARATION OF OLD SURFACES

Stripping of previously papered surfaces

When repapering old walls it is most desirable, both on technical and hygienic grounds, that the old paper should be removed. Failure to do this might result in the whole of the paper becoming loose and developing large blisters, and it will certainly mean that the joints of the old paper will be visible through the new decoration in a most unsightly fashion.

The paper should be well wetted several times with warm water and allowed to soak thoroughly; it is then removed cleanly with a stripping knife or broad scraper. Care should be taken to avoid digging into or "stabbing" the plaster with the scraper, since some old surfaces are very easily damaged in this way. Where more than one paper is present and removal is difficult, it is far better to spend time in repeated soaking rather than in persistent but ineffectual jabbing with the scraper, which invariably results in damage to the plaster.

Varnished papers are the most difficult to remove because, of course, water will not penetrate the varnished surface and so it is impossible to soak the paper. Fortunately these materials are not much used nowadays, and it is only on older properties that they are likely to be encountered. It is first necessary to soften and remove the varnish; this can be done by applying a proprietary varnish-remover, or alternatively by means of a home-made composition of stiff flour paste with a pound of caustic soda added. The paste "stays put" when brushed on to the surface and it should be allowed to remain there until the varnish is softened. Repeated applications are necessary in stubborn cases, such as where more than one layer of varnish paper is present, and the process is both tedious and unpleasant. Floors, paintwork, hands and clothing must be protected from the caustic stripper, and only vegetable fibre brushes should be used.

Where impervious materials such as varnished papers, Lincrusta or Anaglypta are to be removed, or when papers coated with water paint, emulsion paint or plastic paint have to be stripped, it is better to use the Sanderson wallpaper stripping machine described in Chapter 4. This is easily the most effective method, and as well as speeding up the work there is less risk of damaging the plaster surface and therefore less time to be spent on the wearisome job of "making good". As an added bonus the surface is sterilized by the steam.

With so many vinyl papers and washable papers in use a different situation has arisen, and for stripping these papers another technique is called for. In the case of vinyl papers the method is to lift a corner of the material near the base of the wall and, by pulling the vinyl, to detach it from the base-paper. It is sometimes suggested that the base-paper be left in position and used as a lining for subsequent papers, but it is much better to remove it, and this is easily done by soaking it with water and stripping in the usual way. In the case of washable papers the surface can be scored with coarse sandpaper to allow water to penetrate. Pre-pasted wallpapers are stripped by lifting the two bottom corners of the paper and pulling the length steadily off the wall from bottom to top.

Various kinds of wallpaper stripping compounds are now on sale, to accelerate the process. These are usually supplied in sachets, the contents of which are mixed with warm water. By decreasing the surface tension of the water they enable it to penetrate the paper more quickly.

After stripping, the walls should be thoroughly washed down to remove all traces of old paste and size and small pieces of paper. This is important, as old paste and size left behind on the surface may putrefy and set up a mould infection in the newly decorated surface, whilst any particles of old paper will show up unpleasantly as large nibs under the new paper. It is a useful economy of time if, whilst the wall is still wet, the holes and cracks are made good, especially along the top of the skirting board and around the architraves (see "Making Good", page 281). The surface is allowed to dry and is then sandpapered down and dusted off prior to being coated with weak size.

Previously distempered surfaces

It is essential that all surfaces previously treated with whiting or size-bound distemper should be thoroughly washed off down to the bare plaster, otherwise any paper hung upon them will very soon become loose. Size distemper is generally easy to remove, all that is necessary being to soak it well with warm water, then to scrub it off with the side of a half-worn flat brush, rinsing it several times and finishing it off with a swab or sponge. The water soon becomes charged with the distemper and needs to be changed very frequently; only clean water is used for the final rinse. The holes and cracks are then made good, the surface sandpapered, and a coat of weak size applied.

Water-painted surfaces

Oil-bound water paint cannot be washed off, since it is insoluble. Provided the surface is sound it only needs sandpapering down to remove the nibs, followed by a coating of weak size. If any parts are loose and flaking they should be thoroughly removed by washing and scraping, and it may be advisable to bind down the remainder with a penetrative oily sealer and allowed to dry before sizing. If for any reason the existing coatings are suspect — such as, for example, when it is apparent that there are so many accumulated coats that the adhesion is becoming weak — use of the Sanderson stripping machine should solve the problem. Alkaline strippers are not recommended for this purpose.

Emulsion-painted surfaces

These present a real difficulty, especially when a fairly heavy paper is to be hung. Emulsion paints, being completely insoluble, cannot be washed off, but unfortunately because of their composition they do not penetrate very deeply into a plaster surface, and consequently their adhesion is not so good as that of oil paint or even of water paint. This means that very often an emulsion paint coating is in perfectly good condition and yet when wallpaper is hung upon it the strong contraction that occurs as the paper is drying causes it to loosen and break away from the plaster — which of course means that the paper also becomes loose. It is desirable that this fact be anticipated, and even when the emulsion paint is in completely sound condition it should be coated with a highly penetrative oily primer, allowed to harden, and then coated with weak size, followed by a thin pulp lining. As an alternative, the emulsion paint should if possible be removed with the Sanderson stripper, though it should be recognized that some varieties of emulsion paint cannot be removed in this way.

This problem is one of the commonest causes of trouble in present-day practice. Far too often we have seen whole rooms and in some cases whole houses where the entire decoration has had to be renewed at enormous expense after a matter of weeks or months, because every single length of wallpaper has been falling away from the walls.

Oil-painted surfaces

These are probably the most troublesome surfaces upon which to hang paper; the hard, non-absorbent nature of the ground allows moisture to condense behind the paper, and this may weaken the adhesive to such an extent that the paper tends to lose its grip.

It is necessary first to wash down with sugar soap or soda solution in order to kill the gloss, rubbing down simultaneously with pumice block or coarse flint waterproof abrasive paper to provide a key, the surface being finally rinsed down with vinegar and water to remove or neutralize the alkali. When dry, the surface should be glasspapered and dusted off to remove any grit and coated with weak size mixed with a handful of whiting or plaster of Paris. It should then be cross-lined (i.e. hung transversely to the final paper), using a thin pulp lining paper.

Scrimming

It is sometimes desired to hang paper on matchboarding (i.e. tongued-and-grooved woodwork) in order to match other papered surfaces. In such cases it is necessary first to scrim the surface, that is to cover the woodwork with a thin strong muslin or "paperhanger's scrim". Some decorators soak the scrim in flour paste, wring it out and spread it over the surface, the edges being secured with tacks which may be removed when the scrim dries; others paste the woodwork first, then lay on the dry scrim, using a paperhanger's wide roller to secure good contact, and finally coating the surface with thin paste. The scrim tightens out to a level surface when dry; it is then cross-lined with butt joints before hanging the final paper.

The above method is suitable for old and seasoned woodwork. New tongued-and-grooved boarding is rarely properly seasoned, however, and is capable of considerable "drying-out" movement which causes the joints to open and split if the scrim is pasted down. To avoid this, some authorities recommend that the dry scrim be stretched over the surface, secured with copper tacks, then coated with good jelly size, laying it *on* the surface rather than brushing it in. The scrim then tightens up on drying, when it may be cross-lined as above and finally hung with the desired paper.

SURFACE DEFECTS

In all cases the appropriate action must be taken in addition to the normal preparatory processes.

Efflorescence

This, as described in Chapter 19, is readily recognizable as a white furry deposit on the surface; it is caused by soluble salts, present in the structure, coming to the surface as the moisture dries out. Although it usually occurs in new buildings it can also be found in old buildings where moisture or damp are present. It affects the adhesion of wallpaper and may also stain the face of the paper.

Sufficient time must be allowed for moisture to dry out completely, and any efflorescence which forms during this period must be brushed off *dry* at frequent intervals. On no account must the efflorescence be washed off or the salts will be dissolved and re-absorbed by the plaster.

Mould growths

On any surface where these occur it is essential to kill the mould and sterilize the surface before papering. No surface preparation, and no stripping of old wallpaper, must be carried out before sterilization, otherwise live spores will be dispersed into the atmosphere, causing other parts of the fabric of the building to be infected.

Where existing wallpaper is infected, a suitable fungicide should be mixed into the water used for stripping. After stripping, all refuse must be collected and burned immediately. A fresh fungicidal solution is then mixed, and the wall surfaces washed down to remove all traces of old paste and size; the surfaces are then allowed to dry. A new solution is made up and the wall surfaces liberally coated with it; they are then kept under observation for a week, and if necessary washed again with new fungicidal solution.

In the case of new surfaces or previously painted surfaces, the affected area is washed down with fungicidal solution and allowed to dry. The surfaces are then treated with a freshly mixed fungicidal solution and kept under observation for a week as already described.

Damp walls

It is most important that the cause of the damp be diagnosed and the proper measures taken to cure it at its source, and that the surface be allowed to dry out completely before papering.

Where for any reason a complete cure cannot be undertaken, there are various forms of damp treatment available, but these can only be regarded as a temporary measure and cannot be expected to provide a permanent cure.

Pitch paper is applied by pasting the pitch-coated side and hanging at right angles to the direction of the finished wallpaper.

Laminated lead and metal foils should be applied in accordance with the manufacturer's instructions. The metal surface should be cross-lined before papering.

Damp-proofing solutions are applied in accordance with the manufacturer's instructions. When they are dry, a coat of weak size is applied before papering.

Condensation

Here again the cause of the fault should be ascertained and the appropriate measures taken to eliminate it, usually by improving the ventilation; otherwise there will be a danger of the paper discolouring and loosening and there will also be the risk of mould growths forming.

In situations where condensation may occur, a lining of expanded polystyrene may be applied to form a buffer and improve the thermal insulation of the structural surfaces. The polystyrene is hung in accordance with the manufacturer's instructions.

PASTES AND ADHESIVES

The traditional adhesive for paperhanging is *flour paste*, freshly made up from plain wheaten flour, and many expert paperhangers still prefer to use this for all normal work in spite of the wide range of proprietary materials which is available. The recipe for making it is as follows.

Place 1 kilogram of plain (not self-raising) flour in a clean pail, and gradually add tepid or lukewarm water, beating it up vigorously with a flat wooden "stir-stick" until a smooth creamy batter is obtained. Boil five litres of water, preferably in a pan so that it can be poured quickly, and while it is *still boiling* pour it very rapidly but steadily into the pail, stirring all the while; continue stirring until the paste has thickened. When properly scalded the paste will appear slightly transparent. Allow it to stand for a few minutes, and then gently pour a cupful of cold water over the paste to prevent it from skinning. Allow the paste to cool before using it. Use it while it is still fresh; after a day or two it will begin to turn sour and in this state it may affect certain colours, besides losing its strength. It is sometimes recommended that alum or soda should be added to the paste, but this is not advisable because it may discolour the paper.

There is no doubt that properly made flour paste is an excellent adhesive, strong enough for even the heaviest kinds of paper including relief materials; indeed, when flour paste was the only type of adhesive available, and in the days when householders wanted Lincrusta panelling to be hung as a permanent wall-covering, it was quite common to see rooms decorated with this material standing up to a lifetime's wear without any deterioration or loosening. One sometimes wonders why the present-day decorator is so ready to buy expensively packaged proprietary materials when the advantages they offer over the cheap home-made article are so slight. Flour paste is of course a very versatile material, which can be thinned as required to suit any weight of paper.

When exceptionally delicate papers are to be hung, a completely transparent paste is desirable, and here again the traditional *starch paste* is an excellent adhesive with considerable strength. The starch is mixed to a batter with cold water and scalded with boiling water as described above.

Hot-water paste powders consist of either flour paste or starch paste prepared as a powder which keeps indefinitely; they are mixed in the same way as normal flour paste, the exact details regarding quantities depending on the brand. *Cold-water paste powders* are also based upon flour paste or starch paste, reconstituted by the addition of water, and are useful in situations where hot water cannot be obtained or where a supply of paste is urgently needed for immediate use; they are prepared by sprinkling the powder into a quantity of cold water, stirring all the time, and are ready for use within a few minutes of mixing. Hot- and cold-water paste powders are also obtainable with a mould inhibitor included for use with vinyl papers. It is of course advisable that *any* paste used for vinyl papers should contain a mould inhibitor; the vinyl coating on the face of the paper being impervious to moisture, the paste tends to remain wet for a considerable time, which may cause it to putrefy. For this reason the adhesive recommended by the manufacturer is to be preferred.

Flour and starch pastes are generally of high solids content when made up ready for use.

Cellulose pastes have the advantage that even if they are not used for some time after mixing they do not deteriorate or putrefy. They are supplied in powder form in sachets, and the paste is prepared by sprinkling the powder into cold water, stirring all the time. It is widely believed that because cellulose paste is completely colourless it will not stain the face of a wallpaper, but the belief is groundless; if paste of any description is allowed to encroach on the face of the paper it will loosen the colour used in the printing, so that when the papering brush is passed across the surface it is bound to smear. It is just as important for the paperhanger to be scrupulously clean when using cellulose paste as when using any other kind of adhesive.

Cellulose pastes are of comparatively low solids content when made up ready for use.

A recent development is a *starch ether-based* adhesive which combines the easy-working properties of cold-water starch paste with the quick-mixing properties of cellulose paste. This again is supplied in sachets, and is prepared by stirring into cold water. This type of paste is of medium solids content when made up ready for use.

All present-day ready-made paste packs normally contain a mould inhibitor. Some firms produce a range of adhesives of various strengths, the strongest being often described as *heavy-duty* paste, and this generally has a higher proportion of mould inhibitor, recommended for materials such as Vynaglypta. Other firms supply a single general-purpose powder which is mixed in varying proportions with water to produce the strength and consistency required for any specific type of wallcovering.

Extremely heavy materials such as Lincrusta can, as already mentioned, be applied with thick flour paste, but the manufacturers supply a specially formulated *Lincrusta glue* which is admirably suited to the purpose.

Dextrine is a yellow maize starch with exceptional adhesive qualities and is sometimes added to starch or flour-based pastes to improve their adhesiveness when heavy materials are being fixed. Dextrine always used to be supplied as a powder, in which form it was mixed with cold water to the required consistency (becoming a dark reddish-brown in colour when mixed) and allowed to stand overnight before use; today it is supplied in 500 ml or 2·5 litre containers ready for immediate use.

In addition to the paste powders, a very extensive range of adhesives has been developed in recent years for use with modern decorative materials. They are based upon synthetic rubber, acrylic resin, PVA, etc., and are carefully formulated for specific purposes. Some of these materials are fairly slow-drying, and are supplied in semi-liquid paste form for use with vinyl papers, PVC-coated fabrics, metallized panels and the like; others, intended primarily for the fixing of plastic sheetings, etc., or for rigid materials, are known as "contact adhesives"; these set immediately when the sheets coated with them are placed in the selected position. A great many firms are engaged in the manufacture of these materials, and they all supply detailed information about their various products, indicating the precise uses for which they are suitable, together with the correct method of application.

PREPARATION OF THE PAPER

Shading

Wallpapers often exhibit slight differences of tone between one roll and another, and it is therefore necessary to "shade" them before they are cut up into lengths. If the discrepancy is going to lead to any difficulty in hanging the paper, the merchant who supplied the paper should be informed at once, and the faulty rolls will be replaced, provided they have not been cut into or damaged in any way. But even when it is possible to use the paper without having it replaced, the risk of having any variation occurring part way along one wall must be avoided, and this means carefully comparing the rolls of paper whilst they are dry; once the paper is pasted the variations are impossible to detect, but they become very obvious when the paper has dried out on the wall, and at this stage it is too late to rectify the error.

The method of shading is to lay the rolls face upwards on the board, unrolling them to a distance slightly more than the length of the board and spreading them out, one behind the other, so that about 150 mm of each roll is visible. When viewed in good light any slight difference of tone is readily detected. The careful paperhanger will repeat the process when the rolls have been cut up, shading the individual lengths.

Semi-plain or mottled papers are sometimes slightly darker down one edge than the other, and this should be regarded as a legitimate ground for complaint.

Trimming

Practically all wallpapers are now supplied in ready-trimmed form, and in most cases the rolls are individually wrapped and packaged. This spares the paperhanger the task of trimming off the selvedge, and makes perfect matching and butt-jointing very much easier. It must be remembered, of course, that the purpose of the selvedge was to protect the edges of the paper during transit, and this means that wallpaper, whilst being transported from one place to another, must be handled with care and not thrown around in a way that might cause damage.

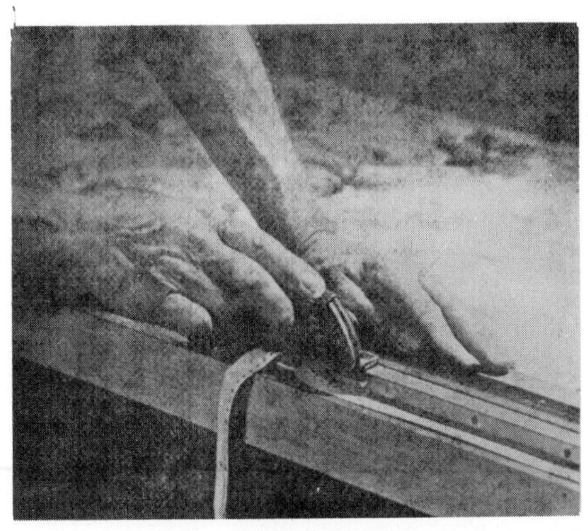

Courtesy: Crown Decorative Products Ltd

Fig. 30.6 Trimming on the table

Using the Ridgely trimmer and straightedge

In the case of any paper still supplied with a selvedge, the method of trimming depends on the type of paper. For normal commercial work the pocket trimmer is adequate. This consists of an adjustable guide plate and self-sharpening circular cutters through which the paper is drawn, the selvedge being trimmed off one side along the entire length of the roll and the process then repeated along the other side.

On high-class work and in the case of good quality hand-printed papers, the paper is generally trimmed *wet*, i.e., after pasting and folding. For this purpose the paper is folded very accurately and both edges are trimmed with knife and steel straightedge, using a zinc strip to prevent cutting the table. The Ridgely trimmer (Fig. 30.6) is a patent device for doing the same thing, the trimmer being a sharp circular steel blade which runs along a brass track on the straightedge. This may also be used for dry trimming when several lengths may be cut at one stroke. A rigid table such as the Ridgely, or a paste-board on two trestles (Fig. 4.4), is always necessary when trimming with knife and straightedge.

Cutting up

The paste-board should be 2 metres long by about 600 mm wide, but since the vast majority of the paste-boards in use were made before the days of metrication, it follows that for many years to come paperhangers will be working from the old 6-foot boards. The main point, however, is that the paperhanger uses the board as a measuring guide when cutting up. With a 2-metre board, for example, if the length of paper required is 3 metres, he

Fig. 30.7 Matching up prior to cutting required lengths

(Fig. 30.7–30.21: Courtesy: Crown Decorative Products Ltd)

unrolls one full paste-board length and then adds one metre. Similarly a 4-metre length would be two full lengths of the board. A good plan is to paint a series of narrow lines with the lining fitch across the top surface of the board at 300-mm intervals; this enables the paperhanger to dispense with the rule whilst cutting up.

Fig. 30.8 Matching a "set" pattern

Before cutting a patterned paper it should be ascertained whether the design is a "set" pattern or a "drop" pattern. In the "set" pattern, the unit of repeat appears at the same level at both edges, whereas a "drop" pattern appears at a different level; this is usually half-way between the vertical repeat, when it is then called a "half-drop" pattern. When the design is a drop pattern, it is usual to cut each length alternately from two rolls in order to avoid undue waste in cutting. Arrange to have a full repeat at the top of the length and let the pattern run out at the bottom where it is not likely to be very noticeable. Do not cut the length "neat", but allow about 75 mm at top and bottom for cutting.

Measure wall or ceiling to ascertain the length of paper required, then cut this length, face uppermost, on the paste-board. Having cut the first length, use this to measure all the other full lengths. Leave the cutting of short lengths over doors and under windows, etc., until required, then cut them from suitable "short ends". When all full lengths are cut, roll them up together, then unroll them face down on the board, allowing an equal amount to hang over at each end. Push them well back, leaving 125 or 150 mm clear space along the front edge of the board; shift the top length to the left until the end of the paper coincides with the end of the board, then draw it forward until the front edges of both paper and board are flush. The paper is now in correct position for pasting.

Pasting

The paste should be thinned with cold water to a "round" consistency free from lumps. If the paste is too thin, the paper will have no "slip" and will tend to "stay put" as soon as it touches the wall; overthin paste also makes the paper too limp and causes undue stretching. Lay the stir-stick across the top of the pail to act as a rest for the paste brush, which should be a half-worn flat brush. Apply a brushful of paste along the whole length of paper about 25 mm off the back edge, then insert the fingers of the left hand under the back edge to raise it while passing the paste brush over it; this is to avoid touching the next length whilst pasting the back edge. Next paste the front edge (always drawing the brush *towards* the edge to avoid any paste getting on the face side) and finally paste the centre portion.

Fold the paper by taking hold of the right-hand corners with both hands, moving towards the left and allowing the paper to fall into position. The beginner always finds this a tricky business but soon becomes proficient with practice. The expert paperhanger folds his paper by lifting the right-hand

Fig. 30.9 Folding the pasted length

In this view, the greater length of paper is pasted and carefully folded for cutting, as described in the text. The shorter length is then pasted and folded to meet the edge of the first fold

front corner with his left hand, merely using his right hand to guide the paper over into position. After folding the paper, draw it along to the right, paste the rest of the length and fold likewise. Turn about an inch of the folded ends back to facilitate unfolding.

Many paperhangers do not lift the back edge when pasting, but allow the paste brush to sweep over on to the next length. This is undoubtedly the quicker method but has the disadvantage that where paste is allowed to encroach on the next length, such parts receive a double coating which causes uneven expansion down one side of the paper; thus, when the paper is being hung, it tends to run off to one side instead of running true along the edge of the previous length. It is desirable that the same period of time should elapse between the pasting and final hanging of each length, otherwise differences in expansion may interfere with securing a true match.

There are three important points to observe in pasting. Firstly, paste the *edges* well and apply sufficient paste to give the necessary "slip" on the wall. Secondly, keep paste off the front face by careful brushwork. Thirdly, keep the hands clean by frequently wiping them on a hand towel or clean rag; the utmost cleanliness is essential at every stage of paperhanging, but it is at the pasting stage that most contamination is likely to occur unless clean habits are cultivated.

HANGING WALLPAPER

It is assumed that all removable fixtures have been removed so that a clear run is obtained. Have all requisite tools — scissors, rule, papering brush, etc., in the apron pocket ready to hand. Commence at the window side and work away from the light. It is better for the beginner to hang his first length to a plumbed line rather than to rely on the window edge as a guide. Therefore, snap a plumbed chalk line 533 mm (21 in.) away from the window, then, with the folded paper over the arm, mount the step ladder and carefully release the top fold. This is done by taking the corners between the forefinger and thumb of each hand and allowing the top fold to open out very gently by its own weight (taking care with thin papers to avoid tearing) and thus exposing the pasted side to the wall. Place the top corner of the matching edge against the chalk line at the correct height, letting a small amount at the top fall over the fingers towards you (so as not to soil the cornice or ceiling) and keeping the off-edge well away from the wall. The paper should now be in contact with the wall at one point only, so that by raising or lowering the hand holding the off-edge the paper may be swung pendulum-wise until the matching edge registers with the chalk line; then, and not till then, should the paper be allowed to settle against the wall. If the paper is allowed to fall naturally into position in this way it will not require any coaxing to get into position. Brush down the centre with the papering brush and out towards each edge, expelling the air from behind the paper and brushing well to ensure good attachment. In the case of satins and similar finishes, the felt-covered roller is used instead of the papering brush to avoid defacing the delicate lustre effect.

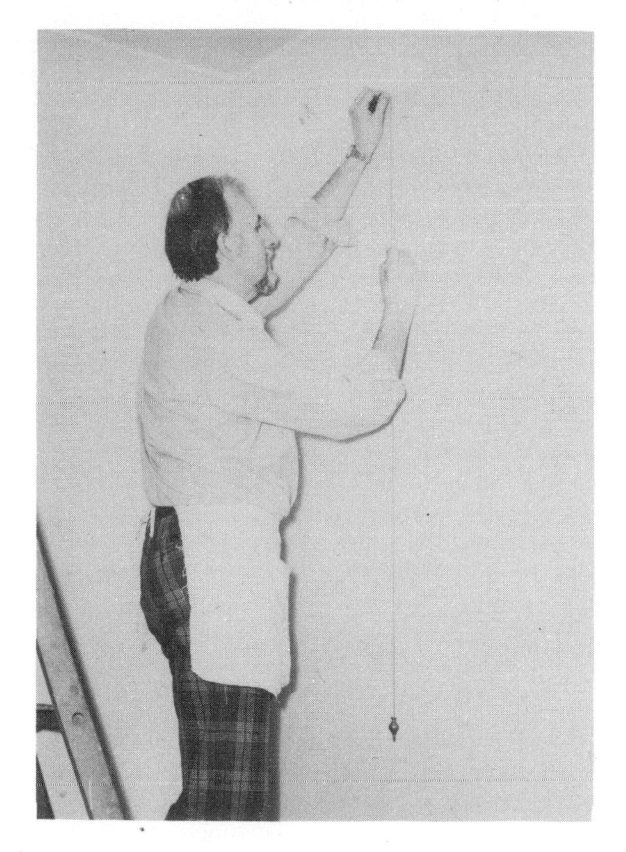

Fig. 30.10 Using the plumb line to obtain a true vertical

While on the steps finish off the top by holding the top edge slightly away from the wall and gently running the back edge of the scissors into the angle to mark for cutting. Cut along the line, then sweep the paper upwards into the angle, brushing well to secure firm contact. Then dismount the steps and attend to the lower portion which is still folded and free of the wall. Carefully unfold, brush down the centre and out to the edges, then mark and cut the bottom edge in the same way as the top. Do not throw pasted cuttings on the floor or stick them to the steps (a dirty and dangerous habit), but fold each piece *directly it is cut* to prevent it sticking to anything and throw it into a spare pail or a cardboard box. This keeps the place tidy and actually saves time in that there is less cleaning up to do. Always sponge picture rails, skirtings, architraves, etc., to remove any paste which may get on the surface, otherwise the paint is liable to crack badly.

The next length is hung similarly, carefully raising or lowering the paper until the pattern is matched, then making contact at that point while holding the off-edge well away from the wall until the matching edge is swung into correct position. The off-edge is then brought to the wall and the paper brushed down the middle and out towards the edges as before. If the first

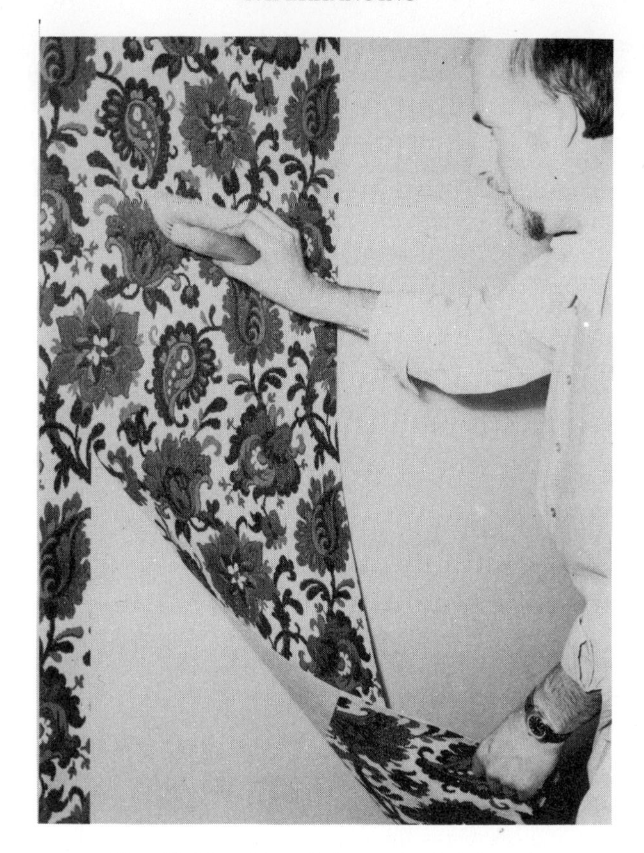

Fig. 30.11 Use of the papering brush

length is well and truly hung the matching-edge of the next length will meet it perfectly along its whole length. If there is any tendency to overlap or to run off, the paper may be coaxed into its right position by brushing the paper to right or left as required instead of down the centre; it should be remembered, however, that when this becomes necessary the succeeding lengths will be similarly affected and the degree of divergence is likely to increase rather than diminish. Greater accuracy is always necessary when joints are butted, since any overlap or open joint will show up badly. Plumb each length to prevent any divergence from the vertical.

When approaching a corner, it may happen that the space between the corner and the last length hung is less than a full width; this means that the next length will have to be cut, since it rarely happens that the angle is true enough to allow the paper to be taken round in one piece. In fact, the angle may be hollow in the centre or so much out of square that there may be considerable differences in width. It is therefore necessary to measure at top, bottom and centre to find the greatest width and add 5 or 6 mm extra to allow the paper to turn the corner. The next length should then be pasted and "folded for cutting", that is to say, folded with greater accuracy than is normally required, the edges being brought flush along the entire length and any ripples or bubbles being smoothed out before cutting. Bring the edge of the folded length flush with the paste-board, and then, using the metric rule as a guide, run a line with the back edge of the scissors; or if greater accuracy is required, use the straightedge. Cut along this line, then hang the respective lengths separately, one each side of the angle and making sure to plumb the length which turns the corner. Beat the paper well into the angle with the papering brush, or use the angle roller. Where necessary cut for door and window openings in the same way.

At the completion of each wall the joints may be *lightly* rolled with the narrow seam roller without much fear of paste being squeezed from the joints on to the paper face. Short lengths over doors and under windows, etc., will be cut as and when required. When there are many of these, as on a dado, or between shelves, it is more convenient to paste two or three pieces to start with and hang the first one, then paste another and hang the second and so on; this gives a little more time for each length to soak and become pliable before hanging.

When working from each side of the window, the pattern is almost certain to be out of match when the two sides of the room are joined up, so this must be arranged in the least conspicuous corner. Where the pattern is large or very pronounced, it should be centred on important walls such as the chimney breast. Where the walls are to be panelled with a stiling border, the pattern should be centred in each panel to give a symmetrical effect.

In high-class work, any panelling is first set out on the wall (previously lined) before hanging the filling paper; if the setting-out is done with a soft lead pencil, the lines will be automatically transferred to the back of the pasted paper when the filling is hung and so act as a guide for cutting tops and bottoms. The stiling border is then butted carefully to the edge of the filling paper and mitred at the corners. This method of procedure ensures that stiling and filling present a flush finish; it should always be used when hanging wood-grain papers where inlaid borders and cross-banded effects are desired.

Hanging lining papers

Lining papers are hung to cover uneven walls, hair-cracks, etc., to provide a surface of equal absorbency prior to the application of paint, or to make a good foundation where a good quality paper is to be hung. In the latter case it should preferably be *cross-lined*, i.e. hung in a transverse direction to the final paper, or, alternatively, hung so that the joints do not correspond with those of the final paper. Joints should be butted and, if the finish is to be paint, the edges should be trimmed if at all damaged or uneven. When hanging horizontally, the paper should be folded in short loops as for ceiling work.

It is sometimes desired to line only a part of a wall or ceiling, such as over a bad stain or other defective patch. In such cases the lining paper is applied to cover the desired portion, the edges being turned up about 25 mm all round and left to dry. When quite dry, the turned-up edge is torn off, leaving an irregular "feather edge" which does not show when coated with emulsion paint or papered over.

Hanging ceiling papers

When pasting ceiling papers they should be folded concertina-wise in a series of short folds or loops, each about 450 mm, so that each loop opens out automatically as the work progresses across the ceiling. Commence near the window, working away from the light. Snap a line 500 mm away from the angle to provide a true guide line for the first length; this is better than using the angle or cornice as a guide, since these rarely provide a true line. It may be necessary to start with less than a full width if the pattern is such that it requires centring, or where it is desired to arrange for a ceiling rose or patera to fall at the edge of the paper instead of in the middle.

Fig. 30.12 Snapping a chalk line for the first length on a ceiling

Arrange the steps and plank so that the whole distance across the ceiling can be reached comfortably; the ceiling should be about 25 mm above the head for comfortable working, otherwise the arms will become unnecessarily tired through over-reaching. Mount the steps, holding the folded paper in the left hand and using a roll of paper as a support. Open out the first fold with the right hand and place the matching edge on the chalk line, holding the folded part in the other hand close to the ceiling. When in the right position, brush across the paper so that the first section is firmly attached before unfolding the next loop (Fig. 30.13). If the first section has been placed right, the next and succeeding folds will run true along the chalk line, the paper being brushed away from you, first in the middle then out to right and left, until the opposite side is reached. The ends should be trimmed off as in dealing with tops and bottoms on the wall. If there is a cornice, trim neat to it; otherwise allow the paper to turn the angle, with a 10-mm lap on the wall.

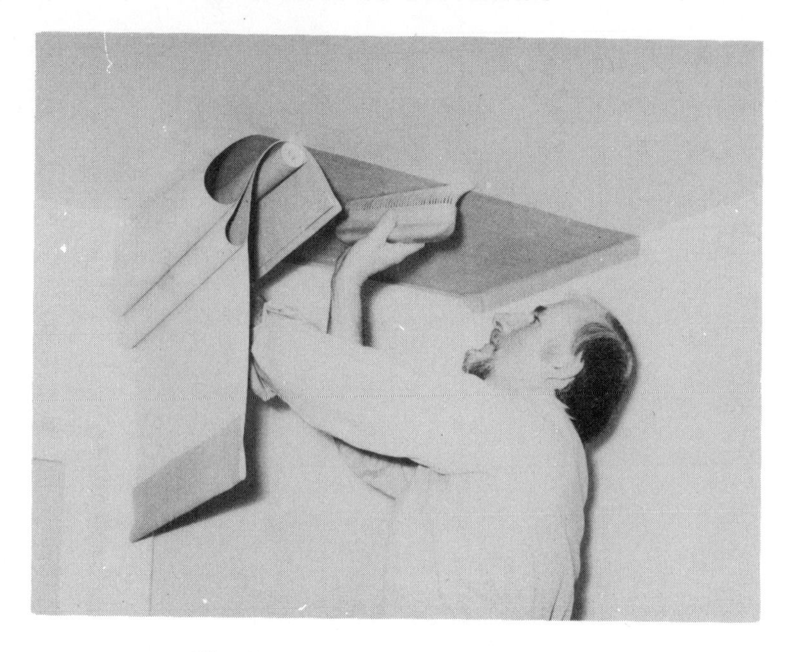

Fig. 30.13 The first fold in position

It is brushed out smoothly while the remaining folds are supported with a roll of paper close to the ceiling

The next length will be hung in the same way except that the pattern will need to be accurately matched at the first point of contact or it will be wrong all the way across. Papering ceilings is, at best, a strenuous job and while lengths of 4 to 5 metres can be comfortably managed single-handed, longer lengths may require some assistance. It is usual to have someone hold the paper, at least for the first half of each length, since the weight of the pasted paper can make it very tiring for one man. Paperhanging is greatly expedited when there are two to do the work — one pasting while the other hangs the paper. Pasting is usually done by the apprentice, and he can either help or hinder the paper-hanger according to his proficiency and cleanliness in handling paste and paper. The well-trained boy will anticipate the paperhanger's every requirement instead of always waiting to be told what to do next. He should fold accurately for cutting where necessary and should always be ready with the next length when and where it is wanted.

Where possible, try to arrange for an obstruction, such as a light patera, to come at the *edge* of the paper to facilitate working round it. Proceed to hang the paper until the obstruction is reached, then cut in from the nearest edge of the paper through the centre of the patera, followed by short cuts at right angles, also through the centre. Continue past the obstruction and finish hanging the length, then return to the patera and make diagonal cuts so that the eight triangular ends can be pressed round the patera with the point of the scissors to obtain a cutting line; then trim off and beat well into position with the papering brush.

All paintwork must be dry before papering is commenced and all edges that come in contact with the paper should be finished, otherwise the paint would "strike" into the paper when cutting in. The parts that do not touch the paper can then be finally varnished or enamelled after the paper is hung. Always take great care to sponge off *immediately* any paste which gets on to the paint work, for if allowed to dry the paste is liable to crack the paint badly.

HANGING SPECIFIC TYPES OF PAPER

While the basic principles of hanging wallpaper are the same in every case, each individual paper has its own distinctive features, which may call for some modification of the technique or some special care in handling. Any paper that is embossed, for instance, should indicate to the paperhanger that each length must be given the same length of time for soaking after pasting and before application. The reason is that the embossing tends to flatten out as the paper becomes moist, and the longer it soaks the further the paper stretches. Unless the lengths are treated uniformly, it will not be possible to match the pattern on adjacent lengths as one will be stretched further than the other. Furthermore, the papering brush should be used lightly and with a dabbing rather than a sweeping action, otherwise the pattern will be further distorted. When an embossed paper is being hung on a staircase it is advisable, especially on the long pieces, that two men should work together, one to take the top part and control the matching and hanging, while the other man supports the weight of the lower end so that the paper does not sag unduly. To some extent the good-quality embossed papers known as duplex papers are less troublesome in this respect, as the backing paper takes a lot of the strain and reduces the tendency for the emboss to flatten out. Duplex papers should be hung fairly soon after pasting and should not be allowed to become oversoaked; the paste should be of a thick consistency. On no account should the paste be used until it is completely cool, because hot paste makes the paper stick firmly together when it is folded and this may lead to the backing paper separating from the face.

Metallic papers It is important that perfectly fresh paste should be used when hanging papers with a metallic finish or with some parts of their pattern printed in bronze. Pastes, especially the flour-based types, develop organic acids as they ferment, and for this reason a paste which is turning sour may cause the metal to blacken. In the case of the very expensive metallic papers like Tekko, preparation of the surface is of foremost importance, as any grit or unevenness shows through and is emphasized by the smooth metallic lustre of the finish. All trace of grit must be removed and the walls made perfectly smooth. A lining paper is essential; a fairly stiff paste should be used, and it is desirable that the paste both for the lining paper and the finishing paper should be strained to ensure that it is free from lumps.

Some papers need particularly careful handling, especially at the stage of pasting and folding. If satin papers are creased or folded abruptly, the china clay coating cracks badly; and the coating is readily scratched if there is any dirt or dried paste in the bristles of the papering brush — although of course it is better that a felt-covered roller be used in their application.

Plastic prints need very careful handling; they must not be oversoaked because the plastic material becomes crumbly when moist, and the paper must be folded very gently and smoothed down without any hard brushing, because the friable plastic paint is very easily dislodged from the surface. Moirés and soirettes are easily marked, and so are mica ceiling papers; the silvery mica is very loose and comes away in a powder if the papering brush is used too harshly. *Wood grain* and *marble papers* with a wax-polished surface need gentle handling, with especial care during hanging to avoid creasing, which spoils the finish; any paste which gets on to the face must be wiped off with a cloth, and the surface should be polished over with a soft duster after hanging.

Vinyl papers With these it is essential to use a paste that contains a fungicide, because the surface is completely impervious to moisture, so that the paste remains wet for a long time, encouraging the formation of mould growth. If a lining is hung, the same kind of paste should be used for the lining. After pasting, the paper is hung immediately and not left to soak, and the surface is smoothed down with a papering brush or damp sponge, particular care being taken to avoid the formation of air bubbles. Any paste which strays on to the face must be sponged off, and on no account allowed to dry on the surface. It is most important that the edges should be accurately butted and not allowed to overlap; the vinyl paper will not stick to itself, so an overlapping joint stands out loose and cannot be stuck down. This leads to difficulties in the corner of a room, where some overlap is inevitable; in these places the edges are touched in with one of the tube adhesives which are stated to be suitable for the purpose. Sometimes when the paper is being cut to fit round light switches, etc., it tends to de-laminate — in other words, the surface coating becomes detached; in this case a little of the paste is placed at the back of the surface film, which is then pressed down to make an invisible repair.

Hand-printed papers obviously need careful handling, if only because customers who have paid for an expensive and highly individual wallpaper are entitled to expect that it will be properly hung. But there is another factor. Whereas the pattern on machine prints is very often supplemented with a light embossing after printing in order to give added interest and texture to the paper, screen-printed papers are not supplemented in this way to anything like the same extent, simply because the production process does not lend itself to this treatment. It therefore follows that any slight defects in pasting, slight smears of paste on the surface, and traces of fingermarks, which would barely be noticeable on the slightly textured machine print, are very conspicuous indeed on the flat level surface of the hand-printed paper.

The essential feature about hanging *flock papers* is absolute cleanliness at every stage. Two pasteboards are used; on one, the whole of the paper as cut to the required lengths is placed, while the other one is reserved for pasting, only one piece of paper being on the board at any one time. Both boards are kept covered with clean lining paper, and a clean dust-sheet is laid on the floor beneath them so that no paper ever comes in contact with the floor. A felt-covered roller is used to smooth the paper down. The wall surface is always cross-lined, and if the flock paper has a selvedge it must be trimmed with knife and straightedge.

Japanese grasscloths and *oriental silk cloths* are hung on a coloured lining paper; trimming, if required, is by knife and straightedge. A thick paste with low water content is used, because a thin paste would cause staining, and over-soaking must be avoided. A felt-covered roller is used for application. No attempt can be made to match up the weave along the joints, and in fact the random effect is considered to be one of the attractive features of these materials.

Wood effects Realwood and Flexwood are fixed according to the manufacturer's directions, using the special contact adhesive which they recommend. Great care is needed in fixing, because the two surfaces adhere as soon as they are brought into contact, and it is therefore not possible to manoeuvre the material into position. Accurate setting-out is essential, and fixing consists of placing only one corner or one edge on the chalk line, while the rest of the panel is held well away from the wall. When contact is made in the right position, the panel is allowed to go to the wall and a wide roller used to press it down. The strapping or stiles and rails are fixed afterwards with the recommended adhesive, taking care to remove any of the latter which accidentally gets on to the face of the Realwood, otherwise such parts will reject the subsequent stain and appear lighter than the rest of the work. Normal paperhanger's pastes are not suitable for Realwood panelling because the inevitable soaking of the material sets up expansion and contraction forces which may, in certain cases, result in the plaster skimming being pulled away with the backing.

For the hanging of *Rollywood* a lining paper is essential, usually a stout brown lining. When this is dry, a latex glue is brushed on to the surface of the lining. The wood material is cut exactly to size before hanging, and is placed in contact with the wall, unrolled slowly, and pressed into firm contact with the glue. The cut ends of thread are sealed with a touch of quick-setting glue to prevent them from fraying.

Hanging relief materials

The hanging of relief materials involves various modifications in technique according to the type of material, whether high or low relief, hollow-backed or solid. If the materials require trimming, a knife and straightedge should be used. What *is* essential in all cases is that the wall or ceiling surface should be cross-lined. A plastered surface never gives the right degree of porosity or the uniform absorbency needed for such heavy material; we have seen countless jobs where the whole of the decoration has become detached because the trifling extra expense of lining was omitted through a false sense of economy.

Low reliefs

Machine-embossed materials of the familiar Anaglypta or Supaglypta type, supplied in rolls, are hung with a fairly stiff paste with low moisture content, or preferably with the proprietary pastes recommended by the manufacturer. Two lengths should be pasted before the first is hung, then a third length pasted and the second hung, and so on, thus giving each length time to soak and become flexible. Over-soaking causes too much expansion and may distort the pattern if the material becomes too limp. For the same reason, the paste is brushed well out of the hollows. Avoid stretching the material during hanging, as the subsequent contraction may cause the joints to open; the paper should be gently beaten into position with the papering brush, rather than brushed to right and left as with ordinary wallpapers. Fold carefully to avoid making kinks in the pattern.

Vynaglypta has an impervious vinyl coating, and it is generally recommended that this material should not be soaked but hung immediately after pasting, using the adhesive suggested by the manufacturer (see Fig. 29.1–29.3).

Hanging Lincrusta

This is a solid-backed material which requires a good smooth wall to show it off to advantage, especially in the wood-grain finishes. The surface should therefore be well glasspapered and dusted off to remove all nibs, etc., and any uneven plaster should be cut out and made good prior to sizing and lining. A stout brown lining paper is used.

For hanging this material the manufacturers supply a special Lincrusta glue which is strongly recommended for the purpose. The Lincrusta is cut to length and laid face downwards on the board; each length is sponged over liberally with warm water which is allowed to soak for 20 or 30 minutes; two or three applications of water may be necessary. The purpose of soaking with water is to cause full expansion of the material, thus preventing the formation of blisters. At the end of this time the moisture is wiped off with a dry cloth and the Lincrusta glue is applied, being brushed out firmly and evenly. Each length, as it comes to be hung, is placed in position and pressed down with a pad of cloth, after which it is rolled down firmly with an uncovered roller, squeezing from the centre outwards and pressing out any surplus glue. The edges are butt jointed. As an alternative to Lincrusta glue a very stiff flour paste reinforced by the addition of dextrine may be used.

Lincrusta tends to harden up in storage. In the ordinary way the act of soaking it with warm water will make it soft and pliable again, but if the material has been kept for a very long time it may have become so rigid that it cannot be unrolled without cracking. In such cases warm water is of little use, and in fact soaking in water may cause the brown paper backing to separate from the face; a better plan is to place the rolls for a short time in a warm place such as a warm airing cupboard or close to a radiator when the heat will generally make it become pliable again.

Lincrusta is generally supplied in a light putty tint, but if a different colour of ground is required it should be finished with sharp oil paint and allowed to dry hard before scumbling. On no account should it be sized before paint or stain is applied.

Hanging fabrics

Cheap hessian and canvas which is to be painted after application is not used to any great extent today, but in cases where the decorator is required to deal with it, it is advisable to hang it dry — in other words, the wall surface is pasted with a good stiff paste and the fabric is then unrolled on to the surface and well brushed out. Unless the seams are to be covered with a moulding or slatting, the method of butt jointing the fabric is to run the next length so as to overlap the previous length by about 50 mm; the overlap is then cut down the centre with a very sharp knife and steel straightedge, the trimmed-off edges being pulled away to allow the butt edges to come together. Roll down with a wide roller or beat down with a papering brush.

Decorative hessian and canvas, when supplied as rolls of cloth without any backing, are applied in much the same manner, the wall being coated with adhesive and the fabric rolled out dry and pressed into contact with it. If they are very wide — and sometimes they are well over a metre in width — the handling of them is a two-man operation. Care is needed to avoid squeezing them into distorted shapes when pressing them down, as this makes the lines of the weave sag, and it can also cause undue stretching, so that they shrink a great deal when drying, which leads to the joints opening. When they are supplied on a backing they are much easier to handle. Those on a latex backing can be pasted with normal paperhanger's paste and hung like a wallpaper, or alternatively a latex fixative can be brushed on to the wall and the fabric rolled on in dry condition. When they are backed with paper they are hung exactly like a wallpaper. Sometimes they are supplied in self-adhesive form; in this case there is a backing of protective paper to keep the adhesive in good condition until the material is required for use. When the fabric is being hung, the top part of the protective paper is peeled back a little way and the top of the fabric is carefully placed in position; the rest of the backing is gradually peeled away, while at the same time the fabric is smoothed down; since there is no wet paste, the whole operation is perfectly clean.

PVC-coated fabrics The surface to which these are applied must be perfectly dry. Whatever adhesive is used must contain a fungicide. A latex adhesive is usually recommended, although various paste powders are claimed to be suitable. On non-porous surfaces the adhesive is applied to the wall and the dry fabric unrolled and pressed into contact with it. On porous surfaces, however, where the adhesive might be partially absorbed before the fabric is placed in position, it is better practice to coat the back of the fabric with adhesive and hang it wet. As the material is smoothed down, care is taken to eliminate air blisters. The edges must be butt-jointed because the vinyl surfaces will not bond to one another.

Hanging pre-pasted wallpapers (Fig. 31.14–31.21)

A plastic trough is supplied with the paper. The trough, filled with water, is placed against the skirting board beneath the area to be covered; the length of paper is rolled up from the base with the pattern side inwards and is immersed in the trough. After soaking for one minute (or two minutes for vinyls) the top edge is gripped and slowly withdrawn in a vertical direction till the required height is reached; the paper is then smoothed down with a damp sponge.

To hang long lengths, as on a staircase wall, the length is back-rolled with the pattern side outwards, so that the free end is the base or skirting-board end. The roll is immersed in the trough and immediately re-rolled under the water. It is taken from the trough and allowed to drain for a moment before hanging — starting at the top, holding the roll in one hand and matching the pattern with the other. By feeding from the roll as it comes down the wall, the whole operation becomes more manageable. This method is illustrated in Fig. 30.20. Fig. 30.21 shows the same technique applied to hanging pre-pasted paper on a ceiling.

Many paperhangers, however, prefer to hang the paper by the conventional method; they ignore the presence of the adhesive backing on the paper, and make a separate adhesive in the form of a paste, which is applied in the customary manner on the back of the paper. The paper is then folded, either with a large fold at the top and a small fold at the bottom for hanging on a wall, or with concertina folds for hanging on a ceiling. There is a widespread feeling among paperhangers that the conventional method of pasting and folding gives a much greater measure of control over the whole operation. Some of the firms manufacturing pre-pasted papers acknowledge this fact by giving information about both methods on the instruction slip supplied with the paper. A heavy-duty vinyl adhesive containing a fungicide is usually recommended.

In the case of materials such as Novamura, the wall or ceiling surface is pasted with the adhesive recommended by the manufacturer, and the material is fed directly on to the surface from the dry roll.

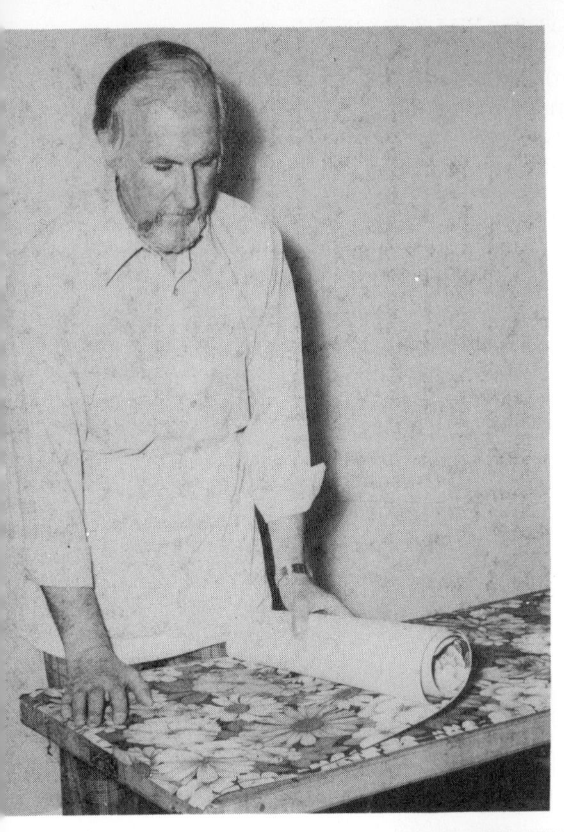

Fig. 30.14

The cut length is *loosely* rolled up with the pattern inwards

Fig. 30.15

'he roll is completely immersed in the ·ater trough for one or two minutes, as equired, then withdrawn vertically and pplied to the wall

IANGING PRE-PASTED VALLPAPERS

Method A: Normal method

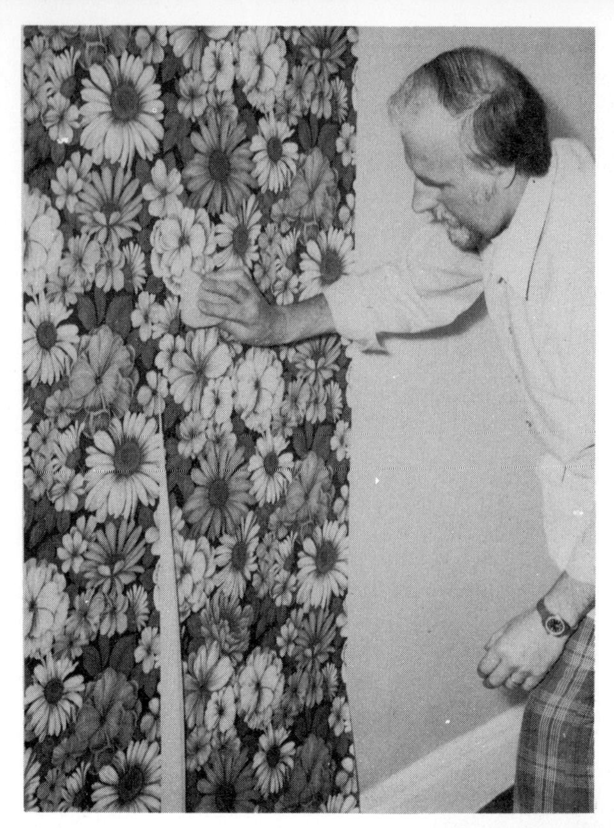

Fig. 30.16

The pattern is matched and the pape smoothed down with a sponge or dam cloth to exclude bubbles and wrinkle then trimmed at top and base

Method A (*continued*)

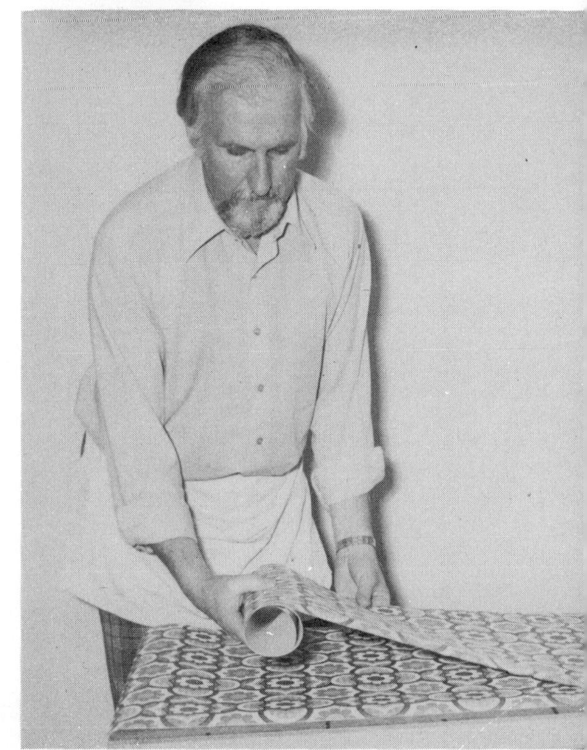

Fig. 30.17

The required length is rolled up with the pattern *outwards*

HANGING PRE-PASTED WALLPAPERS

Method B: Hanging long lengths

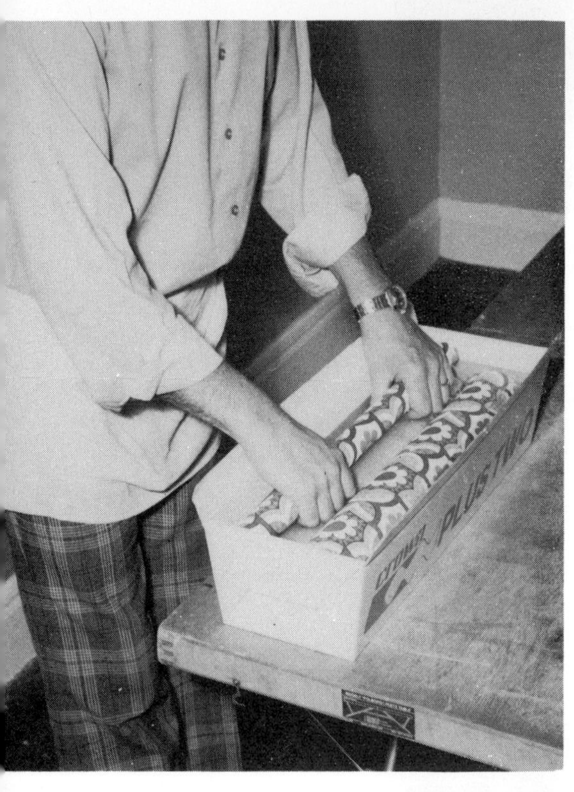

Fig. 30.18

The roll is completely immersed, then back-rolled in the water

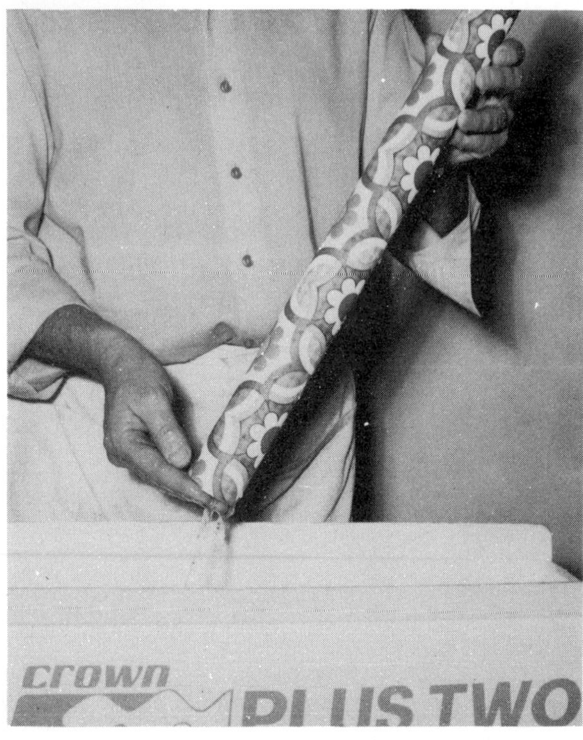

Fig. 30.19

he roll is withdrawn from the water and
llowed to drain before hanging

IANGING PRE-PASTED VALLPAPERS
Iethod B (*continued*)

Fig. 30.20

Holding the roll in one hand, a sho
length is applied to the wall and smooth
into position with the other hand, feedi
from the roll as it comes down the wa
Long lengths, as on staircase walls, can
easily managed in this way.

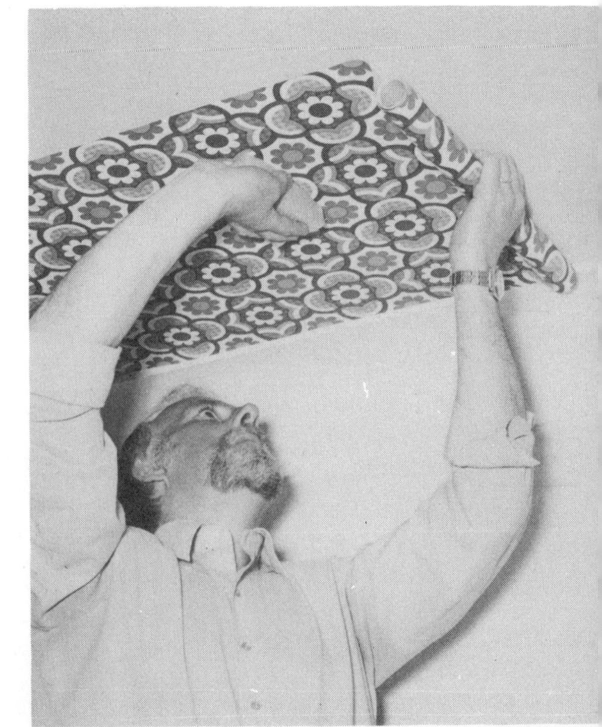

Fig. 30.21

The same procedure as in Fig. 30.20 can be used on ceilings

HANGING PRE-PASTED WALLPAPERS
Method B (*continued*)

31

Drawing and Design

The value of drawing ability to those engaged in painting and decorating can hardly be overestimated. It is not likely that every student will become his own draughtsman and designer, or that he will desire to do so, but the ability to indicate by a few lines the plan and elevation of a room, to modify the proportions of ornament, or to enlarge or reduce a given pattern is a very desirable qualification for every good craftsman. The student who possesses more than average drawing ability, however, should make the most of his natural talent by diligent study and practice at a good school if possible, where, in addition to the guidance of a teacher, he will greatly benefit by seeing the work of other students. Those unable to attend an art class can do much by private study and observation, assisted by a cultural magazine such as *Studio International* or an instructional magazine like *The Artist* which deals with the *technique* of drawing and painting. In any study it is important to keep the ultimate goal firmly in mind, and the decorator will therefore look upon everything he sees with a strong *decorative* bias and will try to cultivate a sense of *design*.

Object and memory drawing

The first essential in drawing is the correct use of the eye in observing the form of objects. The student is apt to mistake what he *knows* for what he *sees*. For instance, he knows that this page is rectangular, but from different angles he will see that it appears to be an irregular diamond shape. His first lesson, then, must be to *see* correctly and his second to draw what he has seen. But he must also learn to discriminate between mass and detail. He knows that a tree has many leaves, but seen at even a short distance the tree becomes a mass, a shape, and the individual leaves are no longer distinguishable.

To take another example — on entering a room papered with a wallpaper that is full of detail the eye is not conscious of more than the general masses of form and colour. Even the colour is not what you imagine it to be, but a visual intermixture of the actual colours of ground and pattern. In a sketch, the student would try to generalize the colours and indicate the pattern by showing the shape of the masses. It would be quite impossible to put in all the detail that close inspection reveals, and just as difficult to put in every leaf of a tree, and neither course would carry a true impression of what the eye had seen. In studying the actual appearance of objects seen at a distance, or in various positions, the use of one eye at a time, and the use of the half-closed eyes, will both assist the beginner to see correctly and also help him to delineate shapes and broad masses better.

The art of drawing, however, does not consist in putting down *all* that the eye sees, for even if that were possible, it would be as futile as trying to put in all that a microscope would further disclose. But having trained the eye to see correctly, it is necessary to cultivate the art of *selection*, to be able to put down essentials while subordinating or ignoring the less important details. As soon as the student starts this process of selection and rejection, he has unconsciously begun to *design*, for design is nothing more than fashioning or arranging something in accordance with one's ideas.

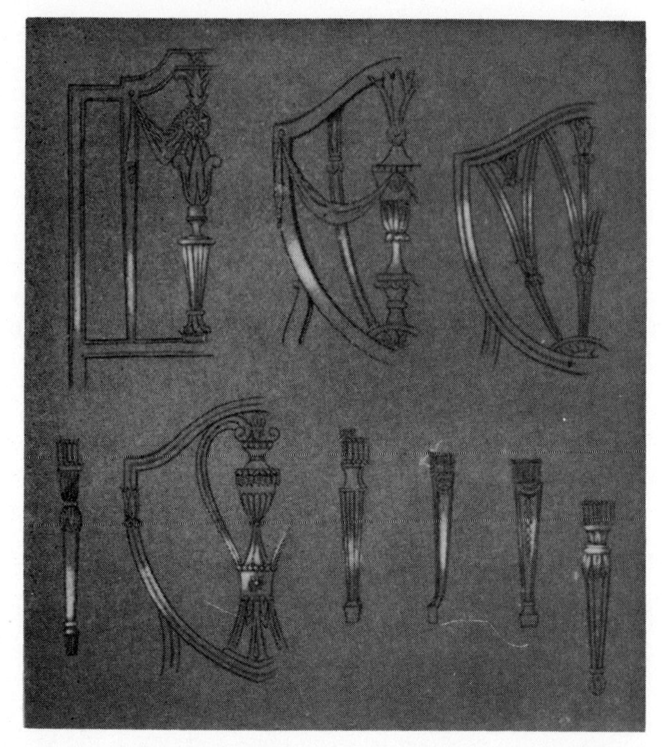

Fig. 31.1 Details of Hepplewhite furniture by Erik Foote

A common mistake with most beginners is the tendency to choose far too elaborate a subject to draw, as though there is some special merit in the *amount* of work or detail put in a drawing. They find little inspiration in the common objects of everyday life — the teapot, cup and saucer, table and chairs, etc. Consequently, when these simple things are shown in a student's design, or a perspective, they are, more often than not, very badly drawn. It is useless to attempt ambitious decorative motifs, or to start drawing the human figure, etc., before one can render simple elemental shapes satisfactorily; the basic shapes in simple objects are only multiplied and complicated in the more elaborate forms. The student is therefore urged to get a good grounding in common object and memory drawing.

Fig. 31.2 Studies of medieval illumination by J. T. Howson

Just as the eye requires training to see correctly, so does the "mind's eye", or memory, need training to register correctly what has been seen, since all drawing is a form of memory test. You look first, then you set down your impression on paper. The eye cannot be engaged in both at the same time, so you have to memorize for a brief period what you have seen. Some people have almost photographic memories and are able to retain a clear impression which enables them to *see* the object on paper as they draw it. Others retain only fleeting impressions and have constantly to shift their gaze back and forth, from object to paper. The best way to cultivate the memory is to gaze at the object *intently* for a few minutes, then draw it without looking at it again. You will soon want to refresh your memory and will look at the object more keenly next time you see it. It is also a good plan to draw from memory something seen earlier in the day, or even several days before. An excellent exercise, for instance, is to jot down one's impressions after visiting an exhibition, or to recall some particular item and make a sketch of it. The great value of memory drawing is that it teaches close observation.

FREEHAND DRAWING

Practice in freehand drawing should preferably be done on an A1 or A2 sheet so that the drawing can be fairly large. The drawing board should be supported on the edge of a table or a drawing trestle, with the lower edge resting on the knees; in this position the pencil may be held almost at arm's length, the arm moving freely from the shoulder. For small drawings and fine detail work the pencil may be held in a writing position, but this position will be found needlessly cramping for larger work and the pencil is best held loosely between the four fingers and thumb, with the knuckles resting lightly on the paper for support. Thus the drawing is kept well away from the eye, and this makes it considerably easier to compare the general proportions of the drawing with the subject being drawn.

Detail or shading should never be put in until one is quite satisfied that the drawing is in correct proportion; this should be checked frequently by holding out the pencil at arm's length, using the thumb as a gauge to measure widths, heights and main subdivisions and comparing these with the drawing. The *pencil* used depends largely on personal preference and the type of subject, but it should be remembered that the paper also plays a big part in the result; an HB, for instance, which might be too hard on some cartridge papers, would be quite soft and black on hand-made, "not" surface Whatman. In general, however, B or 2B will be suitable on cartridge for object and memory drawing, while HB or H would be appropriate for ornament or plant form on hot-pressed Whatman.

Drawing, of course, is not confined to working with the blacklead pencil. There are times when one looks at an object and realizes instinctively that it could be better represented by some other medium. The painter and decorator will, by the nature of his training, very often feel happier with *brushwork* in water colour or poster colour. But even when colour is not required, it is a good thing to practise drawing with a variety of different techniques, and we are fortunate today in having so many useful alternatives to the pencil at hand. The *ball-point pen* is a wonderfully fluid instrument for both freehand drawing and for detail work as well. and is especially valuable for architectural studies, where it gives the crisp details of pen and ink without the need for slow and laborious workmanship. Any student would be well advised to use the ball-point pen regularly for freehand sketching. because it lends itself ideally to the production of quick and accurate studies. One of the most important advantages it offers is that it compels the student to draw directly without a lot of preliminary setting-out. It would in fact be fatal to the freshness of the drawing to do any preliminary pencil work; it is far better to draw directly on to the paper or on a smooth white cardboard, because this encourages a bold approach and helps the student to overcome the timidity that he so often feels.

Another most useful tool available today for drawing is the *felt pen*. This is a most satisfying thing to draw with; for one thing it is so very quick, and the speed of working, together with the fact that rubbing out is impossible, very soon builds up the student's confidence. It is excellent for bold vigorous effects, and here again it is an ideal medium for architectural studies as well as for sketching and for quick representations of light and shade (Fig. 31.3). Once confidence has been won, it is a good thing to experiment with colour felt pens.

One should try to draw with a definite purpose in view and should carefully consider at the start what that purpose is. It may be to render the particular effect of light and shade on a plaster cast or a piece of drapery; if so, it must be remembered that light and shade effects are transitory and that by moving the light source, a totally different effect may be obtained, though the actual subject remains the same. Thus, while light and shade and cast shadows may have magnificent pictorial value, this does not necessarily show the *true* form of an object. One must therefore consider whether one wishes to record the particular appearance of an object at a given moment under certain conditions of light, or whether it is desired to express the *true form* or contour of the object, irrespective of lighting conditions. It is possible, for instance, to express flatness or roundness, or to indicate a change of plane, by judicious shading or hatching, without direct reference to local lighting or incidental cast shadows, and such a drawing may prove to be a more useful and informative record to the designer than a carefully worked-up photographic rendering of light and shade. If the student bears these points in mind, his drawing will have more purpose and value than it might otherwise have.

The student is sometimes urged to avoid working from printed copies and we are entirely in agreement when this refers to the slavish and laboured copying of photographs or printed illustrations. Used in the right way, however, prints and photographs are legitimate sources of reference and every student should keep a scrapbook or a reference file for newspaper and magazine cuttings. These may provide authentic details of pose or costume, buildings, trees, etc., or may inspire a decorative rendering of the subject. Where an illustration cannot be cut out and kept, a quick sketch may be made in pencil, pen and ink, or wash, noting only what is required and leaving out irrelevant details. Alternatively, a tracing may be made, preferably direct in ink. It is a curious fact that some people regard tracing as a rather dubious practice, or even something to be ashamed of. It can only be such when it is used dishonestly, as in passing off some other person's work as one's own.

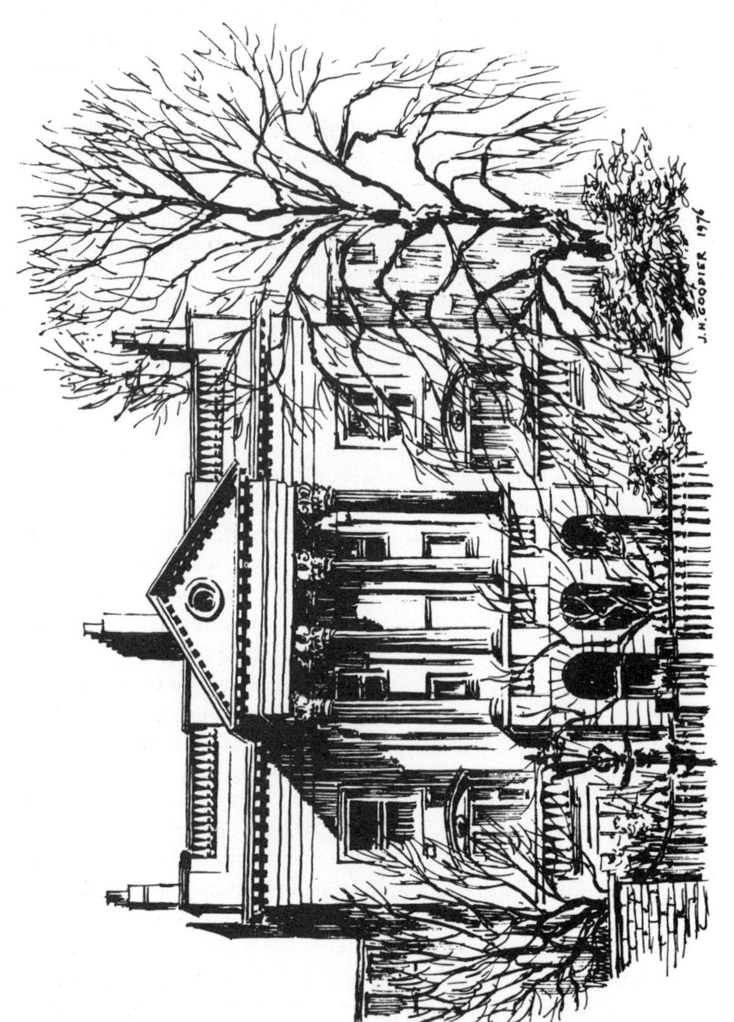

Fig. 31.3 Felt pen drawing is very rapid

It is the ideal medium for the painter and decorator needing to make a quick drawing.

The illustration is a freehand sketch completed on the spot in less than 30 minutes. It shows one of the houses designed in the Classical style by the eminent Regency architect John Nash about the year 1820; the house is in the Regent's Park area of London.

Tracing paper should be used whenever it will serve a legitimate purpose. Apart from its convenience in making a quick record, it is invaluable for purposes of *analysis*, such as in studying lines of composition, disposition of masses, etc., for by placing a sheet of tracing paper over an illustration, the main lines or masses can be boldly drawn in with brush or pen and Indian ink, thus encouraging direct methods of working and promoting confidence in draughtsmanship. In this way it is possible to approach nearer to the spirit of the original than by any slavish copying. Also, in working out experimental design ideas, modified or alternative arrangements can be tried out on tracing paper without disturbing the original conception.

Drawings often require enlarging or reducing to fit certain spaces; the enlargement of drawings by the square method is described on page 448 and drawings may be reduced by reversing the process, or by using the proportional scale method as described on page 428.

LIGHT AND SHADE

In order to study the principles of light and shade, one cannot do better than work from simple geometrical solids such as the cube, cone, cylinder and sphere. If unable to attend an art class where these are provided, the student may easily construct the first three in white cardboard, whilst a white-painted ball will serve just as well as a larger sphere. Place the objects separately against a black background for preference, with the light falling from one side only, and observe the effect of the light on the different planes. Then place a large sheet of white card under the model, and another in a vertical position near the shadow side. Observe how the parts of the model previously in deep shade are now lit up by *reflected* light. In the cube, the strongest contrast of light and shade will be at the corner where the dark plane meets the light plane, but the dark plane gets increasingly lighter as it recedes from the corner, due to the influence of reflected light. Notice too that the shadow cast by the model also contains some reflected light which gets lighter as it recedes from the edge of the shadow. Thus we find that shades and cast shadows will need to be rendered stronger at the extreme edges, graduating to a lighter tone within the shadow itself.

When we observe the effect of light and shade on curved surfaces such as the sphere we find that, roughly speaking, one half is in light and the other half in shade, the light merging gradually into the dark. When the dark half is lit with reflected light from the white card, we perceive that the darkest part now forms a diffused ring round the circumference of the sphere; this is generally referred to as the *line of shade*. In rendering the sphere, therefore, the line of shade must be correctly indicated in order to express the contour convincingly. When the sphere has a glossy surface, the source of light is reflected at one point and this is called the *highlight*. This too must be carefully placed in your drawing. In the case of a cylinder, the highlight will appear either on the top edge or along the side of the cylinder, depending on the position of the object in relation to the source of light and the eye.

Studies illustrating the above points may be carried out in pencil or wash, or with chalk and charcoal (or conté crayon) on tinted paper, the latter being a particularly convenient method. The lessons learnt will prove

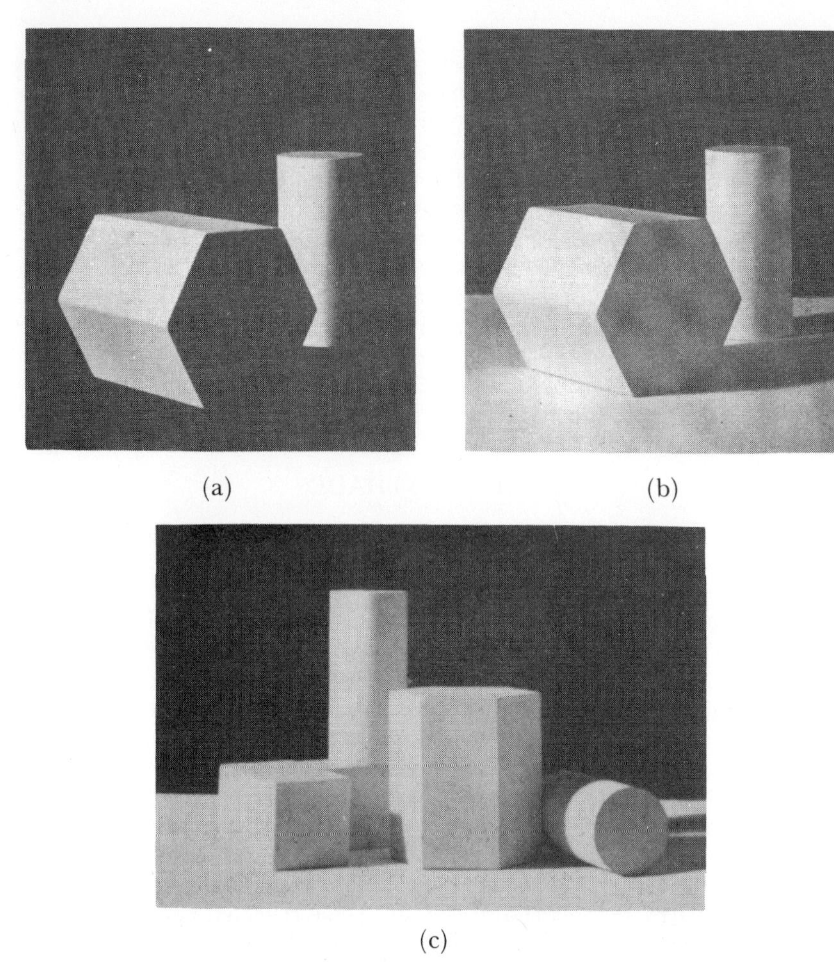

(a) (b)

(c)

Fig. 31.4 Light and shade

Simple cardboard models used for light and shade study showing (a) models lit from one side against a black background; (b) models placed on a white card and a white sheet used to reflect light on the shadow side; (c) a group of models seen under the same conditions as (b) — note the subtle interplay of reflected light

invaluable later on when rendering interiors in perspective, for though the problems of light and shade, cast shadows, highlights on polished furniture, etc. (Fig. 31.1), are much more complex, the basic principles are exactly the same.

A knowledge of the principles of light and shade is also necessary in the representation of relief ornament and drapery, etc., as in all three-dimensional objects. In conventional monochrome and polychrome painting, for instance, the range of tones is usually limited to three or four tints or shades, and these therefore require very careful placing if a convincing effect of contour is to be achieved (Fig. 31.5).

Cast shadows are of considerable use in rendering architectural elevation

Fig. 31.5 Painting in monochrome

Two examples of ornament painted in light, medium and dark tones of the ground colour. This type of painting provides excellent practice with the sable pencil

drawings and in defining projections such as pilasters or overhanging cornices, etc. The shadows should be drawn very carefully, then rendered in a *graduated* wash with the deepest part along the edge of the shadow; this gives an attractive suggestion of reflected light within the shadow and looks much better than a uniform flat wash of colour.

GEOMETRICAL DRAWING

This form of drawing is of great importance to the decorator, for by its use all forms of panelling and the subdivision of ornament are set out. There are many books dealing specially with the theory and practice of geometry, but for the average student *Geometrical Drawing for Art Students* by I. H. Morris (Longmans, 1958) can be recommended. The student is advised to pay particular attention to those parts which deal with the construction of polygons, the ogee, trefoil and other foils, the several geometrical arches and the ellipse; two methods of drawing the latter are shown in Fig. 31.6.

For all geometrical drawing and designing, a drawing board is absolutely

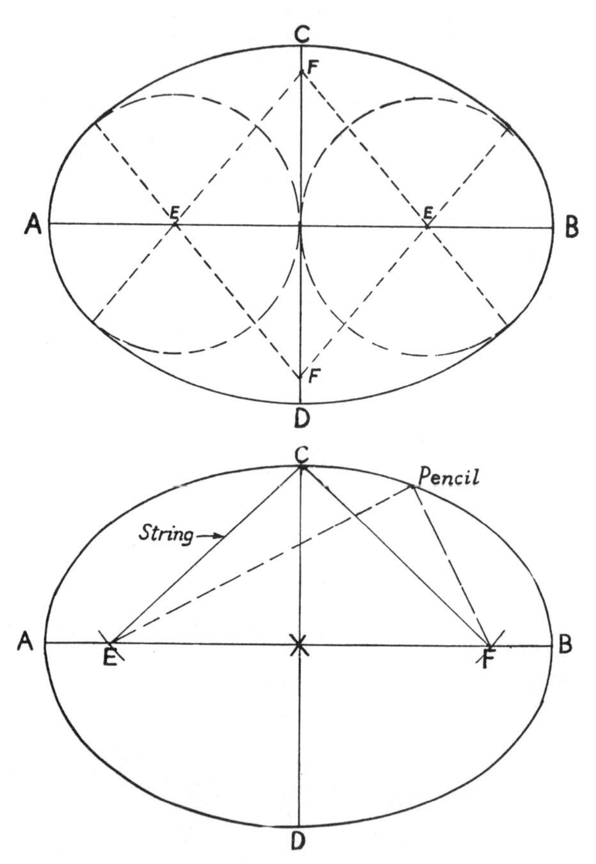

Fig. 31.6 Two methods of constructing an ellipse

Top Divide line AB into four equal parts and describe circles from centres E, E. Find points F, F on line CD and describe arcs to complete the ellipse.

Bottom Set out centre-lines AB and CD, then with distance AX describe arcs from point C to intersect line AB at E and F. Fix pins at points E, C and F around which stretch tightly some strong thread. Remove pin at C and replace with pencil point, moving same (with thread held taut) to describe the ellipse as indicated with dotted line. This method (sometimes called the "gardener's method") is a practical way of setting out an ellipse on ceiling or wall.

necessary. An A1 board with a T-square nearly a metre in length and two set squares, one 60° and one 45°, a metric rule and scales will be required; also a good set of drawing instruments comprising compasses (for pencil and ink), dividers, ruling pen, and a spring bow compass for small work; the compasses should have fine needle points to avoid making large holes in the paper. The main requirements in all geometrical drawing are *accuracy* and *neatness*; these aims can be achieved only by using a fine-pointed H or 2H pencil for all construction work (which should be drawn *very lightly* and left showing in the finished drawing, not rubbed out), the figure or design being drawn stronger with an HB pencil, or ruled in ink. The drawing paper should be well pinned to the board, or secured with self-adhesive drafting tape, all horizontal lines being drawn with the T-square and all perpendiculars with the set squares. As long as the working edge of the board is true, it is immaterial whether the board be square or not, as rectangularity should be secured entirely by using the set squares against the edge of the tee.

In the practical application of geometry in the workshop or on the job, there are many handy methods that are worth noting. For instance, in setting out large circles on a wall, a strip of stout paper pinned at one end, with a pencil point stuck through at any given place, not only produces a true circle or arc, but also keeps a record of size, so that the same curves may be described again and again by placing pin and pencil in the respective holes each time. The workman who has a sheet of paper can easily make a true square, since a correct angle of 90° results from folding any such sheet exactly into four. Any weight suspended on a cord gives a true *vertical* line (a plumb line) which, together with the paper square, will enable one to obtain a true *horizontal* line when required.

An *ellipse* can be struck fairly accurately on a ceiling or wall with two pins or nails and a length of cord. To do this, set out two lines representing the length and breadth of the required ellipse and crossing each other in the centre at right angles (Fig. 31.6). Take a length of cord, measure off the distance of the longer line and make a loop at each end. Place a pin at one end of the shorter line, over which place the loops. Place a pencil in the loose part of the cord, extend it until the doubled cord is tight and mark off with it two points E and F on the longer line. Then place a pin at each of these points, slip one of the loops over each pin and with the pencil in the loose cord describe the ellipse.

Proportional scale It is frequently necessary to divide a given space into a number of equal parts. For example, a wall 1438 mm wide is required to be divided into 5 equal parts. If a 1500 mm rod is placed across the space diagonally so that the ends of the divided edge of the rod exactly touch the limits of the 1438 mm space, and the 1500 mm are ticked off from the rod, the division is accomplished more accurately and quickly than by calculation. This principle is of great value and may be applied in many ways. For example, a signwriter has written a large sign, say 4 metres × 3 metres, and he wants to set out a smaller one proportionally, say 3 metres × 2·25 metres. If the spacing and heights of the various lines of lettering are ticked off on a strip of paper placed vertically on the larger sign, the proportions will be reduced accurately in the small sign by being ticked off from the same paper

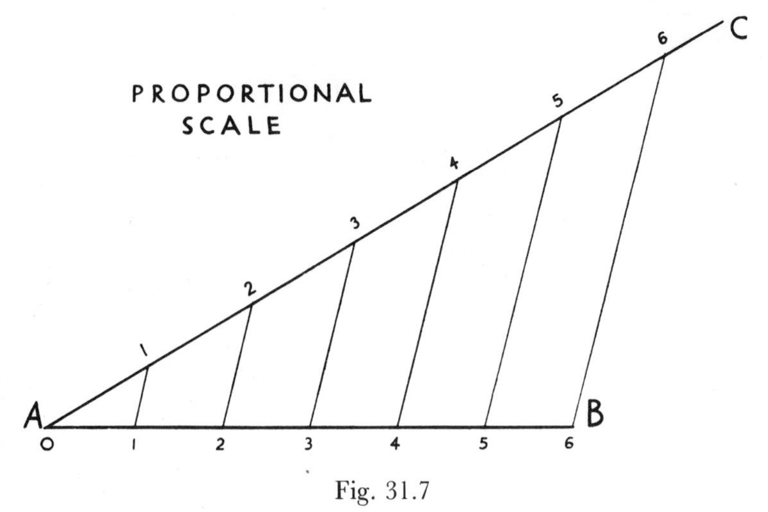

Fig. 31.7

To divide a line AB into any number of equal parts, draw a line AC in any angle and step off the number of equal parts required (in this case six). Join 6–B and draw lines parallel to it from the respective points as shown.

placed at the angle at which it fits the height of the smaller sign. The proportions of ornament may be reduced or enlarged by application of the same rule (Fig. 31.7).

Plain scales

A drawing to scale is a drawing made smaller than the actual work, in which every part is proportionate to the full size. The proportion is represented by some definite comparison which is termed the "scale"; a scale of 1/100 of the actual size is commonly used for drawings of large buildings and of 1/10 the actual size for furniture, decoration and details.

To set down a drawing to scale is fairly simple. The full-size measurements are taken in metres, and set out on paper to the reduced scale by ticking off the same number of metres on the reduced scale. The student is recommended to study BS 1192, Building Drawing Practice.

Plans, sections and elevations

In looking at an object directly from above, one sees it in *plan*; when looking at it horizontally, it appears in *elevation* (Fig. 31.8). If one could look down on a building with the roof removed, the floor *plan* would be seen bounded by the walls in *section*; this is actually seen when a new building is in the early stage of construction. Likewise, a vertical cut through a building would disclose a *sectional* view of walls and floors.

The correct drawing of sections is very important in all working drawings, since all constructional details and projections such as mouldings are shown in this way. Sections are generally shown hatched with lines ruled diagonally, or tinted with a light wash, or filled in solid black with Indian ink. In architects drawings, the sections are often tinted in various distinctive colours

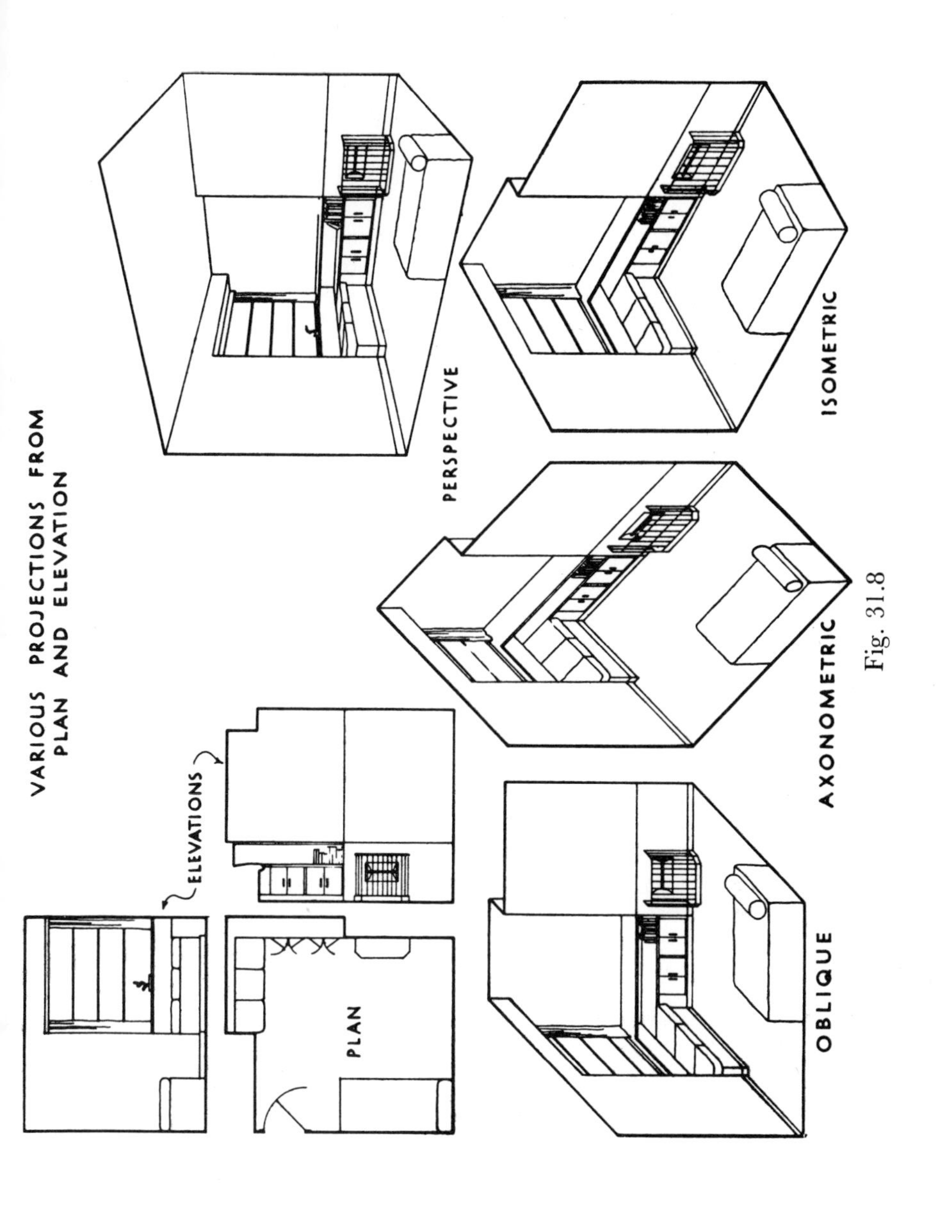
VARIOUS PROJECTIONS FROM
PLAN AND ELEVATION
ELEVATIONS
PLAN
PERSPECTIVE
ISOMETRIC
AXONOMETRIC
OBLIQUE

Fig. 31.8

to denote the different materials used in construction, e.g. brick, wood, steel, concrete, etc.

In preparing schemes for interiors, the positions of all furniture, fitments, carpets and rugs, etc., are shown on the floor plan, while ceiling panelling or decoration is shown in a separate ceiling plan. Wall treatments are shown on the elevations to the four walls, either set out round the four sides of the floor plan, or in one continuous strip, i.e. in *extended elevation*.

The student should closely study any architects' drawings that come his way in order to learn something of the technique of architectural drawing; the standard of draughtsmanship is usually high and the decorator-student should note particularly the variety and quality of line work. The use of different thicknesses of line makes all the difference between monotony and liveliness; important wall divisions, for instance, will be indicated in a stronger line than less important divisions, whilst boundary lines to the complete plan or elevation are best shown in a double-thick line to make the general shape clear. In showing the different members of a cornice or panel moulding, the lines should be drawn *lightly* with a slight strengthening at each end or at mitres. This avoids the heavy appearance which results when such closely ruled lines are drawn with a uniform thickness. Again, in drawing details such as a Corinthian capital, the detailed carving of volutes and acanthus should be put in very lightly with a fine line, a stronger line being used to delineate the general shape of the cap. In this way, the drawing "reads" much more easily than when a uniform line is used throughout.

This *variety* in the quality of line is a mark of good draughtsmanship, whether the subject be architectural, landscape, plant form, or the figure. It is a matter of giving emphasis where needed, and only when the student becomes aware of this quality can he hope to do really sensitive drawing.

Projections

Plans, sections and elevations are two-dimensional, i.e. they can only show length and breadth. They can be linked together, however, to show three-dimensional form, i.e. length, breadth and height, by means of solid geometry or pictorial projection such as oblique, isometric, axonometric and perspective views. These methods are not quite so forbidding as their names would imply; they are briefly described below and illustrated in Fig. 31.8 with different projections from plan and elevation of a day nursery.

Oblique projection is the simplest form of solid geometry, produced simply by drawing an elevation of one wall with lines drawn at 45° from each corner to correspond with the length of the side walls (to the same scale as the elevation) and lines drawn from these points to complete the view (Fig. 31.8). Thus, a three-dimensional appearance of the room is seen at a glance. There is inevitably a sense of distortion in such drawings because one is used to the foreshortening effect of receding planes, as in perspective. Sometimes the oblique lines are drawn at *half scale*, however, to reduce distortion and obtain a foreshortened effect; this type of projection is favoured by carpenters in showing mortise and tenon joints, etc.

Axonometric projection consists of drawing a plan at an angle to the horizontal (usually at 30°, 45° or 60°) and from each corner, drawing verticals (at the

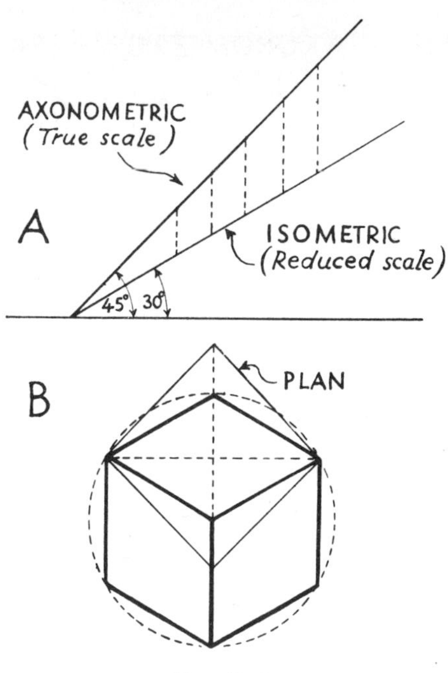

Fig. 31.9

A shows construction of isometric scale, and B an isometric cube in relation to its plan

same scale as the plan) to form the walls (Fig. 31.8). This method of drawing is particularly useful for showing the interiors of rooms, since floor and walls, furniture and fitments may be drawn to scale and measurements of panelling or fitments, etc., may be taken off direct by the craftsman on the job. Many working drawings showing plan, elevation and sections of carpentry and fitments are prepared in this way and the clarity with which constructional details are shown greatly facilitates the work of the craftsman. The positions of chairs and other furniture are indicated on plan and projected vertically to show the solid appearance, while wall subdivisions such as frieze and dado are measured off on the vertical heights and shown by lines drawn parallel to the floor plan. The heights of doors and windows are similarly indicated. Circular objects are shown as true circles on plan, but where these occur on the vertical planes (as, for example, an arched recess in the wall), the curve must be drawn freehand within the lightly indicated construction lines as shown in Fig. 31.10.

Isometric projection is very similar to axonometric except that the lines of the plan are drawn at 30° to the horizontal. Thus, instead of being a *true* plan as in axonometric, the isometric plan is of *diamond* shape. Verticals are projected to the same scale as the plan, exactly as in axonometric projection, and all the work may be done entirely with the 60° set square and T-square. Strictly speaking, all isometric drawing should be done to a reduced scale known as the *isometric scale* as shown at A and B (Fig. 31.9) to avoid distorted appearance. In practice, however, it is possible to ignore the reduced scale and to work to the same scale as one would normally use for an axonometric drawing.

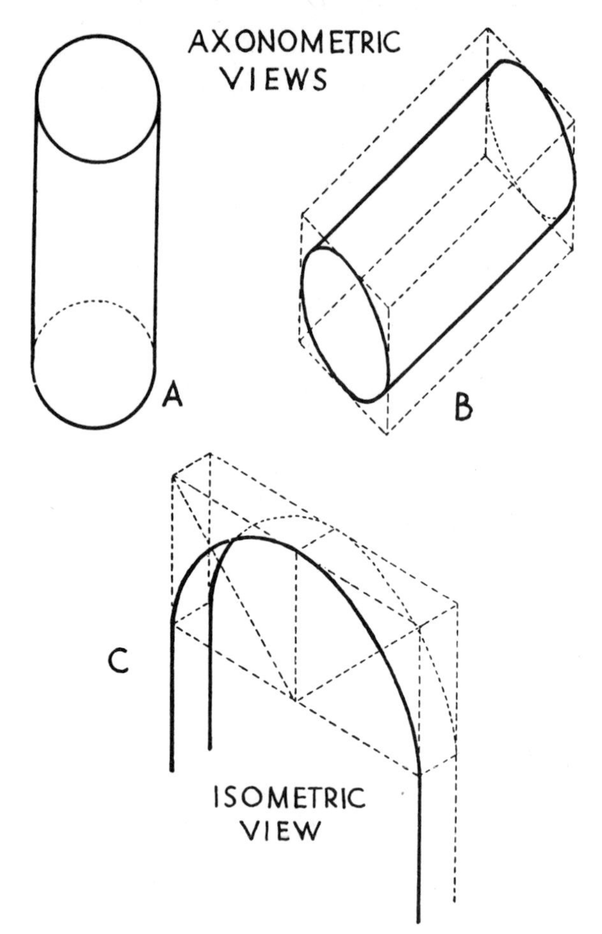

Fig. 31.10 Projection of curved objects
In B and C the curves are drawn freehand within the construction lines

A comparison of axonometric and isometric projections shows that the less acute angle of the latter tends to give a better feeling of perspective, though the lines do not actually converge.

Perspective projection is the nearest we can get to the representation of solid objects as the eye sees them. If we look at an object, such as a flat drawing board, directly from above, we see a true plan. If we move to one side the shape of the plan becomes distorted and the sides seem to converge instead of appearing truly parallel. Moreover, the nearer the plan is raised to eye level, the nearer a straight line it becomes. When it is raised above eye level, we then see the underside (Fig. 31.11). From this we deduce that the converging lines, if produced, would meet at some point at eye level, that is, on the *horizon line*; this point is called the *vanishing point* and its position is determined by (a) the distance of the object from the *station point* (i.e. the point where the viewer is standing) and (b) the angle which the plan makes in relation to the *line of sight* (Fig. 31.12).

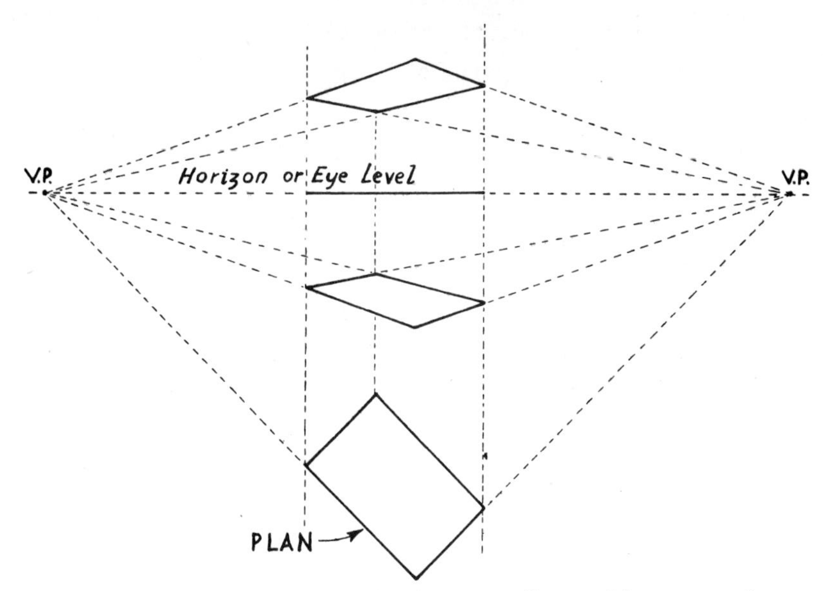

Fig. 31.11 How a regular plan becomes distorted in perspective

Its sides appear to converge upwards or downwards accordingly as it is placed below or above eye level. At eye level it appears as a straight line

When the plan is set *square* with LS (line of sight) there is only one vanishing point, since all lines parallel to LS will vanish towards CV (centre of vision); all lines at right angles to LS appear parallel to the horizon and can be drawn with the T-square. This is called *parallel* perspective. When the plan is set at an angle to the line of sight, however, all lines inclining to the left of LS will have their VP (vanishing point) on HL (horizon line) to the left of CV, while those lines inclining to the right of LS will have their VP on HL to the right of CV. An object so placed is seen in *angular* perspective. Examples of parallel and angular perspective are given in Fig. 31.12.

While the principles of perspective are fairly simple once they are understood, it is not an easy subject to teach without numerous illustrations showing the methods of working step by step, and it would be useless to attempt to describe the setting up of even a simple interior in perspective within the confines of this chapter. There are many books on the subject but very few which deal adequately with interior perspective. In our view, a good one for the decorator is *Applied Perspective* by John M. Holmes; this is clear, concise, well illustrated and interestingly presented, unlike so many books on perspective which make extremely dull reading.

Rendering

Architectural drawings are usually washed or *rendered* in water colour, and the student who wishes to specialize in interior decoration will need to attain some proficiency in the handling of water colour in order to present his colour schemes in a convincing and attractive manner. The basis of all water colour rendering is the *flat wash* and this should be practised diligently until one is able to lay in large areas with confidence. Requirements for success are

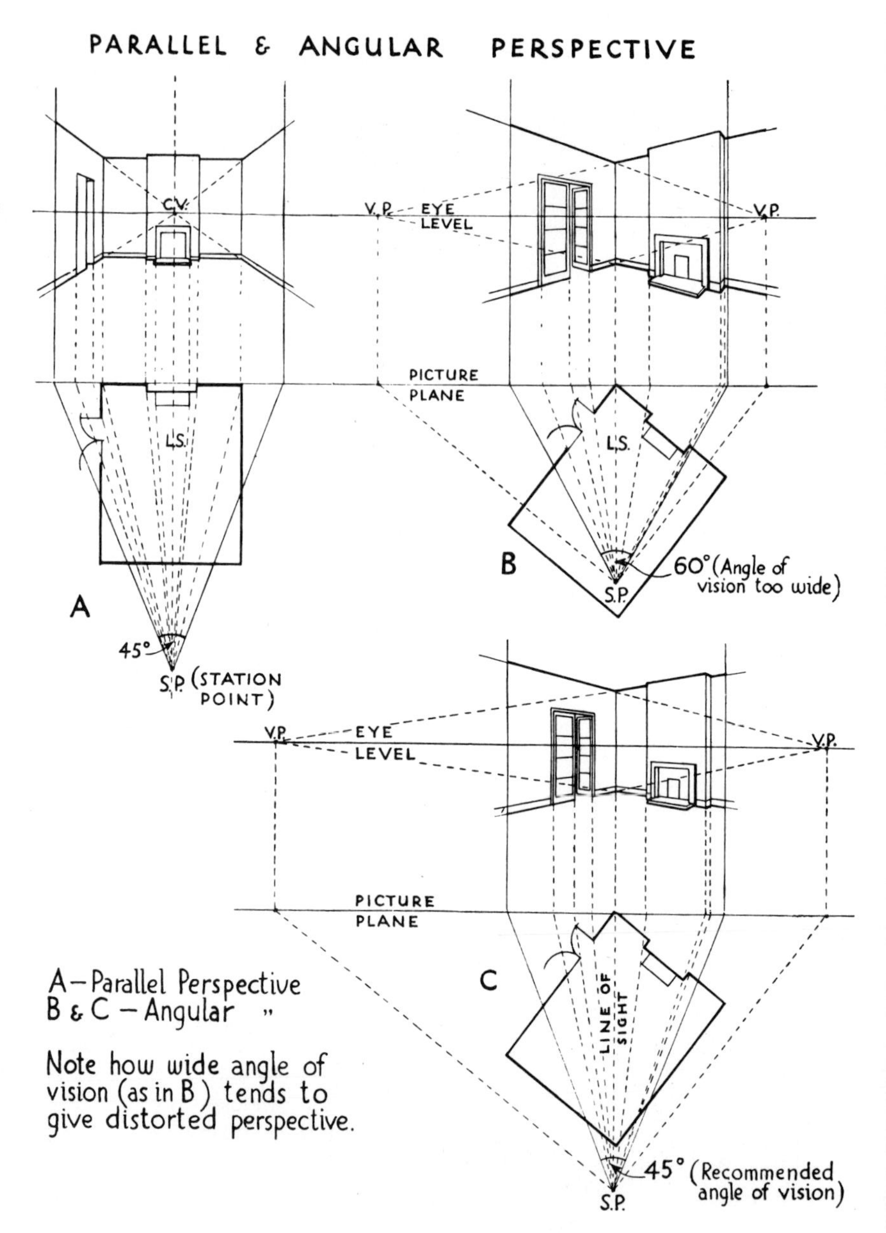

Fig. 31.12

It should be noted that the true PP (*picture plane*) would be represented by a line drawn at right angles to LS (*line of sight*) at the point where the angle of vision intersects the ground plan. In the above examples, however, the *picture plane* has been moved forward for convenience in setting up and also to increase the size of picture.

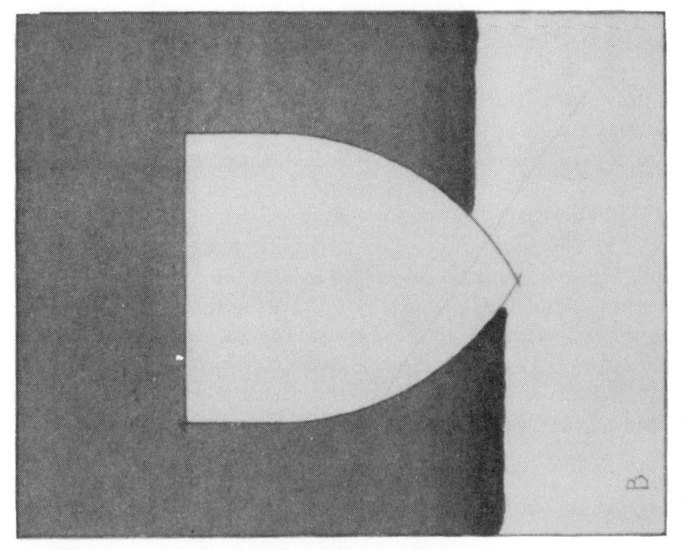

Fig. 31.13 Method of laying a flat wash in water colour

A The colour should always be laid on horizontal bands
B If a shape is to be left clear, work in horizontal strokes down each side simultaneously

(1) properly strained water colour paper or mounted board to avoid cockling; (2) plenty of colour mixed in a large saucer, and (3) a large sable water-colour brush, size 12 for preference. The board should be tilted at a slight angle (15° or so) and the colour laid with fully charged brush in horizontal bands, commencing at the top of the paper and allowing each succeeding band of colour partly to overlap the previous one; in this way the ridge of wet colour is caught up and flows into each band of colour as it is applied, the aim being to apply the colour swiftly (but without flurry) so that the colour flows down the paper with as little interruption as possible (Fig. 31.13). This is fairly easy in an unbroken flat wash, but becomes much more difficult when there are intricate shapes to be worked round. In such cases it is sometimes advisable to damp the paper with a wash of clean water, blot off and follow immediately with the required colour, always maintaining the horizontal bands while working round obstructions, and avoiding the temptation to "cut round" the shape and fill in afterwards.

In order to gain experience in handling water colour, it is good practice to wash in simple cubes, prisms, cylinders, cones and any other examples of solid geometry, rendering these in light and shade with cast shadow and reflected light; these exercises make interesting monochrome studies in sepia or other colour, but they also form the first essential steps to be mastered before going on to the rendering of an interior with its more complex problems of light and shade, colour and texture.

Fig. 31.14 A good conventional treatment of the heraldic Scottish lion

DESIGN

The term *design* is very wide in its application, but to the decorator it may be taken to mean not merely the drawing of ornament or pattern, but *planning* or *arrangement* generally. Thus, one may plan the interior of a room, including general colour scheme, motifs for wall decoration, arrangement of furniture and fittings, etc., all of which require the act of *designing*. In designing a wall decoration, the arrangement of motifs or pattern is considered in relation to the wall space; in designing a colour scheme, the distribution of colour in carpets and furnishings is considered in relation to the colour of walls and woodwork; in signwriting, the placing or layout of lettering must be considered in relation to the shape of the sign. Thus it will be seen that *composition* is of fundamental importance in all design work.

Decorative design, generally, calls for some simplification in the treatment of outline or form, a good instance being the *silhouette*, where the whole shape is painted in flat colour. Another instance is the simplified treatment of the lion in heraldry, where the general outline, details of mane, tail and claws, etc., are rendered in a simple decorative form. These are known as *conventional* treatments. In a good design, the conventional treatment is dictated by the material and the means by which the design is produced. Thus, a painted heraldic lion would clearly show the influence of the brush in the flowing

Fig. 31.15 "The King's Head": design for an inn sign by Reg Ball

Fig. 31.16 Heraldic painting and lettering — design by Jesse Kendal

lines of mane and tail, etc.; if it were stencilled, the details would be expressed in the form of ties and the whole design would have a *pierced* quality. If the design is carved or modelled, it will show the influence of the chisel or the modelling tool.

Decorative spraywork is another instance where the obvious influence of the spray technique largely dictates the result; the limitations imposed by templates, for instance, where these are used in carrying out the design, necessarily call for a simple conventional treatment.

A good design should "read" well, that is to say, all detail should be subordinate to the main shapes which in themselves should be easily recognized. In other words, the units of a design should have good *silhouettes*. This is demonstrated perfectly in good examples of heraldry (e.g., the Windsor Stallplates) and in all good poster and display work; in fact, some modern posters display a strong heraldic quality in their utter simplicity of treatment.

The decorator's aim, of course, differs from that of the poster designer, but nevertheless he cannot do better than study some early examples of heraldry (not the debased forms adopted by Civic Authorities and corporate bodies); for the principles of blazoning, which govern legibility, conform to the principles of good functional design. These same principles applied to mural decoration will ensure good arrangement of masses, avoidance of unnecessary detail, and good definition of the individual figures or units of the design which otherwise might be overwhelmed by detail or background.

In the design of ornament or pattern the motif or unit of repeat is sometimes naturalistic, but more often is highly conventionalized or even abstract (i.e. having no direct reference to nature). Wallpaper designs and appliqué cut-out motifs often show quite natural-looking flowers, birds or landscape effects, but even so the influence of block and stencil, or machine printing, is always apparent. In any hand-painted decoration, the effect will depend a lot on the type of brush used; the writing pencil, for instance, will of necessity produce a *linear* effect, the fitch being more suitable for broader work, whilst a larger mass may require a 50- or 75-mm paint brush or even a small flat wall brush. Stencilling will be limited in colour by the number of plates used, and the unit of repeat limited by the size of the stencil plate (see also Chapter 33, "Applied Decoration").

We have only been able to indicate in general terms some of the principles of design as applied to decoration, but it is a vast subject requiring specialized study. Two books by Lewis F. Day, *Nature and Ornament* and *Pattern Design*, are excellent for students of design.

32

Heraldry

For the complete study of heraldry the student is referred to the standard works by Fox-Davies and Boutell (see Appendix 6), since it is not within the scope of a textbook like this to deal with the subject at any great length. For the signwriter or decorator, however, who may occasionally be called upon to depict armorial bearings, as in municipal work, it may be useful to give a few general guide-lines as to the correct treatment of the shield and its charges.

Shields are used, not only in true heraldry, but in decorative emblems of all kinds — civic, ecclesiastic, domestic, or commercial. There is no set form for the shield, nor are there any absolute proportions. In the earliest times the shield was of great length, but in the 13th century it began to decrease and continued to do so until, in the later Queen Anne period, it became only slightly taller than its breadth. A pleasing proportion is about 9 parts height to 7 parts breadth. The surface of the shield is called the *field* , the right side as borne by the owner being the *dexter* side and the left side the *sinister* (these positions are reversed, of course, when viewed by the spectator).

Divisions of the shield

The principal divisions of the field are —*fesse*, *pale*, *bend*, *saltire*, *chevron* and *quarterly* (usually written "party per fesse", "party per pale", and so on). Further subdivisions are called *Honourable Ordinaries* and *Sub-ordinaries* (see Fig. 32.1).

'Ordinaries" are usually one-third of the width of the shield in breadth. The *chief* does not vary in size, but the other ordinaries have narrower or diminutive forms under different names. The *chief* alone is often used "imposed upon" the other divisions, as on the arms of Manchester, in which the field is divided by diagonal bands (termed *bendy*) and upon which the *chief* is overlaid.

Divisions of the shield and ordinaries are often shown with wavy, zigzag or scalloped edges, as shown in the "lines of partition" in Fig. 32.1.

Charges

Heraldic symbols and devices, referred to as *charges*, include all manner of animated creatures, celestial bodies, flowers and ornament, the treatment being highly conventionalized, except where they are required to be shown *proper* or natural. It is in the rendering of charges that the decorator has most freedom, for he can use his own discretion in the actual drawing so long as the heraldic description is adhered to. For instance, if he is required to paint

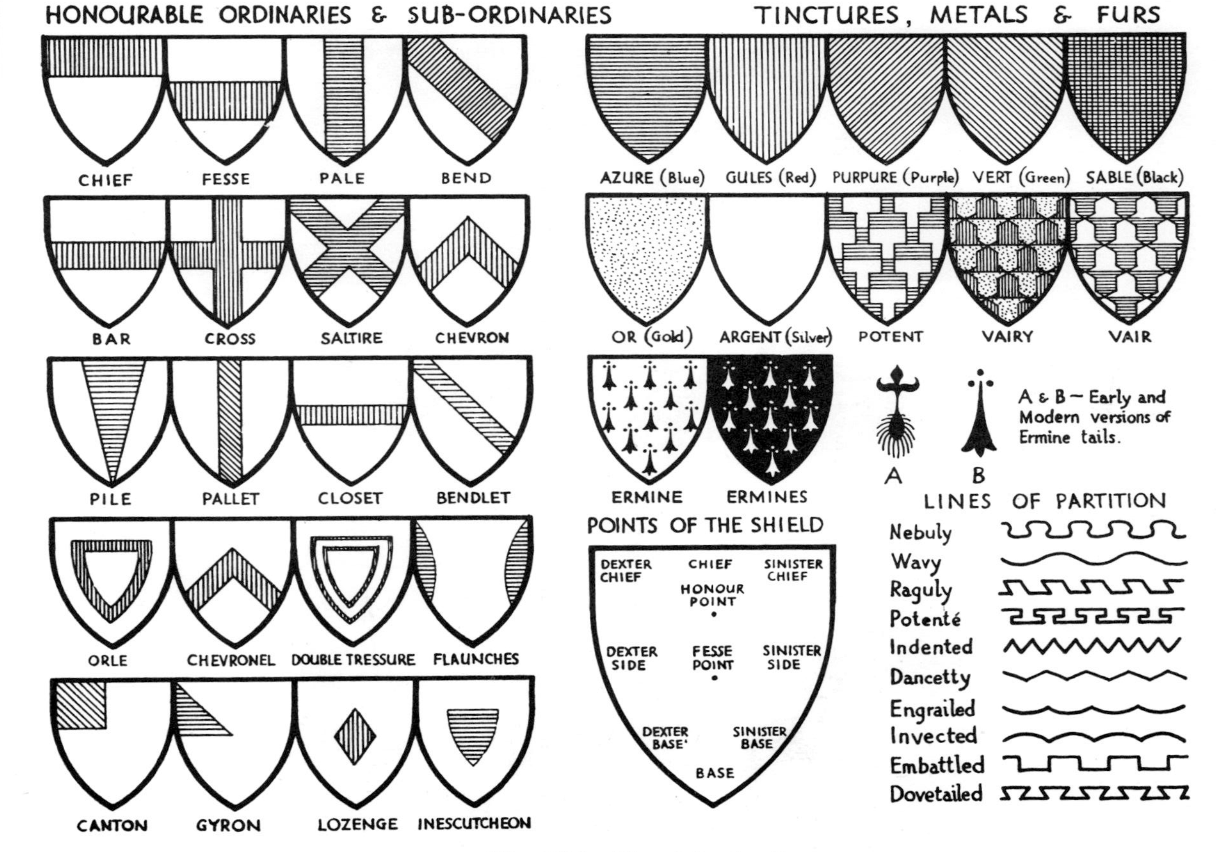

Fig. 32.1 Heraldic details

a lion *rampant* from a printed copy which he considers is poorly drawn, he can re-design the lion, providing he shows it still *rampant*. A good instance of this occurred when the Scottish Lion was used as a symbol for the Empire Exhibition at Glasgow. Instead of the "rag and bobtail" lion so often depicted on heraldic shields, the new design was an excellent example of conventional treatment and of good heraldry (Fig. 31.14). The heraldic beasts shown in Fig. 32.2 are designed essentially for flowing brushwork and attempt to conform to the true spirit of heraldry in maintaining a good *silhouette*. Broad flat treatment of this type is much to be preferred, we feel, to the highly worked up, shaded and outlined charges usually seen.

Fig. 32.2 Painted heraldic charges
Showing influence of brush technique

Achievements

Complete armorial bearings (usually referred to as an *achievement*) include the *shield*, *crest*, *helm* or helmet, *wreath*, *mantling* and *motto*. The space taken up by the crest and helm should not exceed two-thirds the height of the shield.

The rules as to heraldic colour must be strictly observed. No colour should be shown upon a colour, no metal upon a metal nor fur upon fur; for example, a red rose may be depicted upon field of gold or silver but not on blue or green. White may be for silver and yellow for gold, but the rule still applies; thus a red rose may appear on a white field (for silver) or yellow (for gold), but not a white rose on a yellow ground. The origin of this rule was purely functional, for it was essential that in battle the arms of one knight should be distinct from those of another without any possibility of confusion. In very strong sunlight, for instance, it would sometimes be impossible to distinguish one metal from another and therefore impossible to discern the shape of the design; and in certain lights the tonal difference between one colour and another would be insufficient to make the design stand out prominently.

The only instance in which the colour and metal rule does not apply is when the objects are described as *proper* or *natural*. Thus a leopard *proper* must be painted in natural colours, black and yellow, even though on a gold or yellow ground.

Tinctures and furs

The usual colours, or *tinctures*, and the names used for each in a written description, are as follow —

Silver or white	Argent
Gold or yellow	Or
Red	Gules
Blue	Azure
Green	Vert
Purple	Purpure
Black	Sable

Less frequently we find —

Blood red	Sanguine
Orange	Tenne
Sky blue	Celeste
Red-brown (or mulberry)	Murrey

In black-and-white illustrations, engraved surfaces and carved work (which the decorator often has to work from) the colours are indicated by

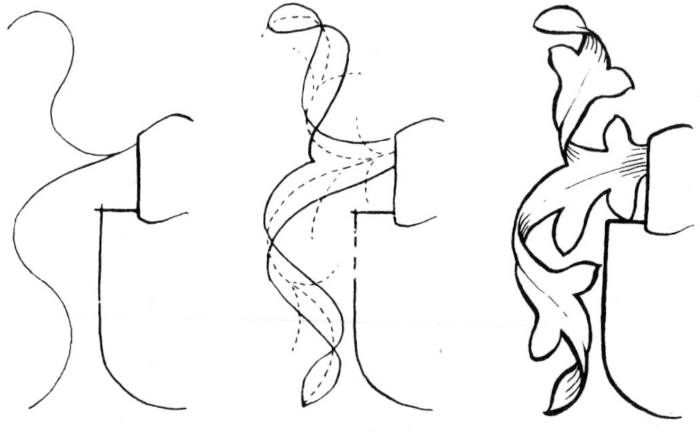

Fig. 32.3 Simple method of drawing mantling

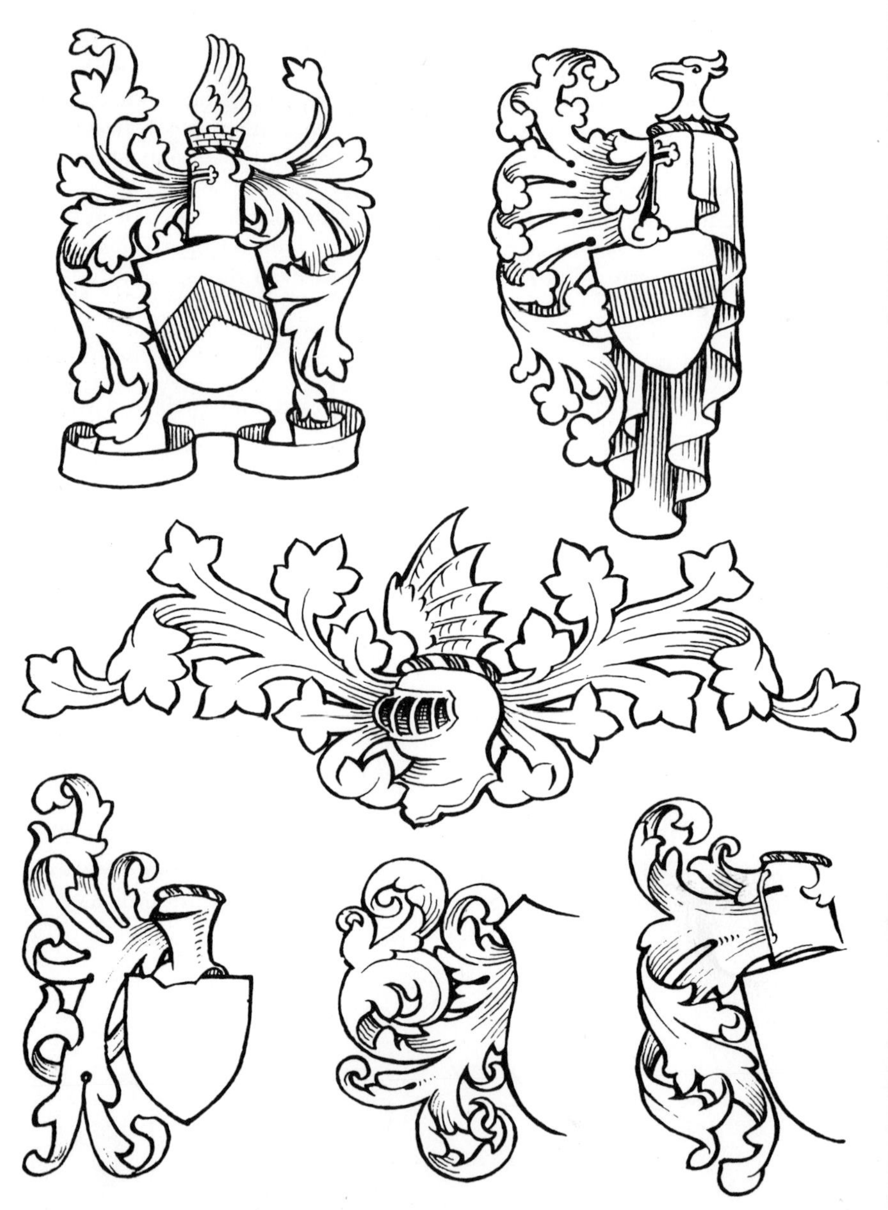

Fig. 32.4 Various types of mantling

lines or hatching, as follows —

Silver	A plain white surface
Gold	A dotted surface
Red	Lines drawn vertically
Blue	Lines drawn horizontally
Green	Lines diagonally left to right
Purple	Lines diagonally right to left
Black	Lines vertically, crossed by lines horizontally

The above are shown in Fig. 32.1 together with the different furs.

Mantling (Fig. 32.3, 32.4) should always be shown the colour of the shield, with the reverse turnovers the principal colour of the charge; thus, the mantling to a gold shield on which there is a black cross would be gold with black lining or reverse turnover. Mantling is often very elaborate in treatment, especially when shown as richly foliated acanthus, as in the heraldic plates by Dürer; in a modern setting, however, we prefer to see it treated as a simple flat decoration, without realistic light and shade effects.

33

Applied Decoration

ART AND CRAFT

Many keen students of painting and decorating aspire to excel in one or other of the higher branches of the craft. Some will aim at becoming proficient in graining and marbling, others will want to specialize in signwriting, while those with a strong sense of design will be chiefly interested in the various forms of *applied* decoration, a wide term which covers all forms of surface decoration ranging from simple stencils and lining to hand-painted ornament, motifs and murals. The painting of mural decorations — by which is meant the design and execution of a pictorial or decorative composition as a wall painting — has attracted man from the earliest times, when he first adorned his cave with simple engravings and paintings of animals and men.

The student apprentice should know and come to appreciate that the training he receives as a painter and decorator is something to be prized; that his craft has a long and honoured tradition and is, in fact, essentially the same as that practised by the greatest painters of all time. If he has the opportunity to study some of the early Italian and Flemish altarpieces in the National Gallery, he will there recognize many familiar craft processes — finely modelled gesso, gilding, glazing, wiping, stippling, graining and marbling — and he will experience that glow of admiration which every craftsman feels on beholding a job of work superbly done. Thus may begin his first steps in art appreciation, finding in craftsmanship a common bond — something he can understand and admire. And if he bears constantly in mind the fact that some of these early church paintings were done by men who practised the same craft as himself, he may come to feel a justifiable pride in his own calling and resolve to uphold its fine traditions.

Our main purpose in the above preamble is to show that the world's greatest paintings are the result of *craftsmanship* wedded to *art*, the important point being that the artist cannot express himself to the full until he has first mastered the technique or craft side of his job. The apprentice who has strong leanings towards art and who feels an urge to rise above the general run of painting and decorating should firmly resolve to become proficient in the various branches of decorative technique and expert in the handling of the tools of his craft. The following sections deal briefly with the practical application of ornament and decorative wall painting, from the preparation of the preliminary small-scale drawing to the various decorative techniques employed in the full-size design. Decorative spray work is considered in Chapter 6.

THE PRELIMINARY SKETCH

Where a proposed scheme involves any hand-applied decoration it is usual to submit a drawing for the client's approval. Generally, the drawing is in perspective, since the layman often finds an elevation drawing unattractive and even difficult to understand. From the decorator's point of view, however, elevation drawings showing the exact layout of the proposed decoration to accurate scale are of more practical use, especially if the design is to be of an elaborate nature.

A plan of ceiling and elevations of the four walls is therefore set up on a strained water-colour sheet (if the paper is strained *after* the drawing is done, the scale is liable to be inaccurate owing to the stretching of the paper).

The preparation of scale drawings has been discussed on page 428 and need not be enlarged upon here except to say that in drawing any decorative details to scale, the size of the finished work must be continually visualized. This comes naturally to anyone who is used to working to scale, but one often sees students' drawings which plainly show that the importance of *thinking* to scale has not been sufficiently realized. The space occupied by the decoration should always be judged in relation to the area of the whole wall (or ceiling) so that it looks perfectly appropriate when enlarged to the full sze. If the question of colour is largely settled at the sketch stage it may obviate much "trial and error" experimenting on the wall. A colour which looks unsatisfactory on the sketch will probably look worse on the wall.

When the sketch is completed to satisfaction, the question of enlargement comes up. The usual method is by "squaring-up", that is, by drawing a "grid" of squares over the scale drawing, numbering the top row and lettering the left-most vertical row, then drawing a similar grid of enlarged squares on the cartoon or detail paper, numbering and lettering each square likewise. Thus, if the scale drawing is 1/10, a grid of 10-mm squares would be enlarged to 100-mm squares on the full-size drawing. The details which appear in the small squares are then drawn in the corresponding large squares, the letters and numerals assisting in locating the appropriate square. If it is desired to preserve the small sketch, it need not be defaced by drawing the grid actually upon it, since it may be covered with thin cellophane on which the grid can then bc set out with a glass-marking pencil and afterwards wiped off with a dry cloth if necessary.

Enlargement is, at best, a rather laborious method and some designers use various ingenious devices to expedite the work. If an epidiascope, or better still an overhead projector, is available, this will be of considerable help, for then the design can be projected up to the required size on to thin detail or tracing paper, on which the design can easily be traced.

It will be found that when the design is enlarged it will need careful correcting and strengthening, and more detail adding, for what appears adequate in a small sketch often appears very unsatisfactory full size. In an important mural, for instance, the *cartoon* (as the full-size working drawing is called) is often worked up to a considerable degree of finish and used as a study for the final painting.

SETTING OUT

A knowledge of elementary geometry is essential for setting-out work, for ignorance of geometrical principles can lead to a great waste of time, even on simple jobs. The use of diagonals in centring ceilings and in checking up the squareness of a room should be understood, also the use of a spirit level and plumb-line. The setting out and drawing of ellipses should be mastered; two simple methods are shown in Fig. 31.6, page 426.

The intelligent use of the proportional scale method (Fig. 31.7, page 428) in subdividing awkward spaces can save a lot of time (and headache) for those who, like the writer, have no instinct for arithmetic. For instance, to divide a wall filling measuring, say, 2·616 metres high into five equal horizontal bands — take out your chalk line and pin one end at the foot of the wall, then measure off exactly 5 metres, pull the line taut until the 5-metre mark just touches the top of the wall and snap a line (Fig. 33.1). Step off your five spaces at 1 metre intervals along this oblique line, then snap horizontals through these points and you have your wall divided into five equal bands. If a tape measure be used instead of a chalk line, the five points may be ticked off direct on to the wall. In the same way, smaller spaces can be centred or divided into any number of parts merely by running the metric rule obliquely across the space until a convenient dividend can be read off.

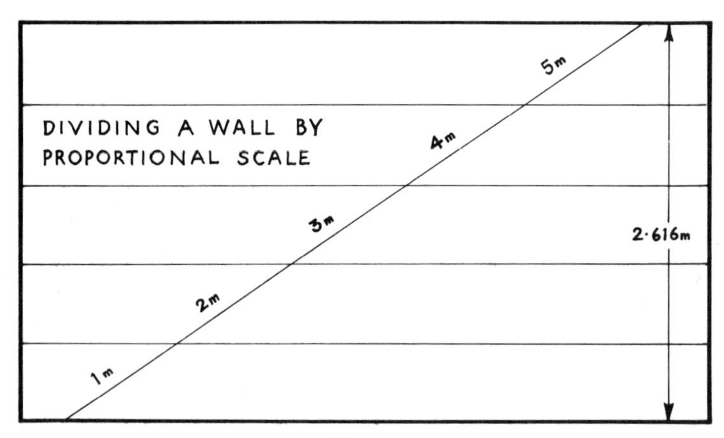

Fig. 33.1

The requirements for setting out are — metric rule, 2-metre and 1-metre bevelled straightedges, large-size 45° and 60° wood set squares, wooden compasses, chalk line, plumb-line and a few stencil pins, preferably those with brass heads and fine needle-points which do the least damage to the plaster. Fine crochet cotton makes the most satisfactory chalk line for lining or similar precision work, but twine may be used for setting out broad work, e.g. scene painting, etc. Self-chalking lines are a great convenience. Pale blue chalk is probably the most useful for setting out work since it can be seen equally well on light or dark grounds. Soft vine charcoal is also necessary on occasion, but avoid the use of any sticks which tend to scratch the surface.

Simple ornament or decoration may be sketched directly on the wall surface, but where the work is of an elaborate nature and is to be painted by hand a cartoon is generally prepared first. This is especially necessary where the figure is involved, for there are very few decorators competent to render the human form without something of a struggle. The cartoon is prepared on thin detail paper (obtainable in wide rolls), first roughing-in the design with charcoal, then drawing more carefully with thick lead or even outlining the subject with a sable and thin sepia ink. When completed, the back of the cartoon is rubbed with dry umber to enable the design to be transferred to the wall; the cartoon is then pinned into position (or fixed with self-adhesive tape to avoid making pinholes), and the design is traced through with a blunt point, such as a knitting needle. A quicker transfer method is to outline the design heavily in very soft charcoal on the *reverse* side (the design can easily be seen through detail paper), then pin up the cartoon in position on the wall and rub over the lines with a spoon or other smooth article; if the ground is textured, rub vigorously with a cloth, folded into a tight wad. The design will then be transferred to the wall surface, but, being charcoal, will require either fixing or pencilling-in to prevent the drawing being dusted off and lost. A third method is to prepare a *pounce*; this is especially useful where the design is to be repeated. The cartoon is laid flat on a soft ground, such as lino or cardboard, and the design is pricked through with a needle point set in a handle, or with a special pricking wheel made for the purpose. The punctured design usually forms a burr on the reverse side of the paper and this should be removed by rubbing with fine glasspaper, otherwise the design will not pounce through distinctly. If the design is symmetrical the paper may be folded accurately on the centre line and both halves pricked through at once. The pounce is then fixed in position on the wall and the design is pounced through by rubbing or "pouncing" over it with powdered chalk or dry colour tied up in a small cambric bag, known as a "pounce-bag". Alternatively, the pounce may be rubbed over with a fitch or sash tool dipped in dry colour or powdered whiting.

In all setting-out work, avoid the use of strong powder colours or highly coloured chalk, since these are sometimes very difficult to remove afterwards and cause trouble by "working up" when painting begins; black lead is especially troublesome in this way and we need hardly mention that a copying ink pencil should never be used on any account because of its liability to bleed.

STENCILS AND TEMPLATES

A stencil is a cut-out pattern or "pierced" design, usually cut in paper, thin card, or lead foil. By application of colour through the cut-out portion, the design may be quickly reproduced a number of times. The design must be very carefully considered so that the stencil holds together well and is capable of withstanding a fair amount of usage without breaking apart. The pierced portions are held together by "ties" which should form an integral part of the design. Weak delicate ties must be generally avoided since they cannot stand up to more than one or two repeats. A *positive* stencil is one in which the ornament is cut out; a *negative* stencil is obtained when the background is cut out and the ornament itself forms the ties. A multi-plate stencil is a design cut from two or more plates by means of which "outline" and other effects may be produced and ties filled in (hand-printed wallpaper borders and appliqué motifs are generally produced by multi-plate stencils).

Stencil paper may be purchased ready oiled, in which case the design is usually set out on lining or detail paper and traced down on to the oiled paper through a carbon sheet; ordinary pencil does not function well on oiled paper, but if desired the tracing can be dispensed with and the design set out direct on the oiled paper with a wax glass-marking pencil. Many prefer to oil their own paper, in which case the design is first set out on stout cartridge paper (purchased in wide rolls) and the paper is coated with boiled linseed oil plus a little goldsize. When quite dry, the design is cut out and the stencil may then receive a coat of thin knotting before putting in use. Oiled paper is much easier to cut than unoiled paper. Cutting is generally done on a large square of plate glass, but some decorators prefer sheet glass because it does not get defaced so readily as plate. Others prefer to cut on a soft bed such as lino, but this tends to leave an undesirable burr on the underside of the stencil. Small holes may be punched out or cut out with cork borers (as used by chemists), using a piece of thick lino or thick card as a bed; the burr on the reverse side may be removed by lightly glasspapering.

Stencil knives are shown in Fig. 33.2, or a Stanley knife (Fig. 33.3) or good pocket knife with stainless steel blade may be used; cheap soft steel is useless

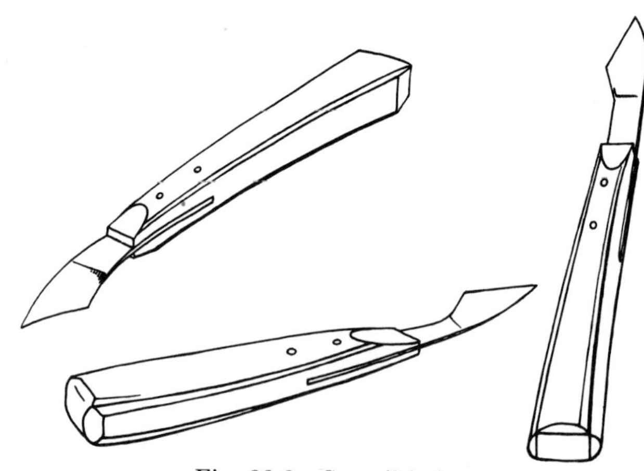

Fig. 33.2 Stencil knives

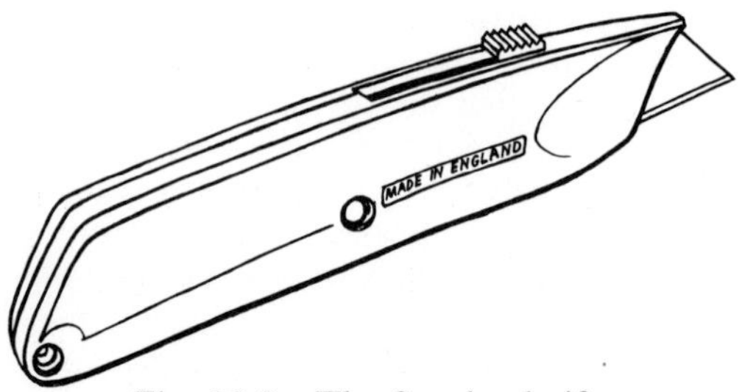

Fig. 33.3 The Stanley knife

since it requires sharpening after every few strokes. An oil-stone should be kept handy since a sharp point is really essential for clean cutting. Small intricate shapes should always be cut out first, as the plate is considerably weakened once the large areas are cut. Another point — if the stencil paper is taken from a roll, it should always be back-rolled to take out the curl *before* cutting the design, otherwise every cut edge will tend to curl up and thus make clean stencilling very difficult, if not impossible.

In designing stencil borders, a full repeat should be cut at each end in order to secure perfect register. Avoid bending the stencil round corners if possible, either by cutting a special corner stencil, or by designing a suitable stop. Stencils may be cut in duplicate by fastening two sheets of stencil paper together with paper clips or adhesive tape. Where a long run of border pattern is required, as in church work, two or more sets of stencils should be provided for replacements and also to allow the work to proceed simultaneously on different parts of the job. Single and multiple "line borders" may be applied very speedily by stencil (as an alternative method to running painted lines with straightedge and lining fitch), but great accuracy is required both in cutting the stencil and in applying it; otherwise the ties will show when joining up.

On ceiling work, stencils may be cut from thin card (oiled and knotted) and if necessary reinforced or supported by plasterer's thin laths. Lead foil is useful for stencilling small mouldings and coves since it lies snugly to the surface contour.

Templates are shapes cut from stencil paper, cardboard or thin plywood round which colour is applied by means of brush, sponge, or spray gun. Templates can be used for shields, cartouches and borders such as wavy bands and zigzags, either alone or in conjunction with stencils.

STENCILLING

The tools required for stencilling are — stencil brushes, palette, dippers, and stencil pins or adhesive masking tape for holding the stencil plate in position. Guide lines are first snapped on the wall with a chalk line and corresponding guide lines and centres should also be indicated on the stencil so that it can be placed immediately in position. The stencil plate may be held with stencil pins (Fig. 33.4) if the work is well above eye level where the pinholes will not show, but for dado borders and other work subject to close scrutiny it is advisable to use self-adhesive masking tape instead since this will not deface the wall surface. Templates may be similarly held with masking tape as indicated in Fig. 33.9.

Colour should be of fairly thin consistency and should not work "sticky", as this causes the stencil plate to be lifted away from the surface by sticking to the brush. Flat oil paint thinned with white spirit works well, and the addition of a little glaze medium will retard the set sufficiently to prevent the colour "piling" on the stencil plate. Where light colour is stencilled over a dark ground a proportion of titanium white may be added to the colour to secure maximum opacity, otherwise two coats will be necessary.

Do not dip the stencil brush in the paint pot, but pour a little colour on to a palette, lightly dab the brush tip in it and rub out well on the palette before applying to the stencil; this avoids excess colour in the brush and makes for clean stencil work. If the brush is overcharged, the colour will splurge under the edges and will look very ragged and unsightly.

Fig. 33.4 Stencil pin

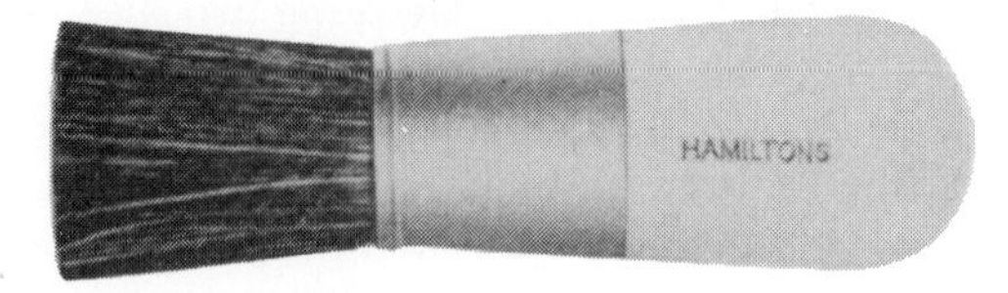

Fig. 33.5 Stencil tool

Colour gradation or blending is readily obtained with the stencil method, but we prefer an even or uniform tone, seeing that stencil work is essentially a flat decoration (except, of course, where actual relief effects are produced with plastic paints). Unusual and highly decorative effects may be obtained by double stencilling, that is, by stencilling first in, say, cream on a medium blue ground and, when dry, overstencilling in a tinted glaze colour such as sienna; if the plate is placed slightly out of register, a thin line of cream will be left on one side of the pattern and a dark line of the glaze colour on the other, thus producing quite a striking effect of bold relief. If gold or other metal is used instead of cream, the effect is even further enhanced.

Stencilling need not always be done with stencil brushes; fitches, sash tools and sponges may all be used for special effects. The spray gun lends itself ideally to the execution of rapid stencil and template decoration, though care must be exercised to avoid blowing colour under the mask; this is likely to occur when the gun is held at an angle, or when the air pressure is too high.

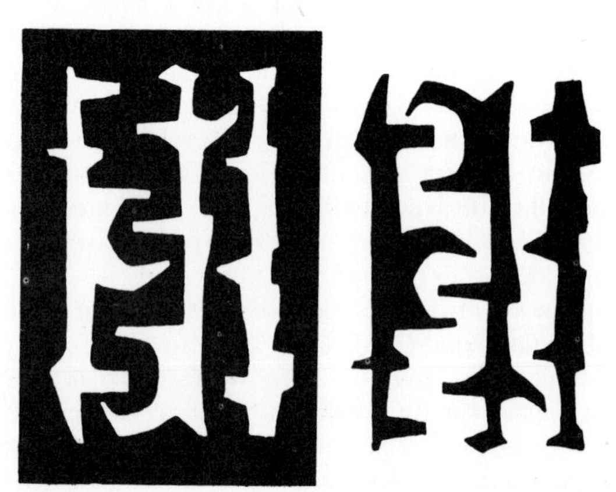

Fig. 33.6 Abstract stencil design by Reg Ball

Stencilling is essentially a repetition process; it is not economic to design, cut and apply only one stencil motif, though, of course, if the same motif can be used on other jobs the preparation of a stencil is justified. In practice, the decorator usually carries a stock of stencils, ringing the changes on them by using a different colour scheme on each job, or by slight adaptations and combinations with other stencils. The application of stencils is regarded by many as being a mechanical process, and while this may be true in that a very careful workman can produce good clean stencilling, the method is capable of a great variety of decorative effects in the hands of a designer. For instance, colour may be removed through a stencil plate, as in what are termed "wipe-out" effects.

DECORATIVE SPONGE-STIPPLING

Hitherto sponge-stippling has generally been confined to the production of broken colour or background effects, but the technique is one that has great possibilities in the rendering of decorative motifs and murals, especially when so much interest is being displayed in unorthodox treatments in contemporary decoration.

Sponge-stippling can be used as an alternative to large-scale spraywork where this would involve the use of costly or wasteful templates. Sponging, however, should not be regarded merely as a substitute for something else; it will stand on its own merits as a legitimate decorative technique. It may be used either alone or in combination with other methods, such as brush stipple, dry-brush dragging, and so on, with or without templates, and once the knack of manipulating the sponge has been acquired it is a fairly quick and easy method.

The facility with which gradations and "blends" are produced with the spray gun has made these effects widely popular, and during recent years we have become all too familiar with a slick type of sprayed decoration in café and cinema, often grossly overdone. The excessive use of any medium or style ultimately produces its own antidote, and a surfeit of the mechanically smooth gradations of the spray gun seems to have resurrected a new interest in the character and spontaneity of handwork. Sprayed decoration, of course, has come to stay, and will no doubt be used more and more in the future, but the wise decorator will use it with greater discrimination, balancing the merits of spray, brush and other methods with a sense of the appropriate.

Sponge-stippled gradation cannot compete with the spray gun for speed or perfection of finish but, on the other hand, it often gains on a large job by cutting out the necessity for templates which (unless there is much repetition) use up much time and material; moreover, the sponge finish, if carried out nicely, has its own attractive character.

When a continuous band of colour is required, graduating from a hard edge into the background, the edge can first be run in with fitch or 25-mm tool in solid colour, the gradation being achieved by sponge-stippling away from this — heavily at first, then more lightly — into the ground colour. If a solid painted edge is not desired, a slip of oiled stencil paper can be used as a

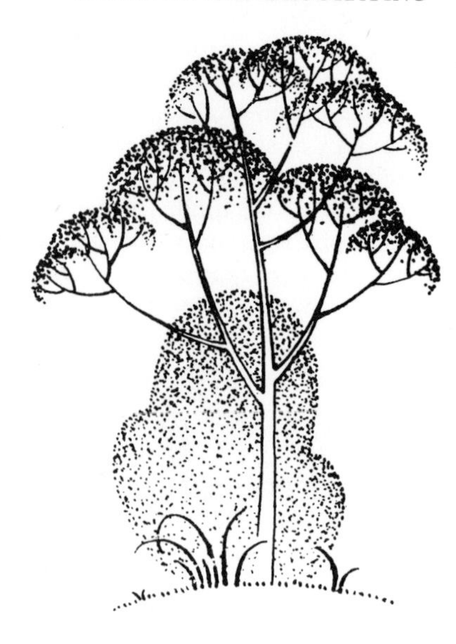

Fig. 33.7 Tree motif in sponge-stipple and pencil work

mask from which to commence sponging. Gradual curves can be negotiated in this way by shifting the paper along instead of cutting a template for the whole curve. With practice, however, on broad work one can sponge away from a pencilled or chalked line without the aid of a mask and still obtain a fairly clean well-defined edge. The sponge can be trimmed to a straight edge to facilitate this; it also helps in sponging close to awkward corners and mouldings.

The actual execution of any stippled decoration is, of course, a highly individual affair, and the decorator will be guided by his own experience and artistic convictions. It would be presumptuous to lay down dogmatically how this or that should be done, but one can sometimes usefully suggest ways and means. The illustrations, therefore, must be regarded as an attempt to show some possibilities in decorative sponge-stipple, rather than as good examples of decoration.

Sponge-stipple motifs look best, we feel, on a plain or very slightly textured ground, because the sponging then provides a pleasant contrast to the background. When it is desired to work on a sponged background, however, this should be kept as unobtrusive as possible, so that the oversponged design has sufficient contrast to prevent its getting lost in the background treatment.

One useful way of making a simple motif more interesting is to create a sort of field or surround, either by sponging a "vignette" on which to paint the motif (as shown in Fig. 33.8 (a)), or a nebulous shape sponged round a template (as in Fig. 33.8 (c)).

Apropos the use of a template, we have found it a good plan to cut small holes or V-shapes round the edge, over each of which is placed a strip of masking tape (Fig. 33.9). This holds the template or stencil close to the wall and prevents colour from creeping under the mask.

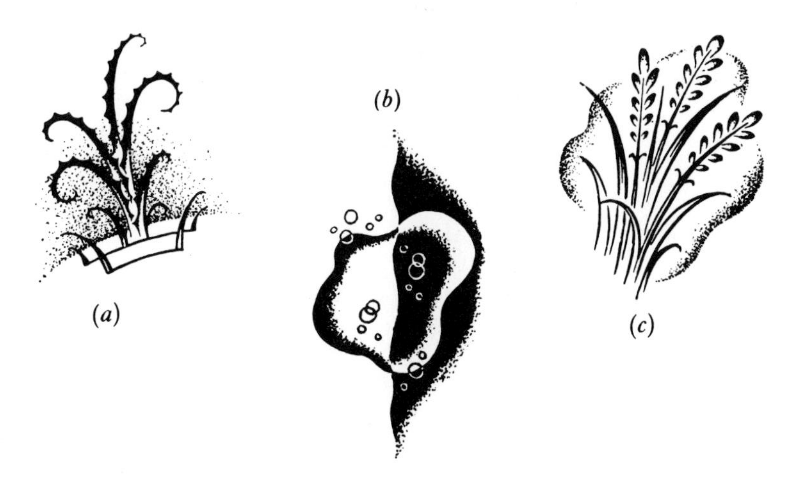
(a)
(b)
(c)

Fig. 33.8 Sponge-stipple and pencil work

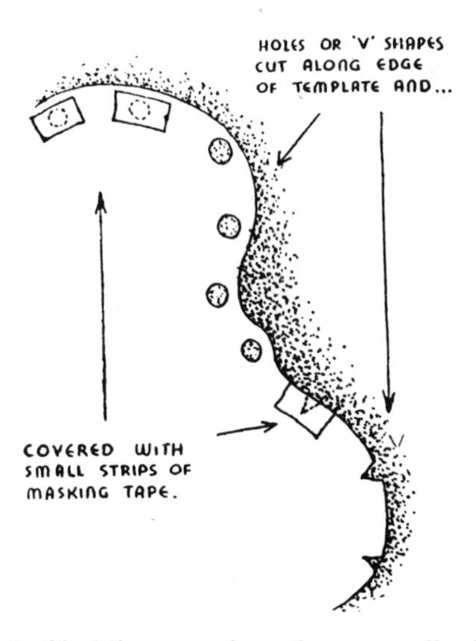
HOLES OR 'V' SHAPES
CUT ALONG EDGE
OF TEMPLATE AND...
COVERED WITH
SMALL STRIPS OF
MASKING TAPE.

Fig. 33.9 Method of holding template close to wall when spraying or sponging

The use of masking tape in this way avoids defacing the wall with stencil pins

Fig. 33.10 Brush stipple technique

Fig. 33.11 Brush stipple and "dry brush" dragging on a rough-textured ground

Fig. 33.12 A design in water paint and crayon

After Verrocchio's "Raphael and Tobias"

When attempting to portray the figure in sponge technique it should be kept as simple as possible, with accent on the purely decorative — treated in this way, features and other details are not missed if left out.

BRUSH TECHNIQUE

There are many alternative methods of producing gradation effects without having recourse to the spray gun. For instance, the required shape may be outlined very broadly and freely with fitch or sash tool and the gradation obtained by softening the wet colour into the dry ground, either by pouncing with a large dry stencil brush, or by rubbing out the wet colour with a dry fitch or sash tool. To achieve the same gradation effect on a large-scale mural, the colour may be softened into the ground with a dry 100-mm wall brush. These effects are best achieved on a low-textured flat oil paint ground, and if a proportion of glaze medium is mixed with the colour to retard its setting, the gradation is greatly facilitated and kept under control.

Low-textured grounds such as stippled water paint, flat oil paint and sprayed plastic finishes lend themselves admirably to "dry brush" effects. The size of the brush depends on the size of the work in hand, anything from a hoghair fitch to a 75-mm flat brush being suitable on occasion. The colour

Fig. 33.13 Litho crayon drawing by Leslie Wood

The technique can be employed to good effect in mural work, especially on a slightly textured ground

To face page 458

PLATE 2

Lo Tung, Chinese poet of the T'ang dynasty (A.D. 620-907)

A fine decorative painting by Eric Fraser at the Tea Centre
Lower Regent Street, London

Reproduced by kind permission of the artist

To face page 459

PLATE 3

"Spring and Summer"

A mural decoration in a private house in Cheshire, painted by F. H. Baines F.B.I.D.

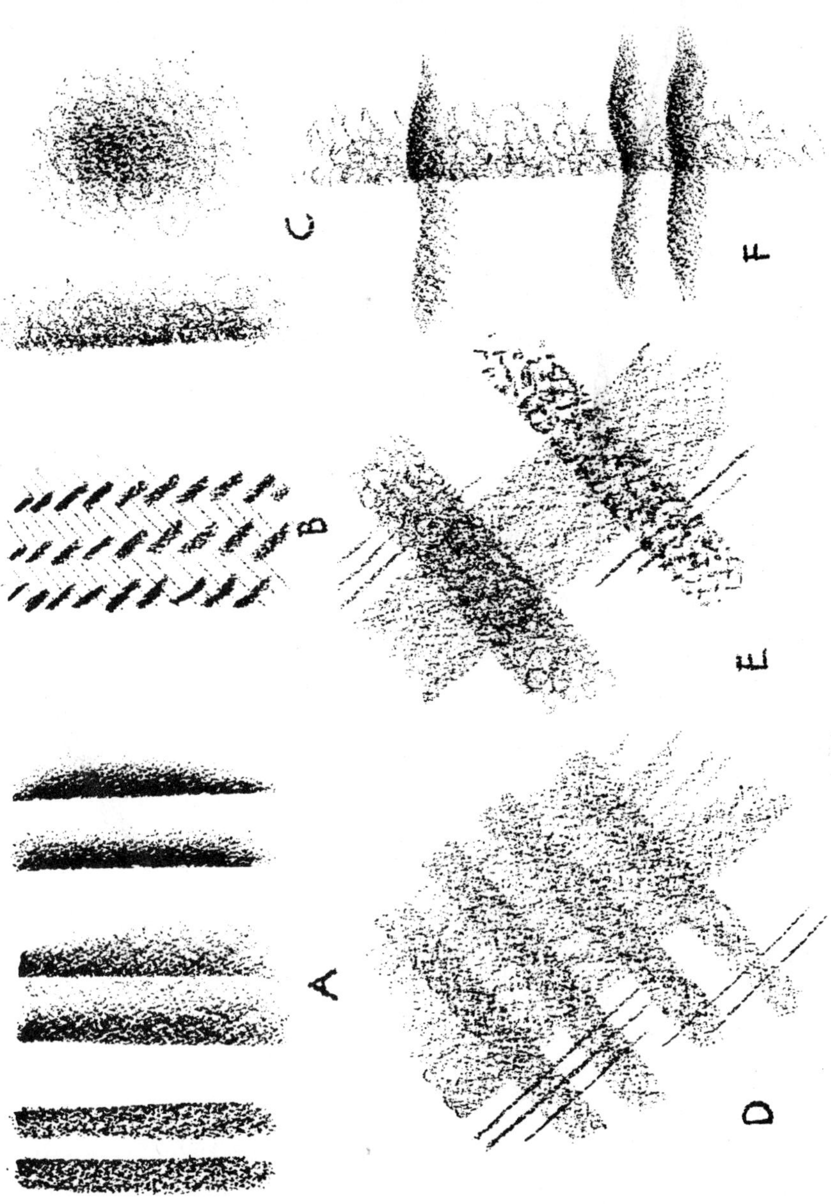

Fig. 33.14

(A) Plain and gradated chalk strokes on a stipple ground, the third pair showing the effect of softening-off with the thumb; (B) herring-bone effect of thick and thin strokes; (C) scribble texture; (D), (E) and (F) contrasting tones and textures produced with flat and point of the chalk respectively.

needs to be on the "round" side, a little being taken up on the brush and then rubbed out on a piece of corrugated paper to remove any surplus. The brush is then lightly *dragged* over the textured surface so that only the relief parts catch and retain the colour.

Or again, decorative gradation can be achieved by hatching and cross-hatching or by feathered-off brush strokes. Such treatments have much more character and liveliness than any smoothly sprayed air-brush gradation.

CRAYON AND PASTEL

The attractive feature about this dry decorative medium is that so much can be done in different ways, ranging from a complete mural decoration carried out entirely in crayon or pastel on a painted or emulsioned ground, to a few sparing strokes applied as a finishing touch to a painted decoration. The figure, for instance, may be painted in flat colour and the form or modelling put in afterwards with crayon as in Fig. 33.12, or the subtle modelling of the nude may be suggested by using chalk or pastel and gently softening off with the finger.

A slightly textured ground is desirable for the crayon to bite on and a hair-stippled water paint provides a very pleasant surface on which to work. For broad effects, a more pronounced texture, such as that produced by an imitation stone paint, is very satisfactory; on such a ground the author once carried out some quite large mural decorations in a café. The design was first painted in, very broadly, then blended into the ground with a dry brush; when dry, it was worked over with large wax crayons which emphasized the texture by colouring only the relief and missing the hollows. The finished effect closely resembled a spray spatter, demonstrating how this freehand method can achieve a spray effect without the laborious preparation of large templates which an actual spray job would involve. Another example of wax crayon technique is shown in Fig. 33.13.

Fig. 33.14 demonstrates some of the effects and textures that can be obtained by varying the chalk strokes. Using the "flat" of the chalk will give broad strokes of an interesting texture, especially when working on a rough ground. By varying the pressure of the chalk, a gradated stroke may be produced as in example A; folds in drapery may be readily suggested in this way, and "softening off" may be done by running the pad of the thumb along the edge of the stroke. Detail, of course, will be done with the point of the chalk or crayon which may be sharpened with a pocket knife if necessary.

Water paint seems to be the ideal ground to work on with chalk or pastel, which blends beautifully into the matt ground. Wax crayons have a slight "sheen" of their own and are therefore more suited to working over flat oil paint.

On ceilings and friezes which are normally out of reach, this type of decoration need not be "fixed", but where it is necessary, chalk and pastel may be sprayed with charcoal fixative to prevent rubbing off; if the decoration is on a water paint ground, a mist spray of clear cellulose lacquer may be used without "glossing up". There is one important point to bear in mind, however, when fixing with lacquer, namely, that certain colours — notably reds — are spirit-soluble and tend to run or bleed (doubtful colours may be tested before using by dipping the chalk into cellulose thinners; if the thinners become tinted, that particular colour should be avoided). Wax crayons, of course, do not require fixing to the same extent as chalk or pastel, but may be sprayed if so desired.

LINING AND PICKING OUT

The running of lines and "picking out" of architectural features are operations which every painter should be able to perform efficiently. They call for skill and confidence in the handling of tools, and should therefore be practised diligently by the young painter until proficiency is attained.

Lining on ceilings and walls is carried out with the lining fitch and straightedge. Lining fitches (Fig. 33.15) are of hog-hair set in tin, the bevelled edge enabling the brush to be held at an angle to facilitate its run along

Fig. 33.15 Hog-hair lining fitch

the straightedge. Some fitches are made with very short hair and, though they are easier for the beginner to manage, the long-haired type hold more colour and so require less frequent charging. For narrow lines, a single chalk line is snapped, but for lines of 20 mm or more, double chalk lines should be struck; the superfluous chalk is lightly dusted off before lining, otherwise the thin colour tends to spread into the chalk. Lining colour should be fairly thin, but not turpy or the bristles will tend to spread out; it should contain a

Fig. 33.16 Round and flat hog-hair fitches

proportion of goldsize or varnish which helps to hold the bristles close together and makes for a clean edge. "Ready-bound" colour thinned with equal parts flat varnish and white spirit runs well and is excellent for practising on lining paper, since it does not "strike" into the paper as would oil colour.

The lining fitch should be well charged with colour but only the tip is used, the brush being drawn lightly and swiftly along the straightedge; it is worth while doing a fair amount of practice on lining paper until confidence and the knack of using the fitch is attained. Some decorators glue pieces of cork on the straightedge to hold it away from the wall, but the writer prefers to use the bevelled edge fairly close to the wall as he finds that this gives greater control at the bristle end.

When lining out a series of panels on ceilings or walls it is quicker and more convenient to stencil the mitred corners, particularly when multiple lines are being run, or when the corner is "stepped" or otherwise embellished. In an emergency, if no lining fitch is available, an ordinary flat fitch may be used quite satisfactorily; the short artist's fitch known as "Brights" is capable of producing quite thin lines.

Lining on woodwork, such as on doors, shop fronts and signs, is generally done with a sable writer or liner (Fig. 33.17), the pencil being guided by running the middle finger along a straightedge, or along a convenient moulding or beading. Gold lines are invariably run this way, the goldsize being slightly tinted with a little lemon chrome to act as a guide, and eased if necessary with a little white spirit. Short, fine lines may be run with the aid of a bevelled set-square by holding the sable penwise and running the middle finger quickly along the smooth edge of the set-square; lines as fine as those produced by a ruling pen may be done in this way after a little practice.

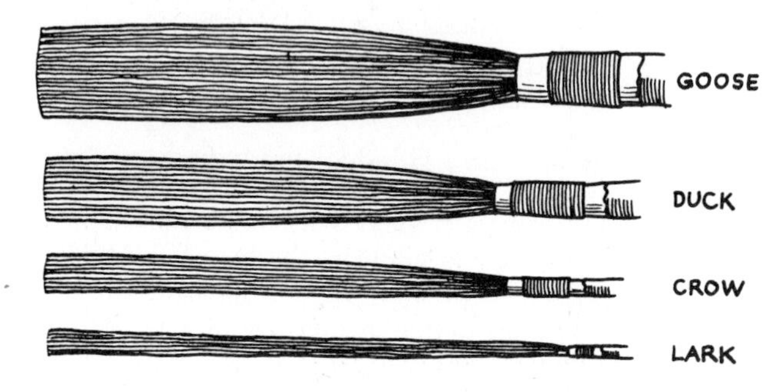

Fig. 33.17 Sable liners in quills

Picking out

In the past, a great deal of painter's and decorator's work consisted of picking out mouldings and enrichments in several colours. Every domestic cornice had its half dozen or more different tints, the result often being unbearably fussy, judged by present-day standards.

Picking out is nowadays generally confined to the decoration of public buildings such as the theatre, leisure centre, or the municipal assembly hall, especially where there is elaborate plaster ornamentation. The purpose and true function of picking out mouldings and enrichments is to add colour and emphasis to existing architectural features which require enhancement. The details of a heavily panelled and ornate theatre ceiling, for instance, may be substantially lost when viewed from the stalls unless judiciously picked out in colour; when, however, the panels, stiles, mouldings and relief ornament are treated separately in different tints, the design of the whole ceiling at once becomes discernible and easily "read".

In painting mouldings and relief ornament, some thought must be given to the appropriate *placing* of the colour so that unimportant details are not brought out too prominently. The effect of relief can be either enhanced or destroyed by the use of colour. For instance, if colour is applied only to backgrounds or recessed portions, the lighter relief parts will be thrown up and enhanced, but if the relief ornament itself is picked out in strong or dark colour on a lighter ground, the effect of relief will be largely lost and it will tend to look like flat painted ornament (Fig. 33.18). Some colours are retiring

Fig. 33.18 Picking out

The effect of relief is enhanced when ornament is painted lighter than the ground

whilst others are forward or assertive and this needs to be considered carefully if the natural contour of relief decoration is to be preserved. We shall have more to say on this particular subject in the chapter on Colour.

Picking out is generally done with round or flat hog-hair fitches, or with 25-mm tools on broader work. In the case of low-relief work, it may be found convenient to put in the whole of the work first in the lighter colour, then cut in the background with the darker colour, using either a fitch or the softer sable or ox-hair flat pencil, according to the scale of the work. Ox-hair poster brushes are excellent for such work, being stronger and less expensive than sable. Small beads and fillets on woodwork are best run in with a sable pencil; these are often laid in with goldsize and afterwards gilded. Narrow quirks can be done very conveniently with the short Bright's fitch. Only small amounts of colour are generally required and signwriter's dippers or finger pots are very suitable for the purpose.

SGRAFFITO

The term "sgraffito" refers, strictly speaking, to the Italian method of coloured surface decoration in plaster. A thin coat of tinted plaster is laid over a differently coloured plaster ground, the design being scored or scraped out from the top coat to expose the ground colour beneath. This method is readily adapted to a paint technique — in fact, one might regard colour combing and certain graining processes as a form of paint sgraffito.

The technical requirements are (a) that the ground should be hard and non-absorbent, and (b) that the finishing colour should "stay put" so that the scored-out design remains crisp and clean. Eggshell paint makes an excellent ground. The finishing coat may be flat oil paint with a proportion of transparent glaze medium to retard the set and flow; this remains "open" long enough to enable the design to be scored out cleanly with a piece of hard rubber or felt. Steel and rubber combs, the graining horn, shaped pieces of plastic or the rounded end of a palette knife may also be used for securing different effects.

Where translucent effects are desired, tinted glaze may be used on a light ground colour; the scored-out design then appears as a light colour on a dark ground. By the use of opaque flat oil paint on a dark ground, however, the reverse may be obtained, i.e. a dark pattern on a light ground. An attractive effect can be produced by using the flat paint over a hard glossy ground so that the design appears in gloss on a flat ground; if the colours are the same, or fairly close together, an effect similar to *damask* is produced in this way.

Paint sgraffito must necessarily be carried out in a direct manner, using firm confident strokes; a timid, hesitant approach can only result in a smudgy mess. Therefore, as in graining, patient practice is needed to gain experience and confidence. Good drawing ability is essential and one must have a clear picture in mind of the finished design. Make a rough pencil sketch (on paper) of the design, then try it out on a spare panel in order to work out the technique. When satisfied with the preliminary try-out, lay in the surface to be decorated, using either flat oil paint or a tinted glaze medium, and stipple to obtain a uniform colour. If, desired, of course, the glaze can be first manipulated with crumpled paper, rubber stippler or combs, etc., to produce a broken colour effect. Then with the chosen tool (wedge-shaped rubber, felt, or palette knife, etc.), the design is drawn or scored on the wet surface, using firm pressure and wiping the tool clean after each stroke. One advantage of this process is that if the design is not satisfactory, part or whole of it can be stippled out and re-drawn.

When the colour is quite dry, the design may be tinted by applying glazes to appropriate parts with a sable pencil or a fitch, stippling such parts with a small stencil tool or scoring out further detail as desired. The design should be simple and direct and should *look* as though it has been scored from wet paint. Any attempt to over-elaborate will look fussy and therefore less successful.

Fig. 33.19 Paint sgraffito decoration by D. S. Dalton, A.T.D.

The design was drawn on wet paint with a piece of hard rubber, exposing the light ground colour

MOTIFS AND MURALS

The subject of mural painting is so vast that we can touch only the fringe of it here, and that mainly from the technical aspect. The problems of design and composition involved in a large mural of any importance can be handled only by one who has had a sound art training after years of patient study and experience, during which time the artist forms preferences and convictions which dictate his own outlook and style. The writer's own accomplishments in this direction are on such a modest scale that he is unqualified to discuss the higher aesthetics of mural painting. Nevertheless, the range and scope of applied decoration is so wide that there is much that can be done by

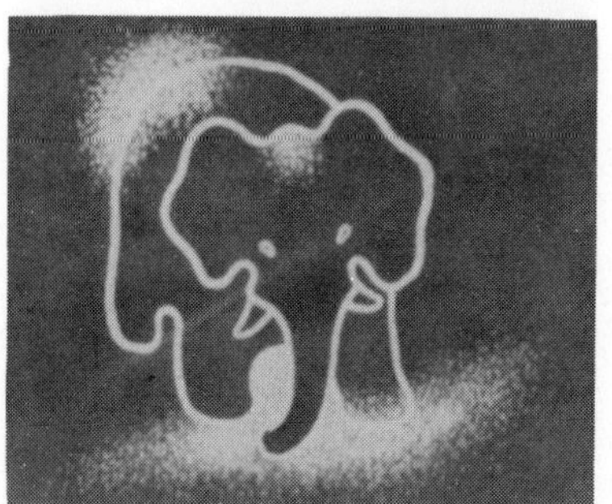

Fig. 33.20 Painted nursery motifs

These motifs show an interesting variety of line and stipple techniques

Courtesy: F. H. Baines, F.B.I.D.

the decorator of average artistic ability — much indeed that is his own special province.

We will confine our remarks, then, to the consideration of applied motifs and murals of a light decorative or impermanent character, appropriate to the domestic interior, the café, the cinema, etc. We use the word "impermanent" here to differentiate between the type of applied decoration that is usually painted out each time decoration is undertaken, as against work which must be preserved on account of its historic value or high artistic worth. First let us define the terms "motif" and "mural". A *motif* is generally regarded as a single decorative device or ornament which forms one or more focal points of interest or embellishment in a scheme. A *mural*, however, is a complete composition or picture occupying a part or the whole of a wall, or even extending round the four walls of a room.

The growing popularity of motifs and murals is no doubt due to the natural reaction from the stark simplicity of bare walls which the modernists demanded in their insistence on absolute functionalism and which, in turn, was a healthy revolt against the decadent Victorian and the even more deplorable jazz periods. Thus, the wheel has turned full circle, with a renewed interest in surface decoration and pattern. If we are to avoid the worst errors of the past, however, we must never forget that applied decoration should always be subservient to architecture. It is here that the average decorator needs to exercise most restraint, for he is generally so enthusiastic and anxious to display his prowess that he, more often than not, does not know when to stop. Much of the beautifully executed stencil work of the past would have been so much more acceptable if it had been confined to smaller areas and not allowed to sprawl willy-nilly over walls and ceilings in an undisciplined manner. Unfortunately, the same can be said of modern spray decoration in some of our cinemas and restaurants.

To preach restraint is not to preach timidity, however. Bold design and execution can be quite acceptable, even in domestic work, provided the *colour* is right. A design in soft pastel tints might be quite delightful, whereas the same design in strong blatant colour would be unbearable.

Colour and treatment will largely depend, however, on the *scale* of the work and the type of interior. In domestic work, for instance, any applied decoration will be under fairly close observation for long periods; colour must therefore be appropriately reticent and "livable-with". In the cinema and theatre we have the other extreme, the decoration being a good distance from the eye and only seen for a relatively short time. Moreover, when we visit places of entertainment, we expect to be stimulated. Design and colour must therefore be bold and exciting or we shall be disappointed. This is a fundamental point of difference between public and domestic decoration.

Not all public decoration calls for such stimulation, of course; the place of worship, for instance, calls for decoration which is large in scale yet quiet and serene in colour, whilst the municipal assembly hall requires something appropriate to civic dignity. In the hotel, we may require a little of both, such as a quiet note of colour in writing room and lounge, as against a gay sparkling treatment in the cocktail bar.

Choice of subject may be suggested by the type of interior or indicated by the client's wishes. Where the decorator has a free hand in domestic work, he can often hit on a happy idea by making a few discreet enquiries as to his client's interests and hobbies. Sport, music and flowers are good subjects. In a bedroom scheme, an interesting design can often be originated by incorporating the occupant's birthday sign (i.e. the appropriate sign of the Zodiac). In the dining-room, a still-life motif may provide a suitable note.

The public house often provides a fine opportunity for the decorator and there is generally no lack of subject. The main topics of conversation in such places — cricket, football, racing, etc. — can be freely drawn upon, but the decorator should take great care to get his details correct for there are no greater critics than the habitués of "the local". Sometimes the name of the inn will provide the subject for a design, e.g. the "Fox and Hounds", "Dog and Partridge", etc., just as they do for the inn sign. Local history or, in the case of a country inn, a decorative map of the district, may provide interesting subjects; these generally entail a fair amount of research into historical costume, architecture, topography, etc., and this is a valuable experience which should not be shirked. No designer can do authentic work without adequate references. And in searching through old prints and engravings, one unconsciously absorbs the atmosphere of the period which tends to come out in one's own work. The influence of 18th and 19th century prints shows unmistakably in the work of some of our present-day designers.

It is often thought that the subject of a motif must of necessity be of a decorative nature to be suitable as wall decoration, but this is not so. It is the *treatment* that makes the subject decorative. Dancing figures, leaping deer, galleons in full sail and so on have become rather hackneyed subjects; it is refreshing to break away by choosing some familiar common object and giving it an unusual treatment. A cup and saucer, for instance, or a wine glass and bottle, paint can and varnish bottle, table and chairs — any group of common objects can be turned to decorative use by a little imaginative treatment, sometimes merely by an axonometric view (looking down on them) or tilting at an unusual angle, or by using unfamiliar colour. Such subjects usually give pleasure because they are easily recognized and the unusual treatment intrigues the beholder.

Abstract design, on the other hand, needs careful and expert handling; the shapes may be meaningless and unintelligible, but if the shapes make good "pattern" and the colour is pleasant, the beholder is again rather intrigued. It is not enough, however, to throw together jagged triangular shapes, circles and rectangles and call it abstract design; there must be some feeling of congruity in the shapes themselves and in the colour used. If, after studying good examples of abstract design one feels out of sympathy with it, leave it alone. Good abstract design can only come from an inner conviction, never from copying a style just to be in the fashion.

A *mural* painting is generally a much more ambitious affair than a decorative *motif* and usually occupies a much greater area of wall space; the colour should therefore be less assertive — in fact, the larger the mural, the quieter and more retiring should its colour be. A wall painting, unlike an easel picture, should remain essentially a background decoration, a flat two-dimensioned treatment which does not compete for attention with anything *within* the room. Bold light and shade or stereoscopic effect should therefore be avoided, the accent being on the decorative rather than the realistic. The treatment should be as broad as possible, consistent with scale and distance from the eye. Light, airy, pastel tints are in tune with modern requirements since murals no longer have to harmonize with heavy, elaborate ornamentation in the architectural setting.

In the more intimate café or domestic mural, the vignette or "frameless" treatment is probably the most satisfactory, for then the mural *looks* like a wall painting instead of perhaps an easel picture merely stuck on the wall. When the general wall colour is allowed to play its part in the mural, so much the better — it then looks even more a part of the wall. Preferably a little of the wall colour should be mixed with all the mural colours to obtain harmony; if, for instance, the wall colour is a deep cream or beige, use this for reducing your mural tints rather than pure white; your design will then sit comfortably on the wall and will not look too high in key or chalky, as it otherwise might. When in doubt about strength or choice of colour, etc., it is useful to fix cellophane over the design set out on the wall (using self-adhesive tape) and roughly paint in the design with the trial colour. The effect can thus be judged *in situ* and any modifications or alternatives can be tried out before finally committing oneself to the finished painting.

In executing your motif or mural, any or all of the decorative techniques previously described may be used, as and when appropriate. Hence the necessity for mastering the various technical processes before embarking upon a work of any importance. An essential requirement in decorative wall painting is the confident use of the fitch in *freehand* brushwork, that is, the ability to draw freely with the brush held at arm's length, without the use of the mahlstick to support the hand. The mahlstick may, of course, be necessary in drawing fine detail at eye level, but otherwise it cramps free movement of the arm and should be dispensed with. There is nothing more enjoyable or exhilarating than to wield the fitch freehand, once confidence has been gained.

A desirable feature about any applied decoration or mural is that the finish should be uniform with the ground colour, being neither more glossy nor yet flatter. If the ground is in flat oil paint the same material should be used for the decoration, thinning either with white spirit only or with a mixture of half and half flat varnish and white spirit. It should be remembered, however, that if oil stainers or tube colours are added, the material may dry with a higher gloss. This may be minimized by tempering the oil colour with a little china clay before adding to the flat oil paint. Where there is much working of one colour into another, as in mural painting, parts of the work may tend to gloss-up, due to fresh paint being worked into partially set paint; in other words, a "flashing" effect occurs. This uneven finish greatly mars the appearance of the work and it may be necessary to apply a coat of flat varnish over all to obtain a uniform matt finish. Alternatively, a coat of transparent glaze medium (well thinned with white spirit) may be applied and hair-stippled. A glossy finish is generally unsuitable for motifs and mural painting, since the reflection of lights, etc., may interfere with seeing the design. This may not be objectionable, however, in a bathroom where the design is generally at eye-level and therefore easily seen.

Water paints or emulsion paints are ideal for decorative painting and mural work, drying as they do with a luminous uniform flat finish. The strong stainers made specially for tinting water paints provide an ample palette for most needs; in fact, the writer has carried out scenic work with only white and the three primaries — bright red, yellow and blue — and has found this limited palette quite adequate. The great advantage of a strictly limited palette (as will be discussed further in the chapter on Colour) is that greater harmony is likely to be achieved than when a larger number of different colours is employed. When carrying out any large-scale decorations in water paint, it is advisable, if the surface is absorbent, to apply a coat of stabilizing fluid first. This makes for cool working and enables colours to be blended together more easily.

34

Signwriting

The lettering of signboards is generally referred to in the trade as "signwriting", though strictly speaking "letter painting" would be more correct, since the term "writing" implies the use of the pen. The term is common trade usage, however, and even the sable pencil used for lettering is known as a "writer".

Signmaking is a highly specialized industry, involving not only the painting of signs on all types of surfaces, but also the manufacture of cut-out letters in wood, metal, plastics, etc., and various forms of neon and other electric signs. We propose to confine our remarks here to *painted* lettering, since this is the form which chiefly concerns the painter and decorator.

The good all-round craftsman can generally wield a sable pencil almost as dexterously as he wields the flat brush, but good lettering does not depend on dexterity alone. A high standard of draughtsmanship is required to produce really well-shaped letters, and a sense of design and taste is needed to place lettering satisfactorily within its allotted space.

The first requirement of lettering is that it should be easily read. This means that lettering should be of a simple type, devoid of any excrescences that detract from its legibility. The plain block letter known as *sans serif* (i.e. without serifs) was chosen by London Transport as being the most legible, and Eric Gill in designing "Gill Sans" in the 'thirties showed that it could also be beautiful. The similar fount, "Univers", has also achieved wide popularity as a basic design for sign painting. Antique Roman lettering, as depicted on the base of the Trajan Column in Rome, is generally held to be the most beautiful style ever produced and is accepted as a standard by which others are judged (Fig. 34.1). The student of lettering should make a deep study of this style, practising it until he becomes conversant with its many subtleties, for the man who can give an intelligent rendering of Antique Roman can generally give a good account of himself in almost any other style.

The preliminary study (and practice) of pen lettering, particularly of the early historical manuscript types, helps the student to appreciate the development of the thick and thin strokes in brush-formed Roman letters and the slightly inclined axis on which the round letters are based. There are many books dealing with this aspect of the subject; some of these are listed in Appendix 6. An excellent one is L. C. Evett's *Roman Lettering*, published by Pitman, in which a detailed analysis of each letter is given.

ABCDEFGHIJKL
MNOPQRSTUV
WXY&Z, 123456
abcdefgh 7890
ijklmnopq
rstuvwxyz

15°

CENTRES

Fig. 34.1 Alphabet based on Antique Roman

Note method of constructing the "O" and similar letters, the thickness of the letter being built upon a central skeleton line which is a true circle (shown dotted)

ABCDEFGHIJKLM
NOPQRSTUVWX
Y&Z.1234567890
abcdefghijklmnop
qrstuvwxyz
1234567890

Fig. 34.2 Block alphabet
Capitals, lower case and numerals

The point system and the construction of Antique Roman

The ingenious point system, devised by Ernest Sanderson, of Bradford, provides the beginner with a useful method of constructing the Antique Roman alphabet (it should not be confused with the printers' system of measuring type in multiples of the "point", i.e. 1/72 of an inch). The "point" here is a unit of measurement determined by the width of the thick down-stroke or limb of a letter, the thin horizontal stroke being half a point. Points are allotted for the width of each letter (which varies so much in the Roman alphabet) and the same point is used in measuring the height; thus the width and height of the letter bear a definite relationship to the thickness of the down-stroke, carefully based on the correct proportions of the Antique Roman. The latter is referred to as an "11-point" alphabet because the thick stroke goes into the height eleven times. A heavier letter is obtained by using "10-point" or "8-point" alphabets and so on. Block alphabets may be set up in the same way.

The system has certain limitations in practical use, particularly when setting out several lines of lettering, and since it entails a little arithmetic it can be confusing to the young painter. Nevertheless we consider it to be a reasonable method of introducing the Antique Roman alphabet to young students because it teaches them the relative proportions of each letter in an easily understood manner. But it is only an introduction, and the student should be encouraged to dispense with it as soon as possible. Roman lettering is far too intricate and subtle to be reduced to a series of mechanical measurements, and indeed, while a system of numerical mnemonics can be helpful in indicating the comparative widths of the letters, it is of no value whatsoever in conveying the individual qualities of each letter form. Once the student has become familiar with the appearance of well-formed letters and has grasped the principles underlying their construction, he will subconsciously adapt the system to his own needs. Even then, it is to be hoped that he will continue to study the Trajan inscription; the more he learns about it, the greater will be his understanding and his proficiency in all branches of lettering.

The endless fascination of Roman lettering is due to the extreme care with which it was worked out by the artists of the ancient world. In the first place, the Romans as a race were brilliant engineers, and both they and the Greeks before them were fully aware of the limitations of our human eyesight; they knew that the inevitable result of placing numerous straight lines in close proximity to one another was that the eye would play tricks, and that because of optical illusions the lines would appear to be distorted. They corrected these optical illusions by making sure that all the lines which appear to be straight were actually slightly curved. For this reason it is impossible to produce good Roman lettering by ruling the lines with drawing instruments; the letters can only be formed properly by working freehand. Similarly the serifs cannot be described with a pair of compasses; far from being circular in shape, they are actually a very subtle elliptical shape growing out from the curve of the main part of the letter; furthermore the line between any pair of serifs is curved into a delicate fishtail shape.

These are just a few of the general factors affecting the whole of the inscription. Beyond this, each individual letter was very carefully worked out to make it as effective as possible. To take just one example, the letter R is a remarkable construction, with many features that most people have never even noticed. The curved part of the letter, for instance, does not curve uniformly throughout its distance; the top half is a fairly regular curve, but the next quarter flattens out so as to meet the right-hand sloping stroke (the tail) at right angles; it then swerves round sharply into a short horizontal stroke which again is practically straight. The slope of the tail itself varies considerably; if the letter following the R is straight-sided, such as a D, L, N, etc., or the even more difficult wedge-shape of the letter A, the slope of the tail can be very slight, whereas if the next letter is a T or a V the tail can have a greater slope to help to fill the extra space. Similarly the tail can be finished with a serif or can be curved round at its extremity. It is an illuminating exercise to draw or paint a letter R with the rounded part shaped in a regular curve, and then place it beside a properly formed letter based on the Trajan example — it will appear quite unmistakably weak and formless in comparison with the Roman example. And there are similar subtleties of form to be observed in every letter of the alphabet.

LEGIBILITY

To be really legible, lettering must be well shaped, well spaced and have good *tone* contrast with its background, i.e. light lettering on a dark ground or vice versa. Colour contrast is not so important as tone contrast; for instance, bright red letters on a bright green ground of equal tone value would probably be unreadable at a little distance because, although the *colour* contrast is violent, there would be no *tone* contrast and the letters would tend to merge into the ground.

The space within and between letters is as important as the shape of the letters themselves and the aim should be to have the space between letters as equal in area as possible; thus one should allow a wider space between straight-sided letters such as H, I, N, than between such curved letters as D, O, C, or open-sided letters such as E, S, Y, etc. Combinations such as LA, LT and TWY require closing up, and straight-sided letters should be opened out in order to equalize the areas of space between them. Lettering should read as an *even tone* except, of course, where it is necessary to emphasize important words or headings. It is sometimes necessary to shock the passer-by into attention by eye-catching headings (as our poster hoardings amply demonstrate), and once the attention is gained the message must be easily and quickly read.

Marginal space is of utmost importance and bears a great influence on the aesthetic appeal and legibility of lettering, whether on a sign, a poster, or a book page. There can be no greater mistake than that of totally filling the available space with the idea of getting "value for money". Plain space can be more eloquent than large type, as any newspaper or magazine will show. Satisfactory margins cannot be arrived at by rule of thumb; it is largely a matter of taste and sense of proportion, and the student should study street

ABCDEFGHIJKLMN
OPQRSTUVWXYZ
abcdefghijklmnopqr
stuvwxyz
1234567890

Fig. 34.3
Albertus, a modern type-face which is particularly suitable for brushwork

ABCDEFGHIJKLM
NOPQRSTUVWXYZ
abcdefghijklmnopq
rstuvwxyz
1234567890

Fig. 34.4 Rockwell Bold, a modern development of Egyptian which reads well on facia signs

signs and press advertisements and try to discriminate between good and bad layout.

Lettering may be purely functional, in which case the prime object is to get the name or message over, or it may be used solely as a decoration, in which case legibility may be of secondary importance. An instance of this would be an inscription on a classic entablature where the lettering is used as a thing of beauty in itself. Where lettering is used as a decoration, the letter forms should retain their identity and should not be debased by distortions or vulgar excrescences. There is probably nothing more decadent than letters formed from the human figure in various grotesque attitudes, or from twisted branches of trees (the so-called "rustic" alphabet), or from contorted ribbons and so on. Such misshapen freaks should never be permitted to disgrace a signboard, even for the sake of novelty. This is not to say that decorative styles are not permissible, however, for there are many historical types which are really fine when used appropriately. Where a rich decorative effect is required, for instance, letters may be embellished with *inhabited* ornament, that is, by applying a fanciful filigree ornamentation to the thick stroke of the letter without altering or distorting the essential letter shape in any way. Some Victorian playbills provide charming examples of this type, though the bill itself might be a horror owing to the prevailing habit at that period of using so many different type faces on the same bill. The latter should be an object-lesson to the student, of what *not* to do; the mixing of too many styles is showy and ostentatious and should be avoided. When different types *are* used together, they should be limited to two, or three at the most, and should be carefully chosen so that they do not look incongruous.

The representation of solid letters, showing imitation projection by means of shading and cast shadow, is often criticized, quite rightly, on the grounds that such "three-dimensional" letters cannot possibly be in correct perspective from all angles, but we would not go so far as to say that such lettering should never be done. There are times when it is quite appropriate and there is no gainsaying the fact that it *can* look extremely decorative when well executed. Some of the glass stall-plates, popular with provision merchants early in the century, were marvellous examples of craftsmanship, embodying as they often did every known artifice — gilding, embossing, outlining, shading and blending, diapered grounds, etc. — but the magnificent richness of effect was often gained at the expense of legibility. What we would say, therefore, is that when shaded letters are called for, take care that the shading does not distort the *essential letter shape* and thus impair legibility. For instance, white or cream lettering on a medium-toned ground might be enhanced by shading in black or dark colour, whereas a medium-toned letter similarly shaded on a white or cream ground would tend to be less legible, because letter and shading would, at a distance, tend to blend together and read as a silhouette on the light ground, thus distorting the letter shape.

Where lettering needs some form of embellishment, outlining is much safer than shading since the letter shape is always preserved and even emphasized. For example, gold lettering on a light or medium-toned ground tends to lose itself when viewed at certain angles and a black or dark-coloured outline is therefore advisable to obtain greater definition and legibility.

ABCDEFGHIJKLMN
OPQRSTUVWXYZ
abcdefghijklmnopq
rstuvwxyz
1234567890

Fig. 34.5 Univers Bold, probably the most widely used sans-serif fount from the later '60's onwards

There are many variants — in light, medium, extra bold, condensed and expanded versions

Needless to say, the suitability of letter type for its position has an all-important bearing on legibility. Such styles as "Gothic" or "Old English" were primarily intended for manuscript to be read at close quarters and are generally unsuitable for sign work, except in the case of a trade name (such as the name of a newspaper) with which the public is already familiar. Antique Roman, originally carved in stone, is eminently readable even though it is light in construction. Bolder types are generally required where the sign is placed high up on a building, in which case "footed" Roman, block or heavy "Slogan Script" (Ultra Bold) may be used.

A study of the various books on lettering (see Bibliography, Appendix 6) and of printers' type-faces will help the student in his selection of good types (Fig. 34.3–34.5). He should also collect any cuttings from newspapers or magazines, advertisements, headlines, bookplates and title-pages that appear worthy of study, pasting them into a scrapbook for future reference.

GROUNDS

Many large signboards are now constructed of heavy laminated plyboard, or composite blockboard, the preparation and painting of which should follow sound painting practice as outlined in the appropriate chapters of this book. Metal-faced plywoods are generally of non-ferrous metal or galvanized finish, and it is always advisable to etch the surface with a chemical mordant solution prior to applying the first coat of paint (see page 260). Where old signs are to be repainted it is generally advisable to burn-off and body-up afresh, but if it is desired to avoid this on grounds of economy the surface should be well rubbed down with pumice stone and water to remove all traces of the old lettering which otherwise would show (in relief) through the new finish.

In glass work, the ground is painted last, that is, after the lettering is completely dry and hard. Colour for glass painting should be made from colour in turps (or ready-bound colour) thinned with varnish and a little turps, in order to provide maximum adhesion, ordinary oil paints being insufficiently tenacious. Some signwriters still grind all their colours for glass work by hand, tempering with varnish and turps and sometimes adding a spot of boiled linseed oil to ease the working. But the more normal practice today is to use alkyd gloss paint, which is tenacious enough to adhere to the glass and sufficiently opaque to give good obliteration in one coat. It is most important that the glass be thoroughly cleaned and all trace of grease removed, otherwise the paint will peel off.

Streamers, banners and temporary signs are generally written on prepared sign cloth, using sharp colour, i.e. colour in turps thinned with varnish and turps. Lettering upon silk banners, etc., is set out by means of a "pounce" and written in sharp colour; alternatively, the lettering may be laid in with emulsion paint mixed with a proportion of clear emulsion medium for added flexibility. Upon this the lettering may be written in sharp colour or gilded, as required. Posters, window bills and valances executed on paper (including showcards) may be carried out in poster colour or latex emulsion paint, tinted with water paint or emulsion paint stainers which are excellent for this purpose (see page 312). Ordinary oil paints should be avoided because they strike badly into fabric or paper, leaving an oily stain round the letters.

TOOLS

Whilst brushes are the most important of the signwriter's tools, his kit should also include the following items — metric rule, mahlstick, snap line, dippers and small palette or handboard. Straightedges will be required as occasion demands. Hand rests or bridges are used for flat work on the bench.

The mahlstick should be long enough to give freedom in working; it is easily made by covering the tip of a rigid stick with a small tight wad of chamois leather to prevent marking the sign, or a properly jointed three-section stick may be purchased.

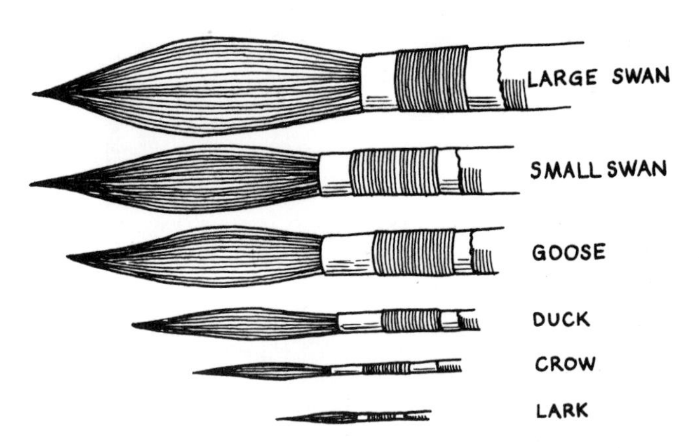

Fig. 34.6 Sable writers in quills

The best-quality writing pencils are made of red sable; they have a lively spring and are much superior to brown sable in this respect. Ox-hair is suitable for work on coarse surfaces such as cement and brick, etc., but for large lettering on gable ends, the hog-hair fitch or inch tool is required. The so-called camel-hair (actually squirrel or pony hair) has no spring and is useless for writing on vertical surfaces, but its floppy nature is sometimes an advantage in glass work done on the bench since the loaded brush tends to cling to the surface.

The chisel edge pencil is, to our mind, the ideal tool for most types of lettering but many prefer the pointed variety. Similarly, some writers prefer their pencils in quill, others in metal ferrule; the former need very careful handling to avoid splitting the quill, especially when fitting the handle, which should only be done after dipping the quill in warm water to expand it. A point to watch with metal ferrules is that, after washing, the pencil should not be twirled vigorously between the palms, as this tends to cut the outside hairs on the sharp edge of the ferrule. Sizes in quill range from *lark* (the smallest) to *large swan* (or *eagle*) (Fig. 34.6), whilst metal ferrules are generally numbered 1 to 12 (Fig. 34.7). Writing pencils are generally supplied in two sizes of hair-length, whilst the so-called "one-stroke" brushes, designed for poster and ticket work, are considerably shorter to allow for short quick strokes (Fig. 34.8). Sable and ox-hair blenders are shorter still; they are principally used for blending light and dark "shading" colours in "projected" lettering.

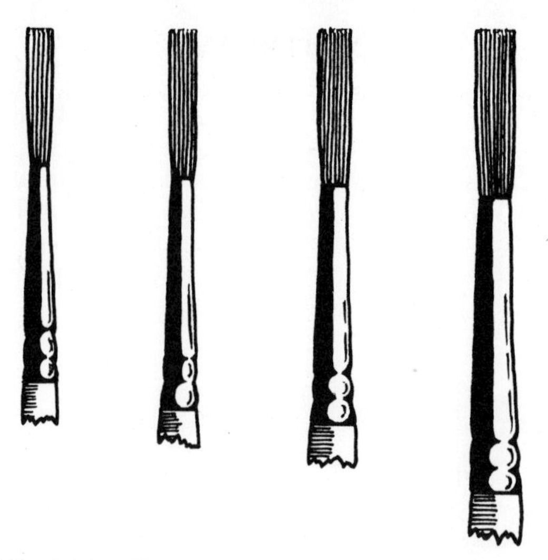

Fig. 34.7 Short sable writers in metal ferrules

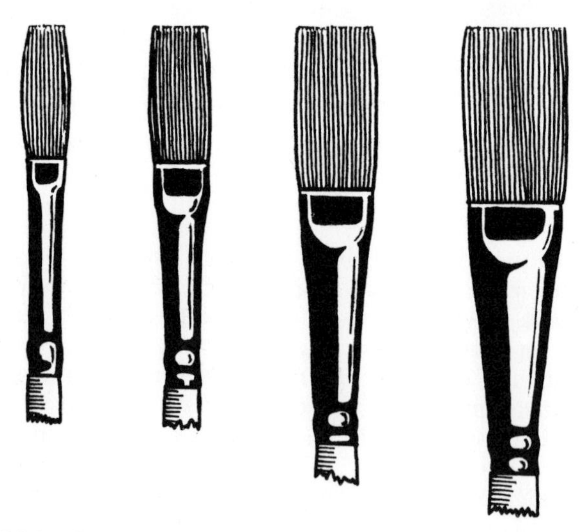

Fig. 34.8 Short sable "one-stroke" writers in metal ferrules

Sable pencils are delicate tools, easily damaged; they should be used with great care and should be washed out thoroughly after use. If in daily use, they should be washed in white spirit and well greased with Vaseline or tallow, but if used only occasionally they should be rinsed out in white spirit, then washed in warm water and mild soap, paying particular attention to the root or heel where the hair emerges from the ferrule or quill. Pencils should be kept in a case, secured to a strip of card or tin by means of rubber bands. If the hair inadvertently becomes bent it can often be straightened by dipping in warm water and holding it close to a hot surface (such as a hot-water pipe) until the heat pulls it straight.

MATERIALS

All paints for lettering should be smoothly ground and well strained to eliminate "bits" or other undesirable particles; one cannot expect to write smoothly with "bitty" stuff. Use only permanent pigments on jobs which are to last and avoid the too-liberal use of goldsize in working over newly painted grounds, otherwise cracking may be expected.

Lettering colour should be about the consistency of cream, but only experience will teach the writer to recognize the best working consistency. The colour should brush easily and yet have sufficient "cling" to steady the pencil.

SETTING OUT

Before the student can paint good well-shaped letters he must learn to draw them really well, both in lead pencil on paper and with chalk on blackboard. Thin detail paper is the best for setting out lettering since it can be used either as a "trace" or a "pounce". Having drawn out a satisfactory alphabet, the back of the sheet may be rubbed with powdered whiting or chalk and the lettering traced through on to a dark ground by going over the outlines carefully with a blacklead pencil sharpened to a good point. On a white or cream ground, pale blue chalk may be used. Alternatively, the sheet may be laid on cardboard and the outlines pricked through to make a pounce as described on page 450; this method is very useful where more than one copy is necessary, since the inscription may be pounced through quickly instead of being laboriously traced each time.

The blackboard is specially useful for practice in setting out lettering and layouts in chalk, for this is the method used in practical setting out on the signboard. The chalk should be scraped down to a point or narrow chisel edge, so that clean crisp lines can be drawn; thickly chalked smudgy lines are not conducive to good clean lettering. Besides, in setting out a newly painted sign, a very light touch is needed to avoid damaging the newly painted surface with the chalk. Pipeclay is better than chalk in this respect. As the student becomes more adept at lettering, his setting out will become more sketchy, sufficient only to indicate the correct position and proportions of the letters, the actual drawing of the letters then being done directly with the brush. Long and patient practice is needed before this expert stage is reached, but not until this expertness is attained can the student regard himself as a really practical signwriter.

In setting out any sign, the student should first work out his ideas on paper. After roughing out several thumbnail sketches to decide the best layout, he should set out the lettering to a 1/10 scale or larger, taking care to work out the spacing and proportions of letters accurately and leaving nothing to chance. He will then be able to "step off" the letter widths and spaces from the scale drawing on to the signboard, placing each letter exactly in position, knowing that if the set-up looks right on the drawing, it will look right on the sign. As the student gains in experience, he will no doubt find his own short cuts and will be able to set out more sketchily and expertly. Even so, the scale drawing should not be lightly dispensed with, for it is far better to work out one's problems on paper than to indulge in trial and error on the actual signboard where the surface can be so easily defaced by excessive chalking and rubbing out.

The *point system* (page 474) gives a carefully apportioned number of points for each letter width, but for those who are not familiar with this method, the following may be taken as a rough guide to the proportions of a good Roman alphabet. Taking the height of letter as a unit of measurement we find that certain letters occupy a square (i.e. width equals height) whilst others occupy three-quarters and some half a square. W is slightly greater in width than height whilst I and J occupy the least space.

Table 34.1

Proportion of width to height	*Letters*
Width equals height	M, O, Q
Seven-eighths	C, D, G
Three-quarters	A, H, K, N, T, U, V, X, Y, Z
Half	B, E, F, L, P, R, S
Miscellaneous	I, J, W

Ideally, these proportions should be adhered to as far as possible, giving and taking with the spaces between the letters rather than altering the letter shape; Antique Roman cannot well be expanded or condensed without losing something of its character and fine proportion. If, therefore, a facia is not long enough to accommodate comfortably the required lettering, a style other than Antique Roman should be chosen, preferably a type which lends itself to condensing or closing in.

The *proportional scale* method may be used to advantage when setting out a small-scale drawing. Suppose, for instance, we have to letter a facia 5000 mm by 450 mm with the words INFORMATION BUREAU in Antique Roman (Fig. 34.9). First rule two horizontal lines across your paper, say, 25 mm apart. Then set out the words carefully (using either the point system or the proportions given in Table 34.1), spacing the letters nicely and ignoring for the moment the length of the sign. When quite satisfied with the lettering and spacing, tick off at each end what you consider to be a suitable clear space; this could be twice the letter height, i.e. 50 mm. Mark these ticks A and B.

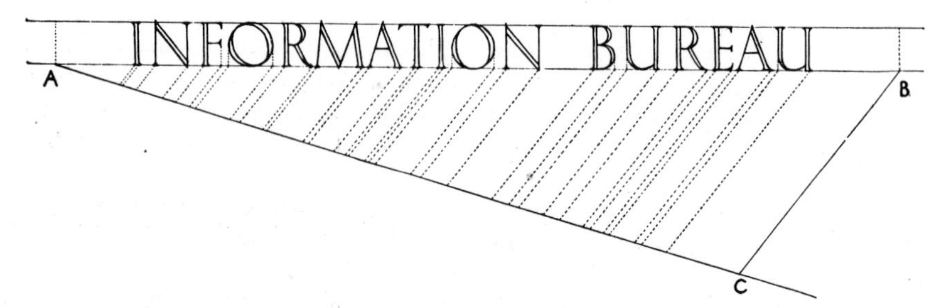

Fig. 34.9 Setting out by proportional scale method

From A, draw a line at any angle and measure off the length of the facia at 1/10 scale, i.e. 500 mm, calling this point C. Connect point C with B, then draw lines parallel to CB, marking off each letter width and space from AB on to AC. This gives the correct setting out of the lettering to 1/10 scale, the height of the letters being determined by the width of the letter O. Chalk lines are then snapped on the facia and the position of the letters ticked off from the scale drawing. Each letter can then be lightly sketched in chalk in its correct position.

The ratio of height to length

One of the things that often puzzles the aspiring signwriter is to determine what height of lettering to use in any given situation, and there is another method of scale drawing which automatically gives the correct height. It is particularly useful on long signboards and shop facias. The method is to draw two parallel lines 15 mm apart on a piece of paper, and within these lines to set out in blacklead pencil or ballpoint pen the required lettering, paying careful attention to the shapes and forms of the letters and the spacing between the letters. Make no attempt to regulate the length or to squeeze the letters into any given length; just let the length develop naturally from the correct spacing of the letters. Suppose, as in Fig. 34.10, that the words NATIONAL THEATRE when set out in this manner and with a suitable space at either end, occupy a length of 300 mm; this immediately shows that the height of the lettering is 1/20th of its length (300 divided by 15). Thus, if the signboard is 3 metres long the correct height for the letters will be 150 mm, if the signboard is 4·8 metres long the letters will be 240 mm high, and so on. The problem is solved.

NATIONAL THEATRE

Fig. 34.10 Rough layout sketch to determine the correct height of letters in any given situation

But this is not all. The next step is to mark off the line of lettering on the paper at 15-mm intervals as shown (reduced) in Fig. 34.11. When the signwriter is setting out the actual signboard he strikes the top and bottom lines with the chalk line the appropriate distance apart, and then marks off horizontally at intervals of the same distance, which means that he is virtually squaring up the drawing. It is then an easy matter to sketch in the positions of the letters lightly, knowing that they are bound to work out correctly. The beauty of this method is that the piece of paper with the preliminary sketch on it is only narrow and is small enough to be handled and used when working on a scaffold at a considerable height above the ground, even in windy weather.

NATIONAL THEATRE

Fig. 34.11 Adapting the same layout sketch to assist in setting out a signboard accurately

The author has used this method for many years in many situations and can recommend it as being completely practical. At one time he was called upon over a long period to write the eaves-level signboards on hotels and public houses in the Manchester district, and was frequently faced with the task of setting out signs of such great length that the ladder scaffold had to be dismantled and re-erected further along the street ten or twelve times in the course of the job. In these circumstances it would have been impossible to chalk out the whole of the sign before beginning to paint in the lettering; this would have been a mammoth task involving an enormous amount of scaffold shifting, very laborious and time-consuming. By the method described it was possible to plan the work at ground level, taking only a few minutes. The work was then done piecemeal; the scaffold was placed in the first position with a plank of some 5 metres long, and the first part of the lettering was set out, written in goldsize, gilded, washed off and burnished, before moving the scaffold along and carrying out the second part of the work. There is no other method by which this could be done with the confidence that it would work out exactly right over the whole distance. The method has also been used on large street banners made of signcloth, and on one occasion was adopted for lettering corrugated iron sheeting along the side of a factory nearly a hundred metres long, with letters 5 metres high.

Some people would ask why such an awkward size as 15 mm was chosen for the preliminary sketch when 10 or 20 would work out so much more easily. The answer is that experience has shown that 15 mm is the ideal height for working out letters and spacing; 10 mm is too small to give an accurate enlargement, and 20 mm is just that little bit too big so that it is difficult to space the work properly.

Measuring by letter width

A rough-and-ready method used by some signwriters is to count the number of letters, reckoning each space between words as a letter also. If, for instance, there were twenty letters to be fitted into a 6-metre length of board, with half a metre space at each end, this would mean a horizontal width of 250 mm for each letter (5 metres divided by 20). Of course the letters would not necessarily be 250 mm wide because of the space between them and because the width of individual letters varies. This is a matter of adjustment as the setting out proceeds, but the height of the letters would be 250 mm and the lines snapped out accordingly.

The advantage of all these methods is that one is certain from the start that the whole of the lettering will fit in the allotted space, whereas if a sign is set out by guesswork, it is often necessary to make several trials before the lettering is satisfactorily placed, all of which entails more chalking in and rubbing out than is good for a newly painted surface.

When the lettering has been lightly sketched in, a few verticals might be drawn with a set square to make sure the letters are really upright, for most people have a tendency to draw uprights leaning slightly to one side or the other. If centre-lines are drawn down the vertical strokes of the letters, this will be sufficient to check the freehand work.

Any superfluous chalk should be lightly dusted off with a painter's dust brush before commencing to "write", this being specially important when goldsize is being laid-in for gilding; the goldsize tends to spread into (and be absorbed by) superfluous chalk and this subsequently interferes with the adhesion of the gold leaf, usually causing ragged edges.

WRITING

The beginner invariably finds the writing pencil rather awkward and difficult to control at first and this tends to increase his nervousness. It is therefore advisable, before attempting any actual lettering, to practise a little with the sable, learning to hold it correctly and trying out various brush strokes. Thick and thin strokes, curves, swirls, wavy lines, leaf and petal forms may be tried out on a blackboard, using a creamy mixture of whiting and gum water which can be readily washed off. The student gains confidence in this way, knowing that anything he paints can easily be removed and a fresh start made; the chief cause of nervousness is the fear of making a slip and spoiling the sheet of paper or painted board. When this fear is removed the student can forget about spoiling anything and so can concentrate on mastering the handling of the pencil.

It seems to be the time-honoured practice in many places to start a student off with a Block alphabet. This, we feel, is too exacting for a beginner, since the production of straight lines and square corners is not easy. We have found it better to give the student some free-flowing script or simple ornament to paint at first, as this allows freer brushwork in which inaccuracies are not so apparent. When the student makes a "passable" job, he is encouraged somewhat and is thus able to tackle the more exacting Block or Roman with a little more confidence and control of the pencil than he would otherwise command.

It is a good plan to practise early exercises with white or light colour on a dark ground because in this way every brush stroke may be clearly seen and the need for proper laying off as in normal painting practice will be appreciated. Black should be avoided in the early stages because when lettering in black on a light ground it is extremely difficult to see the position of the pencil, especially when correcting a stroke.

Take up the writing pencil, holding it as you would hold a pen, and charge it with colour from the dipper; work it about on the palette to remove surplus colour and bring the hairs to a square working edge. Then, resting the wrist firmly on the mahlstick to steady the hand, draw a long stroke down one side of a serif letter, starting with a thin stroke on the tip of the serif, increasing pressure slightly on the down-stroke and finishing off lightly at the tip of the bottom serif. Try to do this in one steady stroke; too many bites only induce niggling and timidity whereas practice in long bold sweeps soon creates confidence. Except when starting or finishing a stroke, do not fix the eye on the tip of the pencil, but look a little ahead of it and allow the pencil to follow the eye.

Fairly large letters (i.e., 100-mm or over) should be practised first, using a large well-charged pencil and letting each stroke do as much as possible. It is not good practice to outline the letter first and fill in with a larger brush, except in very large-scale work. The habit of using brushes too small for the letter size tends to encourage "fussing"; moreover, a large pencil is much easier to control. The aim should be literally to *draw* the letters boldly with the brush rather than merely to fill-in a chalked outline. The expert sign-writer does this so well that he only needs to indicate the skeleton of each letter in his setting-out. When negotiating curves, the pencil should be slightly twirled between the fingers to assist its passage round the curve; this movement becomes quite easy once the knack is acquired and it helps to develop a supple wrist.

When using oil colour, the consistency of the paint should be just right; if it "slips", a little varnish or goldsize should be added to give the pencil more cling to the surface, but if, on the other hand, the colour works slow or drags too much, it should be eased with a drop of white spirit as required. This question of experimenting to find the right balance between too free a consistency and one which is too heavy is most important to the student. Lay off each letter evenly and avoid fat edges. Light lettering on a dark ground requires two coats to gain solidity, but even two coats may look patchy and uneven if the colour is not properly laid off.

Letters which terminate in a point (A, M, N, V, W) or a curve (C, G, O, Q, S and U) should be carried slightly beyond the horizontal chalk lines, otherwise they will tend to look shorter than the other letters.

When painting very large letters (that is, in the region of 600 mm or over) the mahlstick gets in the way and cramps one's style if used in the ordinary way; in such cases the lettering is best painted freehand.

Excellence can only be attained by dint of hard work and diligent practice. Concentrate always on the *quality* of the work; facility and speed will develop naturally with constant practice. The best writers are usually speedy ones.

When in doubt as to type, choose the simpler; never sacrifice legibility to elaboration. Do not mix slanting and perpendicular letters indiscriminately and beware of mixing periods and styles without careful consideration. The development of good Antique Roman, sans serif, and copperplate script will cover practically any style that may afterwards be attempted. Copperplate is just as disciplinary as Antique Roman and is a good basic style for script and lower-case italics, etc.

Use the best sables procurable and, for preference, those with long flexible hair; when the mastery and control of a long sable is acquired, its value is soon apparent in the rapidity with which long sweeping curves can be effected. The greatest error a beginner can make is to use too small a brush; use the largest with which it is possible to do the work in hand.

GLASS WRITING

The first point to note about glass writing is that the work is done *behind* the glass, which means that the lettering must be written in reverse. The beginner finds this rather disconcerting at first and it requires some practice to accustom oneself to writing backwards and at the same time to produce well-shaped letters. Another difficulty is caused by the thickness of the glass itself; this thickness comes between the setting-out on the front of the glass and the actual writing on the back and must always be allowed for, otherwise the lettering may easily get out of line and appear ragged and untidy from the front. The eye must therefore be kept constantly at the same level as the writing pencil to ensure that the lettering on the back truly corresponds with the setting-out on the front. It is generally advisable, especially with small lettering, to strike top and bottom lines on the *working* side of the glass to ensure correct sighting with the face side.

Setting out

This is usually done in lead pencil on lining or detail paper which may be affixed to the front of the glass with self-adhesive tape or by pasting down with weak paste; this latter eliminates any trouble due to the paper sagging. In the case of bench work, the setting out sheet may be placed upon carbon paper and the outlines traced through so as to appear in reverse on the back or, alternatively, treated with paraffin to make the paper transparent so that the setting out can be clearly seen through it. Expert writers, however, often set out directly on the face of the glass by snapping the guide lines in chalk and using either pipeclay or glass-marking pencils for setting out the lettering.

Materials

As already indicated, the paint for glasswork should be tenacious to ensure good adhesion. The colour of the glass, being of a decided greenish hue, will modify any colour painted behind it and all colours therefore must be tested behind the glass when matching up. White, cream and stone colours may be tinted slightly pink to offset the green of the glass.

Writing on glass

The writing pencil tends to slip on the smooth glass surface if the colour is too oily or thin; the addition of varnish gives the colour a certain amount of "cling" and so steadies the pencil. Avoid breathing directly on the glass in cold weather as this condenses into moisture and thus impairs adhesion. An electric hair-drier is invaluable in driving moisture off glass. Needless to say, the glass should always be scrupulously clean and dry at time of painting.

When working on the bench a bridge rest is required on which to steady the hand; this is a long narrow piece of wood having cork or wood supports at each end to lift the wood off the glass in the manner of a bridge. When working flat in this way many writers prefer camel-hair pencils because their lack of spring allows the hair to keep in contact with the surface and thus under better control than the livelier sable.

Any gilding will be done as described under "Glass Gilding" (Chapter 35). Sometimes, when gold lettering is required, the outline and the background are painted in first, leaving the lettering clear. When the background is thoroughly dry and hard the letters are then gilded, burnished and backed up. This method of putting in the background first is also adopted in the case of illuminated signs, the letters being afterwards backed up with translucent white or colour, and stippled.

Normally, however, the ground colour or "backing up" is usually put in last. It should be of the same tenacious composition as the lettering colour, and should always be stippled with a hog-hair stippler to produce a uniform finish, and finally varnished. Where the lettering is to appear on clear glass (i.e. without a ground colour), the letters should be backed up and finally pencil-varnished, allowing the varnish to extend about 5 mm beyond the boundary of the letters as a protection against condensation and cleaning, etc. A good boat or spar varnish is the best for this purpose since it has better water resistance than ordinary oil varnish, and two coats should be given for preference.

GLASS EMBOSSING

Glass embossing is an etching process which produces sunk letters or ornament upon glass by means of dilute hydrofluoric acid; the process is done on the back of the glass and the deep etching therefore appears as a raised or *embossed* letter or pattern when viewed from the front. It is sometimes used upon "flashed" or coated opal glass (as in "sheet ruby") to remove the coloured glass film and leave a white letter, or vice versa; at other times to produce a clear letter upon a ground glass surface, or a slightly obscure letter upon a clear surface. There are two distinct contrasting "whites" obtainable on glass by means of acid. First, the slightly dulled surface produced by etching with ordinary acid; second, a matt or dead white produced by the use of "white" acid. Ground glass is sometimes introduced by rubbing the portions of the glass whch are left raised after the use of the acid with a copper-lined wood block or a glass slab, using fine emery powder and water for abrasive. By gilding upon embossed and clear glass, a beautiful matt and burnished gold effect is produced. Combinations of "frosting" and aciding are used for glass which is required to remain uncoloured, as in doors and windows where it is desired to obscure the view yet retain the light.

Glass embossing requires great care at every stage of the work. The glass is first cleaned thoroughly and placed face downwards upon a reverse tracing of the setting out, in which, of course, the letters all appear backwards. The parts of the glass which are not to be etched by the hydrofluoric acid are usually protected by being painted over with two coats of Brunswick black, but there is always a serious risk of the acid finding its way under the edge of the "resist", and the following is therefore the safer and surer method.

First, coat the glass *all over* with Brunswick black and allow to dry. Then rub the surface well with beeswax and on this film lay a sheet of thin stencil foil (lead foil), smoothing out with a rubber squeegee or a piece of boxwood having a rounded edge. Next, trace down the design in reverse, using a hard pencil so that the foil is slightly indented; the design is then cut through the foil with a sharp stencil knife (only light pressure is required) and the parts to be embossed are peeled off. The parts thus exposed are gently scrubbed with a short stencil tool damped with white spirit or paraffin to remove the beeswax and Brunswick black. This cleansing must be done thoroughly and slowly, using a minimum of solvent, which, if used too liberally, will tend to find its way under the protective foil. When the mask has been thoroughly cleaned up, leaving the design in clear glass, lay a sheet of brown paper over the surface and go over it with the rounded boxwood to press the cut edges of the foil into close contact with the glass (they sometimes tend to be lifted in the course of stripping out). The plate is then ready to be subjected to the action of the acid. But first it is necessary to erect a low wall of tallow round the edge of the glass to form a shallow "tray" in which to pour the acid. The glass is then laid upon a flat bench thickly covered with sawdust to form a *perfectly level bed*; this is very important and a spirit level should be used to ensure that the glass is truly level.

The room should be well ventilated so as to allow the acid fumes to escape. The acid is then poured carefully on to the glass to an even depth of 10 mm or so. The strength of the solution is generally 1 part acid to 2 parts distilled water; it is prepared in a shallow dish or tray of sheet lead having a lip from which to pour the acid on to the glass. There should be a leaden lid to the tray. The same acid may be used again and again, until it has been reduced by waste and evaporation, when new acid can be added. The acid should be strained or filtered occasionally through linen or filter paper, to remove any particles of stopping wax, glass sediment, etc. Hydrofluoric acid must be kept in airtight gutta-percha or lead bottles, or the diluted acid may be kept in the lead tray if the lid is stopped round with Russian tallow to prevent leakage and escape of fumes, otherwise windows in the room are liable to be affected. Rubber gloves should be worn as a protection against possible acid splashes.

The time necessary for the exposure of the glass to the treatment will vary according to the strength of the acid and the make of glass. A test should be made on a strip of waste glass of similar kind to that about to be used. When the acid has bitten deep enough, it is poured back into the tray through an aperture made in the wax "walling", by gently tilting the glass; the face of the glass near this point must be smeared with tallow to prevent damage to the front of the glass should the acid run underneath. When the acid has been poured off, the glass is rinsed thoroughly with water to remove any remaining acid, the tallow walling is removed and stored in a pot for future use, the foil is stripped off and the Brunswick black resist is cleaned off with paraffin, white spirit or petrol. The glass is then thoroughly washed with soap and warm water, followed by vinegar and water, and polished ready for gilding or whatever other method of finishing is intended.

The beauty of embossed glass is seen to the best advantage when it is combined with burnished gold, for then the matt effect produced by acid embossing makes a pleasant contrast with the bright polished or burnished part. In lettering, the centre is usually embossed while the rest is left burnished, thus giving the effect of a matt centre and burnished outline.

Imitation embossing

Where cost will not permit the actual embossing of glass, an imitation emboss effect is usually carried out. The effect obtained is not an actual emboss, but merely a matt finish which, in combination with burnished gold, gives the "matt and burnished" appearance associated with embossing.

The matt effect is obtained with clear varnish in the following manner. Assuming that the letter or ornament has already been outlined in black and is quite dry, the setting-out paper is removed from the front of the glass to make it possible to see where the varnish is to be applied. The back of the glass (i.e. the working side) is then rubbed over with a clean damp sash tool on which is the merest trace of whiting or French chalk. When this dries it should have a thin white chalky or "frosted" appearance; this acts as a guide when putting in the matt parts with clear varnish since it becomes transparent as soon as the varnish touches it.

When all the matt parts are in (usually the centres of letters, leaving clear outlines) the varnish is allowed to dry; then, with clean dry fingertips, *very gently* rub the whiting into the clear varnished parts to obtain an even matt effect. The glass is then polished off with a clean dry cloth and is now ready for gilding as described under "Glass Gilding" (Chapter 35). When gilded, the finished work bears a superficial resemblance to embossed work, in that the lettering has matt centres and clear burnished outlines.

35

Gilding and Bronzing

The term *gilding* may be understood broadly as referring to the application of gold or other metals in a thin leaf form by means of an adhesive or "mordant". *Bronzing* generally refers to the application of *imitation* gold or other metals in powder form, also by means of a mordant. The various metals used for decorative purposes are the following —

Gold (in various tints and thicknesses) ...	82 mm × 82 mm in books of 25 leaves
Silver	115 mm × 115 mm in books of 50 leaves
Platinum	82 mm × 82 mm in books of 25 leaves
Aluminium	82 to 150 mm square in books of 25 leaves
Copper	82 mm × 82 mm in books of 25 leaves
Tin	82 mm × 82 mm in books of 25 leaves
Abyssinian gold (alloy of copper and tin)	90 to 115 mm square, in bundles
Ducat gold (alloy of copper and aluminium)	82 to 120 mm square, in books of 25 leaves
White metal	140 mm square, in books of 25 leaves
Dutch metal (alloy of copper and zinc) ...	90 to 115 mm square, in bundles
Bronze powders	In jars or packets by weight

GILDING

Gold leaf

Gold is beaten to a leaf of extreme thinness, the ductility of the metal allowing as many as 2500 leaves, measuring 82 mm square, to be obtained from 28 grams (one ounce) of fine gold. The best quality is 23 to 24 carat English gold leaf, but in judging quality, thickness of metal and freedom from blemishes must be taken into account. Gold leaf can be supplied in double thickness, or even four times the ordinary thickness where resistance to severe atmospheric conditions is required or where extreme durability is desired, as on church steeple weather-vanes which can be reached only at infrequent intervals. It is also supplied in various shades, such as pale, medium, deep and extra deep. There are two types of gold leaf, namely, *loose* and *transfer*. Loose gold leaf is simply interleaved between thin paper in the form of a book, the pages being dusted with a red earth powder called *Armenian bole* to prevent the gold sticking to the paper. Loose gold is extremely delicate and must be handled with great care; it should be stored in a dry place and may with advantage be slightly warmed before use to prevent its sticking to the pages of the book. Transfer gold is much more easily handled, the gold leaf being loosely attached to a thin waxed tissue which extends about 25 mm beyond the gold to form a margin by which it is held during application. Transfer gold is also supplied in ribbon form in rolls of about 20 metres and in widths varying from 4 mm to 75 mm, although the supply position is uncertain partly because of fluctuations in the price of gold and partly because there is very little call nowadays for ribbon gold.

Foreign gold leaf is cheaper than English gold, but is much less dependable and varies greatly in quality. Imitation gold, copper alloys and silver leaf quickly tarnish unless protected by clear lacquer. The alloys are generally much thicker and more brittle than gold leaf and if necessary may be cut to the required shape with scissors.

MORDANTS

Gold leaf is generally attached to a painted or varnished surface by means of a tacky adhesive such as Japanner's goldsize or oil size; in certain cases, as in glass gilding (which will be discussed later) it is applied by floating on a water size.

Japan goldsize

Japan goldsize is a quick-drying short-oil varnish which is made to dry at speeds ranging from one hour to twenty hours, but the drying time may be further modified by the addition of a little boiled linseed oil. The goldsize should be brushed out to a firm even coating, for if flowed on too liberally it may wrinkle badly when the gold leaf is applied. If the goldsize is lightly pigmented with finely ground colour in oil (preferably tube colour) this acts as a guide and shows up any excess or uneven laying off. When the right state of tackiness is reached it should be gilded immediately since it holds its tack for only a comparatively short time; once it dries it quickly becomes hard and tack-free and thus impossible to gild without laying-in afresh. To test for tackiness, the *back* of the forefinger should be stroked lightly over the goldsize; as soon as this can be done without marking or smearing the surface it is ready for gilding.

A practice which is growing rapidly and is now widespread is that of using an ordinary decorators' gloss paint as a mordant instead of goldsize. Generally a yellow paint is used. The arguments advanced in favour of this dubious practice are (a) that the paint works more easily than goldsize, and (b) that because it is opaque it makes any blemishes in the subsequent gilding less apparent. In fact, although nobody would admit it, the real reason for the growth of the habit is sheer laziness, in that it is easier to use a little of the gloss paint that is already on the job instead of bothering to obtain the right material. In our opinion this is a thoroughly bad and slipshod practice, not worth a moment's consideration by any real craftsman, and one that gives unsatisfactory results. Admittedly when the gilding is first completed it is not easy to detect anything wrong, but we have seen work which after less than two years' exposure has lost its lustre and become badly cracked. Quite obviously, it goes against all the canons of proper painting procedure to trap a paint film which has not completely hardened underneath an impervious film of metal sheeting, however thin.

Japan goldsize is used for lettering, lining, etc. in all cases where the area to be gilded is comparatively small, and for all outdoor work irrespective of area. It is obtainable in various drying times, varying between 1 hour and 20 hours, and generally speaking the slower the drying time of the goldsize the more brilliant and lustrous the final gilding will appear. But for outdoor sign work the best policy is generally to use a goldsize with fairly rapid drying time, otherwise there is a serious risk of the material being spoiled by a shower of rain before the gold can be applied, or by dust blowing about in the atmosphere which causes the surface to become gritty. Large areas on indoor work are best gilded with oil size.

Oil size (*sometimes called old-oil goldsize*)

Oil size is a thickened or oxidized oil originally prepared by exposing linseed oil to air and light for about six months in a wide shallow vessel, covered with fine gauze to exclude dust, until it became "fat"; it was then pigmented with yellow ochre, litharge added for drier, and thinned to usable consistency with boiled linseed oil or varnish, the latter providing hardness and additional lustre. It is now more convenient to purchase ready-prepared oil size made from "stand oil", which may be similarly thinned with boiled oil or varnish if necessary. Prepared oil size is available in various gilding speeds from 4 to 24 hours; the quicker oil sizes retain their tack for 3 or 4 hours, but 24-hour size will generally retain its tack for several days. The longer it holds its tackiness, the better is the result, provided that the size ultimately dries firm and hard throughout. Once the oil size is laid it must be protected from contamination by dust, etc., until it is ready for gilding.

Oil size is generally used where large areas of surface are to be gilded, as for example large wood or metal letters for exterior signs, or where a solid gold background is required either for lettering or for interior decoration, the lustre and burnish being greatly superior to Japan gilding. It must be carefully applied and evenly brushed out since it has a pronounced tendency to flow or "curtain" into ridges or fat edges. It must be allowed to assume the correct degree of tackiness before the gold leaf is applied. It is not suitable for other leaf metals and bronze powders, which are liable to discoloration by interaction with the oil.

Some manufacturers have now ceased to market oil size, and there may therefore be some difficulty in obtaining it. In certain circumstances a slow-drying Japan goldsize may be used as an alternative, but it should be emphasized that there is no satisfactory substitute for oil size, and nothing else can take its place for top-quality work.

GROUNDS

The ground for gilding with Japan goldsize or oil size should be non-absorbent; it may be finished in oil paint, varnish or enamel, flat or glossy, the important point being that before laying-in with the goldsize the ground should be hard and *free from tack*, otherwise the gold will adhere where it is not wanted. Gold leaf will attach itself to a surprising degree on surfaces that may feel quite hard to the touch, and on new varnish or enamel it is generally advisable to apply a coat of "glair", a weak size made by adding a pint of lukewarm water to the white of an egg; on very small signs, this should be laid evenly with a soft brush because otherwise it tends to form into ridges, but on normal signboards the only practicable way of applying it is with a soft clean sponge. If a pinch of sifted whiting is added, the egg glair will dry with a slightly chalky finish upon which any laying-in with clear goldsize will show quite plainly, but this can lead to grittiness which detracts from the lustre of the gold. The same objection applies to a practice adopted by some signwriters of lightly pouncing the glaired surface with whiting or French chalk. Pouncing with chalk in any shape or form, however, is better avoided because quite apart from the risk of damaging a newly painted surface, the goldsize always has a tendency to "creep" or spread into the chalk, producing a blurred and ragged edge to the work. Much the better plan, to make sure that the goldsize can be seen, is to pigment it with a little chrome yellow tube-colour in oil. The egg glair must be washed off immediately the gilding is finished, otherwise the paint may soon be affected by fine cracking or crazing because of the contractile properties of the glair. After washing, the surface is sponged down and leathered off. Some signwriters today are too lazy to use egg glair, and pretend that it is not necessary with present-day hard gloss paints; and when the gold sticks around the edges of the lettering they take it off by abrading the surface with whiting and water. This is a thoroughly disreputable practice; it leads to the surface of the paint being scored by abrasion so that it attracts dirt, and a short time after the work is finished the letters are seen to have a dark shadow all round them.

LAYING THE GOLD

Transfer gold

Only when the goldsize has assumed the right tack should gilding commence. If the size is too soft or wet, the gold will be dull or lustreless and may soon discolour. Never test the surface with the finger *tip*; always use the back of the forefinger or knuckle, as this is less likely to mark the surface of the goldsize.

The book of transfer gold leaf is generally placed in the apron pocket with the clear margin of tissue protruding. Take a leaf with the left hand, holding the margin between finger and thumb, and place it over the sized part; then pass the thumb or the first finger of the right hand over the hair (to lubricate) and gently rub the back of the tissue. The gold will leave the waxed tissue and will attach itself to the tacky surface. Move the leaf to another part, rubbing gently as before, and repeat until there is little or no gold left on the tissue; then take another leaf and continue the operation. It is possible to see through the semi-transparent tissue any parts that are missed, the leaf being manipulated accordingly until the work appears solid and free from pinholes. When rubbing, take care that the leaf does not slip across the part just gilded as this will burnish that portion and will not be uniform with the rest of the work. Slipping is sometimes caused by an excess of red powder on the tissue or by perspiring hands; in the latter case the fingertip or thumb should be covered with a handkerchief, or rubbing may be done with a small wad of cotton wool (in fact, some gilders prefer the latter always).

Start by gilding the portion first laid-in with size and progress methodically in the same direction so that the degree of tack is fairly uniform. The gold should adhere only to the sized portions and should leave the tissue cleanly; if the gold leaves the tissue too readily, it is left in loose ragged edges beyond the sized portions and this causes needless waste of gold leaf especially on small lettering. To remedy this, the books of transfer gold should be put in a press or under a heavy weight for a short time. It should be remembered that transfer leaf is supplied "loosely pressed" or "firmly pressed" to order, the former being especially suitable for large surfaces where whole leaves may be rapidly laid in quick succession.

When gilding is completed, the surface may be *gently* polished with cotton wool to remove all loose gold or "skew". If the gold is damaged by this treatment it has probably been applied too soon, that is, before the size has reached the right degree of tack. Once the leaf is applied the surface is sealed from the air and the goldsize is thus prevented from hardening properly. It is therefore very important that the size should be left until it is quite ready for gilding.

The used tissues should be collected along with any gold "skewings", as these can be sold back to the goldbeaters, as mentioned later.

Ribbon gold

Ribbon gold may be applied by means of the *gilding wheel*, a little instrument consisting of two wheels, one of which contains the gold leaf upon thin paper in a narrow band or ribbon, the other being covered with soft felt. A handle is attached. The wheel is simply rolled along the course of the goldsize or tacky varnish surface and the tissue paper falls clear. The felt wheel presses the gold as firmly in place as required, the pressure being regulated by the operator at will according to the condition of the goldsize or mordant. The gilding wheel was much used by coach and carriage painters for lining direct upon varnished surfaces, the ribbon gold being applied before the varnish hardened off, i.e. while there was just sufficient tack to hold the gold. The method was also useful for lining long runs on flats, beads and mouldings. However, the use of the gilding wheel is rare nowadays.

Laying loose gold

Loose gold is an extremely delicate and sensitive material, requiring great care in handling. It clings to the fingers or to anything greasy, tears easily, and becomes airborne with the slightest breath of air. It is generally applied by means of the gilder's tip and cushion, but many gilders prefer to lay direct from the book. Both methods are therefore described.

Laying with tip and cushion

The gilder's cushion (Fig. 35.1) is a small flat board about 200 mm by 125 mm, the top of which is padded with felt and covered with a tightly stretched chamois leather. A screen of parchment is fitted around one half to protect the loose gold from draughts. There is a leather thumb strap beneath by which the cushion is held in the left hand like a palette; another leather loop holds the gilder's knife (Fig. 35.2), a long smooth blade set in a balanced handle which keeps the blade clean by holding it away from any surface on which the knife rests. Gilders' tips are of badger hair set between two thin

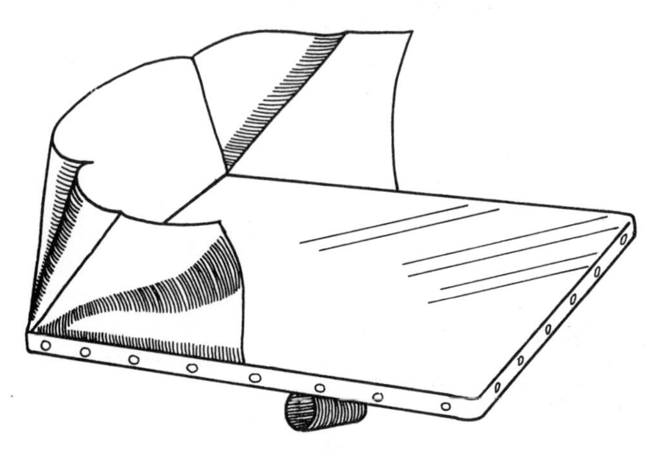

Fig. 35.1 Gilders' cushion

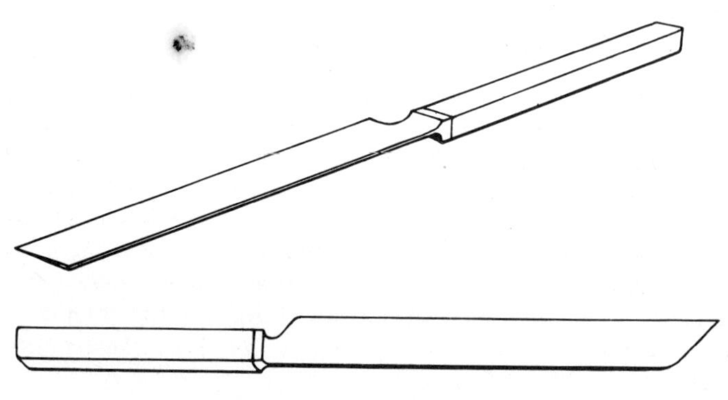

Fig. 35.2 Gilder's knives

The handle is weighted so that when the knife is laid down the blade remains clear of the working surface, and is thus prevented from coming into contact with anything that might contaminate it with grease

cards (Fig. 35.3) about 75 mm wide and in different hair-lengths from 30 to 60 mm. A camel-hair mop or dabber (Fig. 35.4) is also required to press the gold into quirks and enrichments, etc. All tools should be scrupulously clean and dry; the cushion is occasionally dusted with French chalk to prevent the gold from sticking to the chamois leather. The knife blade should not be touched by the hand as it must be kept free from the slightest trace of grease; if the gold tends to stick to the knife, the blade should be rubbed with whiting or French chalk.

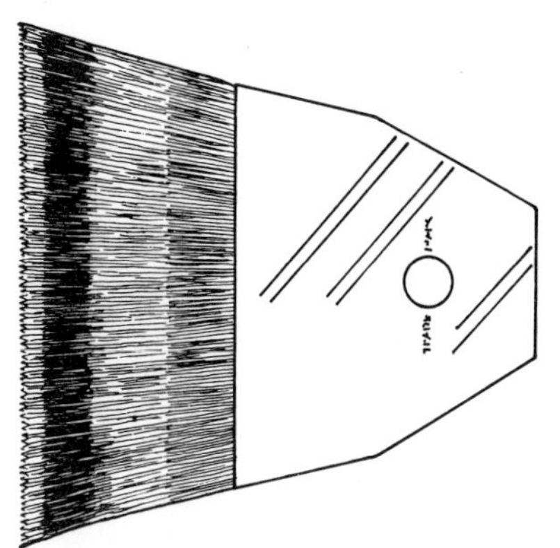

Fig. 35.3 Gilder's tip

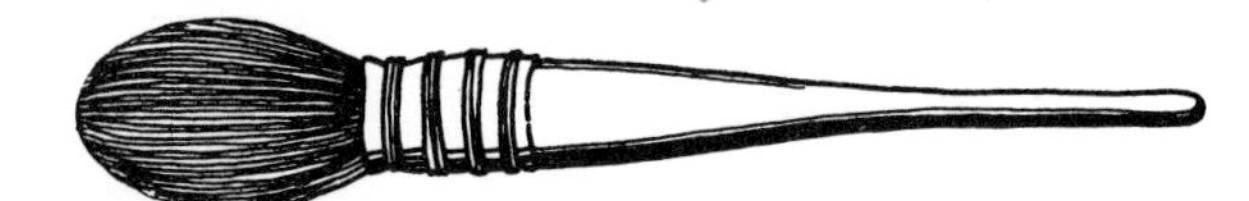

Fig. 35.4 Camel-hair domed mop in wire-bound quill

When the size is ready for gilding, turn out about a dozen leaves of gold into the back part of the cushion, i.e. that part surrounded by the parchment screen; this is done by merely opening each page of the book in turn and gently blowing the gold out on the cushion. The tip and cushion are then held in the left hand and the knife in the right. With the knife, a leaf of gold is gently separated from the rest and brought to the front of the cushion, laid squarely and flattened out with a gentle puff of air directed to the centre of leaf. In doing this, the mouth is placed directly over the leaf and the gentlest "puff" given (as though making the letter "P" sound with the lips only). The leaf should then flatten out smoothly on the cushion, ready for cutting. If, however, the leaf does not lie properly, or if a corner is folded under, the cushion may be tapped lightly with the flat of the knife, thus causing a little draught which raises the leaf at that point and enables it to be smoothed out with another gentle "puff".

The leaf is cut to size with a quick saw-like movement of the knife (actually the gilding knife is not ground to a sharp cutting edge, but is quite smooth, so that the gold leaf is not so much "cut" as divided or separated). The knife is then slipped into the loop provided for it under the cushion or transferred to

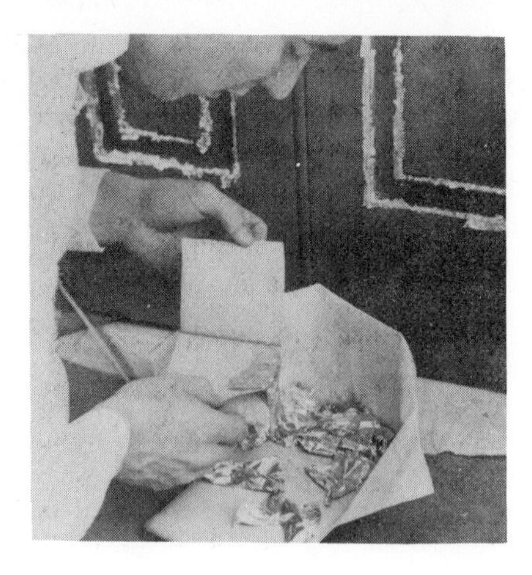

(1) Shaking out leaves of gold from the book on to the leather cushion

(2) Flattening out a leaf of gold on the leather cushion

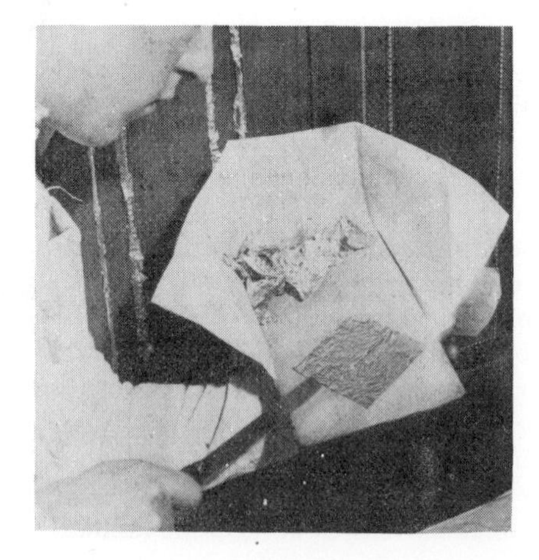

(3) Cutting the leaf of gold on the leather cushion and picking up the cut section with the knife

Fig. 35.5 THE "TIP AND CUSHION" PROCESS OF GILDING — (I)

Courtesy: Campbell, Smith & Co. Ltd

(4) Picking up the cut section of gold leaf with the gilder's "tip"

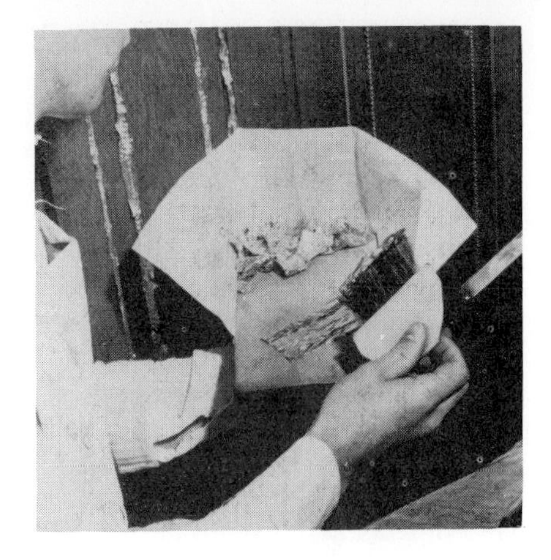

(5) "Laying" the cut section of gold leaf on to the moulding, which has previously been treated with goldsize

(6) "Skewing" off surplus gold leaf with a badger-hair brush into the gilder's home-made skew bag

Fig. 35.5 THE "TIP AND CUSHION" PROCESS OF GILDING—(II)

Courtesy: Campbell, Smith & Co. Ltd

the left hand and the tip to the right. The tip is drawn across the hair or the cheek to give it the *merest trace* of grease, and the gold is then picked up from the cushion with the tip and laid upon the tacky surface. The process takes longer to describe than to carry out and once the technique is mastered it is fairly simple. No matter how dexterous the gilder becomes, however, he must necessarily work quietly, breathe gently, and move about with caution, avoiding any sharp jerky movements that may cause the gold leaf to float off the cushion. Should a leaf become airborne do not attempt to catch it but allow it to settle before gently picking it up with the knife.

The tip must not be too greasy or the gold will become too firmly attached and will tear in transferring itself to the tacky surface. Lay the gold methodically, allowing each piece to overlap slightly at the join. Use whole leaves wherever possible and cover all holes and crevices before dabbing down with the camel hair mop or cotton wool. Gild the high parts of mouldings and enrichments first and fill in the hollows last. When the gilding appears to be solid, press well down with cotton wool before "skewing off" (i.e. removing the surplus loose gold) with a wad of cotton wool or a soft hog-hair tool. The surplus gold or "skew" is collected carefully and put by, along with used transfer tissues, until there is sufficient to send to the goldbeater who pays for this according to the value of the gold recovered.

Fig. 35.5 shows the main stages in the tip and cushion process.

Laying from the book

When gilding a large flat surface, much time can be saved by gilding direct from the book, either by using a long-haired tip to apply the whole leaves, or by discarding the tip altogether and taking the book in the left hand, opening it with the right to turn the gold leaf straight on to the tacky surface. When sufficient dexterity has been attained, large letters and flat areas of oil-size gilding can be executed with great rapidity in this way.

Another method of laying from the book is that originating in America and practised by some specialist signwriters in this country. A flat board, such as a piece of plywood about 100 mm × 150 mm (i.e. slightly larger than the book), is held in the left hand and the book of loose gold laid upon it. The pages of the book are turned over and held down by the thumb while the gold leaf is picked up with the tip and transferred to the surface. If only half a leaf is required, the page is folded half way in a sharp straight crease and the finger nail is run along the crease to cut the gold leaf, using the folded page as a shield and a straightedge. The exposed half of the gold leaf which has been cut by the finger nail is then picked up by the tip and transferred to the work in hand. Any width or portion of the gold leaf can be cut in this way by folding the page as required, the great advantage being that only part of the gold leaf is exposed, the rest being held down by the thumb under the folded page. Thus, loose gold can be used in situations that would otherwise be most unfavourable, such as in shops or stores where passers-by are continually creating air disturbance. The loose gold is under control all the time and the method is much quicker than using tip and cushion.

It is important that the finger nail be kept fairly long and smooth, and it should be dusted with French chalk or whiting to prevent the gold sticking to it. When the gilder becomes really adept in this method, the gold leaf is cut as the page is folded and creased, all in the same movement. Of course, the gilding knife may be used for cutting, if preferred, but the finger nail method means one tool less to handle and makes for greater convenience and speed. It is specially convenient in glass gilding compared with the tip and cushion, and once the new technique has been mastered there is no question of going back to the traditional method. In fact, it is definitely worth while using a book or two of gold leaf merely to practise and gain proficiency in gilding from the book.

Sizing and varnishing gold

Oil gilding and Japan gilding do not normally require any protection, but on interior work a coat of pure gelatine or parchment size is sometimes given to even up the lustre and to protect the surface from dirt deposits; the size can be washed off when renovations are carried out, thus removing the accumulated dirt and restoring the gold almost to its original condition.

When gold is required to be varnished (such as gold lettering on commercial vehicles, etc.) it should first be sized with a weak gelatine size applied evenly with a flat camel-hair or a soft hog-hair brush. Allow the size to dry thoroughly before varnishing with a pale finishing carriage varnish.

GLASS GILDING

Gilding upon glass is generally done with isinglass size prepared as follows. Take a pinch of the best Russian isinglass (this is a gelatinous substance obtained from the sturgeon), dissolve it in a pint of almost boiling water (preferably *distilled* water) and strain it through white filter paper, or through a loose wad of cotton wool placed in a funnel. Alternatively, size for glass gilding can be made by dissolving a gelatine capsule in almost boiling water. It is important that glass for gilding be scrupulously clean and free from grease; it should be cleaned with a little whiting and water which should be allowed to dry, then polished off with a clean cloth.

If the gilding is to be done on a window *in situ* the lettering or design should be set out in pencil on lining or detail paper and fixed in position on the front of the window, so that from the inside the outline is seen in reverse. If, however, the glass is to be done on the bench, a reverse tracing should be made by laying the setting-out sheet upon carbon paper and drawing over the outlines which will then appear in reverse on the back of the paper; this is pasted on the front of the glass which must be carefully propped in a vertical or slightly tilted position on the bench. If the lettering is to be outlined, this should be done *after* gilding, using one of the proprietary blacks specially prepared for the purpose; allow to dry hard. If the letters are outlined *before* gilding it will be found that an extremely thin line of gold leaf sticks around the outside edge of the outline, even after the surplus gold has been cleaned away; nothing will dislodge it, short of scrubbing the outline with a stiff brush such as a damp stencil brush, using a touch of whiting as an abrasive; this of course damages the edge of the outline and may well damage the gilding itself. The line of gold, although so very thin, shows up with remarkable clarity, and spoils the appearance of the work.

When all is ready for gilding, turn out about a dozen leaves of loose gold into the cushion, bring one leaf forward with the gilding knife, flatten it out and cut it to size in readiness for laying as described for oil gilding (or, if preferred, work direct from the book as described on page 502). The isinglass size is applied to the glass with a wide flat camel-hair brush, a flowing coat being given only to that part immediately to be gilded. The size should flow perfectly over the glass; any tendency to ciss indicates that the glass is not sufficiently clean. The gold is picked up from the cushion (or from the book as the case may be) with the tip and carried to the glass *but without actually touching it*, for it will be found that when the tip is held about 3 mm from the glass, the wet surface appears to exert a sort of magnetic "pull" which causes the gold to jump from the tip to the glass. The tip must be kept quite dry. Before picking up the gold, the tip is lightly passed over the hair or cheek to remove any rouge or bole, but if the tip is made too greasy the gold will not leave it so readily and may tear. If the glass is sufficiently flooded at the required part, the flowing size will cause the gold to stretch itself out smoothly on to the glass, the whole art of glass gilding being to lay each piece of gold leaf in such a way that it lies perfectly flat and free from wrinkles or cracks. Thus the gold must be laid as soon as the size is applied, and the expert gilder holds the size brush and tip in the one hand, flowing on the size and laying the leaf almost in one operation.

Allow each piece of gold leaf slightly to overlap the next and make no attempt to confine the gold strictly within the outlines of the lettering or design; any surplus will be cleaned off afterwards. The size must always be used freely and allowed to run off quickly to secure the brightest result. Gilding should generally proceed in vertical bands, working from top to bottom so that the size is kept wet until the gold is laid; if the size happens to dry on any portion of the glass, it must be thoroughly re-wetted before laying with gold, otherwise it will cause cloudiness. Too strong or dirty size will also cause clouding since it will be seen between the gold and the glass.

Faulting (second gilding)

When the first gilding is finished, it must be allowed to dry thoroughly; this may be expedited by blotting the surface carefully to absorb some of the moisture, or by using, where possible, an electric hair-dryer. When quite dry, the loose edges of gold must be skewed off with cotton wool. This will expose many faults in the gilding (such as fine cracks, pinholes or bad joins) and these must be "faulted" by laying small pieces of gold over the faulty parts, or the whole may be double-gilded as in the best work. When faulting or double-gilding, the size brush must be passed lightly over the surface only once, so as not to disturb the first gilding.

When faulting or second gilding is complete, inspect the result from the front of the glass. If the gilding appears cloudy or streaky, it will have to be cleared by "scalding". This is done be applying *almost* boiling water to the glass with the size brush and allowing the water to run down over the gilding, or better still, by pouring the water direct from the spout of a kettle over the whole of the work to wash away surplus size. Allow to dry, then inspect the results; if still cloudy, repeat the scalding process until the gilding clears. The gold should then dry clear and with a bright burnish which is enhanced by lightly polishing with cotton wool. Scalding must obviously be done very cautiously in cold weather and not at all in frosty weather owing to the danger of cracking the glass.

The gilding is now ready to be backed up with a mixture of black Japan and red lead or with a proprietary black; the former is better as it is harder and more protective. The lettering may be pencilled in direct, since it is generally possible to discern the setting-out through the gold (which appears a translucent green colour when viewed by transmitted light). If the lettering is double gilded, however, it will be difficult, if not impossible, to see the setting-out lines through it and a pounce must be used to define the precise shape of the letters. A pounce is also necessary in the case of bench work.

When the lettering colour is quite dry and hard, the surplus gold may be removed with a moist cotton wool pad. Trim up the corners where necessary, using a sharp 6-mm joiners' chisel, then back up and finally varnish as detailed in "Glass Writing".

WATER GILDING

Water gilding on *matt* or *burnish* size is really a specialist job, though in the past the method was frequently used by high-class decorators on Rococo compo ornament, cornices and similar gilded enrichments. It is now rarely used except by picture-frame gilders, but the method has a certain interest for decorators, hence the following brief outline.

Burnish size has pipeclay and blacklead in its composition with parchment or gelatine size as binder and is capable of taking a high polish from an agate burnisher. *Matt size*, which contains pipeclay, Armenian bole and

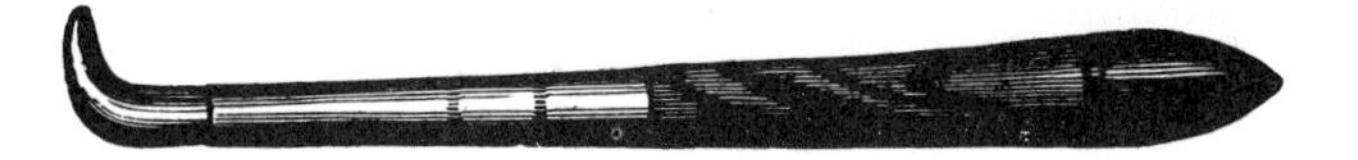

Fig. 35.6 Agate burnisher

other ingredients, is used where a matt gold effect is required in contrast with burnished gold. Both matt and burnish size are supplied, ready prepared, by goldbeaters. In matt and burnish gilding, the work is bodied-up with several coats of gelatine size and whiting, the respective parts then being treated with five or six coats of matt or burnish size as the case requires. The size is laid with a camel-hair brush, each coat being allowed to dry hard and rubbed down with very fine glasspaper. When sufficiently bodied-up, the surface is gilded with water only, that is to say, the size is well wetted with the camel-hair brush fully charged with water and the loose gold laid immediately from the tip as in glass gilding. This must be accomplished very quickly as the water is soon absorbed; the expert gilder

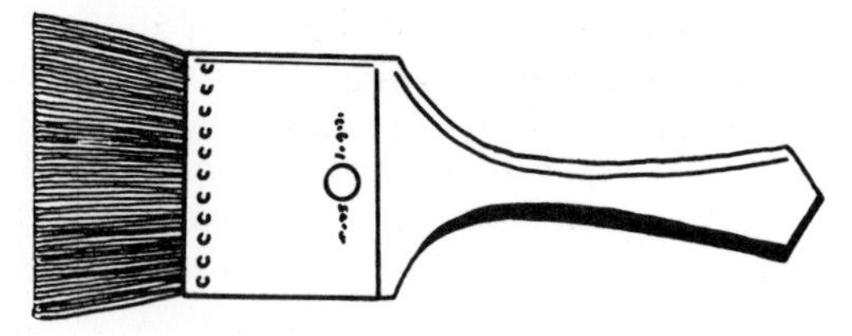

Fig. 35.7 Camel-hair flat

holds the water brush and tip in the same hand, laying the gold as soon as the water brush has passed over the surface. When dry, the matt portion is coated with clear parchment size and sometimes coloured with ormolu. The burnished portion, however, is polished with an agate burnisher by rubbing lightly over the gold as soon as it dries. The gold thus takes on a high polish and retains it, but it must not be sized.

Ormolu for matt gold is prepared from white shellac dissolved in spirits of wine tinted to the required colour with dragon's blood; a few drops are added to the gelatine or parchment size to produce a rich, even, lustreless surface of any desired depth.

OTHER LEAF METALS

These are thicker and more robust than gold leaf and may be handled and cut to shape with scissors. With the exception of platinum and white metal, they quickly tarnish and must be protected with lacquer. Silver even tends to tarnish at the edges whilst in the book. Platinum is scarce and is rarely used.

Only Japan goldsize must be used as a mordant since oil size may stimulate oxidation of the metal. The size should generally be tackier than for gold leaf and, for the thicker and more brittle qualities, it is necessary to use "bodied" size, i.e. Japan goldsize that has been exposed to the air in a shallow tin until it has thickened. Alternatively, the goldsize may be bodied-up with a little Venice turpentine.

The metalled surface may be protected from tarnishing by coating with clear lacquer (one or two ounces of white shellac dissolved in spirits of wine). This should be done as soon as possible after the metal leaf is applied. It is important to remember that the application of metal seals up the surface more or less hermetically, which makes it desirable that the painted or varnished ground should be quite hard and the condition of the goldsize just right when the leaf is applied.

BRONZING

Metallic bronze powders in imitation gold and various colours have the same limitations as the alloy leaf metals in that they readily tarnish unless protected with lacquer. They can be applied either in powder form, by dusting over a tacky Japan goldsize with a camel-hair mop or a hare's foot, or in liquid form by mixing with a suitable bronzing medium. The latter may be either a cellulose medium or a low-acid Japan goldsize. Ready-mixed metallic paints tend to lose their lustre on storing, due to interaction between the bronze and the medium, and they are therefore better mixed freshly as and when required. It is for this reason that makers often supply the bronze powder and medium in separate containers.

Metallic paints are sometimes glazed to produce oxidized and other decorative effects, and in some cases are even gloss varnished. It is important, however, always to apply a coat of thin shellac lacquer first, to protect the bronze from the oxidizing effects of the oil in the glaze or varnish (see page 356).

Coloured lacquers may be made by tinting a shellac solution with spirit soluble dyes such as Saffron, Turmeric, Sanders, dragon's blood, etc. These are sometimes useful for finishing small fittings which have been bronzed or covered with leaf metal.

36

Colour

Colour exerts so great an influence in our lives that our first impressions are largely dominated by it. When we enter a room for the first time, or look at a new dress or a picture, we experience a sensation of pleasure or otherwise, depending on whether the colour is to our liking or not.

Everyone with normal vision is colour conscious, though not by any means to the same degree; children and primitive peoples, for instance, show a marked preference for strong, bright colour and vigorous combinations which may appear gaudy and unpleasant to more cultured tastes. Some people are so extremely sensitive to colour that they may suffer acute mental discomfort in the presence of clashing or incongruous colours, whilst others may be quite undisturbed in similar surroundings.

The decorator should be able to discuss colour problems in a knowledgeable and convincing manner, especially as it is sometimes necessary to warn a client against an unsatisfactory choice of colour. This may require some diplomacy to avoid any appearance of running counter to the client's wishes, but if the decorator has a sound practical and theoretical knowledge of colour he will generally inspire confidence in his opinions.

COLOUR THEORIES

Colour has been the subject of a great number of textbooks and the student is often bewildered by apparently conflicting and misleading theories. The chief cause of confusion is probably due to the totally different results obtained by mixing coloured *lights* as compared with *pigment* mixtures, and we shall therefore attempt to make this distinction clear, as well as to offer suggestions for the practical application of colour to buildings.

The advanced student should acquaint himself with the three-dimensional conceptions of colour as expounded by Munsell, Ostwald and Lawrance. As an introduction to the subject, the beginner cannot do better than to read *Colour* by H. Barrett Carpenter, followed by *The Art of Interior Design and Decoration* by John M. Holmes, both of which are extremely lucid and well illustrated in colour. Another most useful work is The Department of Education and Science's Building Bulletin No. 9, *Colour in Schools* and, as a supplement to this, A. E. Hurst's *Colour in Buildings* might be studied (see Appendix 6). But the student should not only read about colour; actual experiments should be carried out so that he gains in practical experience as well as in theoretical knowledge.

COLOUR TERMS

Before we can begin to discuss colour we must define the expressions used, to ensure that reader and writer are thinking alike. Much confusion is caused by the absence of any commonly accepted terminology; in America, for instance, some of the Munsell terms are quite different from ours, and since these have been adopted by certain writers in Britain we include them here.

Hue is simply another name for "colour". We speak of a "reddish hue", a "yellow-green hue", and so on.

Purity refers to the strength of colour, i.e. its intensity or, as the scientist calls it, "saturation". The colours of the rainbow, for instance, are of maximum purity. When a pigmentary colour is mixed with another, or with white, black, or grey, its purity value is reduced or weakened. The Munsell term *chroma* (the Greek word for colour) is synonymous with *purity*.

Tone value refers to the lightness or darkness of a colour. When white is added to a colour a lighter tone is produced, whilst the addition of black produces a darker tone. Thus a colour may be raised or lowered in tone value. In Munsell the word *value* is used by itself to express the same meaning.

When a colour is lightened or reduced with white it is referred to as a *tint*. When it is darkened by the addition of black it becomes a *shade*. The term *shade* is often loosely applied, as in "shade card" (meaning "colour card"), or a "light shade" (when a light *tint* or pale colour is meant).

The *Chromatic Scale* refers to the whole range of colour which, for convenience, is arranged in the familiar colour circle. Ideally the colour circle would be a continuous band of colour graduating or blending from yellow through green, blue, violet, purple, red and orange, back to yellow, but it is generally shown stepped in separate distinct hues, the number of which varies with different authorities. Carpenter, for instance, shows *fourteen*, Holmes *twelve*, Munsell *ten* and Ostwald *eight* basic hues.

Harmony Colours such as yellow, orange and red; yellow-green, green and blue-green, etc., which lie adjacent or fairly close to each other on the colour circle are said to be *harmonious* or *analogous* colours, since, in combination, they are pleasing to the eye.

Contrast Colours which lie directly opposite or nearly opposite on the colour circle are *contrasting* colours. Carpenter defines contrasting colours as being "those which when placed side by side intensify each other but do not change".

The contrast gained by the juxtaposition of two complementary colours is said to be a *true* contrast. These pairs of colours are termed "complementary" because, when added together as coloured lights, they produce *white light*; thus yellow and blue, purple (or magenta) and green, red and blue-green are pairs which produce white light and are therefore true contrasts or complementaries. Colour circles which show such pairs lying directly opposite are therefore based essentially on the intermixture of coloured light.

Contrast as applied to *pigmentary* mixtures is said to be true when the two pigments mixed together produce a *dark neutral grey*; thus yellow and violet, orange and blue, red and blue-green are pairs which are capable of producing a neutral grey, *but only when the hues are correctly related.* For example, to find the true contrast to a particular scarlet it is necessary to make trials with various blue-greens until one is found that will, when mixed with the scarlet, produce a neutral grey (i.e. a grey which inclines neither to red nor green). This applies to every pair of true pigment contrasts, and if the opposite hues on any colour circle do not produce a neutral grey when mixed together, they are not correctly related according to the principles of pigment mixture. (Incidentally, very few published colour circles are satisfactory in this respect.)

Discord When colours of maximum purity are arranged in a circle, we find an orderly progression from light to dark in this order —

Yellow (lightest colour, nearest to White)

Orange	Green
Red	Blue-green
Purple	Blue

Violet (darkest colour, nearest to Black)

Rood has called this "the natural order of colour". When this natural tone order is reversed, i.e. when black is added to the light colours and white to the dark, we get a circle of *discords*, with violet now the lightest hue and yellow the darkest. It will be noticed that red and blue-green, being midway between light and dark, retain their natural tone relationship and do not become discordant with each other until one is made lighter or darker than the other (see Plate 4).

THE MUNSELL COLOUR SYSTEM

The publication in 1953 of the Department of Education and Science's Bulletin No. 9, *Colour in Schools* (already referred to), introduced the *Archrome* range of 47 colours specially recommended for schools; later, in 1955, the Paint Industry Colour Ranges Committee, in conjunction with the Royal Institute of British Architects and various Government Departments, agreed on a standard range of 101 colours which incorporated the *Archrome* range and was adopted by the British Standards Institution as BS 2660:1955. This, in turn, was superseded on January 1st 1973 by BS 4800:1972, Paint Colours for Building Purposes (see below).

All the colours in the Archrome and BS 2660 ranges were selected from *The Munsell Book of Colour* published in America in 1942, and although this was based on the work of Albert Munsell whose colour system was first published in 1905, the Munsell system was little known in Britain until the above-mentioned Bulletin No. 9 first appeared. *The Munsell Book of Colour* was revised in 1966 and is published in a glossy edition containing 1450 detachable colour chips, and a matt edition containing 1115 colour chips mounted permanently. It is published by the Munsell Color Division, Baltimore, Maryland, U.S.A. The United Kingdom agents are the Tintometer Sales Ltd., Salisbury.

The Munsell Colour Atlas shows the colours arranged in the form of a three-

PLATE 4

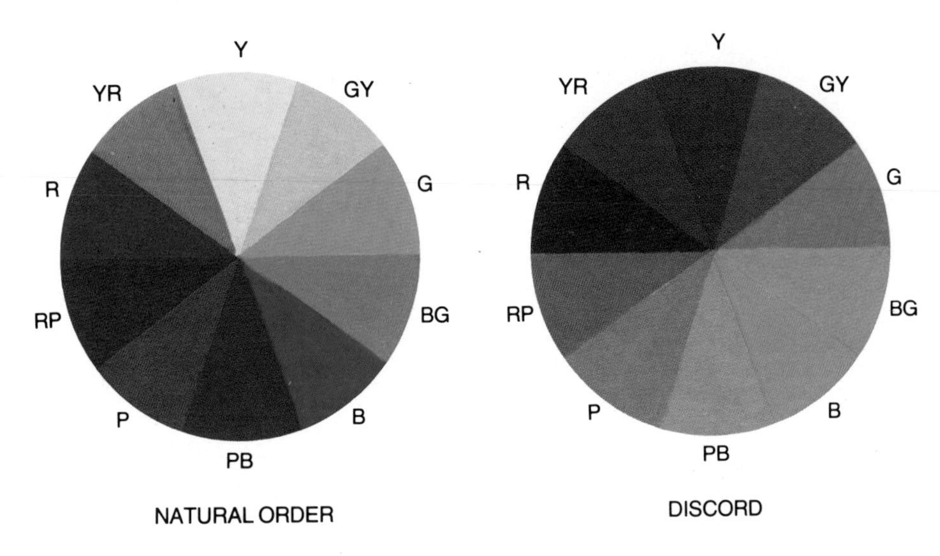

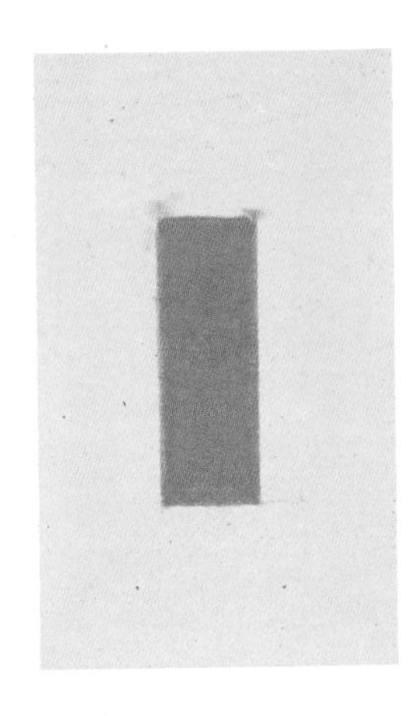

COLOUR MODIFICATION

Colour circles and modification

Courtesy: Crown Decorative Products Ltd

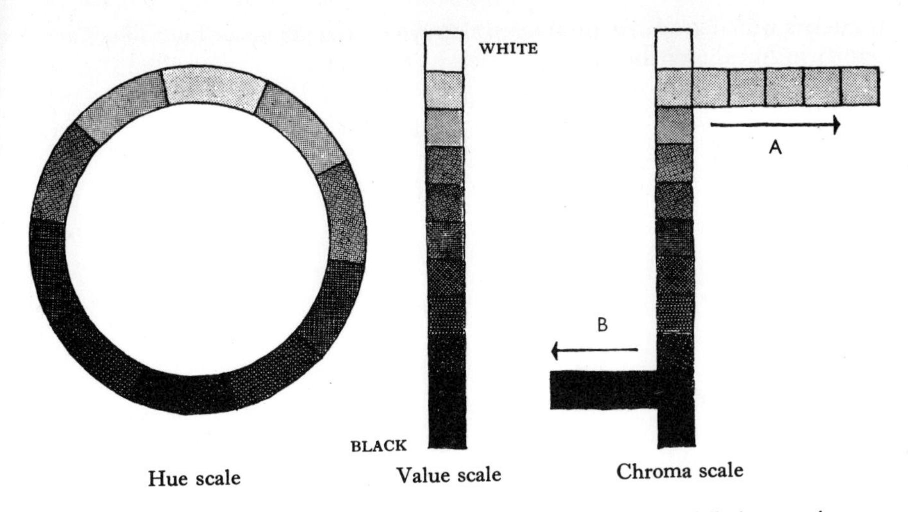

The arrows at A and B indicate the graduation from neutral grey to full chroma, the tone value for each branch being constant

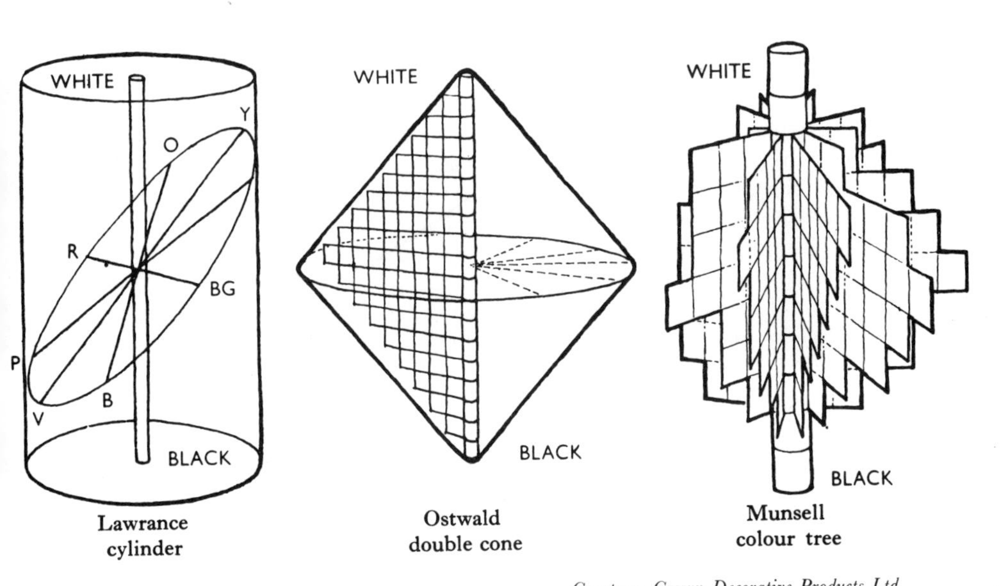

Courtesy: Crown Decorative Products Ltd

Fig. 36.1 Colour solids and the Munsell colour system

dimensional "solid", usually referred to as the "Munsell Colour Tree" (Fig. 36.1). The vertical trunk or centre pole is a scale of greys graduating from black at the base to white at the top, and radiating from this are horizontal branches which graduate from neutral grey to a full colour or hue. There are ten main hues, namely —

Y (Yellow)
GY (Green Yellow)
G (Green)
BG (Blue Green)
B (Blue)
PB (Purple Blue)
P (Purple)
RP (Red Purple)
R (Red)
YR (Yellow Red)

Each of the 10 hue sections is subdivided into 10 parts, thus forming a circular scale of 100 subtly graded hues. In Munsell terms *hues* are the colours in the circular scale — yellow, green, blue, purple, red and the intermediate hues; *value* refers to the vertical dark-to-light scale from black to white; *chroma* refers to the horizontal scale graduating from neutral grey at centre to the purest available hue at the end of the branch; thus a colour of weak chroma is a greyed hue whilst one of strong chroma is a much purer colour.

In specifying a Munsell colour the hue reference is given first, followed by the value and lastly the chroma; thus 5.0 R 6/12 (a strong pure red) of 7.5 GY 8/2 (a pale grey-green).

The Munsell notation is a very precise and logical method of colour description, in contrast with colour names which are often meaningless, and for this reason it was adopted with enthusiasm by the architectural profession. Any conceivable colour can be described exactly by Munsell notation, but unfortunately the cost of the Munsell Atlas is so high as to make it almost prohibitive, and for general purposes one has to refer to the selected range shown in the BS 4800 Colour Cards or BS 5252 Colour Matching Fan.

The complete Munsell Atlas, shown in the form of a colour solid, is, however, the only way in which one can visualize simultaneously the three-dimensional aspect of colour, i.e. hue, value and chroma. A useful comparison between the Munsell and Ostwald systems appears in the Department of Education and Science's Bulletin No. 9.

BS 4800:1972

This British Standard entitled "Paint Colours for Building Purposes" consists of 88 colours, including black and white. Each colour is identified by a code consisting of three parts.

The first part of the code signifies the hue, and consists of an even number with two numerals.

The second part signifies greyness and consists of a single letter.

The third part signifies weight and consists of odd numbers with two numerals from 01 to 55.

For example, the BS 4800 colour equivalent to what is popularly known as magnolia is classified as 08 B 15.

It will be seen that colours with the same letter belong to the same greyness grouping, those with the same first part are of similar hue, and those with the same final part are of similar weight.

For many years the building and associated industries have recognized the need for a standardized range of colours, offering a reasonable amount of choice within the limits of economical production, and catering for all types of materials including paint and other building products. When BS 4800 was published in 1972, the British Standards Institution also published a document called "Draft for Development 17" which envisaged a "Basic range for the co-ordination of colours for building purposes", the next stage of which was implemented four years later. It is therefore necessary that BS 4800 should be considered not in isolation but in the wider context of a further British Standard, BS 5252, which was introduced in 1976.

BS 5252:1976

This is the first British Standard based on and superseding DD 17 and is entitled "Framework for Colour Co-ordination for Building Purposes"; it provides 237 systematically related colours from which specific ranges may be chosen. It is linked with three new standards giving colour ranges for vitreous enamels, plastics, and flooring (sheet and tile) which join the existing BS 4800 range of paint colours.

The framework is co-ordinated in terms of three visual attributes, namely hue, greyness, and weight.

BS 5252 has 12 hues or colours.

The *greyness* scale shows any hue graduated from "clear" (such as bright yellow) down through shades containing increasing amounts of grey. In BS 5252 even the greyest of colours are seen to be based upon one of the parent hues, and a range of greys as evenly distributed as possible for that hue is provided. There are five groups of greyness in BS 5252.

Weight distinguishes the apparent lightness of any hue or greyness. In each of the blue/greyness categories of BS 5252 a range of up to eight "weights" is chosen. Each greyness group is designed to contain eight weight columns.

In BS 5252 the colours are set out to show these relationships clearly, and the key numbers to be used by the manufacturers in addition to their own chosen brand name for the colour are supplied beneath each of them. The range includes seven completely neutral greys, as well as black and white.

The fact that each colour relates to its neighbour in a systematic way is of significant importance in providing a reasonable measure of choice. A designer choosing colour by weight (i.e., by lightness or darkness) can take his choice from an equal degree of weight over twelve hues and several greynesses. Since colours in all degrees of greyness, no matter how remote from the clearest group, are based upon a known parent hue, chosen effects of harmony or contrast can be achieved with much greater accuracy.

The Standard is called a "framework" because it remains the basis for further colour selection. It has been specifically designed in its present form so as to serve the needs of the building industry, with an exceptionally wide range of warm colours to blend with wood and brick, but it does not reflect the limitations of any particular branch of the industry.

There are now four sections of the building industry with specific ranges taken from the co-ordinating framework. This does not mean that these sections of the industry need in any way to restrict their overall ranges to the chosen colour standards; it simply means that the standard range is included with their own.

Other British Standards on Colour

The other three standards which, together with BS 4800, are included in the co-ordination framework, are as follows:

BS 4900:1976 "Vitreous Enamel Colours for Building Purposes". This supersedes the former BS 1358 in respect of colours applied to metals for use as interior and exterior claddings. There are 199 colours in the standard, all of which are available in gloss or semi-gloss finish.

BS 4901:1976 "Plastics Colours for Building Purposes". This contains 124 colours covering opaque plastics products such as laminates and moulded and extruded components. The type of finish is not specified.

BS 4902:1976 "Sheet and Tile Flooring Colours for Building Purposes", which has 143 colours from the co-ordinated framework.

The existing BS 4800 "Paint Colours for Building Purposes" consisting of 88 colours was based on the restricted range of the framework as it stood in the year 1972.

LIGHT AND PIGMENT

Although the painter and decorator is dealing with pigment intermixtures, it is desirable that he should understand the differences between coloured *light* and coloured *pigment.*

Colour is the effect produced on the eye by the different wavelengths of light. When there is no light there is no colour — only blackness. The colour of pigments (and of all "coloured" matter) is determined by their chemical composition and their capacity to absorb or reflect certain coloured rays from the light in which they are seen. A white surface, for instance, *reflects* all the light rays whilst black *absorbs* them all. A red surface reflects only red rays and absorbs the rest; similarly a green surface reflects only green, and so on. When a red surface is illumined solely by blue light or green light (in which red rays are absent), the light rays are totally absorbed and the surface therefore appears black.

White light, or sunlight, is made up of innumerable different wavelengths or, in other words, colours of light; when it is passed through a glass prism and projected on to a white screen, it is split up into its constituent parts and the familiar *spectrum* or "rainbow band" is produced. Conversely, white light may be reconstituted by adding together the spectrum colours as lights. It is not necessary, however, to use the whole of the spectral band since the three light primaries, i.e. red, green and blue, will suffice. When these three light primaries are projected separately on to a white screen and allowed to converge until the beams overlap, *white light* is reproduced. When only two light primaries are mixed, a *secondary* colour is produced; thus red and green produce yellow; red and blue produce purple (or magenta); green and blue produce blue-green (see Fig. 36.2). It is important to note that light *secondaries* are of ligher luminosity (i.e. nearer to white light) than the *primaries* which produce them; this is due to the "additive" effect of mixing coloured lights. When a light primary is mixed with its contrasting secondary the result is *white light* again because all three primaries are then involved. Thus projected light combinations of blue and yellow, red and blue-green, green and purple, all produce *white light.*

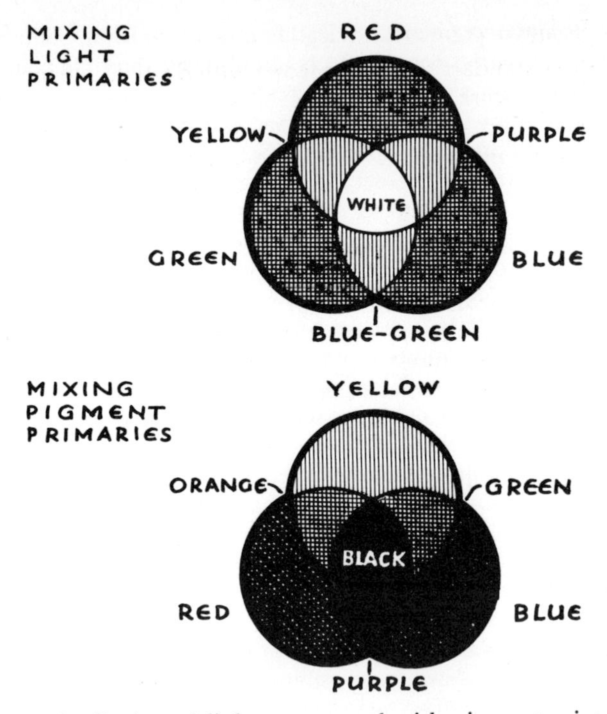

Fig. 36.2 Projected lights compared with pigment mixtures

Showing the "additive" effect of mixing the three light primaries (producing white light) compared with the "subtractive" effect of mixing the three pigment primaries (producing black, or near black)

When we consider the effects of *pigment* mixture, however, we find many important differences. First, the pigment primaries are red, blue and yellow; these are the three basic colours which cannot be produced from any other combination, but which, in theory, are capable of producing all other colours. A mixture of two primaries produces a *secondary*, thus — red and yellow produce orange; yellow and blue produce green; blue and red produce purple. Two secondaries produce a *tertiary* colour; thus green and orange produce citron or citrine; orange and purple produce russet; purple and green produce olive. When coloured pigments are mixed together, a certain amount of darkening and loss of purity occurs; thus secondary colours are less pure than primaries and tertiaries are lower still in purity and tone value. When all three primaries are mixed together the result is a dark neutral grey (almost black) and the same occurs when two complementaries are mixed as noted previously. This darkening or "subtractive" effect (which is directly opposite to the lightening or "additive" effect of mixing coloured lights) explains the limitations of the three-colour theory, for whilst it is true that almost every hue can be produced from red, yellow and blue, the results of such mixtures cannot compare in purity or brilliance with those of available manufactured pigments.

It will be seen from the foregoing that failure to distinguish between the mixing of lights and of pigments can lead to much confusion. Some authorities base their theories on the optical effects obtained by spinning coloured discs, and whilst this method is extremely interesting, it can also be very misleading if one attempts to apply its principles to the mixing of pigments, because the *visual intermixture* produced by spinning discs is, in effect, the intermixing of *reflected* coloured light. For instance, we have noted that blue and yellow lights produce white light; if we paint the two halves of a disc in a similar blue and yellow and spin it rapidly the effect produced is a neutral grey.* Thus, when projected or reflected light is being considered, yellow and blue are complementary. Yellow and blue *pigments*, however, produce *green*, and it requires a mixture of yellow and *violet* to produce a neutral grey. These distinctions must always be borne in mind.

Optical colour effects are highly important to the weaver (since he can usefully exploit the effects of the visual intermixture of closely juxtaposed coloured threads), but to the decorator their chief interest probably lies in the production of sponge-stipple, spray-spatter and colour combing effects in all of which visual intermixture plays its part in providing a more "vibrant" quality than is possible with plain colour. The main concern of the painter, however, is with the *actual* mixing of pigments, and he therefore requires to study colour theory mostly from that angle. The method described by Holmes in *Colour in Interior Decoration* is both practical and convenient. By selecting twelve basic coloured pigments of maximum available purity (as he advocates) and adjusting each pair of contrasts until they produce a neutral grey when mixed, the colour circle becomes of great practical value instead merely of theoretical interest.

COLOUR APPLICATION

All colour is modified by its environment or surroundings. Its effect can only be judged in relation to other adjacent colours in the scheme, the area occupied by it, and the amount and quality of illumination.

We have noted that when *true contrasts* are placed side by side they intensify each other without changing their hues. A small spot of one on a background of the other will glow with great brilliance, but when distributed in large amounts or in equal areas the effect is clashing and disturbing; this may be avoided by *reducing* the contrast, i.e. by adding a little of each colour to the other and producing what is sometimes called a "broken" colour.

Colours which are widely separated (but not truly opposite) on the circle tend to induce their contrasts in each other and therefore do not make good combinations; orange, for instance, tends to make purple appear bluer whilst purple makes the orange look slightly greenish. The effect of an *induced contrast* is clearly seen when a small patch of light neutral grey paper is placed in the centre of each of larger patches of say red, yellow and blue and viewed simultaneously; the grey will appear to incline towards blue-green, violet and orange respectively (Plate 4).

---

*By spinning *pale* blue and yellow together, a greyish "white" is produced. Pure white cannot be obtained with disc mixtures because reflected light is never as brilliant as projected light.

It may be noted here that when we are looking at applied colours, we are, in effect, seeing coloured *reflected light*, and any results of colour modification will therefore conform to the effects of light mixture and not to pigmentary mixture. For instance, if we paint bright yellow lines on an ultramarine ground and view from some distance away, the yellow will tend to appear white, due to the additive mixture of blue and yellow reflected light (whereas a mixture of blue and yellow pigments produces green). Similarly, if we look closely at scarlet spots on a bright green ground, each colour will appear quite intense and brilliant, but if we move away and view at some distance the green will mix optically with the scarlet and will incline towards yellow (due to the additive mixture of green and scarlet). Thus we see that the effect of juxtaposed colours at a distance may be quite different from that at close quarters (see Plate 6). The student should not be content to accept these statements but should test them out for himself.

In practice, of course, the decorator rarely uses colours of maximum purity, except perhaps in very small areas. He certainly would not use an all-over pattern in complementary colours without first greying them off considerably; in fact, most of the decorator's palette is made up of greyed hues or pastel tints. Nevertheless, although pale tints will not show the effects of colour modification so noticeably as the examples quoted above, the *tendencies* will be the same. We have already indicated that when contrasting tints of fairly equal tone-value are intimately juxtaposed, as in spray-spatter and similar effects, they tend to neutralize each other due to visual intermixture, though the resultant "grey" has a distinctly vibrant quality.

Distribution and choice of colour

The art of applying colour lies in its satisfactory distribution. In general terms, the largest areas in a scheme should be occupied by the least insistent hues. In a domestic interior, for instance, ceiling and walls may be painted a fairly pale tint with woodwork to match, stronger tones being used in the furnishings; the deepest tones could be introduced in carpet or floor coverings, but the strongest and purest colours should be reserved for the smallest areas or focal points, such as flowers, pictures, cushions, and so forth.

Dark picture and dado rails, skirtings and architraves tend to "outline" a room and emphasize the wall subdivisions in a rather insistent manner; moreover, the picking out of such features in strong colour may attract more attention. Linear features such as these are best in a neutral tint, preferably white, and this applies also in particular to beams and cornices which form a framework to a coloured infilling.

For simplicity in a small room, woodwork may be painted to match the walls unless it is specially desired to create a bright gay atmosphere, as in a nursery or in a kitchen, where a stimulating effect is desired; in such cases a note of contrast may be introduced. Choice of wall colour should always be appropriate to the type of room — restful in lounge and bedrooms, warm and cheerful in entrance hall and dining-room, light and cool in a warm room such as the kitchen. It is generally conceded that red is exciting and stimulating, that yellow and orange are both warm and cheerful, that green is pleasant and restful, that blue and grey are cool and tranquil, and that purple, though rich and "regal", can also be depressing. Whilst such generalizations may be useful, much depends on the particular *quality* (i.e. purity and tone) of a colour and especially on the influence of other colours which may be in close juxtaposition. Green, for instance, can be pleasant enough, but it can also be very unpleasant in certain circumstances; green and yellow, for example, would be most inappropriate in a ship's dining-room or an aircraft interior.

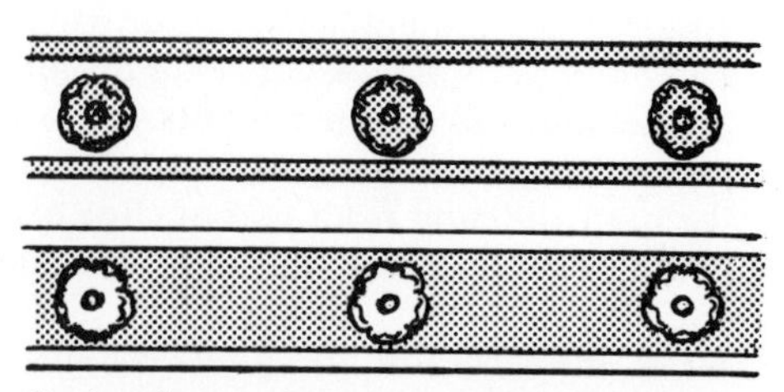

Dark colour applied to raised parts tends to read as a silhouette & hides the detail. When applied to the background, however, the relief ornament is enhanced.

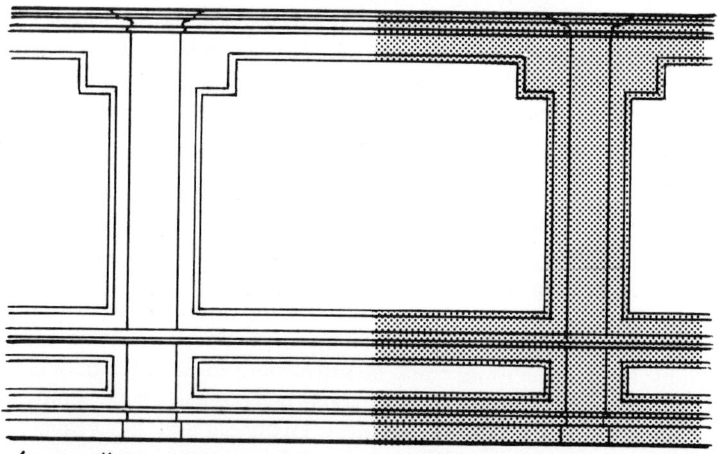

A panelled room usually has sufficient surface interest provided by light & shade of the relief mouldings; when stiles are painted a darker colour the "framing" effect may be unduly emphasized.

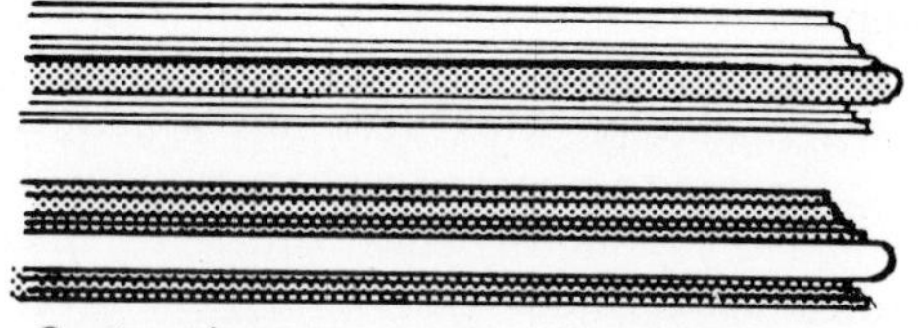

Dark colour on recessed members of cornice mouldings strengthens the natural light & shade. Dark colour on projecting members does the opposite.

Fig. 36.3

Light colours tend to make a room look bigger, whilst dark or strong colours have the opposite effect. Colours in the yellow to red range are of a forward, assertive nature, whilst the cooler greys, greens and blues are quiet and retiring, suggestive of distance. Thus, *advancing* or *receding* colours may be usefully employed to modify proportions, as for example, to lower a high ceiling, or to shorten a long room or corridor (see Plate 5; also "Colour", Chapter 29).

Whether or not panelling and enrichments should be picked out in colour is largely a matter of taste. In the case of "period" rooms it should be done in a manner appropriate to the period. In fairly small rooms it should be avoided if possible, since it may give undue emphasis to panels and mouldings and tend towards fussiness; if such mouldings or enrichments are painted uniformly with the wall or ceiling in a light colour, the light and shade of the relief work will usually provide sufficient interest.

A large room or assembly hall is rather a different matter. Applied colour may be very desirable in order to avoid monotony, or to provide additional richness and emphasis to relief ornament and mouldings which might otherwise be lost because of height or inadequate lighting. In picking out cornices, mouldings and enrichments, the colour should be applied in such a way that the relief contours are enhanced, since careless placing may easily result in *hiding* the modelling. For instance, if mouldings or relief ornament are picked out in a colour darker than the ground, the sense of relief is lost and the ornament will tend to appear, from a distance, as a flat painted silhouette. Reverse the situation, however, and the ornament is at once enhanced, appearing as a light relief on a darker ground with the modelling now plainly visible (Fig. 36.3). When modelled relief and projections are thus painted *lighter* than the ground, with hollows and recesses *darker*, the colouring becomes complementary to the natural light-and-shade and the effect of relief is strengthened. The same thing happens when relief work is scumbled and wiped; the colour left in the hollows and interstices emphasizes the higher parts that are wiped clean. (See also "Picking Out", page 462.)

Selective wall treatments

When the four walls of a room are painted a uniform colour, each wall is given equal interest. If we treat one or more walls in a different colour, we tend to give them a special interest which may or may not be justified. It is not enough to depart from the orthodox for the mere sake of novelty; walls should receive a different colour treatment only for sound logical reasons (Fig. 36.4). On the Continent, where the vogue for selective wall treatment first appeared, walls continually lit by a brilliant sun were painted a dark tone to reduce glare, whilst the darkest wall, i.e. that containing the window, would be painted a light colour.

Rooms having panelled walls or other architectural features are not suitable for this sort of thing, of course, but on the other hand, there are modern box-like rooms, devoid of cornice, picture and dado rail, etc., which simply ask for such treatment. In some cases the apparent height or shape of a room may be modified by careful placing and choice of colour. For instance, in a bedroom the bedhead wall could be a deep, warm tone with the opposite wall in a pale cool tint to suggest distance, the remaining walls and ceiling being white. In a

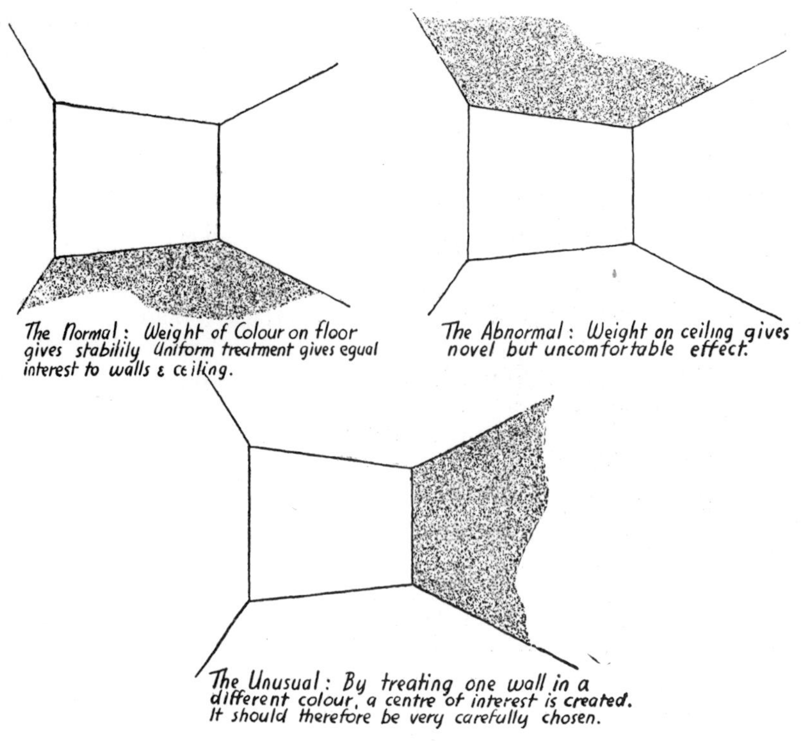

The Normal: Weight of Colour on floor gives stability Uniform treatment gives equal interest to walls & ceiling.
The Abnormal: Weight on ceiling gives novel but uncomfortable effect.
The Unusual: By treating one wall in a different colour, a centre of interest is created. It should therefore be very carefully chosen.

Fig. 36.4

very long room, if the two end walls are painted a strong deep tone of advancing hue with the side walls in white or a very pale receding tint, the room will tend to appear shorter. In a long narrow corridor, the "tunnel" effect can be offset by painting one wall in a strong dark colour and the opposite wall in a very pale blue or white (Plate 6).

Again, in a lofty room, the ceiling may be apparently lowered by painting it a strong tone, the walls being preferably a pale tint or white. Ceilings in black or Midnight Blue can be very effective where appropriate, but may be too oppressive in a living-room. Pale blue is a quite appropriate colour for ceilings since we are used to seeing it overhead. A low room may have a coloured ceiling, provided a very pale tint is used.

With selective wall treatments the attention or interest may be diverted to any desired part of the room simply by means of suitable colour. A wall may be selected as a background for a divan or a bedhead, or in a dual-purpose room the dining-nook may be treated differently from the rest of the room; thus a "focal point" may be created at will. We have mentioned elsewhere (page 383) how pattern and texture in the form of paperhangings may also be used in this way, in conjunction with paint and distemper.

When walls are treated in different colours it is generally desirable that they should have a decided change of *hue* in order to provide a clearly marked difference, for if there is only a slight variation in colour (as, for instance, white and pale grey) the effect may be totally cancelled out by the varying amount of

PLATE 5

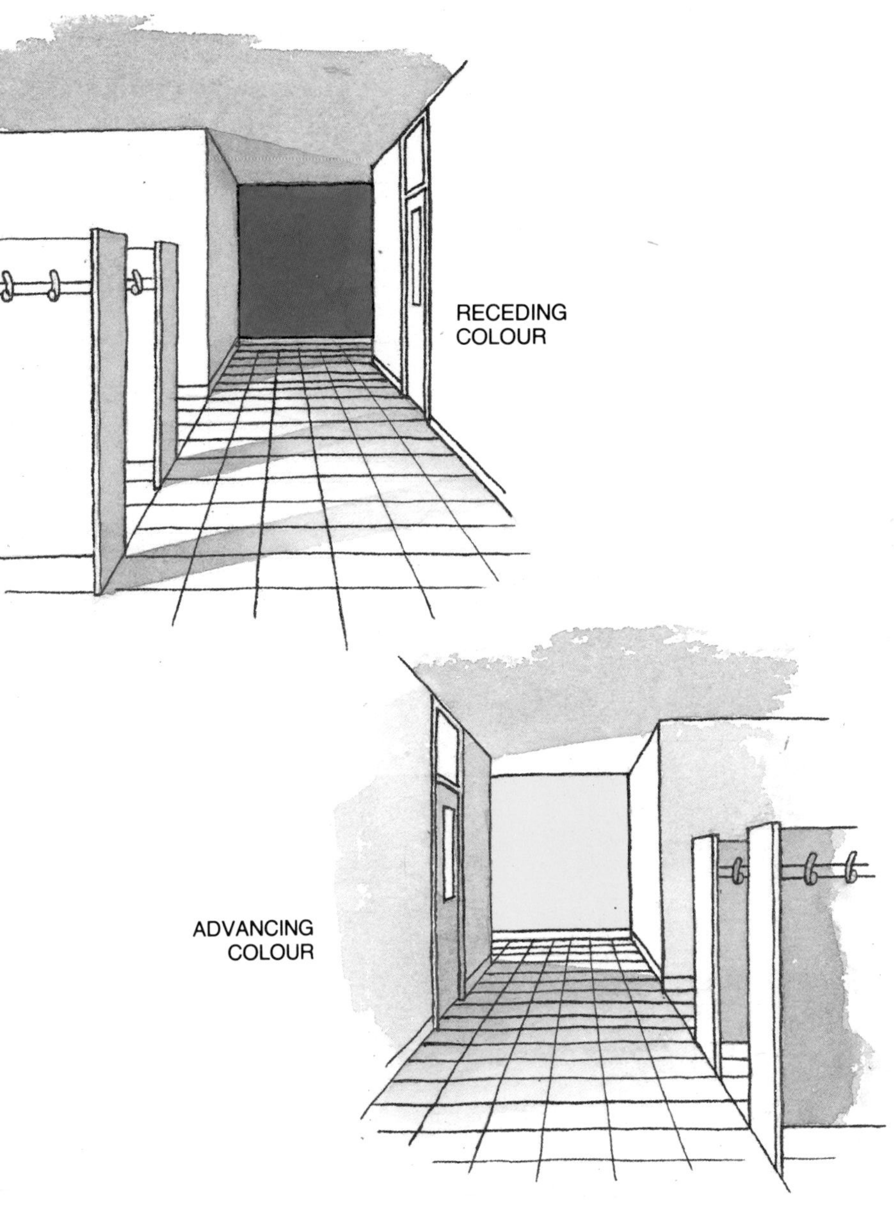

Receding and advancing colours

Courtesy: Crown Decorative Products Ltd

To face page 521

PLATE 6

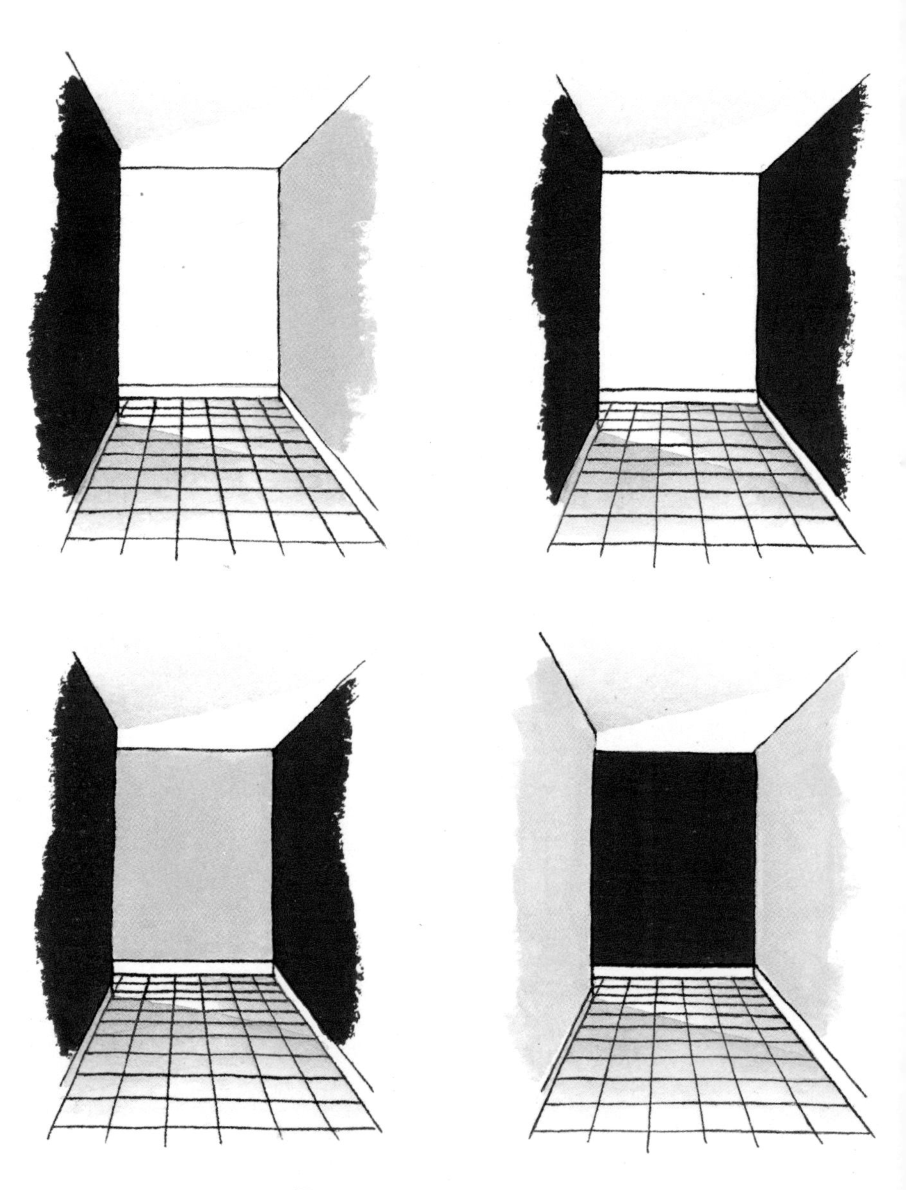

Tunnel effects in corridors

Courtesy: Crown Decorative Products Ltd

daylight falling on each wall. The most satisfactory effect is secured when one wall is in a fairly decided colour such as pink, blue, primrose, rust, etc., and the adjacent wall in a neutral tint as, for example, white, off-white, or a pale grey. When all four walls are evenly lit by means of concealed or diffused lighting, however, a more subtle colour change is then permissible between one wall and another.

In the schoolroom the decorator can employ these unorthodox schemes with every chance of success. Children respond to bright clean colour and desire it in large quantities. A sunlight yellow wall adjacent to white or off-white walls, with ceiling in pale blue, the woodwork generally in a pale grey and doors in bright blue or signal red — such a scheme provides a stimulating and happy atmosphere for children to work and play in. The chalkboard need not be in black but may be in very dark dull green, blue, brown or maroon. The wall space surrounding the chalkboard should also be a deep shade to avoid brightness-contrast between board and wall.

It is good to note that factory owners are now realizing the importance of colour in workshop and canteen. Clean wall tints of high light-reflective value induce high spirits and increase working efficiency. The colour of machinery need not be the usual depressing dark green or brown, since modern synthetic finishes and machine enamels in light colours stand up well to cleaning and hard wear. The high-lighting of working parts is now more widely used. For instance, the general framework of a machine may be in a mid-grey or grey-green with dangerous moving parts, start and stop levers, etc., picked out in orange to make them clear yet not too distracting; any parts in shadow or difficult to see are made more visible when picked out in light grey; and pulley guards and fences, etc., may be in a medium blue. Thus machinery can be made good to look at, while increasing the safety factor and general efficiency. In any factory scheme it is highly important that all colours used as symbols should be consistent; signal red, for instance, should be reserved exclusively for all fire appliances, fire buckets, fire doors and escapes, etc., so that these are instantly visible in an emergency and confusion is avoided. Similarly, first-aid cabinets may be a standard bright green (for safety), and pipelines, electric conduits, etc., may each have their distinctive identification colours as recommended by the British Standards Institution (BS 1710:1971, Identification of Pipelines, and BS 2929:1957, Safety Colours for Use in Industry). Colour may thus be used functionally while adding to the general decorative effect.

Public buildings

Bright strong colour is very stimulating and exciting, but the eye quickly tires of it and seeks relief when over-stimulated. Neutral grey and greyed hues are the most restful to the eyes. In places of entertainment, cafés, cocktail bars and similar places which are seen for comparatively short periods, we expect and enjoy more vigorous colour than we could tolerate in our home surroundings. Warm colours are generally preferred, with notes of sharp contrast introduced in the form of decorative motifs, drapery, seating or flooring. The principle of using paler or slightly greyed hues for the largest areas must still be observed, however, for large masses of strong primary colour tend to be overwhelming.

In municipal or civic halls, more restrained and dignified colour is called for, but even here the ceiling may carry rich strong colour, especially if it is fairly high and has elaborately ornamental plasterwork. Deeply coffered (sunk) panels, for instance, may be painted vermilion, or an intense blue, relieved with gold and touches of contrasting colour; the supporting stiles, mouldings and lower walls being in more subdued tints and greyed hues. Such a ceiling seen from the ground will glow with an intense richness which has something of a heraldic quality about it.

In places of worship, any rich colouring should generally be reserved for the focal point — the chancel or sanctuary — the rest being quiet and restrained. The frontispiece shows a scheme of this nature applied to a modern church interior.

Discordant colour

When the term *discord* is used, one instinctively thinks of music — atonal music, perhaps, for discords tend to jar; yet music, and life generally, would be unbalanced and incomplete without occasional discords. While subtle discords in music lend a note of piquancy which is pleasant to the ear, loud and prolonged discords are to most people unbearable. In the same way, discordant colour can be violently unpleasant in large amounts, yet very pleasing and even beautiful when used with restraint and discrimination.

The important difference between aural and visual discords is that, whereas the musician uses discords intentionally and the result is instantly recognized by the hearer, the painter and decorator generally uses discords quite unwittingly and he may find the result pleasing or displeasing without knowing why. This suggests that our musical sense is more developed than our knowledge or appreciation of colour.

We have previously noted that discord is produced when the natural order of colour is reversed. We shall therefore know what to expect, for instance, when we make orange or green lighter than yellow; crimson or vermilion lighter than orange; blue lighter than green, etc., and use them adjacently. We shall know why pale mauve or lilac (pale violet) is the most difficult of all colours to use as the basis or background of any scheme, throwing as it does all colours darker than the ground out of their natural tone order. Nevertheless, many beautiful schemes (e.g. Persian wall paintings and manuscripts) are based on pale violet or purple backgrounds, whilst pale blue is also a common ground in oriental decoration; they often have touches of sienna, dark blue and crimson, etc., with a liberal use of grey, black, white and gold which, being neutral, pulls the whole scheme together. These schemes tend to appear somewhat bizarre at first, but their beauty grows as one becomes familiar with them. That is probably true of many schemes based on discords — they have an unusual and subtle quality which may appeal only to a person of cultured or refined taste.

Discordant colour is most unpleasant when used in large areas; for example, pale blue or green walls with brown dado or woodwork (brown being a dark orange). The brown tends to look hot and the walls look pale and washy. Yet a *single note* of brown, such as a sepia print on a pale blue wall, can look most satisfactory. The decorator need not therefore avoid discords, but he should try to recognize them and use them always with great caution.

The limited palette

We have noted that, in theory, any hue can be produced by intermixture of the three pigmentary primaries — red, yellow and blue— but that because of the darkening or neutralizing effect of mixing pigments, the resultant hues are much less pure than pigments already available. It is this very neutralization, however, that is responsible for the unified and harmonious result when a strictly limited palette is used.

If a complete mural or decorative painting, for instance, is carried out solely with (say) lemon chrome, vermilion and ultramarine, using white to lighten the tone and black to darken (though the latter is rarely necessary, since near-black is obtained by mixing all three primaries), the various hues obtained by intermixture are all *related* since they have something in common — a little of each other in their make-up. Such a palette is capable of producing a fairly bright range of hues, but by choosing three duller primaries such as yellow ochre, light red (or Venetian) and ultramarine, a more subdued and subtle colour range is produced, which, if pitched fairly light in tone, gives some lovely "pearly" or opalescent tints, well suited to large-scale murals where quiet and dignified colour is desired.

One is not obliged to use three colours, of course. Two-colour combinations such as Prussian blue and raw or burnt umber, ultramarine and burnt sienna, etc., produce a fine range of related greys when intermixed and reduced with white. A fine range of monochrome tints and shades too may be obtained from almost any of the common pigments when mixed with black and white. Pigments such as Prussian blue, Italian ochre, raw umber, Indian red and deep Brunswick green are each suitable for monochromatic murals, giving an effect similar to the familiar two-colour half-tone magazine illustrations, in which a single colour is overprinted with black.

Incidentally, when white is added to certain colours the resulting tint inclines towards blue. This is probably most noticeable with bright reds which, on reduction, tend towards purple or violet. A little chrome yellow is needed to correct this, or alternatively, tints may be made with cream instead of white.

Colour and lighting

It has already been pointed out in the section on "Light and Pigment" that the sensation of colour cannot exist without light and that pigments can only reflect their true colour when illumined by light which contains their colour. Thus, if red pigment is illumined by white or red light it will reflect red; if, however, it is illumined only by blue-green light, there are no red rays to reflect and it will therefore appear black. For the same reason, violet seen under a strong yellow light and blue illumined by amber or deep orange lighting would both appear black or near-black, since any coloured surface illuminated by a light of contrasting colour will absorb all rays and reflect none. These extremes are not normally encountered in practice, except perhaps in stage lighting, where the principle is sometimes used to produce the illusion of quick-change effects.

It will be seen then that where coloured lighting is installed, as in cinemas, cafés and hotels, the colours for walls and furnishings require to be carefully chosen to avoid being negatived by the lighting. Such lighting normally contains a large proportion of white light, but nevertheless by reason of its colour content it will tend to neutralize any colours in the decorations in contrast to it. Thus a pale blue seen under amber lighting would tend towards russet or warm grey.

Ordinary electric light (tungsten filament) is rich in yellow and red rays (especially that from low wattage lamps) and we therefore find a corresponding deterioration in the purity of blue and violet tints in such light. Yellows are strengthened, though, curiously enough, they may appear paler if the surrounding colour is much debased by the same lighting. *Fluorescent* lighting is supplied in a variety of tints, each having different colour-rendering properties; "warm white", for example, has a decided pinky tint, while "daylight" fluorescent is a cold white which is said to differ very slightly, if at all, from natural daylight. There appears to be no standard as yet, however, and some "daylights" are distinctly bluish (particularly towards the end of their useful life) and this, of course, affects adversely any warm colours in the scheme. It is important to remember that, in addition to the neutralizing effect which occurs when the illuminant is in contrast to the colour of the decorations, there is also bound to be a loss in lighting efficiency since some of the light rays are being absorbed instead of being reflected.

Sufficient has been said to show that in designing any interior colour scheme, the decorator and the lighting engineer must work in close collaboration and that, wherever possible, the decorator should try out all colours under the form of lighting to be used.

37

Measuring and Estimating; Contracts; and Setting up in Business

MEASURING AND ESTIMATING

Most people engaged in painting and decorating will at some time in their lives toy with the idea of setting up in business on their own account. The attractions are obvious, not the least of them being the feeling that the man who runs his own business is reaping the rewards of his own efforts instead of expending his labours for the benefit of some other person. In a great many cases the thought is pursued no further, and there are many excellent craftsmen who are quite content to be employed by a good firm and who ask for nothing more than the intense satisfaction of doing a job of work to the best of their ability. And indeed, a man who knows instinctively that he is not cut out for the rough-and-tumble of commercial competition is happier working under somebody else's direction than he would be if harassed with the worries and anxieties of business life.

But, on the other hand, many decorators do eventually set up in business, and in fact one of the most striking features of the organization of the painting and decorating industry is the vast number of small firms, often consisting of only two or three men, operating on a small scale in the immediate vicinity of their own homes. Indeed, small firms of this kind form the greater part of the industry, and big firms are the exception rather than the rule. Clearly, then, a great many people are fired with this ambition and see it as a satisfying outlet for their abilities.

The keen student-apprentice soon discovers that in every college course he is required to devote part of his training to the subjects of measuring, calculating, and simple costing of work, and that these subjects are obligatory in certain examinations. Ambition, unless it becomes an overweening obsession, is a valuable spur to greater effort, and there is certainly nothing wrong in wanting to build up a reputation as the owner of a first-rate decorating business; after all, that is how every one of the firms already in existence was founded. In some cases, of course, the ambitious student sets himself a different goal, feeling that a highly responsible job in a supervisory capacity with one of the big firms operating on a nationwide basis will provide an exciting career and will give more scope to his energies than the slow process of building a firm from small beginnings. But whatever the student sets his sights on, whether it be a top management job, a business of his own, or perfection as a craftsman, he cannot ignore the economic factor which governs every piece of work he undertakes.

Principles of costing

By far the biggest single factor in the price of most decorating work is the cost of labour, which is considerably greater than the cost of materials. Whatever the size of the job, the customer generally want to know how much it is going to cost before deciding to have it done, and the usual practice is for three or four firms to be asked to submit an estimate, the work being given to the firm whose price is the lowest. It therefore follows that the ability to work out an accurate estimate of cost is a matter of vital importance to anyone concerned with running a business, whilst those engaged on carrying out the work will need to know how the cost was assessed so as to make sure that the job pays its way. It is also plain that the only way of arriving at an estimated cost is to know how long it will take to perform the various operations involved in the work. This calls for a sound working knowledge of all the craft processes, and can only be gained by long experience. The snag is that at the point when they first launch themselves into the business world most decorators have no experience at all of these things, with the result that for a long time they are working on a trial-and-error basis, their prices being nothing more than guesswork. Starting a business and discovering rightaway that it involves working out estimates of prices is tantamount to being thrown into the swimming bath at the deep end before being able to swim. Some may survive, but many don't; no wonder so many aspiring master painters meet with disaster very quickly.

The question that arises at once is, "How do I gain this experience when I've never had the chance of observing office procedure or taking part in the administration of the firm I work for?" The answer is that experience can be gained very easily, that everybody can become skilled and proficient at costing, and that the best time to start acquiring this knowledge is whilst still working as an apprentice and student of the craft. The opportunity is at hand every day and on every job.

Experimental costing

The first requisite is a small notebook with hard covers, small enough to fit in the pocket, and the other essential items are a watch and a metric rule. Suppose the student is given the job of coating a ceiling with emulsion paint, using a brush. It takes only a couple of minutes to measure the length and breadth of the ceiling and work out the area. A glance at the watch when work begins and again when it ends discloses how long it took to paint that area. So an entry is made in the book: the page is clearly headed "Emulsion paint, brush applied" and the area and time are entered, together with a calculation of how many square metres would be applied in one hour at this rate. A further note is made about the condition of the ceiling, i.e. whether absorbent or non-porous; and for further information the time spent on unproductive work such as sheeting up, etc., is added.

Now a single entry is not sufficient, because no two jobs are ever identical. But when five or six entries have been made about the brush application of emulsion paint it is possible to work out an average time in square metres per hour, and this is the basis upon which such work is estimated. Thus, if the average time over several jobs is 10 square metres per hour, it is easy to work out from the hourly rate of pay how much money it costs to apply one square metre.

The student should make a practice of doing this with every job he is given. In this way he will build up a personal price-list for applying emulsion paint by brush, by roller, and by spray on all sorts of surfaces. Similarly he will compile information about the cost of applying priming paints, undercoats and gloss finishes to woodwork and metals, how long it takes to cut in windows, how much it costs to paint downspouts of various diameters, and so on until eventually he has a complete dossier covering all the operations he has ever performed. Apart from anything else, the operation will add a lot of interest to his work, and he will make all kinds of observations; he will notice, for example the difference between applying the second coating on new woodwork, with all that it involves in the way of stopping up nail holes, joints, etc., and the straightforward application of the third coat, the difference in time between applying undercoats and gloss finish to the same surface, and so on. But more importantly when the day comes when he needs to submit an estimate to a customer he will be able to do so intelligently and with confidence, instead of working blind.

Essentials of pricing

When an actual job is being costed, the average figure for each operation is used as the starting point and is adjusted upwards or downwards according to the conditions pertaining to this particular job. Suppose, for instance, that the average figure for applying emulsion paint by brush is 10 square metres per hour, but that the actual job consists of several small areas each needing to be cut in separately and involving several changes of colour; the speed of application on that particular job would fall well below the average figure and the cost would rise correspondingly higher. In every case, the average figure needs to be examined to take account of the nature of the surface, the amount of preparation required, and the amount of unproductive work involved.

Another important point to be remembered is that the average production figure deduced by timing the operation over a measured area will generally be higher than what can be achieved in the normal working day. There are several reasons for this. In the first place, working over a measured area generally takes no account of interruptions. But nobody works at full pressure and maximum pace for the whole of an 8-hour shift; sooner or later the operative tires and the pace flags. Then again, there are all sorts of stoppages to be taken into consideration—the morning and afternoon tea breaks to which the men are justly entitled, the odd pause now and again to light a cigarette, to exchange a few remarks with a workmate about some item in the news, or to discuss some aspect of the job, the occasional visit to the toilet, the pause when the customer comes in to inspect the work, and so on, not to mention the fact that the working day itself is rather less than is entered on the timesheet. It is to be hoped that all the men will be wearing their overalls and ready to start at 8 o'clock in the morning, but there is generally a few minutes' delay while materials are stirred and poured out and while the foreman allocates the work; in the same way, if a section of work is completed 15 to 20 minutes before the mid-day meal break very few operatives will rush to start another job. These are all points that the foreman on the site has to watch in order to keep production up to the required level, but it would be foolish for the estimator to ignore them because they are the facts of life, and if they are not taken into account the estimated time will be too optimistic and the price too low.

It will be noticed that the emphasis has been upon deciding the price by measuring the work — and this is the only satisfactory basis for pricing. Costing by guesswork is futile. Admittedly a decorator after many years of experience on one class of work acquires the knack of knowing roughly how long the normal piece of work will take simply by looking it over; the man who habitually works in small domestic properties can probably make a fairly close guess about the time it will take to paint the walls of a kitchen, for example. But it is only a guess, and if he makes a mistake it can prove to be expensive; what is more, it can be misleading. A man whose prices are consistently too high will not get much work because his competitors will undercut him, and a dwindling supply of work spells disaster; similarly if his pricing is too low he will certainly succeed in getting work but he will never make a satisfactory margin of profit and he may in fact lose money. Both courses are equally bad because they both lead to the bankruptcy court. The fact that many decorators do muddle along by guesswork is exposed by the disparity in the prices submitted for competitive tenders; it is not uncommon to see a range of prices extending over hundreds of pounds with the top price more than double the lowest tender; if all the pricing were properly done there would be very little between them. A useful lesson can be learned from the really big firms, especially those engaged on industrial work where the competition is very keen. All the really important firms insist upon accurate estimating, and the practice invariably is to measure every single item and price it out correctly.

Measuring up

In the actual physical act of measuring up painting work, the techniques of the professional estimator employed by the big industrial firms are worth studying. As already indicated, all the work is measured, and this means accurate measuring with a tape measure and not some slipshod "quick way". Some decorators claim to measure their work when actually they are content with just striding out the length and breadth of the site on the assumption that each pace represents one yard. In fact very few people have a pace exactly a yard long; generally it is considerably lower, and since in any case a metre is larger than a yard the method is woefully inaccurate.

Generally the practice at the big firms is to have two men working as a team; they hold a 20-metre tape-measure between them and one man reads off the measurement and calls it out aloud while the other man writes the figure down. But because they are so skilled at the work they realize very clearly how easy it is for mistakes to creep in, and they double-check every figure to reduce the possibility of error. The man with the book, for example, can easily write down a wrong figure because he has not heard it correctly — this is especially likely to happen when noisy areas such as factory buildings are being measured. To guard against this, he repeats aloud the figure that he thinks the other man has called so that any error can be detected and put right. Then again, to guard against a possible misreading of the figures, the man who is reeling in the tape after taking the measurement is pacing out the line along which the tape was placed; he has trained himself to know the length of his pace and by a quick calculation can check on the accuracy of the figure. He also looks for any repetitive items such as stanchions situated at regular intervals along the building, and uses the distance between them, multiplied by their total number, to verify his overall measurements. A final check is made when all the measurements have been taken.

Both the estimators, having now seen the whole of the site, mentally review the extent of the work and decide on a "spot price" or an approximate figure that they think would cover the job. If when their measurements and prices are finally worked out there is a big disparity between them and the spot price, it could indicate that a serious error has been made and that a significant group of measurements has either been omitted or duplicated to produce a wrong result.

Booking in; progress reports

The measurements written down in the book are compiled in an orderly manner. It is no use covering page after page with a jumbled mass of figures that cannot be interpreted when the estimators get back to their office. The proper method is to put a clearly written heading for each section of the work and to make sure that all the measurements are entered in the appropriate place — thus there might, for example, be a group of pages headed "Workshop No. 6 (Scrap Metal Reclamation Bay)" with sub-headings to cover the ceiling, roof trusses, wall areas, structural steelwork, cranes and gantries, woodwork, etc., and with all the measurements for these various sections entered in their proper place. This enables the estimators to visualize each part of the work when they are back in the office calculating the prices. It is also of the utmost value if and when the firm succeeds in getting the contract, because this method enables the supervisor whenever he visits the site to check off the amount of work completed and compile a progress report.

The importance of such progress reports cannot be too strongly emphasized. Every time the site is visited the extent of the completed work should be compared with the money spent. Without this, it is only too easy to reach the point when practically the whole of the estimated money has been spent and the work is still substantially unfinished. When this discovery is made it is too late to arrest the fault; the job has run into debt, and every day the work continues increases the extent of the loss. The object of the regular progress report is to reveal any divergence between the estimated cost and the actual cost as soon as it begins to appear, so that the reason can be investigated and the leakage stopped.

The work having been measured up accurately, the estimator (who knows the spreading capacity of the various materials specified) is able to work out the quantity of material required and, from this, the cost. Usually the material costs relate to the labour costs roughly in the proportion of one to three, but occasionally the ratio is different. An example of this occurs where a considerable area of work is to be washed down — e.g. where exterior wood cladding which is already finished in clear varnish is to be washed and given a further single coat of varnish for maintenance purposes; most of the time is spent in performing a manual operation in which the material costs are practically nil, and on the complete job the material cost might be less than one-sixth of the labour cost. On the other hand, hanging an extremely expensive paper would present the reverse situation; the actual time spent in hanging the paper would not be markedly greater than normal operations of this kind, and the cost of the paper might be higher than the labour cost.

Overheads

At this stage the actual cost of the labour and materials has been assessed, but there is another significant item to be taken into account, the importance and extent of which is not generally appreciated by the student or the operative painter, and this is the matter of "overhead expenses" — often known nowadays as "establishment charges". These are the costs necessarily incurred in running the firm but which are not directly chargeable to any particular job; they are the essential items of expenditure without which the firm could not operate and which still continue whether the firm is busy or slack. They include the salaries of the office and supervisory staff, the upkeep of the premises (i.e. rent, rates, lighting, heating, cleaning, etc. of the office, workshop, stores and scaffold yard), the cost and maintenance of the firm's transport fleet (cars, vans and other vehicles, together with the drivers' wages), the cost of plant and equipment, the cost of office equipment, telephone charges, printing and stationery, insurances, advertising — it is a formidable list.

Obviously this expenditure has got to be paid for somehow out of the earnings of the firm, and this can only be done by charging up some of the cost to each of the jobs on which operatives are working. But of course it would be foolish to expect a small contract priced at, say, £100 to pay the same amount towards the overhead expenses as a big contract priced at £50,000; the only reasonable method is to make the charge proportional to the size of the job. How, then, is the actual proportion determined? The general practice is to total up the whole of the establishment charges over a complete year and express them as a percentage of the firm's cash turnover during the same period; during the ensuing year, whenever an estimate is submitted, that same percentage is added to the estimated cost of the job. There are various ways of doing this; a percentage may be added to the estimated prime labour cost, or to the combined cost of labour and materials, etc.; the details are worked out to suit individual needs. A careful watch has to be maintained to see that the overhead expenses are trimmed down to essentials and not allowed to get out of hand, and it is worth noting that the more successful a firm's trading operations are, the lower the overheads become by proportion.

Suppose, for example, that a firm's overhead expenses totalled £10,000 during a year; if the firm has a very successful year's trading and its turnover is £100,000 the overheads represent 10 percent of the costs, whereas if the firm finds difficulty in securing contracts and has to cut down its labour force, as a result of which its turnover is only £30,000, the overheads represent a ratio of $33\frac{1}{3}$ percent. This is, of course, an over-simplification, but it remains true that a firm which is prospering usually has proportionately lower overheads than a firm which is going through a difficult patch, and this means that the successful firm can submit a lower quotation when asked to tender for a new contract, with correspondingly higher chances of securing the contract. Success engenders success.

Profit margin

We have now considered the labour, materials and overhead expenses, but when these three are totalled together the estimate still only covers the money which will actually be spent while the work is in progress. A fourth item must be added, and that is the profit which is basically the reason for the firm's being in existence. Some people have the quaint idea that the profit made by a firm is in some way rather dishonest, that it represents money pocketed by a greedy and unscrupulous employer which ought rightly to be shared out between the operatives. Without entering into a discussion about the moral issues, it needs to be pointed out that the profits are not automatically handed over to the firm's principals; a reasonable proportion of the profit is generally ploughed back into the business to replace old stock and to strengthen the firm's trading position and open the way for expansion and growth. The precise amount of profit margin allowed for in an estimate is a private matter decided by the firm's directors as advised by their accountant, in the light of economic conditions at the time. The point to notice here is that sometimes a firm may elect for various reasons to reduce its profit margin. It may be that the directors have their eyes on some wealthy person or some big organization that has never traded with them before and is generally known to be a good customer; by submitting a very competitive tender at the price of cutting their profits to the bone they hope to win the customer. Or it may be that they want to get a contract for decorating some important building in a prominent position because of the prestige it will bring them. Again, it may be there is a general shortage of work and the firm is willing to undertake a contract without making a profit rather than have its workforce dispersed. There are occasions when such a course is justified, but it would be fatal to make a regular practice of it.

CONTRACTS

Types of contract

When a decorator submits an estimate to a prospective customer, it is clearly with the intention of entering into a contract with him. It is hoped that the contract, if agreed, will prove to be mutually beneficial, the customer obtaining a completely satisfactory piece of work and the decorator receiving the due reward for his skill.

A contract is any agreement which the law recognizes as binding, i.e. which the courts will enforce by penalizing a person who breaks it by awarding damages against him.

As a general rule, a contract requires no special form, and can be created by spoken words alone (or even without words, as in the case of items purchased in a self-service store). Some categories of contract, such as for instance those concerned with sale of land, *are* required to be written, and most commercial contracts of any importance are reduced to writing as a matter of practice. Even so, a bare exchange of quotation and order form is enough to create a binding agreement.

There are two main classes of contract relating to construction work, including painting operations, namely (a) *the fixed-price contract*, in which the contract price is fixed in advance, subject to variations and adjustments according to whether or not the specification is strictly adhered to, and (b) *the cost-reimbursement contract*, in which the price to be paid may, at the time the contract is entered into, be left to be determined on the basis of the actual cost incurred by the decorator plus an agreed amount to cover overheads and profits.

There are several recognized forms of fixed-price contracts, including the following.

(i) *The Lump Sum Contract*

This is the simplest form and probably the form under which the majority of painting and decorating work is carried out. The fixed price is a lump sum or overall figure for the complete job, the sum being determined by the decorator on whatever basis he thinks fit, and there is nothing in the agreement to show how the price quoted was arrived at. Sometimes the extent of the work covered by the agreement is very vague. The contract itself may be merely a verbal agreement, or if it is in writing the description of the work may be so brief and imprecise as to be practically meaningless. Sometimes the specification for the work is compiled by the decorator himself after he has visited and inspected the premises, his quotation being based on his own assessment of what is required; or there may be a specification drawn up by the householder or by some other party acting on his behalf. But there are also a vast number of contracts placed in this manner, including some that are on a very large scale indeed, covering every conceivable type of structure — civic buildings, public halls, shops, hotels, business premises, churches, hospitals, etc., where a detailed specification is drawn up by the customer's professional employees or the customer's agents.

It is to everybody's benefit if a detailed specification is used, laying down precisely the extent and the terms of the work to be done, because this means that all the firms tendering for the work know exactly what is required and are tendering on an equal basis. Confusion arises when, as in the case of small domestic work, the specification is vague or non-existent because this leaves the way open for the unscrupulous firm to undercut the reputable decorator and produce a shoddy job that is quite unsatisfactory from the customer's point of view. The same stricture applies when each decorator tendering for the work compiles his own specification; the householder is unable to distinguish between what is a good or a bad specification and only looks at the lump sum price, which again puts the reputable firm at a disadvantage compared with the dubious character.

(ii) *The Bill of Quantities Contract*

This is used on large and important jobs and on works of new construction, where generally the painter and decorator is acting as a sub-contractor. The bill of quantities is drawn up by the quantity surveyor and it details fully and accurately every single item in the entire structure; it is divided into columns in which the dimensions of each item and the number of those items is indicated. The contractor sets down by each item a price (which for the decorator is the price per square metre or linear metre) covering the cost of labour, materials, overhead expenses and profit, and in the last column this is multiplied by the number of such items to give a price for the total number. A bill of quantities is a contract document by which the contractor is fully bound.

The bill of quantities is an important part of the system in construction work, and has a four-fold purpose in that system:

(a) It ensures that every contractor is basing his prices on exactly the same information.

(b) It provides a basis for variations which may occur during the progress of the work.

(c) It gives an itemized list of everything in the structure and this enables the contractor to order his materials and assess his labour requirements.

(d) After being priced it provides a basis for "cost planning" and "cost analysis".

It is generally a very thick, voluminous document with thousands of entries, all to be separately priced,but since painting and decorating occupies only a minor part in new construction work the painting section is usually quite small, consisting of only a few pages. The system works in favour of the big contractor in one of the constructional trades, who is dealing with bulky materials and is thus able to put a realistic price beside each item; it is of less advantage to the decorator who is in something of a dilemma in trying to assess a price covering labour, material, overheads and profit on so small a unit as a square metre of paintwork. On the other hand, there are distinct advantages when for any reason the work is varied from the original specification; there is no ground for wrangling and argument, because the price both of the material deleted and the material added is set down precisely in the bill of quantities.

(iii) *The Schedule Contract*

This is a method sometimes used for work let out by big public undertakings or by Government Departments. The work is itemized, but instead of the contractor working out his own price, each item is priced at rates supplied in a schedule of prices. The schedule may have been specially prepared for the particular job in hand, or it may be a printed schedule published by the Department concerned. The sum paid to the contractor is calculated when the work is completed by applying the given prices to the actual quantity of work done. This type of contract is frequently used to enable the "customer" to put a piece of work in hand promptly before full particulars have been worked out and therefore before a bill of quantities could be prepared.

While the various types of contract referred to in these paragraphs are termed "fixed-price contracts", there is at present so much doubt and uncertainty about material costs, escalating labour rates, and rapidly changing lodging allowances, travelling costs and similar contingencies that contractors are protected by what are known as "rise and fall clauses". Under these they are recompensed for any increases, but the adjustments are usually net and do not affect the profit margin.

In the same way, there are various forms of *cost reimbursement contracts*, including —

(a) *The Cost Plus Percentage Contract* , usually called the "Cost plus". Here the customer and the decorator agree beforehand on a percentage figure to cover overhead expenses and profit, the decorator being paid on the completion of the work a sum equivalent to the wages he has paid out and the materials he has bought plus the agreed percentage. This is often used in emergency situations because it is the quickest way of arriving at an agreement and avoids any delay in commencing operations.

(b) *The Cost and Fixed Fee Contract* Under this system a fixed lump sum based on an estimate of the cost is agreed between the architect, engineer or other appointed agent of the customer, and the contractor. The contractor receives this sum whatever the ultimate cost may be, subject to any agreed variations, and when the work is finished the actual cost is checked against the estimated cost. This system is not often applied to decorating operations but is sometimes used in industrial painting. The advantage of the system is that if it is properly applied it serves both the contractor's and the customer's interests; the contractor's remuneration is not affected and his sole concern is to produce a good piece of work without undue delay and with the knowledge that if he satisfies the customer it enhances his prestige and advances his prospects of future work.

(c) *The Value Cost Contract* Here the contractor's remuneration is calculated as a percentage of a careful valuation of the work actually completed, made on the basis of an agreed schedule of prices. If the final cost is below the valuation the contractor's remuneration is increased and vice versa, so that the contractor has a definite financial incentive to carry out the work efficiently and economically.

Some terms used in contract work

Standard Method of Measurement A document issued jointly by the Royal Institute of Chartered Surveyors and the National Federation of Building Trades Employers to provide a uniform basis for measuring building work generally, including painting and decorating.

Extra-over items These are certain types of processes which involve additional cost over and above normal working practice; for example, paintwork finished by stippling.

Deemed-to-be-included items Work which is implicit and does not need to be specifically mentioned in a job description; for example, the use of the duster brush before applying paint.

Prime cost sum A sum of money to be provided in a bill of quantities for work or services to be executed by a nominated sub-contractor, or for materials or goods to be obtained from a nominated supplier. The contractor is entitled to add a percentage for profit on such items.

Nominated sub-contractor A person or firm specified (usually by an architect) to supply and fix materials or execute a certain part of the work under a prime cost sum; for example, the architect may nominate a decorative artist to carry out mural painting on a site.

Nominated supplier A person or firm nominated (usually by an architect) to supply goods under a prime cost sum, the goods to be fixed or applied by the contractor.

Provisional sum A sum of money provided in a bill of quantities for work or costs that cannot be foreseen; for example, extra work which was not expected and which was revealed when some material has been stripped from a surface.

Interim payment A payment made at some stage of a lengthy contract to cover the cost of work already completed. The extent of the work qualifying for such payment is usually determined by the quantity surveyor.

Working Rules — formerly known as the Working Rule Agreement

These form a document drawn up jointly by the employers' and operatives' associations covering the conditions of employment, i.e. wage rates, overtime rates, rules for the employment of apprentices with provision for their technical education, the allowances for night work, overtime, working at heights, travelling time, etc., procedures for terminating employment and for dealing with disputes, and all other matters relating to working conditions. As well as the national working rules there are regional working rule agreements with variations applicable in certain localized areas. Revisions and amendments to the rules are issued in printed forms at various times during the year.

SETTING UP IN BUSINESS

Costing and estimating are an important part of the day's work to the supervisor, the contracts manager and the master painter, but there are other aspects of management to which attention should be directed. There are, of course, scores of books on business management, not to mention management courses in every technical college and polytechnic in the land, but none of these deal with the subject from the point of view of the painter and decorator. In particular the man who is thinking of setting up his own business is very badly served. As far as we are aware, there is nothing available which tells him the simple things that he wants to know, or which gives him some idea of what to expect. Yet this is the stage at which he is most in need of some advice. His only recourse is to learn the hard way, by trial and error, hoping that he will be able to hold his head above water till he has grasped the fundamentals. But our experience suggests that it is not just a simple matter of learning once for all how to cope with the problems; our belief is that the pattern is constantly changing, and that at various stages in his career he will be facing new situations and having to make decisions without anything to guide him in his choice.

There is no difficulty about starting — in fact it is extremely easy, so much so as to produce a false sense of security. In most cases, what happens is that the young man has been thinking about taking the step for some time and has talked it over with his friends and relatives, who all agree that it is a good idea. Some of them have gone further, and told him that they themselves need some decorating done, and that to launch him in his new career they will let him do the work for them. So there is no shortage of work, and there is any amount of goodwill. He needs little in the way of equipment, he doesn't require any sort of premises, he can work from his own address, and the whole thing is just a change of scene; he goes out each morning just as if he were going to work in the ordinary way, and at the end of the day his time is his own. As yet there is no hint of the extra responsibilities entailed. Most important of all, as soon as a job is finished he is paid, in cash, the full amount that he asks. This is natural because all these people are well-wishers. As one job finishes, there is another waiting to start. And provided he is working from his home address and trading under his own name, he does not need to register the business.

Gradually, however, the supply of ready-made jobs begins to dry up. As more and more of his friends' and relatives' work is completed, what appeared to be an endless source of work recedes into the background. Now he has to start looking round for new customers; now he has to spend two or three evenings a week going around to see prospective clients, and for the first time he finds himself in competition with others. Not all of the people he sees give their work to him. Of course, there is nothing to worry about yet. His friends will by now be recommending him to other people. Or he can advertise in the local newspaper. But now he is working a fresh seam; his customers are people whom he does not know personally. They are not so ready to overlook his shortcomings as his own people; they are more demanding. If he needs to go off during the daytime to see another prospective customer they resent the fact that he has been away for part of the day. Most important of all, when the job is finished they do not pass him ready cash; instead they say "Send me your bill". This is when he discovers that people don't always pay up promptly and that some don't pay up at all. By the time he has done a string of jobs and several weeks have gone by without any money coming in he is feeling the pinch. Unless he has some money in hand to meet this eventuality he can face real hardship and may indeed have to give up the attempt. The business is dead, or to be more precise it was stillborn. Lesson No. 1 — you need some capital before you start in business. Anybody starting on his own needs to understand that the danger period is not right away but when a few months have gone by. Some people, not aware of this, get a false picture of the situation and commit themselves to buying vans, cars and expensive equipment in the first few weeks, and when the bubble bursts they find themselves hopelessly in debt.

Planning for expansion

Assuming that the young man survives this initial period, the next decision he faces is whether or not he is going to expand the business. He may decide that he wants to work single-handed, a course which has much to commend it. He will build up his business on his own reputation, and not have the worry of engaging labour and possibly having unsatisfactory workers. He will not need expensive premises. Admittedly he will not make a fortune, but if he is especially good at some branch of the trade he may become a specialist in this field and carry out graining, signwriting or paperhanging commissions for other people, and this will bring handsome profits. But what he needs to recognize is that if he falls ill or meets with an accident his business comes to a standstill, which can be disastrous. He needs also to look to the future; working alone is fine while he is young and active, but it becomes progressively harder as he grows older. He may realize this and feel that the lack of capital is holding him back, in which case one solution might be to enter into partnership with some other person of like mind. This again is bound to be a gamble; it may work out very successfully or it may turn out to be an unhappy relationship, and there is no means of knowing beforehand which way it will go.

So eventually another decision is needed: is he going to engage two or three more men? If he does, the actual physical work becomes easier with more people on hand to haul the ladders about. He will now have a certain amount of paperwork to do and records to keep, but his firm will still be small enough to operate from home, and he will still be able to work full time on the job himself, thus keeping an eye on the quality of the work.

Once expansion begins, however, he will get the offer of bigger jobs, and will soon realize that with a few more men he will be able to tackle these things more easily. The decoration of a clubroom or small housing estate could be done in a reasonable length of time, whereas with only two or three men it was impossible. Clearly this will lead to greater profits; against which he needs to review the consequences. If he has more than one job running simultaneously he can only work on one of them himself, and in fact will not be able to work on any job for the whole of the day. Part of the time will have to be spent supervising the other jobs and going to the suppliers for materials. He will also need a constant flow of work, or else he will never be able to keep his labour force. His home premises may no longer be adequate. He will need somebody to man the telephone, to take messages from customers (although this is no longer an insuperable difficulty because he could make use of the telephone answering service). Certainly the amount of paperwork will increase considerably, and since he is more likely to be skilled in decorating than in office procedure, he may decide to engage someone to deal with it. Here is a further decision to be made: does he engage an office worker who is looking for a part-time job while her children are at school and who only wants to work in the middle of the day, or does he engage a full-time secretarial worker to take complete charge but who would obviously need much higher pay?

Suppose that having weighed up the possibilities he decides that a policy of expansion will lead to greater prosperity. As the firm continues to expand, the point is reached when he can no longer afford to spend any part of the day in actual manual work but will need to devote the whole time to supervision and organization. How many men can he control before this point is reached? Probably by the time he has ten or twelve men working for him he will be feeling the strain. But if the firm's operations have to cover the cost of a full-time supervisor whose work is of a non-productive nature, he is definitely going to need a fair amount of business acumen. Many people are perfectly capable of controlling a fairly small labour force, but are quite out of their depth with anything larger. What is more, very few people starting in business are farsighted enough to look ahead to this situation, and they never develop a business organization that can cope with it properly. A makeshift procedure of booking materials and reckoning up timesheets may have worked quite well when the firm was tiny, but it is no use expecting it to work efficiently when the volume of responsibility increases. If the decision to expand the firm is made, it must be accompanied by a realization that all the work both in the office as well as on the job will have to be very carefully planned. This is the point at which expert advice is needed, with the financial side in the hands of an accountant.

Ancillary services

As the firm continues to grow the ancillary services must expand as well. Good premises of adequate size are needed, as described in the early chapters of the book. A well-trained and efficient office staff will be required, equipped with all the modern adjuncts to business — electric typewriters, copying machines, a reliable filing system, etc. This however is outside the scope of the present book, and there are plenty of firms specializing in office furniture and offering an advisory service about office procedures. But this does not mean that once a firm has grown in size the details are no longer important. On the contrary, the wider the range of the firm's operations, the more essential it is that the master painter should pay close attention to detail. Take, for example, the matter of stationery. This needs to be attractive and well designed. The man controlling a biggish firm can no longer rely on personal contacts. His stationery, his letterheads, and the way the letters are both phrased and typed, are his introduction to a prospective customer, and they are the only thing on which the customer can judge the firm. It is of no use to say that it is the quality of the firm's workmanship that matters; if the first impression created by dingy notepaper and slipshod presentation is a bad one, the firm never gets as far as first base. This applies in all sorts of ways, and especially in the matter of courtesy. If a customer makes an inquiry, it may arrive when everybody is too busy to give it immediate attention. To put it aside for a few days until there is time to deal with it is not good enough, because by that time the customer will have assumed that the firm is not interested and will have taken his work elsewhere. Every letter, every telephone inquiry should be dealt with at once, and if there is no time for a proper answer it is a simple matter to send a printed card saying that the message has been received and is having attention.

Care is needed, too, to see that the expansion of the firm takes place over a sufficiently wide range of work. It may be that much of the work comes from one direction, that the firm gets the monopoly of the work in its own particular area from a government department, one large establishment, or a chain of hotels, and that without anyone realizing it the main effort of the firm becomes geared to dealing with this one big customer, simply because it proves so lucrative. It can happen — and it has on many occasions happened — that for some reason there is a change of policy in the customer firm, or the customer is taken over by some other organization, and suddenly the decorating firm that was doing uncommonly well finds itself floundering with a large staff and no work. When this happens it is too late to go running around trying to woo old customers that have been neglected for years.

The ultimate stage in a firm's growth is reached, of course, when it has a nationwide reputation in some specialized field. The particular area of speciality depends to a great extent on the personal taste of the principal. There are many large firms which concentrate solely on industrial painting; there are others, like the well-known firm of Campbell, Smith & Co., that have built up a reputation for first-class craftsmanship chiefly in the field of church decoration, but taking in also the decoration of palaces and historic buildings. Here again is an area of decision, because at some stage in a firm's development the choice has to be made: the logical thing is to see that the choice is not haphazard but is based on a critical analysis of what kind of work yields the best returns for that particular firm.

Appendices

(1) HEALTH AND WELFARE

The Health and Safety at Work, etc. Act, 1974, which came into force on 1st April, 1975, is probably the most important statute for industry and commerce ever to have been passed by Parliament. The terms of this Act govern the activities of every company and firm and every individual engaged in any kind of work whatsoever. It affects all employers and employees at every level. Directors, officers, and managers of companies are personally responsible if a breach of the Act occurs as a result of their neglect or through their consent or connivance, and any employee who fails to co-operate with the safety arrangements is liable to prosecution. This is a fact which is not generally recognized; most workpeople are under the impression that if any of the statutory regulations are broken the responsibility rests with their employer, but this is not so and they in fact are equally liable. The Act specifically states that employees have a duty to take reasonable care to avoid injury both to themselves and to others by reason of their work activities, and they must co-operate with employers and others in meeting the statutory requirements. The penalties for a breach of the Act are very severe, amounting in some cases to two years' imprisonment and/or a fine which is unlimited in amount.

It is the duty of every employer to draft a written safety policy code covering the activities of the firm insofar as they affect the safety and welfare both of the employees and of the general public, and every employee should receive a copy of this document. Where the size of the firm permits, a safety committee should be set up, and the employees should be represented on that committee.

Accidents and their prevention

This is of course a matter of the greatest importance to the painter and decorator. The construction industry generally has a very poor record of safety, and within the industry those occupations that are classed as maintenance trades (which include painting and decorating) have a particularly black record of accidents. There are various reasons why this should be so. The industry is dispersed over many thousands of temporary sites, the labour force is highly mobile, and some of the work is inherently dangerous; it is obvious, for example, that a moment's thoughtlessness when working on a lofty scaffold can have serious consequences. But against this must be set the fact that practically all industrial accidents could be avoided if proper precautions were taken. A good motto for any painter to remember is that "accidents don't just happen, they are *caused*". People talk about accidents in a fatalistic manner as though they are simply bound to occur, like bolts from the blue, but actually no accident ever takes place unless there are unsafe conditions on the site or unsafe actions being carried out. Accidents can be prevented by learning what the hazards are, and then making sure that they are eliminated. The commonest accidents of all in our branch of industry are those concerning the use of ladders, yet one rarely sees a ladder in use that conforms with the regulations —either there are rungs loose or missing, or the ladder is not lashed or footed to prevent it from slipping, or it is resting on a pile of loose bricks, tiles or other brittle material, or it is not long enough for the work in hand. All these faults can so easily be checked, and if this were done the accident rate would fall dramatically. A strict code of conduct is laid down in the Construction Regulations, the observance of which would considerably reduce the incidence of accidents.

It is probable that at some time in the future, more safety legislation will be introduced under the terms of the Health and Safety at Work, etc. Act, but at the present time all sections of the industry, including painting and decorating, are bound by the provisions of the Construction (Working Places) Regulations and the Construction (Health and Welfare) Regulations, both passed in 1966. It is the personal concern of everybody working in the industry in any capacity whatsoever to see that the regulations are observed, not only because they are the law but because nobody with any common sense would disregard the measures that exist to protect him from injury.

Insurance

Every employer has certain legal obligations both in regard to workmen's compensation and to third party risks. He is legally required to take out insurance which will provide for the compensation of employees who meet with any accident or mishap, under the terms of the Employers' Liability (Compulsory Insurance) Act, 1969, and quite apart from the legal obligation it is only common sense to insure in this way, because unless it is found by the court that the employee had contributed to the accident by his own negligence the damages awarded against the firm could be crippling. He is also required to insure against any mishap or injury that some member of the general public might suffer. This is to protect the interests of any bystander, or the employees of some other firm, or the occupier of the premises at which the work is taking place, from the results of any activity carried on by the decorating firm.

(2) LEAD PAINTS

The general health of the painter depends primarily on personal hygiene and clean habits but, in addition to applying ordinary common sense in these matters, he is required by law to conform to certain government regulations with which he is expected to be familiar. The incidence of lead poisoning among painters was deemed to be due mainly to the inhalation of lead dust in rubbing down dry lead-coated surfaces or, in a lesser degree, to mixing paint from dry lead pigments; with the passing of the Lead Paint (Protection against Poisoning) Act in 1926, both these practices were prohibited.

A printed copy of the Regulations (obtained from any branch of Her Majesty's Stationery Office) must be exhibited in every painter's workshop or store. The Act requires that every employer must (a) register his name and address with the District Inspector of Factories; (b) keep a register of all persons employed in painting buildings; (c) keep a record of all painting contracts or jobs; (d) notify any cases of lead poisoning, and (e) give to each painter on engagement, and on the first pay day in each year, a copy of the Home Office health leaflet. The essential requirements of the Act are briefly as follow —

(1) Dry rubbing or scraping of lead paint is absolutely prohibited. This means that all rubbing down must be done wet with pumice stone or waterproof abrasive paper, unless the employer or the workman can *prove* that the painted surface does not contain lead. It is not a simple matter to ascertain whether a painted surface contains lead, either in the top coat or in the undercoatings, and the safest plan, therefore, is to rub down all surfaces wet.

(2) Spraying of lead paint is prohibited for *interior* painting only. Nevertheless, in spraying lead paints outdoors, the operative will, in his own interest, wear a mask.

(3) Lead pigments must not be used in dry powder form, *except* where red lead is used in making small quantities of hand-mixed stopping or filling composition.

(4) Washing facilities, i.e. water, soap, nail brushes and towels, must be provided by the employer and used by the workman (a) before meals and (b) before leaving the job. Overalls must be worn during the working period and *removed during meals*; they must be washed at least once a week. Outside clothing must be protected from contamination by lead dust.

The obligation to observe the respective parts of the Regulations rests upon employer and workman alike, and anyone contravening the Regulations is liable to penalty. The incidence of lead poisoning ("painter's colic") among house painters is now practically non-existent and this fact alone is ample justification for legislation which, at the time of inception, was considered extremely irksome. Quite apart from safeguarding health, however, wet rubbing down with waterproof abrasive paper is much more efficient than dry sandpapering, producing a smoother surface in less time. The chief danger is that surfaces rubbed down wet may not be allowed sufficient time to dry thoroughly before paint is applied, thereby trapping moisture beneath the film. Where necessary, old paintwork may be rubbed down with white spirit (or a mixture of one part raw linseed oil and three parts white spirit) as lubricant, using waterproof paper or steel wool as abrasive; this is particularly useful on steel surfaces where rubbing down with water may lead to trouble from rusting. It should be mentioned, however, that some people are extremely sensitive to the effects of spirit solvents on the skin and may develop *dermatitis* if the hands are too much in contact with white spirit. Alkaline detergents are also dangerous irritants.

There are various firms such as Rozalex Ltd., of Manchester, and Century Oils Ltd, of Hanley, Stoke on Trent, etc., which make excellent barrier creams for the protection of operatives performing potentially harmful work of this kind. The barrier cream is rubbed well into the hands before work is begun. The same firms produce industrial cleansing preparations for washing the hands when work is finished, these preparations being far more effective than ordinary soap and water. Materials of this kind are not used nearly so widely as they should be, and when they *are* used they are not always used correctly — it is common, for instance, to see operatives using barrier cream for washing the hands after working, as if it were some kind of soap instead of a protective barrier.

Another source of danger to health is the dirty habit of using lead stopping from the hand instead of from a small handboard, also the use of tools and brushes with unclean handles, by which paint is introduced into the system through the pores of the skin. Handles should be kept clean and should, if necessary, be scraped with a sharp piece of glass or a steel scraper, and coated with knotting. The removal of paint splashes with the finger or thumb nail, possibly followed by scratching the skin or, worse still, biting the nails, is another obvious source of danger. The mixing of paint with the hands now seems unthinkable, yet it used to be a fairly common practice.

(3) CELLULOSE

Very great care is needed when cellulose materials are in use, because of the highly flammable nature of the solvents and the consequent risk of explosion; the solvents are also injurious to health. Cellulose is not used to any significant extent in painting and decorating, so the regulations governing its use are of more importance to those engaged on vehicle painting or on industrial finishing processes in factory work. But if a painting firm sometimes carries out a small amount of vehicle spraying in addition to its normal activities, it immediately comes within the scope of the statutory requirements. It is also necessary that employees should be aware of the requisite precautions because they might be called on to carry out painting work in a factory where cellulose materials are used, and in certain circumstances they might apply cellulose lacquer as a special kind of finish on normal site work.

The decorator who seldom, if ever, uses cellulose does not as a rule realize how potentially dangerous it is. Instances have been recorded of fatalities and serious injuries and of very extensive damage to property caused by some apparently trivial act of carelessness.One example concerned a man wearing steel-tipped boots, which struck a spark when he was crossing a concrete floor; the spark ignited the solvent fumes in the air, causing a violent explosion. Another example concerned a man who was using a chipping hammer to clean the structural steelwork in the paint shop of a factory; no cellulose was in use at the time, but there was a slight leak from a badly-fitting top on a drum of cellulose thinners, and a spark struck by the hammer ignited the vapour with serious results. On one occasion a fair-sized college of art was completely destroyed when cellulose vapour withdrawn from a workshop by an extractor fan came into contact with combustible matter in the flue from a pottery kiln.

The storage and use of cellulose materials are controlled by Statutory Instrument No. 917 (1972), which is called "The Highly Flammable Liquids and Liquefied Petroleum Gases Regulations, 1972"; these regulations have superseded the old Cellulose Solutions Regulations of 1934 which were revoked in two stages in June 1973 and June 1974. The regulations apply to every liquid with a flashpoint below 32°C, and this of course includes cellulose.

The Regulations lay down strict rules about the premises in which cellulose is used or stored and the precautions to be observed in its use. The structure must be built with fire-resisting materials and must provide for pressure relief in the event of explosion, the pressure being vented to a safe place. Cellulose must be stored in a special store-room which is separate from the main workshop and must have a separate entrance. (When the quantity involved is less than 50 litres it can be stored in closed vessels in a fire-resisting cupboard or bin.) Every storeroom, cupboard, bin, tank or vessel must be clearly marked with a warning notice.

Precautions are to be taken so that there is no possibility of any spark or naked flame in any area where solvent vapour might be present and this means that no electrical motor and no electrical switches shall be situated in such an area. Electric-light fittings must be suitably masked and protected against sparking and must be enclosed in a strongly constructed guard to prevent them from being accidentally broken. Extractor fans must be installed to remove fumes and vapour and to conduct them to a safe place. A means of escape is to be provided in case of fire, and suitable fire-extinguishers must be situated in convenient spots. Smoking is strictly prohibited in any work area or storeroom.

(4) PAINT DEFECTS

It has been said that there are more snags in painting and decorating than in any other branch of industry, but whether this is true or not, it is a fact that one can spend a lifetime in the trade and still meet fresh difficulties or circumstances not encountered before. It is probably the realization of this that makes the true craftsman inclined to be cautious and not so ready to take things for granted, as would a less experienced man.

The investigation of complaints regarding the cause of paint defects is a highly technical matter, often involving laboratory analysis. It also requires great tact on the part of the investigator, for the craftsman may resent any suggestion that his ability is questioned. He may have used the same material for years on "similar" jobs without previous trouble and is therefore inclined to assume that the paint *must* be defective when anything goes wrong.

There are many factors to be taken into account, however, since painting conditions vary from day to day (often from hour to hour) through the vagaries of weather etc. The ability to assess the condition of the surface depends largely on the judgment and experience of the painter, who generally has nothing more to guide him than superficial appearance. In many cases, of course, the painter must adhere to a rigid specification prepared by an architect, and this may lead to further complications in assessing responsibility for subsequent paint failures.

The most elusive link in the chain of circumstances is the history of the previous treatment; this can generally be determined only after part of the defective film has been removed and submitted for chemical analysis. For instance, a plaster surface treated previously with water paint may *look* perfectly sound, yet the initial bond with the plaster may be so weak that it is unable to withstand the contraction movements set up by further coats and so flaking occurs. Examination of the flakes will probably reveal a loose powdery backing which an analysis may prove to be (a) calcium carbonate (whiting); (b) calcium sulphate (plaster), or (c) water-soluble salts.

From this may be deduced the history of the previous treatment. The presence of whiting would indicate an improperly washed-off distemper prior to applying the water paint; the calcium sulphate backing would point to a weak plaster skim due, possibly, to a defect such as "dryout" (page 220) or to the premature application of an oil primer. Soluble salts would probably come from the bricks or rendering and would tend to weaken the bond, especially if the surface had been primed before it was completely dry.

The above single instance is given in some detail merely to show some of the possible causes that may lead to peeling or flaking, none of which may be suspected by the painter who may ascribe the trouble to faulty material. In the space at our disposal we cannot go very deeply into this complex subject, but the following is a brief description of the more common defects and their remedies. Information on the treatment of defective surfaces will be found under "Preparing Old Surfaces" (page 263).

Bleaching

The action of strong acid fumes will bleach out certain colours, e.g. ultramarine and any colours containing it. Similarly, Prussian blue and Brunswick greens, etc., are bleached by alkalis (Fig. 19.3). The terms "bleaching" and "fading" are often loosely applied, but they are not strictly synonymous; bleaching indicates a *whitening* of the surface, or a complete change of colour, while fading suggests a reduction in colour strength to a paler tint.

Bleeding

This term is applied to any colour or stain which persists in permeating successive coats of paint or distemper, etc. (Fig. A.1). Materials prone to this defect are —

> Certain lakes and dyestuffs (chiefly reds and purples), cheap black Japans, tar and creosote preparations, knots and resinous timber, mica and metallic bronzes. Sooty cracks on chimney breasts, and copying ink markings, etc., will also bleed through paint and distemper coatings.

Doubtful surfaces may be tested for bleeding by painting a trial patch of oily white paint over a suitable portion of the work; any tendency to bleed will generally show itself within a few hours. Bleeding colours may be insulated by applying two thin coats of "stop-tar" knotting, or alternatively, one coat of patent knotting lightly pigmented with aluminium powder (1 kilogram to 5 litres).

If the ground is soft and elastic, however, serious cracking may result from the application of the knotting and the safest plan in this case is to remove the coating completely.

Blistering

The presence of blisters indicates either moisture or gaseous products beneath the paint film (Fig. A.2). Moisture may be contained within the surface, as in unseasoned timber or damp plaster, etc., or it may be

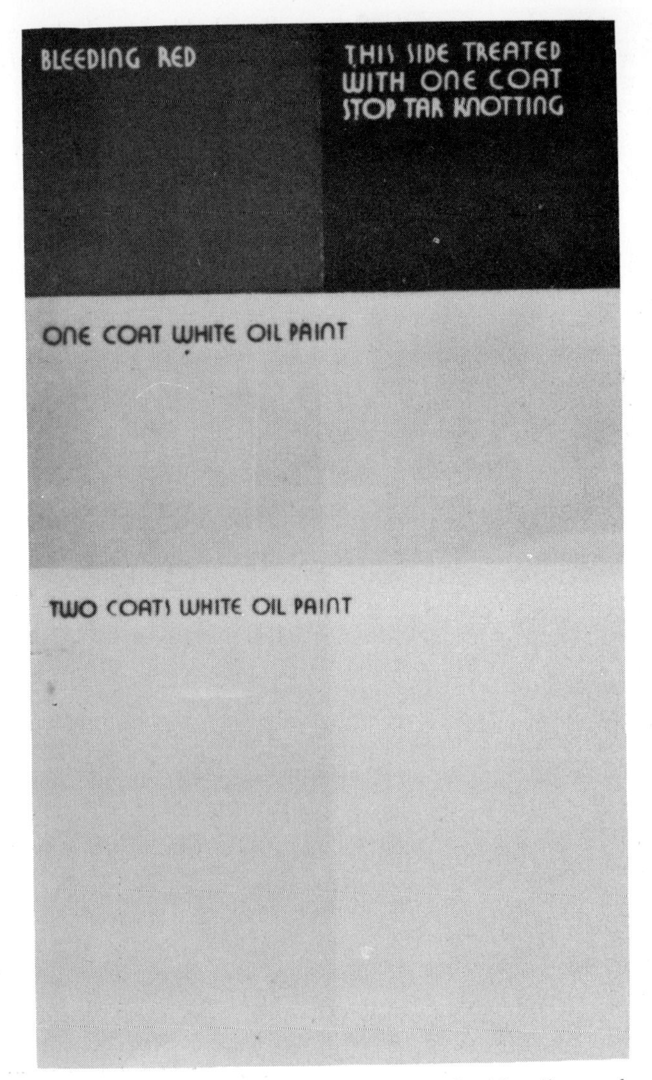

Fig. A.1 Panel showing the effect of a bleeding red

One half is insulated by a coat of stop-tar knotting, but on the untreated portion the bleeding red has turned the white oil paint into a decided pink

imprisoned beneath the film by painting in damp foggy weather, or by failing to allow washed surfaces sufficient time to dry out properly. Water may also be introduced by way of paint brushes and strainers.

Thus, blisters may be caused by the direct pressure of water (as in new plaster and cement surfaces), or by the expanding vapour which occurs beneath the film when the surface is heated by the direct sun, or artificially. Similarly, the volatile resinous matter in timber will generate gases when heated; hence the prevalence of blisters over knots. Soft, oily

Fig. A.2 A close-up view of blistering on painted wood
Courtesy: Paint Research Association

undercoats, and especially old fat colour (smudge), are very conducive to blistering; and dark colours, owing to their greater absorption of heat, will blister more readily than light colours.

The remedy for blistering is to remove all defective paint down to the ground and repaint with fairly "sharp" undercoats. Any large or excessively resinous knots should be cut out and replaced with sound timber, or alternatively cut well below the surface and filled with proprietary filler. Cases of persistent blistering on woodwork have been cured by applying a coat of oil-bound water paint direct to the bare wood (see page 271). Blistering often occurs on galvanized iron and similar zinc-coated surfaces owing to lack of adhesion; for recommended treatment, refer to page 259.

Blooming

Varnishes and enamels are liable to be affected by *bloom*, a defect characterized by a clouding or milky appearance of the work. The defect may be caused by atmospheric conditions prevailing during or after application, or to some impure ingredient or faulty preparation of the varnish itself.

All varnishes are highly susceptible to sudden chilling and condensation of moisture due to temperature change under humid conditions; when subject to such influences during the critical stages in drying, i.e. between the tacky and the dry state, blooming is very liable to occur. Impure air, as in the vicinity of open gas jets, will also tend to promote blooming, and whilst this points to the need for adequate ventilation, direct draughts should be avoided where possible.

It is important that surfaces should be free from moisture during varnishing, and surfaces which have been flatted down wet should therefore be allowed to dry out thoroughly; similarly, water graining or glazing should be given ample time to dry out, as also should sizing on stained woodwork prior to varnishing.

Bloom may appear some days after the varnish has been applied, and in such cases the varnish or the undercoatings may be at fault; the use of cheap patent driers in undercoats or in graining colour is a possible source of trouble since they may contain a percentage of water.

It is sometimes possible to remove bloom by a brisk rubbing with a clean dry wash-leather, or alternatively, by wiping down with a mixture of equal parts linseed oil and acetic acid, afterwards polishing vigorously with a dry leather. If the bloom returns or persists, the surface should be flatted down and re-varnished under good drying conditions.

Spirit varnishes (French polish, etc.) and nitrocellulose lacquers are also subject to blooming or "blushing", especially when applied in cold or damp weather. Best results can be obtained only by applying in a fairly warm, dry atmosphere.

Chalking

A paint coating which has become loosely bound and powdery is described as being "chalky". Ideally, the gradual deterioration of a sound paint film throughout its life should take place by a very slow process of chalking, that is, by a gradual disintegration of the film from the surface downwards. When this is unaccompanied by any other defect, such as cracking or flaking, the paint retains some preservative value until the film finally disintegrates. If a paint film exhibits early signs of chalking, however, it indicates premature disintegration of the coating, which is a serious defect.

Rapid chalking of this kind would suggest that either the paint is faulty, or that it is unsuited to the particular conditions of exposure. Certain pigments, notably titanium and white lead, have a tendency to chalk, particularly when exposed to sea air. Careful formulation, however, may overcome or minimize such tendencies; white lead, for instance, is greatly improved by the addition of zinc oxide, and titanium white is now being made in anti-chalking qualities. Cheap paints often contain an excessive amount of barytes or other extender, which gives rise to early chalking.

The life of a paint film depends mainly on its medium; when this begins to perish it ceases to bind the pigments effectively and powdering or chalking of the surface results. Very absorbent surfaces will rob a paint film of its medium unless the suction is stopped by a suitable primer. Flat paints are generally unsuitable for exterior conditions because the oil content is too low to protect the pigment particles. Paints which have to withstand severe exposure require to be well bound with suitable oil or varnish media to avoid early chalking.

Loss of gloss is the first indication of a mild chalky condition and this is confirmed when colour is removed by wetting and gentle rubbing. A badly chalked surface, however, resembles a bleached loosely bound distemper. In repainting such surfaces, all loose powdery material should be removed by thorough washing and rubbing down in order to secure proper adhesion of the first coat of paint which should be sufficiently oily to bind down any remaining loose particles in addition to stopping suction.

Cissing

This defect is demonstrated in its acutest form when water colour is painted over an oily ground; the water colour refuses to form a continuous film and collects in separate beads or globules, due to differences in surface tension.

Paint and, more frequently, varnish sometimes show a similar reluctance to form a homogeneous coating and the wet film recedes in parts, leaving the undercoat or ground exposed. This may occur as soon as the material is applied, or some hours after application, in which case the wet film tends to thin out in parts and forms a network of low ridges (Fig. A.3).

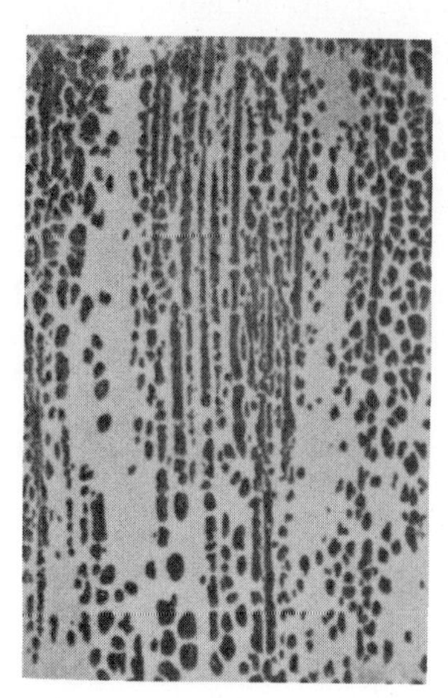

Fig. A.3 An extreme case of cissing

Cissing is liable to occur (a) when varnish is applied over a varnished surface without first flatting down; (b) when undercoats are left too long before following up, or when they become too hard and impervious; (c) when paint or varnish is applied over hard greasy surfaces, e.g. waxed, highly polished, or much-handled surfaces. In such cases, cissing may be prevented by rubbing the surface down with a damp washleather sprinkled with a little fine whiting or fuller's earth, or by sponging down and rubbing with fine waterproof abrasive paper; in either case, of course, all moisture should be allowed to dry out thoroughly before proceeding to varnish.

The second type of cissing referred to above, i.e. delayed cissing, may be due to materials being contaminated by grease, soap, or wax, etc., or to the mixing of different varnishes. Frothing or bubbling caused by white spirit in the brush, or by too brisk application, may also lead to delayed cissing.

Where cissing has occurred some time after application, the only remedy is either to remove the varnish with white spirit, or to allow it to dry and harden thoroughly before rubbing down and re-varnishing.

Cracking (checking and crazing)

Where a hard coating is applied over a soft elastic coating, the top coat (being less elastic) is unable to conform to the expansion and contraction movements of the undercoat, with the result that cracking generally occurs. The greater the difference in the elasticity of the coatings, the more pronounced is the cracking (Fig. A.4).

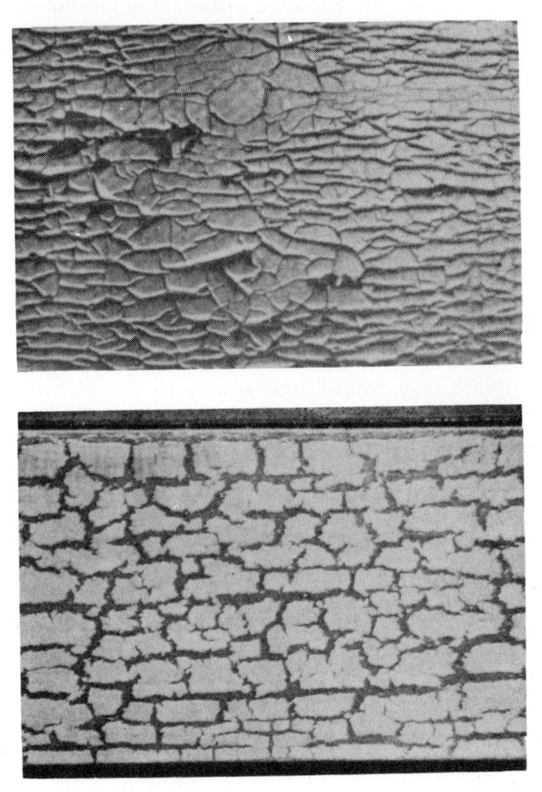

Fig. A.4 Two examples of cracking of paintwork on wood

In (a) the paint edges are beginning to curl and flake, while (b) shows extreme cracking, sometimes referred to as "alligatoring"

Pigments such as zinc oxide and umber tend to produce hard brittle films. Thick undercoats and insufficient drying time between coats tend to produce soft grounds which are too unstable to support hard enamel or gloss paint finishes. Fat oily colour (smudge) should never be used for priming or undercoats on good-class work for the same reason.

Cracking may be expected when knotting is applied over paint (except where the paint is thoroughly aged and hardened), or when paste and size are allowed to encroach on paintwork during (or preparatory to) paperhanging.

Cracking is naturally more common on exterior surfaces exposed to direct sunlight, but if finishing coats are made to dry more elastic than the undercoats, cracking will generally be avoided.

Very fine cracking or *crazing* is generally caused by rapid surface drying of the film. When cracking does not extend throughout the entire film, it is sometimes possible to remove the defective top surface by careful use of a spirit paint remover, or by using a strong solution of Manger's Sugar Soap, in accordance with the maker's directions. Surface cracking of this type is sometimes filled with Japan or water filler, but this can only result in further cracking ultimately, since the cause (i.e. the elastic undercoat) has not been removed. Where cracking is deep seated, the whole of the film should be removed and the surface repainted with properly constituted priming, intermediate and finishing coats.

Damp surfaces

The *cause* of persistent dampness should be traced to the source and any leaky roofs, defective pipes, or exterior pointing, made good and the surface allowed to dry out before attempting to redecorate. If dampness is due to deliquescent salts (as in sea sand), either the defective plaster should be removed and the surface replastered, or the wall should be battened out and covered with wallboard. Damp patches may be covered with lead foil carried well beyond the damp area and fixed with red lead and goldsize applied in a round consistency to the lead foil, which should then be rolled into close contact with the wall surface. Where a complete wall is so covered, commence fixing the sheets from the top and allow each sheet to overlap slightly. The foil can then be painted or papered, but in the latter case it should first be well sized and cross lined.

Pitch paper is often used on damp surfaces, also rubberized fabric hung horizontally with special adhesive. The advantage of the latter is that it is sufficiently elastic to conform to surface irregularities and can be carried over shallow mouldings and cornices if necessary. (See also "Mould and Mildew" page 554.)

Darkening

Chemical action between lead pigments and sulphur compounds usually results in darkening or blackening. For example, white lead paints, lead chromes and chrome greens, etc., will quickly discolour in sulphurous industrial atmospheres (particularly if unprotected by varnish) due to formation of black lead sulphide. Lead pigments will react similarly with sulphide pigments (i.e. ultramarine and vermilion); mixtures of sulphide pigments and copper compounds are also to be avoided.

Some pigments discolour on heating, e.g. lead chromes, etc., and great care should be taken in selecting paints for hot surfaces.

Discoloration

The most common forms of discoloration may be described by the following terms — *Bleaching*, *Bleeding*, *Darkening*, *Fading* and *Floating*, and are briefly dealt with under those headings. Other instances of discoloration are found under "Saponification" and "Mould Growths".

Efflorescence

A white crystalline or amorphous deposit which forms on the surface of new brickwork, cement or plaster as water evaporates (Fig. 19.4). The matter is fully dealt with in Chaper 19.

Fading

Certain colours, notably the lakes, will fade or lose their colour under the action of strong light; they are referred to as *fugitive* colours. Some colours which are fairly fast to light when used full strength, fade badly when reduced with white. Fugitive colours are more prone to fading in flat finishes than when well protected by a varnish or enamel medium, hence the difficulty in providing pale flat wall tints guaranteed fast to light. Pigments which are very fugitive often tend to *bleed* also.

Feeding

Chemical reaction between certain pigments and media may result in *feeding* in which the paint thickens up to an unusable consistency. When paint assumes this condition of "false viscosity" it becomes very difficult to thin down — in fact the addition of thinners sometimes increases the viscosity. Although thick in consistency, the paint works "slippy" under the brush and also loses its opacity. In other words, it becomes useless.

Paints which contain basic pigments such as white lead or zinc oxide are prone to this defect when mixed with varnish media of high acid value. Feeding may also be caused by the loss of solvent by evaporation, which may occur when a paint tin is left unsealed. Another cause is the addition of unsuitable thinners or the mixing of different types of paint. It is important that different brands or types of paint should not be intermixed, and that thinning should only be done in strict accordance with the manufacturer's instructions.

Flaking (peeling, scaling or shelling)

Lack of adhesion is the root cause of this defect, which usually commences with large-scale cracking. The cracked areas curl at the edges, then finally peel or flake away from the surface. Conditions conducive to flaking are (1) loose, powdery surfaces (such as a weak plaster skim or badly washed-off distemper); (2) efflorescence; (3) rust; (4) size or paste under water paint, or over oil paint, and any other circumstance which impairs adhesion. Zinc, lead and other non-ferrous metal surfaces give trouble unless correctly treated before painting (see page 259) as do Keene's and Parian cements (page 220).

When water paint is subjected to steamy atmospheres, as in kitchens and bathrooms, it is continually absorbing and giving up moisture; this sets up expansion and contraction movements which may cause the bond to weaken and finally give way under the strain. Where water paints are exposed to adverse conditions of this kind, without the surface having been properly prepared and treated with a suitable oil primer, flaking may be expected sooner or later.

For re-treatment of flaking surfaces, see "Preparing Old Surfaces", page 220.

Flashing

An unsightly "glossing up" which occurs in patches on flat oil paints, flat varnish, or semi-gloss work. This is caused by the material setting too rapidly, due possibly to being applied over too absorbent a ground, or to working "flat on flat", i.e. applying one flat coat over another. The flashing also occurs where wet paint or varnish is brushed into partly set material instead of into a "live" edge.

Finishing coats in flat oil paint should always be applied over a well-bodied uniform semi-gloss ground. They should be of a creamy but easy brushing consistency, applied *liberally* and quickly, so that each section is joined up while the edge is alive, crossed once, laid off lightly and left. When applied in this manner, flashing will generally be avoided.

Floating

A change of colour may occur during drying of oil paints or water-thinned paints. Certain pigments, notably greens and blues, are liable to separate partly and to rise to the surface during drying, the dry film thus exhibiting a hue different from that of the wet paint. Such colours are said to have *floated*. The thicker the paint coating or the slower the drying, the longer is the period of flotation; thus if the paint is applied unevenly the discoloration may appear in patches. Some paints dry slightly darker, others lighter than the wet state, but when there is a wide difference, such as a green which dries distinctly bluer, the material is seriously at fault and should be replaced.

Grinning

When opaque colour is applied too thinly, or when there is too great a contrast in tone between the ground colour and the top coat, the result is an uneven or unsolid appearance and the ground is said to be "grinning through". When done intentionally, the same effect is termed "scumbling" or "glazing". Gloss paints and enamels are sometimes low in opacity or hiding power and therefore require an undercoating fairly close to the finishing colour in order to avoid grinning. Stencilled or painted ornament will often grin through several coats of paint unless obliterated with an intermediate tone (see page 275).

Livering

A defect in which the paint thickens to a firm jelly and forms a liver-like skin which gradually spreads throughout the mass. This chemical reaction, which takes place between linseed oil media and certain lakes, Prussian blue, etc., is not unlike a vulcanizing process; it usually occurs after long storage, and colours liable to this defect should therefore not be purchased in bulk unless the material can be used fairly quickly. Livered material is useless.

Mould and mildew (fungoid growths)

Moulds are a class of fungus which, under favourable conditions of dampness and humidity, infect and thrive on organic matter contained in flour paste, glue size, casein distemper, oil paint, paper, etc. They favour damp plaster and unseasoned timber and are often very difficult to eradicate (Fig. A.5). Infection may occur by way of airborne spores, or in water used for mixing plasters. The defect is commonly found in such places as breweries, dairies and greenhouses.

Disfigurement may occur in large spots or patches which vary in colour according to the species of fungi; mildew (white and fluffy), black mould

Fig. A.5 Dry rot fungus, showing advanced development beneath a house floor

(musty odour — favours damp paper and size-distemper) and green mould are very common types, while pink or violet fungus is frequently found in greenhouses. The latter seems to favour white lead paint and appears as a stain which permeates the film.

Eradication of mould growths depends partly on the removal of the conditions which favour their development and partly on the killing of the spores by a suitable antiseptic. The first requirement is thorough cleansing of the affected surface, carefully removing all infected paper, paint or other material by stripping, scraping, burning off (as appropriate), and taking care to burn all dry debris to avoid contamination of other surfaces. Plaster surfaces should be thoroughly washed down and the cause of damp should be traced and cured at the source. Adequate ventilation should be provided where possible to allow the fabric to dry out thoroughly.

All the main paint manufacturers produce an antiseptic wash or mould inhibitor which should be well brushed over the surface in accordance with the maker's instructions. Monsanto Chemicals Ltd of Ruabon produce a very effective mould inhibitor called Santobrite which for a long time has been widely used by decorators, but this is no longer available in small quantities. It is still supplied in bulk, when required.

Alternatively, a solution of zinc silico-fluoride in water (170 grams to 5 litres) (6 oz to the gallon) may be applied to the surface and allowed to dry. The sterilized surface should then be left for at least a week or so and kept under observation. If further mould appears, brush it off and apply a further wash of antiseptic, allowing another period of drying before proceeding to decorate.

If size is to be used in the new decoration, or in preparation for papering, add a proportion of preservative such as carbolic acid (about 2 percent of the size content). Where the surface is to be painted, thin the priming coat (but not subsequent coats) with naphtha, as this has good penetrating and fungicidal properties. It is desirable — if rarely possible — to thin subsequent undercoats with pure turps and then to finish with a good hard-drying gloss paint; this treatment will go far to discourage the recurrence of mould stains in greenhouse interior work.

Ropiness and "ladders"

A painted surface which shows brushmarks in a pronounced corded or ridged effect is said to be "ropy". This may be due to (a) insufficient flow in the material; (b) applying paint too liberally and in too thick a consistency; (c) careless laying off, using too coarse a brush, or (d) a combination of these conditions. Paint should always be well brushed out unless it be of a type which requires liberal application, e.g. a gloss finish, in which case it should possess sufficient flow to eliminate brushmarks. Remove surplus paint from the brush before laying off—one cannot lay off properly with a loaded brush.

"Ladders" are caused by carelessness in laying off. Any misses in the final laying off will disclose the transverse brush strokes which appear as ladderlike crossings. These are even more pronounced when the finish is ropy.

Saponification

When a paint based on linseed oil or certain other drying oils is applied over new lime plaster or Portland cement surfaces, the oil is attacked by caustic alkalis in presence of moisture and becomes saponified, i.e. turned to soap. The paint softens up and develops unsightly runs or tears (see page 223). Surfaces stripped with alkaline paint removers are also liable to saponify any subsequent paint coatings unless all trace of alkali is removed and the surface neutralized with a vinegar or acetic acid wash.

Saponified paint, varnish, or enamel should be thoroughly removed by washing and scraping and the surface allowed to dry. Emulsion or water paint may be applied as a temporary measure until the plaster becomes thoroughly dry, when it may be safely treated with an alkali-resisting primer followed by a normal oil or gloss paint finish.

Alkalis will also attack certain colours, notably Prussian blues, lead chromes and any combination of these, i.e. Brunswick greens, etc., and will result in unsightly discoloration. For this reason, only lime-fast colours are used in emulsion and water paints which are so frequently applied on alkaline surfaces.

Sheariness

Flat or semi-gloss paints which exhibit local differences in gloss or uneven finish are said to be "sheary". Over-brushing, or inexpert application, usually results in this defect (see "Flashing").

Sleepy gloss

When a newly applied gloss paint or varnish dries with a dull gloss or lack of brilliance, it is referred to as being "sleepy". This may be due to an absorbent undercoat or to a defect in the finishing material, in which case the maker should be consulted.

Wrinkling (creeping or crinkling)

A puckering of the surface due to either (a) faulty application or (b) faulty material. Too liberal or uneven application of viscous enamels or varnishes, especially in cold weather when the material "works tough", often results in thick films which may wrinkle badly when the film expands in warm weather.

Heavy sags or runs generally show this defect to a marked degree (Fig. A.6). Application over greasy or waxed surfaces may lead to this condition by the underfilm remaining soft while the top dries to a floating skin. Thick soft undercoats also tend to induce wrinkling of superimposed finishing coats.

When the fault lies in the material, it may be due to unsuitable driers or to "fatty" paints which form elastic films, incapable of drying properly throughout.

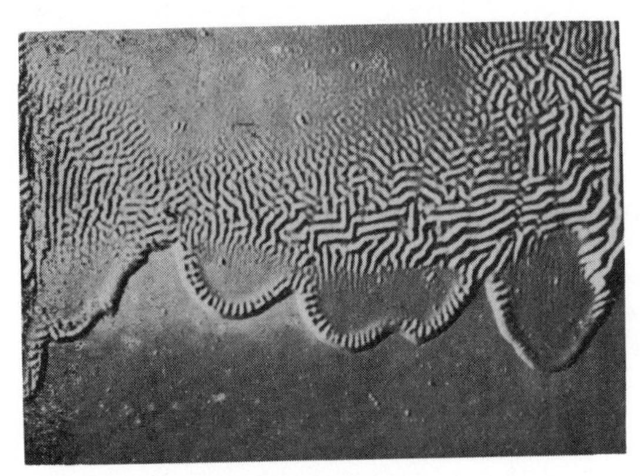

Fig. A.6 Acute curtaining and wrinkling due to uneven application of finishing gloss coat

All wrinkled paintwork should be completely removed, either by burning off or by stripping with paint remover. The work should then be repainted with properly constituted undercoats and finish, allowing ample time for hardening between coats.

(5) STUDY METHODS AND EXAMINATION TECHNIQUE

Most reasonable people employed in an interesting trade or craft will want to study to make themselves more proficient, and having embarked upon a course of study they may well wish to sit for an examination that gives proof of their proficiency. This is a perfectly natural ambition. The examination itself is not the principal goal; the motivating force is the desire for knowledge, and the examination merely sets the seal upon the endeavour. You will very often hear examinations decried on the grounds that many people who have gained no qualifications at all are just as good — or better — at the job than those who have. This may be true, but it is one of the facts of life that if you are applying for a superior post your prospective employer has no other means of judging your capacities. Your qualifications provide a yardstick by which your performance can be gauged. It is perfectly clear that if two men are competing for a senior post, one of them having qualifications and the other without them, all other things being equal it is the qualified man who will get the job. This being so, it is sound common sense to aim at getting some qualifications so that when the opportunity of getting a better position is presented you will be able to apply for it with confidence instead of regretting that the job went to somebody else.

Most examinations at craft level are of the kind known as *multiple choice*, consisting of a number of questions to each of which four or five alternative answers are given; the candidate selects in each case the answer that he thinks is correct. Question papers of this kind are favoured by the examining bodies because they are easy to administer and the whole examination process can be computerized. They do show whether or not a student has achieved a basic knowledge of his subject, but they are not completely satisfactory. For one thing, of course, each question has to be so simple and uncomplicated that only one answer is possible, but the more important drawback is that the really accomplished candidate has no opportunity to display his talent; the mediocre student who can merely say which of the answers is correct gets just as good a mark as the clever student who thoroughly understands his subject and who could, if he were asked, provide the reasons why that particular answer is right. The argument often advanced in favour of this type of question is that a student does not need verbal ability to pass, but this argument is a false one because he still needs verbal ability to understand the question.

Examinations at a higher level than those of craft certificate standard are still very often of the sort that pose a number of problems which the candidate is required to answer. Very little guidance has been given to craft students in the standard textbooks about the techniques of study and examination work. This is a great pity, because there are definitely ways in which you can prepare yourself for a forthcoming examination, and you put yourself at a disadvantage if you don't make use of them.

The first step is to analyse what the purpose of an examination is. This can be stated under three headings, thus: (1) to see whether the candidate has covered the whole range of the syllabus and has a wide enough knowledge of his work; (2) to see whether he can understand a question and realize what is required of him; and (3) to see whether he can set down his knowledge in understandable form so as to convey to the examiner the extent of his learning. We can take each of these three heads in turn and discover how to cope with them.

Reading and notetaking

The best way to be sure of having a wide enough range of knowledge is to get into the habit of reading. All the information you want is available in books and in journals; whether or not you make use of it is your own personal choice. Many people think they can get by without reading, but this is tantamount to throwing away their chances. If you wanted to learn to swim you would not deliberately strap one arm to your body before plunging into the water — so why do exactly the same thing on a mental plane when going in for an examination? Then there are the people who say that books are too expensive, although one very often sees the same people spending far more than the price of a book on a single night out with the lads. When you consider that a book gives you for a small sum of money the whole of a man's lifetime of knowledge and experience you realize that books are actually very cheap. And of course if you are genuinely short of money and cannot afford to buy a book you can still borrow it from the public library.

The most useful thing you can do is to cultivate the taste for reading, and to get into the habit of reading something every day. Buy one or two good standard books on the main aspects of your subject. Find out what books there are that deal with fringe subjects like building science or heraldry or graining or business management, and borrow them. Take one of the trade journals regularly. And don't forget to keep an eye on the daily newspaper; it is astonishing how often you see a paragraph about some new technique that is being developed or some new material that is being launched on the market. Now if you simply read them at the time you will soon forget about them, so it is an excellent idea to start a press cuttings book — a sort of technical scrapbook — using an exercise book with stiff covers. Whenever you see a note about something new, cut it out and stick it in your cuttings book, writing beside it the date when it appeared and in which paper you saw it. Very soon you will find that it is almost like a game; whenever you pick up a newspaper you'll be skimming through it to see if there is any information for your scrapbook. And indeed it is not a bad idea to regard it as a game, because it is far more sensible to make your work enjoyable than to think of it as a penance.

Understanding

We have already seen that if you read widely you will extend the boundaries of your knowledge. But you will not get as far as you might if you don't understand what you're reading. This of course is why so many people do not read much — they get into a panic when they see an unfamiliar word, especially if the word is a rather long one. The answer to this is that whenever you are reading any book or paper, make sure that you have a good dictionary beside you — *and that you use it.* A *good* dictionary, mind; little pocket dictionaries are useless because they don't contain enough information. Now to look up the meaning of a word is a good idea, but you are liable to forget it again quickly, so here is a tip to improve your memory. Make a note of what the word is derived from. Our language is a mixture of many other languages — Latin, Greek, French, Italian, German, Scandinavian, Anglo-Saxon, and others. Suppose, for instance, you look up the word "polychrome"; you find that it comes from the Greek words *polus* meaning "many" and *chroma* meaning "colour". Then perhaps next time you are using emulsion paint you see the word "polymer" on the tin and this will suggest a link; the word "poly" again means "many" and when you look in the dictionary you see that the "mer" comes from the Greek *meros* meaning "a portion"; and when you link this with the explanation of the word you will understand clearly what a polymer consists of.

Once you start doing this you will find it quite exciting and soon discover that you are looking at words with a fresh insight; it is not a task but is as enjoyable as a game of cards and much more useful. And all the while, without realizing it, you will be enlarging your vocabulary so that you express yourself more clearly, and do not feel at a loss when in the company of people who are better educated than you. It will do wonders for your self-confidence.

Handwriting

Now we come to the point of how you convey your knowledge and understanding of a subject to somebody else; you do this by speech or by writing, and in the case of an examination you are limited to writing. And all the knowledge in the world won't help you if the examiner cannot read your writing. Now of course, if you go in for a technical examination you will not lose marks for bad writing or poor spelling. But it is only a short step from writing that it is hard to read to writing that is impossible to read. Clearly, if your writing is so bad that the examiner cannot read it you cannot possibly get any marks. He simply has not the time to spend in trying to decipher a jumbled mass of squiggles and blotches — remember he has thousands of scripts to mark and little over a fortnight to mark them in. So the answer is simple: set about improving your handwriting. It is only a matter of practice, and here again if you realize how attractive a page of neat writing looks you will find that trying to write in the same way becomes a pleasure. As a matter of fact there are other reasons for cultivating your handwriting, but these belong to the next section of our consideration.

How to study

The important thing to notice about study is that it is a private matter, something you work out for yourself, so the first essential is to get away from other people. Choose a time when you can have a room to yourself, or if this is impossible, make use of the quiet room or reference room at your local library. The object is avoid anything that will distract you from your work. For this reason, you should arrange all the necessary books, papers, pens, etc., close at hand because when you are pursuing a line of thought it is distracting to have to break off and search for some missing books or papers.

Notice that we said "Books, papers AND PENS". Simply reading about a subject is not enough. Few people can take in more than a small fraction of what they see or read. Even if the words seem clear at the time, within a few minutes they will be getting blurred in your mind, and probably within a few days you'll have forgotten them completely. This is something you need to be prepared for, and one of the best ways of committing something to memory is to write it down — the physical act of writing helps to impress it on your mind. It is far better to study three or four pages of a book thoroughly than to read through fifty pages without remembering any of them. The technique is this. Read one paragraph or one small section of a book, and then close the book and try to write down in your own words what it was about. Then open the book again and see what you missed or forgot. Don't try to write it out in full, because this would be too laborious; get into the habit of jotting down the important points. Reading a paragraph at a time and writing it down in this way will improve your memory enormously. But there is another stage in the process, and that is to try, two or three days later, to write it down from *memory alone* and then open the book to see how much you had forgotten.

This practice serves two ends — it helps to cultivate one's memory, and it helps one to get into the habit of writing. Many people go into an examination room never having practised writing for more than a few minutes at a time, and yet at the examination they will need to be writing for two hours or even three hours without a break. How can they possibly do this successfully if they have never trained themselves to do it? Be wise, be prepared for what you will need to do, and get yourself ready to cope with the situation.

Preparing for an examination

For some obscure reason most examinations are held in the summer months, in May or June. Generally you will have decided to enter for the examination in the previous autumn, round about September. So you have about ten months to prepare. The first thing to do is to put yourself on to a course of training, just as if you were entering for an athletic event or competing in a race. You know well that if you were training for a sporting event you would do it gradually, building up a little at a time until you reached your peak of perfection. You would not go all out from the start, nor would you neglect to do any training at all in the early stages. It is exactly the same principle here; begin gradually and step it up as you go along. The pace of the training will vary with individual needs, but if you are not accustomed to study, the best plan is to set aside a short time every evening. In addition to your general reading, give yourself a short burst of ten minutes every night of concentrated study. Ten minutes is enough; it is useless to set yourself an impossible target. After two months (i.e. by the middle of November) you should have acquired the habit of spending a regular time every night on your studies, and at this point you can step it up to fifteen minutes each night. Don't try to do too much, because then you run the risk of growing stale and forgetting your work just at the time when you most need to remember it.

After Christmas or New Year push it up to twenty minutes each night, and add to it an hour of solid study every Saturday and Sunday as well. Don't spend an hour watching the clock, of course; set yourself a target of so many pages or so many chapters of a book. By Easter you should be ready to go into real intensive training — something of the order of an hour each night and two hours each Saturday and Sunday, and while you are still gleaning new information make sure that you keep up with the work already done by revising it regularly. Don't make the mistake of thinking that if you have read something you know it and never forget it; that is not true. And remember to keep practising writing.

Three weeks before the examination you come to the final spurt. At this time you put everything else out of your mind and concentrate hard on the coming examination. Tell your mates that you'll not be seeing them for the next three weeks; tell your girl that you won't be meeting her for three weeks but that afterwards there is all the time in the world — you'll find that she is only too ready to co-operate and help you in the effort you are making. It seems a long time when you are looking ahead to it, but afterwards you realize how short it was, and there is no sense in spoiling things at this stage and perhaps failing the examination and wasting a whole year. So for three weeks you spend every spare moment working. Keep it up, however hard it is and however tired you get. Keep it up until three days before the examination, and at that point put your books away. Don't be tempted to go on working right up to the last minute — you will only cloud your mind and get confused; stop now and relax. For the last three evenings go out and get plenty of fresh air and exercise, so that you can go into the examination room feeling fresh, relaxed and rested.

This recipe for preparing for an examination really works. If you follow the plan you will succeed. But you have to start early enough, and do it bit by bit. Many people are always "going to start studying" but never get round to it. But of course there comes a point when it is impossible to do the necessary work in the remaining time. Look at it in this light: suppose you were on a walking tour and you needed to reach a certain spot twelve miles away by 6 o'clock in the evening to catch a train home. If you started off at 10 in the morning you could dawdle along, take a long time over lunch, stop here and there for a rest and explore various little side tracks (8 hours for 12 miles; 1½ miles per hour). If you started at 2 in the afternoon you could still afford to go at an easy pace and pause for refreshment on the way (4 hours, average 3 m.p.h.). If you didn't start till 3 o'clock you could get there but it would be brisk hard walking all the way (3 hours, 4 m.p.h). But if you delayed starting till 4.30 p.m. you couldn't possibly get there (1½ hours, 8 m.p.h.). It's as simple as that.

Examination room technique

When the time arrives for the examination to begin, the way you conduct yourself can make all the difference between success and failure. Some people develop a bad attack of nerves at the very thought of a written test; in a way this is understandable, but on the other hand if you have studied carefully and methodically you have nothing to fear. The best way to counteract nerves is to adopt a positive attitude. Go in with confidence; take strength from the fact that you have worked hard, and determine that you are going to pass convincingly — not just a borderline pass that hovers on the brink of failure but a really good performance of distinction standard. Many people defeat themselves at the outset by believing that they will fail; this is silly. There is no point in taking an examination until you are ready for it; there is no sense in spending money on the entrance fee if you think you are going to lose it. There is nothing to be gained by sitting for the examination just for the experience — just to see what it's like. To fail a test is demoralizing, and the knowledge that you once failed destroys your confidence next time. So go into the examination room with a positive outlook, with the intention of coming out top, and provided you have prepared yourself properly you are bound to succeed unless you make some stupid mistakes.

What sort of mistakes can you make? You should know, so that you can avoid them. First of all, make sure that you arrive in good time. The worst possible way to start an examination is to come blundering in when the test has already started, hot, breathless and dripping with sweat, and with no time to collect your thoughts. Get there 20 minutes beforehand, so that if there is some confusion about which room you are to use you won't become flustered. And make sure that you have all the materials you need. Take at least two pens with you; pens have a knack of running dry at critical moments and it is very disturbing to find out that you have nothing left to write with. But the most important thing of all is to read the instructions and make sure that you understand them — in particular that you know *how many questions* are to be answered, and that you *answer exactly* that number. This, without any doubt at all, is the one thing that causes more failures than any other single factor. It is essential to realize this, so we will give it some consideration.

Some examination papers give you a choice, for instance five questions to answer out of eight; on other papers the questions are all compulsory — you need to answer them all. Now the first thing to do is to place your watch on the desk in front of you, and decide how the time is to be allocated. If you have five questions to answer in three hours that works out at 36 minutes per question; if you have 10 questions to answer in three hours there is 18 minutes per question; ten questions in two hours is 12 minutes for each, and so on. It would be reasonable in the above examples to allow yourself 34 minutes, 16 minutes or 10 minutes for each, so that there is a little time left at the end for checking your work over before handing it in. Don't be tempted to stray beyond the allotted time for any question, otherwise you may not be able to finish the required number. The importance of answering all the required questions lies in the fact that this is the only way of ensuring that you are marked out of the maximum possible marks. If there are five questions to answer and each question carries equal marks, then each question commands 20 percent of the possible marks; if there are 10 questions to answer, each question is worth 10 percent of the possible marks, and so on.

Now if you answer only one question out of five — and it is surprising how many examination candidates do answer only one question — you are making it impossible to pass, because the maximum mark you could get for giving a perfect answer is only 20 percent of the total, and in practically every examination the pass mark is at least 40 percent and is often higher still. There are various reasons why people fall into this trap. A student may see a question about a subject on which he is completely confident; spending all his time in writing a masterly answer to this one question, he gets so engrossed in it that he leaves himself no time for anything else. Another common occurrence is that the student looks at the questions and sees only one that he is really happy about; he hopes that by writing a good answer to this he will conceal his inability to cope with the rest. But whatever the cause, the end result is failure. He will fail if he answers two questions instead of one, because he is hardly likely to score maximum marks for any question. The only sensible course to adopt is to answer first of all the questions you are confident about, but to keep within the allotted time for them, and then to look very carefully at the other questions. However hard they seemed at first sight, you will generally find on a second reading that you can at least write something about them. Supposing you have scored 17 or 18 marks out of 20 for the first answer and about 12 or 13 for the second, you will need about another 10 marks to get a pass; but even if you only score 5 or 6 marks each on the difficult questions you will be comfortably inside the pass-mark bracket.

Another word of warning concerns writing about the subject specified in the question. If, for instance, a question is asked about the preparation of non-ferrous metals for painting, it would be foolish to write an answer — however good it may be — about preparing iron and steel. Many students try this dodge, hoping that the examiner will give them marks for what they have written, even if it is beside the point. What you should know is that when the examination questions are set, the examiner draws up a marking schedule too, in which each question is broken down into small sections and a specified number of marks is allotted to each section. If your answer does not tally with the mark schedule it is bound to fail.

Just as bad as answering too few questions is answering too many. If you answer more than the specified number you will only be credited with the number you should have answered.

Reserve a few minutes at the end of the examination to check your work and correct any silly mistakes you may have made. Ensure that all your answers are numbered; it is sometimes difficult for an examiner to detect which question a candidate is answering. Try to organize your thoughts so that you don't split an answer up into several separate parts dotted all over your answer paper.

(6) BOOKS FOR FURTHER READING*

PAINT TECHNOLOGY

An Introduction to Paint Technology (Oil and Colour Chemists' Association, 4th edn, 1976)

Outlines of Paint Technology: W. M. Morgans (Charles Griffin & Co. Ltd, 2 vol., 1980–)

Paint and Varnish Manufacture, edited by H. W. Chatfield (George Newnes, 1955)

The Science of Surface Coatings, edited by H. W. Chatfield (Ernest Benn, 1962)

Protective Painting of Structural Steel: F. Fancutt and J. C. Hudson (Chapman & Hall, 1968)

Practical Manual of Industrial Finishes on Wood, Metal and Other Surfaces: B. M. Letsky (Chapman and Hall, 1960)

Paint Film Defects: Manfred Hess (Chapman & Hall, 3rd edn, 1979)

PAINTING AND DECORATING

Dictionary of Painting and Decorating: J. H. Goodier (Charles Griffin & Co. Ltd, 1974)

Painting and Decorating: a Guide for Houseowner and Decorator: J. H. Goodier Godwin, 1977)

Painter's Craft Science: L. F. J. Tubb (Macmillan, 1974)

Painting A to Z: J. Lawrance (Sutherland Publishing Co, 1966)

Painting and Decorating (6 vol.): C. H. Eaton (Pitman, 1930)

Practical Painting and Decorating: J. E. Butterworth (English Universities Press, 1930)

Painting and Decorating: an Information Manual: A. Fulcher *et al.* (Crosby Lockwood Staples, 1975)

Paperhanging: J. H. Goodier (Sutherland Publishing Co. Ltd, 1962)

Graining, Ancient and Modern: W. E. Wall (Norwood Press, Mass., U.S.A., 1924)

Graining and Marbling: J. P. Parry (Crosby Lockwood, 1949)

Choosing a Job — Painting and Decorating; J. H. Goodier (Wayland, 1975)

ANCILLARY ART SUBJECTS

Roman Lettering: L. C. Evetts (Pitman, 1938)

Writing, Illuminating and Lettering: E. Johnston (Pitman, 1944)

Lettering for Architects and Designers: Gray and Armstrong (Batsford, 1962)

Manual of Rendering in Pen and Ink: R. W. Gill (Thames & Hudon, 1973)

Nature and Ornament: L. F. Day (Batsford, 1924)

Pattern Design: L. F. Day (Batsford, 1933)

Pattern and Design: N. I. Cameron (Lund Humphries, 1948)

Traditional Methods of Pattern Designing: A. H. Christie (Oxford, Clarendon Press)

---

*Books in this list which are out of print may be consulted in libraries.

Drawing Design and Craftwork: F. J. Glass (Batsford, 1948)
Geometrical Drawing for Art Students: I. H. Morris (Longmans, 1958)
Practical Plane and Solid Geometry: I. H. Morris and J. Husband (Longmans, 1962)
Applied Perspective: J. M. Holmes (Pitman, 1938)
Art of Interior Design and Decoration: J. M. Holmes (Longmans, Green, 1951)
Colour: a Manual of its Theory and Practice: H. Barrett Carpenter (Batsford, 1933)
The Munsell Book of Colour: Prof. A. H. Munsell (revised): Munsell Color Division, Baltimore, U.S.A., 1966: Tintometer Sales, Salisbury
The Ostwald Colour Album: J. Scott Taylor (Winsor & Newton, 1933)
Department of Education and Science, Bulletin No. 9, "Colour in School Buildings" (H.M.S.O.)
Colouring in Factories (H.M.S.O.)
Colour in Buildings (The Walpamur Co. Ltd, 1959)
Interior Lighting Design (British Lighting Council)
The Arts of Mankind: Hendrik Van Loon (Harrap)
Painting and Drawing: A. Daniels (Arco, London)
Mural Painting: H. Feibusch (Black, 1947)
English Art in the Middle Ages: O. Elfrida Saunders (Oxford University Press, 1932)
Complete Guide to Heraldry: A. C. Fox-Davies (Nelson, 1969)
Boutell's Heraldry: revised by C. W. Scott-Giles and J. P. Brook-Little (8th edn, Warne, 1978)
Heraldry: J. Franklyn (Arco, London, 1965)
Heraldry: Sources, Symbols and Meaning: O. Neubecker (MacDonald & Jane's – McGraw-Hill, 1976)

ANCILLARY — GENERAL

English Historic Architecture: Bryan Little (Batsford, 1964)
The Architecture of Britain: Doreen Yarwood (Batsford, 1974)
The Pattern of English Building: A. Clifton-Taylor (Batsford, 1962)
English Parish Churches as Works of Art: A. Clifton-Taylor (Batsford, 1974)
A History of Building Materials: N. Davey (Phoenix House, 1961)
A History of English wallpaper: A. V. Sugden and J. L. Edmondson (Batsford, 1926
A Literary History of Wallpaper: E. A. Entwisle (Batsford, 1960)
Building Quantities Explained: I. H. Seeley (Macmillan, 1965)
What Every Supervisor Should Know: L. R. Bittel (McGraw-Hill, 1968)

B.R.E. (Building Research Establishment) Digests: *Building Materials,* published in book form 1977 (H.M.S.O.)
B.R.E. Digests available separately from H.M.S.O.:
21 New Types of Paint
70 Painting Metals in Buildings, (1) Iron and Steel
71 Painting Metals in Buildings, (2) Non-ferrous Metals and Coatings
106 Painting Woodwork
197 Painting Walls, (1) Choice of Paint
198 Painting Walls, (2) Failure and Remedies

British Standard documents relating to paints, etc. (published by the British Standards Institution)

BSCP 97: 1967–1972 Metal scaffolding [3 Parts]
BSCP 98: 1964 Preservative treatment for constructional timber
BSCP 231: 1966 Painting of Buildings
BSCP 2008: 1966 Protection of iron and steel structures against corrosion
BS 1053: 1966 Water paint and distemper for interior use
BS 1070: 1956 Black paint (tar base)
BS 1215: 1945 Oil stains
BS 2015: 1965 Glossary of paint terms
BS 2029: 1953 White oil pastes for paints
BS 2324: 1966 Red oxide–linseed oil priming paints
BS 2521 & 2523: 1966 Lead-based priming paints
BS 2525–2527: 1966 Undercoating and finishing paints for protective purposes
BS 3416: 1961 Black bitumen coating solutions for cold application
BS 3698: 1964 Calcium plumbate priming paints
BS 3900: 1965–1974 Methods of test for paints [46 Parts, obtainable separately]
BS 4310: 1968 Permissible limit of lead in low-lead paints
BS 4652: 1971 Metallic zinc-rich priming paints
BS 4756: 1971 Ready mixed aluminium priming paints for woodwork
BS 4764: 1971 Powder cement paints

British Standard documents relating to colour

BS 381C: 1964 Colours for specific purposes [93 colours]
BS 950: 1967 Artificial daylight for the assessment of colour [2 Parts]
BS 1611: 1963 Glossary of colour terms used in science and industry
BS 1710: 1971 Identification of pipelines
BS 2929: 1957 Safety colours for use in industry
BS 4727: (Part 4) 1971 Vision and colour terminology
BS 4800: 1972 Paint colours for building purposes [86 colours]
BS 5252: 1976 Framework for colour co-ordination for building purposes [237 colours; colour matching fan available]

Other relevant British Standards

BS 476: Part 7: 1971 Surface spread of flame tests for materials
BS 890: 1972 Building lime
BS 1129: 1966 Timber ladders, steps, trestles and lightweight stagings
BS 1139: 1964 Metal scaffolding
BS 1191: 1972 Gypsum building plasters [Parts 1 and 2]
BS 1192: 1969 Building drawing practice
BS 1397: 1967 Industrial safety belts and harnesses
BS 2037: 1964 Aluminium ladders, steps and trestles
BS 2451: 1963 Chilled iron shot and grit
BS 2482: 1970 Timber scaffold boards
BS 2569: 1955 Sprayed metal coatings
BS 2769: 1964 Portable electric motor-operated tools

BS 2830: 1973 Suspended safety chairs and cradles
BS 3349: 1961 Protective canvas sheets for overhead work
BS 3913: 1973 Industrial safety nets
BS 4232: 1967 Surface finish of blast-cleaned steel for painting

Statutory Instruments

The Health and Safety at Work, etc. Act, 1974
SI 1966, No. 94 The Construction (Working Places) Regulations 1966
SI 1966, No. 95 The Construction (Health and Welfare) Regulations 1966
SI 1972, No 917 The Highly Flammable Liquids and Liquefied Petroleum Gases Regulations 1972
SI 1974, No. 1681 The Protection of Eyes Regulations 1974
Amendment: SI 1975, No. 303

Periodicals

Painting and Decorating Journal
The Decorating Contractor
Design
Interior Design and Contract Furnishing
The Artist
Studio International
Product Finishing
Architectural Review

Index

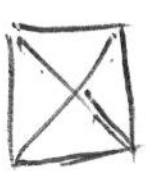